Benchmark Papers
in Geology

Series Editor: Rhodes W. Fairbridge
Columbia University

PUBLISHED VOLUMES

BENCHMARK PAPERS IN GEOLOGY (continued)

Additional volumes in preparation

Benchmark Papers
in Geology / 32

A BENCHMARK ® Books Series

MECHANICS OF
THRUST FAULTS
AND DÉCOLLEMENT

Edited by
BARRY VOIGHT
The Pennsylvania State University

Dowden, Hutchinson
& Ross, Inc.
STROUDSBURG, PENNSYLVANIA

LIBRARY OF CONGRESS CATALOGING IN PUBLICATION DATA
Main entry under title:
Mechanics of thrust faults and décollement
   (Benchmark papers in geology / 32)
   Includes references and indexes.
   1. Faults (Geology)—Addresses, essays, lectures. I. Voight, Barry.
QE606.M4   551.8′7   76-11741
ISBN 0-87933-223-9

Exclusive Distributor: **Halsted Press**
A Division of John Wiley & Sons, Inc.
ISBN: 0-470-98946-7

To Mary Anne, Lisa, Barbara . . .
my companions throughout

SERIES EDITOR'S PREFACE

The philosophy behind the "Benchmark Papers in Geology" in one of collection, sifting, and rediffusion. Scientific literature today is so vast, so dispersed, and, in the case of old papers, so inaccessible for readers not in the immediate neighborhood of major libraries that much valuable information has been ignored by default. It has become just so difficult, or so time consuming, to search out the key papers in any basic area of research that one can hardly blame a busy man for skimping on some of his "homework."

This series of volumes has been devised, therefore, to make a practical contribution to this critical problem. The geologist, perhaps even more than any other scientist, often suffers from twin difficulties—isolation from central library resources and immensely diffused sources of material. New colleges and industrial libraries simply cannot afford to purchase complete runs of all the world's earth science literature. Specialists simply cannot locate reprints or copies of all their principal reference materials. So it is that we are now making a concerted effort to gather into single volumes the critical material needed to reconstruct the background of any and every major topic of our discipline.

We are interpreting "geology" in its broadest sense: the fundamental science of the planet Earth, its materials, its history, and its dynamics. Because of training and experience in "earthy" materials, we also take in astrogeology, the corresponding aspect of the planetary sciences. Besides the classical core disciplines such as mineralogy, petrology, structure, geomorphology, paleontology, and stratigraphy, we embrace the newer fields of geophysics and geochemistry, applied also to oceanography, geochronology, and paleoecology. We recognize the work of the mining geologists, the petroleum geologists, the hydrologists, the engineering and environmental geologists. Each specialist needs his working library. We are endeavoring to make his task a little easier.

Each volume in the series contains an Introduction prepared by a specialist (the volume editor)—a "state of the art" opening or a summary of the object and content of the volume. The articles, usually some thirty to fifty reproduced either in their entirety or in significant extracts, are selected in an attempt to cover the field, from the key papers of the last century to fairly recent work. Where the original works are in foreign languages, we have endeavored to locate or commission translations. Geologists, because of their global subject, are often acutely aware of the oneness of our world. The selections cannot, therefore, be restricted to any one country, and whenever possible an attempt is made to scan the world literature.

To each article, or group of kindred articles, some sort of "highlight commentary" is usually supplied by the volume editor. This should serve to bring that article into historical perspective and to emphasize its particular role in the growth of the field. References, or citations, wherever possible, will be reproduced in their entirety—for by this means the observant reader can assess the background material available to that particular author, or, if he wishes, he too can double check the earlier sources.

A "benchmark," in surveyor's terminology, is an established point on the ground, recorded on our maps. It is usually anything that is a vantage point, from a modest hill to a mountain peak. From the historical viewpoint, these benchmarks are the bricks of our scientific edifice.

RHODES W. FAIRBRIDGE

PREFACE

He who sees things grow from the beginning will have the best view
of them

The historical presentation commonly provides the most penetrating
approach to new subject matter, leading to useful insights not available
from a mere display of currently popular scientific theory. Thus, the historical approach is used in this volume. We wish, in addition, to examine
two contrasting views about the advancement of science. First, there is
the traditional view, which holds that science progresses gradually by an
incremental accretion of factual knowledge; such an accumulation presumably leads to the expansion of theories capable of embracing these
newly discovered "facts." The emergence and clarification of key concepts is thus considered to be quite gradual. New concepts are introduced
where before only vague perceptions existed. Some are forgotten, or
abandoned, only to be subsequently rediscovered decades later. These
concepts would seem to be legitimate scientific benchmarks, and their
subsequent refinement the fruit of hard labor of many, presumably
working with a common philosophy:

In the house humanity has built,
Over thousands of years,
I, too, add one gold nail.

But there is another essential view of scientific advancement. This
competing view is that major changes in science are "revolutionary";
formerly established theories and even "facts" are not simply modified,
but are destroyed. Whole conceptional frameworks are totally replaced
in a relatively short time span. Following the latter view, Kuhn (1962)
contrasts "normal science" with the "paradigm" (i.e., unprecedented,
yet attractive and open-ended scientific achievement). The most obvious
examples of paradigms are associated with such names as Copernicus,
Newton, Lavoisier, and Einstein. But for a more limited professional
group, certain developments in the attempt to comprehend the mechanical problem of overthrust faults may seem as revolutionary as Newton's.

The resolution of the question concerning the appropriateness of
these two contrasting views (i.e., "revolutionary" or "evolutionary"
science) is left to the judgment of the reader. My intention has been to

provide a historically based framework for a specific topic, taken from the special viewpoint, not unbiased, of tectonic geology. In the final analysis, however, it probably must be psychology, rather than philosophy, that will provide the essential key to the understanding of scientific endeavor. This will require a psychological theory of concept formation, transmission, and acceptance and/or rejection in order to adequately comprehend the growth of sciences. Because a generally accepted theory of rational inquiry has yet to be proposed, it is sufficient here to consider a few internal and external factors thought to be significant in the growth of our knowledge of fault mechanics. It is helpful if we follow a theme, and one is thus developed.

This exposition thus begins in the consideration of a key concept developed by an obscure Austrian physicist—the "mechanical paradox of overthrust faulting." We set forth on a trail which leads, ultimately, to the "resolution of the mechanical paradox" as presented by M. K. Hubbert and W. W. Rubey, and beyond: the focus of this volume is thus defined. While the subject of thrust faults and décollement are generally and thoroughly discussed herein, overall perspective to the volume is given by reference to the overthrust theme.

In order to properly examine the quest for mechanical plausibility, however, it also seems necessary to consider early model experiments, deformation experiments with natural geologic materials, and advances in theoretical and applied mechanics, which set the stage for the Hubbert and Rubey 1959 collaborations. A web of continuity is provided by detailed commentary, in which an attempt is made to place the excerpted works in the perspective of geologic discovery. Special attention, where warranted, is given to workers in allied fields: e.g., the civil engineer Karl Terzaghi, who was directly responsible for the discovery of the "effective-stress principle," which later would serve as the keystone of the Hubbert–Rubey hypothesis; Theodor von Kármán, the brilliant mechanicist who developed a methodology for laboratory tests of rock materials so suitable that it has remained essentially unchanged for over a half century; and C. A. Coulomb, who two centuries ago recognized the strength criterion that still bears his name. Readers will discover, if they are not already aware of the fact, that the history of thrust-fault analysis is, essentially, the history of tectonic geology; reference is made herein, with obvious pleasure, to the works of such men as the Alpinists Escher von der Linth, Heim, Bertrand, Schardt, Lugeon, Ampferer; in the Appalachians, to the brothers Rogers, Safford, Keith, Bailey Willis, Rich, Ernst Cloos, Rodgers, and Zen; in the British Isles, to Sir James Hall, Murchison, Geikie, Lapworth, Peach, Horne, and Anderson; in Scandinavia, to Törnebohm; in the Rocky Mountains, to McConnell, Willis, Veatch, Mansfield, Dake, Longwell, Bucher, Rubey, and W. G. Pierce; and to the ingenious mechanicists, Daubrée, Adams, Nádai, Griggs, and their colleagues.

As a consequence, we discover that we must ordinarily choose, from the point of view of overthrust mechanics, between two principal, although not necessarily mutually exclusive, mechanical idealizations: (1) the enhanced-fluid-pressure concept, in which the Coulomb-Terzaghi law is assumed to govern material behavior, and (2) "pseudo-viscous" concepts such as those which have received emphasis by Bucher and Kehle. Viscous idealizations for overthrust faults, as demonstrated herein, had been seriously postulated at the turn of the century; thus, they are not to be considered as particularly modern discoveries. Indeed, from this viewpoint, it is difficult to see on what grounds a mechanical paradox could be substantially founded. The requirement of a paradox, of course, is what scientific revolutions are all about—successor theories are *by definition* theories that deny the validity of their predecessors, for otherwise they would be (merely!) refinements. To what role, then, do we assign the Hubbert–Rubey discovery?

The inertia of mathematical formalism should not be permitted to carry us forward as if nothing but a few formulas and assumptions concerning material behavior and boundary conditions were to be adopted. Like an ancient kingdom, geological science possesses a cultural heritage and a rich history of battles fought and won, of great and ordinary men, of colorful eras, of periods of stagnation. As pointed out further by Walter Jaunzemis, a presentation that is too matter of fact sometimes repels the inquiring mind and deprives us of much of the enjoyment of science! Have not fascination and delight for science been its prime moving forces? In retrospect, I now wish I had room in this volume to include a particular chapter from my well-worn copy of Hans Cloos' magnificent *Conversation with the Earth*.

A list of those who have influenced these pages would practically be co-extensive with a list of my geologic colleagues and students, friends, and acquaintances. Nonetheless, I must necessarily express my personal debt to the following: R. C. Gutschick, Erhardt Winkler, Archie MacAlpin, M. J. Murphy, Bill Fairley, Leroy Graves, and Harry Saxe, at the University of Notre Dame; Fred Donath, Marshall Kay, Walter Bucher, and Rhodes Fairbridge at Columbia University; Wally Cady, Gary Crosby, E-an Zen, J. B. Thompson, P. Osberg, B. Baldwin, J. Bird, L. Platt, Mrs. Selleck, and D. Rumble, in the Northern Appalachians; W. G. Pierce, Hal Prostka, Bill Parsons, W. W. Rubey, Reub Bullock, E. Spencer, J. D. Love, Mel Friedman, David Stearns, and Raymond Price, concerning the Rocky Mountains; my Penn State colleagues, especially Rob Scholten, Gene Williams, Barton Jenks, Dick Parizek, Lauren Wright, Duff Gold, Jon Weber, and B. H. P. St. Pierre; Sam Root, Rodger Faill, and Dick Nickelsen in the Valley and Ridge of Pennsylvania; Jack Currie and Freid Schwerdtner of Toronto; King Hubbert; the Austrians Guntram Innerhofer, O. Schmidegg, and E. Clar; Hans Laubscher, Basle; Ken Hsü, Zürich; Hans Ramberg, Uppsala; Jacques Dozy, Delft; Jean Goguel, Paris; Andrej Werynski, Warsaw; Kazuo Hoshino, Tokyo; Anders Kvale, Bergen; L. Bjerrum, Oslo; Bengt Broms and Nils Hast, Stockholm; H. Illies and L. Müller, Karlsruhe; John

Christie, now at U.C.L.A.; W. R. Hansen, Denver; C. Bruce, Houston; Claire, Agnes, Rose, and George Voight, Mary Raak, Arthur Cheesman and Frank Joseph, of Yonkers; Tony Kamp, Miami; Nellie and George, Croton Dam and Schickshinny; C. Taylor, New York City; J. P. Voight, Togwotee, Wyoming; and especially, the Chairman of the Board, Elmer Enterprises, Inc.

The historically oriented writings of R. Murchison, E. Suess, H. Cloos, E. Cloos, R. Balk, E. Bailey, F. Adams, D. McIntyre, E. Mach, H. Straub, A. Geikie, J. Rodgers, J. Goguel, A. Nádai, and T. Kuhn, and a translation by James Gilluly, have been influential in the development of commentary and deserve specific acknowledgment.

Special mention in the preparation of this volume, among many other matters, is due to Barb Dauria, Dotty Duck, and Judy Bailey of the Geosciences Department, and to Emilie McWilliams and the 1974 Earth and Mineral Sciences Library Staff at Penn State.

BARRY VOIGHT

CONTENTS

Contents

PART V: REFINEMENT AND REFLECTION

Contents

CONTENTS BY AUTHOR

MECHANICS OF THRUST FAULTS AND DÉCOLLEMENT

The Japanese artist tenderly loves the form of things and
respects it; but he loves still more the inner forces
which, emerging from it and frozen for an instant,
have given birth to this beloved form.
"Do not paint created things," instructs an old sage, "Paint
the forces that have created them!"

Le Jardin des Rochers

INTRODUCTION

"Give me matter and motion," said the seventeenth-century philosopher René Descartes, "and I will construct the universe." It seemed
thus, about two centuries later, that a "universe" for tectonic geologists
became illuminated through the discoveries of stone, in arrested
motion, by Arnold Escher von der Linth. His vision pushed aside his
own immediate reactions—"Nobody will believe it. They will call me
a fool"—as he boldly provided evidence of spectacular alpine folds and
overthrust sheets (*nappes*). His evidence seemed clear: layers of rock
thousands of meters thick had moved almost horizontally for scores
of kilometers. Two enormous recumbent folds, one rooted in the north,
the smaller rooted in the south, were reconstructed in Escher's imagination across a Swiss canton, Glarus. This was the *Glarus Doppelfalte;*
with hinges virtually clasped together over a distance of 40 km, two
upper-fold limbs, long since etched away by erosion, were believed to
have once passed through the air above the present mountain summits
(Fig. 1). Thus was laid the cornerstone of *Decken* or *nappe theory*, to
be accepted soon thereafter by a majority of geologists.
   Escher, in the view of Edward Suess (1909), was

> ... one of those possessed of the penetrating eye, which is able to
> distinguish with precision, amidst all the variety of a mountain land-
> scape, the main lines of its structure. He had just come forward with
> the magnificent conception, unheard of in the views of that time, of
> a double folding of certain parts of the Alps, which has since received
> the name of the "double fold of Glarus." Studer opposed him. Such
> movements of the mountains were, he said, contrary to nature and
> inexplicable. Escher did not concern himself with the explanation, but
> with the facts.

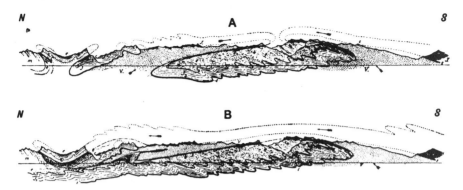

Figure 1 Schematic sections of the folded strata of the Glarner region: (A) the Glarner "double-fold" according to Escher and Heim (1870–1902); (B) the Glarner "nappe" according to Bertrand (1883), Suess (1892), and Heim (1903). m, Molasse (Nagelfluh); e, Flysch; c, Cretaceous strata; J, Jurassic strata; t, Helvetian Trias; V, Verrucano (Permian conglomerates). [From Heim's *Der Bau der Schweizeralpen*, as cited in J. Geikie (1913).]

> A few years later I had the good fortune to make the acquaintance of Sir Charles Lyell, with whom, as with Escher, I maintained friendly relations till the close of his life. On the one side stood Sir Charles, the calm superior philosopher, the lucid thinker and clear writer; on the other dear old Arnold Escher, who entrusted his admirable sketches and diaries to every one indiscriminately, but to whom every line he had to publish was a torment, and who was perhaps only quite in his element up in the snow and ice, when the wind swept his grey head and his eye roamed over a sea of peaks. In characterizing this time I only mention these two important men, because in the contrast of their qualities the whole field of activity in our glorious science is brought into view. Lyell's *Principles*, however, of which the ninth edition appeared in 1853, while investigating at length many fundamental questions, scarcely touched that of mountain-formation: Escher would not enter at all into the discussion of theoretical questions.*

Thus, although to Escher von der Linth we attribute the recognition of Alpine architecture, we must look elsewhere for any consideration of the mechanical question. However, before this could profitably be achieved, the fundamental details of architecture required clarification; Escher's *Doppelfalte* was, of course, subsequently replaced

---

*The editor was immediately attracted to this quotation in the Sollas edition of *The Face of the Earth* for the reason that it seemed a perceptive description of two colleagues at University Park. One of these gentlemen, at home only in the "sea of peaks," almost single-handedly penetrated the secrets of the Beavertooth Range, and most recently produced an important volume on tectonics and gravity; the other, the superior philosopher, successfully inquired into fundamental causes of sand transport, and was most recently honored in the dedication of Arvid Johnson's volume, *Physical Processes in Geology* (Paper 5).

by the interpretation of Marcel Bertrand of France, who in 1884 considerably simplified matters by constructing a single enormous recumbent fold which had moved from south to north (Fig. 1). "Bertrand's vision and the perception of Bertrand, Schardt, and Lugeon regarding overthrust sheets, *"Nappes de recouvrement,"* the so-called "Decken theorie," was correct. It spread in triumphal conquest since 1898, from west to east, revolutionizing the science of Alpine geology. Nothing comparable ever happened in our science. "Impossible, unthinkable, a nonsense, one pulls the ground beneath our feet away—so cried out many people, and even some today. From west to east, it even reached the Carpathians, thanks to Lugeon and Uhlig, within the first decade of our century, eventually penetrating to the Dinarides, Anatolia, and the Himalayas . . ." (Albert Heim, 1919–1922, Vol. II).

For an adequate explanation of the phenomena discovered and explored by Escher and those following in his footsteps, refuge has since been primarily sought in antiquity—in the works of Newton and his successors. In Newton's preface to *Principia,* he presented his work as the mathematical principles of philosophy: "For the whole burden of philosophy seems to consist in this—from the phenomena of motions to investigate the forces of nature, and then from these forces to demonstrate the other phenomena." In assuming that the changes of motion of any object are the result of forces acting on it, the subject of classical (or Newtonian) mechanics was created, a landmark in the history of science because it replaced a merely descriptive account of phenomena with a rational and marvelously successful scheme of cause and effect.*

Newton built on a foundation that included the speculations of Descartes and the experiments of Galileo; indeed, in the inclined-plane studies of Galileo at Padua, and Stevinus at Delft, we have the physical basis for the equilibrium problem on inclined surfaces (e.g., overthrusts induced by gravity).

However, for the student of *Decken Theorie* the light of dawn seemed a long time in coming. More precisely, we should amend this statement by remarking that a few of our predecessors required less "groping in the dark" than the majority. To explain this assertion more fully we need only turn to the question as posed by T. M. Meade and the riposte by M. S. Smoluchowski: i.e., the concept of the *mechanical paradox of overthrust faults.*

---

*Mechanics is often subdivided into three parts: statics, kinematics, and dynamics. *Statics,* which deals with the distribution of forces on bodies at rest, is the oldest of the engineering sciences, with its earliest theories due to Archimedes (250 B.C.). *Kinematics* describes the motions of bodies and mechanisms without reference to its cause; it is, therefore, practically a branch of geometry. Finally, *dynamics* considers motions (accelerations) as they are influenced by forces.

Part I

THE MECHANICAL PARADOX OF OVERTHRUST FAULTING

Editor's Comments
on Papers 1 and 2

1   **READE**
    The Mechanics of Overthrusts

2   **SMOLUCHOWSKI**
    Some Remarks on the Mechanics of Overthrusts

This proposition can be directly illustrated in our first series of Benchmark papers. In an attempt to unravel some of the "weightier problems of geology," T. Mellard Reade, author of an original and systematic treatise, *The Origin of the Mountain Ranges* (1886), in 1908 considered in qualitative terms (Paper 1) a mechanical difficulty concerning the enormous dimensions for the overthrust faults that had been recently postulated in the literature: "Have the authors considered that this means movement of a solid block of rock or rocks of unknown length and thickness of 100 miles over the underlying complex of newer rocks? . . . I venture to think that no force applied in any of the mechanical ways known to us in Nature would move such a mass. . . ."

An adequate answer was not long in coming: its author, M. S. Smoluchowski, an Austrian physicist not only conversant with geological literature, but a fundamental contributor (1909) to the mechanical theory of folding of layered rocks. His reply to Reade (Paper 2) was brief, but is, to my mind, one of the classics of tectonic literature. "It is easy enough to calculate the force required to put a block of stone in sliding motion on a plane bed. . . ," i.e.,

$$F = abcwe,$$

where F is the required force; a, b, and c the length, breadth, and height of the block; w the weight per unit volume; and e the coefficient of sliding friction. The average edge pressure (p) is thus

$$p = \frac{F}{ac} = bwe,$$

which, for large b, is shown to be easily larger than the breaking strength of granite. "Thus," Smoluchowski demonstrated, "we may press the block with whatever force we like: we may eventually crush it, but we cannot succeed in moving it."

But Smoluchowski was not about to condemn theories of Alpine overthrusting, for he thought the above comparison to be, as he put it, "not quite fair." Two reasons were cited: (1) the bed of sliding might not be horizontal but inclined; but, what seems more important, (2) "nobody ever will explain Alpine overthrusts in any other way than as a phenomenon in rock plasticity."

Smoluchowski's work was later to be directly adapted by Hubbert and Rubey in Part I of their classical work "Role of Fluid Pressure in Overthrust Faulting" (Paper 28), in a critically important section subtitled "Mechanical Paradox of Large Overthrusts." An estimate of the maximum length of an overthrust block was presented by Hubbert and Rubey using Smoluchowski's approach, coupled with more recent data on rock strength; second, an estimate of the slope required for gravitational, frictional sliding was made by equating slope angle with the angle of sliding friction as Smoluchowski did. Their purpose was to demonstrate a logical necessity, somehow, for drastically reducing the frictional resistance to block sliding; the remainder of the Hubbert–Rubey and Rubey–Hubbert articles constitutes, as is well known, a magnificent attempt to document the effectiveness of a fluid-pressure mechanism in reducing frictional resistance. The Hubbert–Rubey articles, in fact, comprise the "focal point" of this volume. It therefore may seem ironic that the value of coefficient of friction used in Smoluchowski's rough approximations (0.15, friction of iron on iron) may be nearer to geologic reality for the properties of many thrust surfaces than the value of 0.6 assumed more adequate by Hubbert and Rubey; for example, the reader is referred to the modern literature in soil mechanics concerning the residual-strength concept (see Papers 32 and 39), wherein it is shown that clay-shales of particular mineralogy may commonly have friction values in the range 0.1–0.2.

There is little evidence that Smoluchowski's second and (according to him) most important point, concerning the phenomenon of rock plasticity, has ever been properly appreciated. Hubbert and Rubey (Paper 28) dismiss his alternative briefly on p. 122: "the rocks involved were plastic with a still lower coefficient of friction than assumed" (p. 122); on page 129 they seem to accept Smoluchowski's "plasticity" alternative, but with the substantial qualification, "for such materials as rock salt." However, this seems in part to be a misstatement of Smoluchowski's views. Although he was no doubt aware that Hans Schardt and Albert Heim thought that a thin layer of evaporite rock underlying the western and northern Prealps acted as a "lubricant," a view supported in part by studies of the Jura foldbelts, Smoluchowski was also aware that Heim, Daubrée, and others, even including

7

Reade (1886, p. 140–142, 156–160, 164–177, 208–209), had emphasized well before the turn of the century the general condition of plastically distorted rocks within Alpine and other mountain chains; this was long before the first high-pressure laboratory experiments on rock materials were made (see Paper 10) which proved the capability of many rocks to flow under enhanced confining pressure and temperature.* In citing the appropriateness of "pitch" as a modeling material, Smoluchowski lays stress on his concept of rock as a material capable of rather continuous if inhomogeneous deformation, in contrast to a rigid slide block model on a rigid base.

Another misconception appears to lie in Smoluchowski's use of the term "plasticity"; it is used by him in the sense of resistance associated with deformation by mechanisms of flowage, not in its more restricted phenomenological sense associated today with time-independent mathematical "theories of plasticity." His meaning is made absolutely clear, inasmuch as he refers to the "laws of viscous liquid friction" and suggests that "any force, however small, will succeed in moving the block"; Smoluchowski has a viscous fluid in mind, and "plasticity" as used by him may be considered as "inverse viscosity." In emphasizing the time factor associated with geologic deformation he shows an awareness of the scale factors associated with viscous material; this is hardly surprising in view of his background in physics. Although previous workers (e.g., Sollas, Paper 7) had been aware of the appropriateness of viscous models and others have since pursued similar ideas (e.g., Carey, 1954; see also Papers 44 and 45), it seems to have only been in the last few years that the "viscous model" has been acknowledged as an alternative to the fluid-pressure mechanism for "ordinary" overthrusts. This is aptly illustrated by Kehle's excellent paper on that subject (Paper 47), which begins with the statement: "Existing theories on the mechanics of gravity sliding and orogenic translation impose severe limitations on the size of thrust plates, except in special settings where they are underlain by abnormally pressured sediments . . ."; Kehle then introduces a concept of décollement viscosity—in essence Smoluchowski's model—as an alternative approach to the question. Smoluchowski's work has not, however, been cited in these more recent papers, and may not have directly influenced their development (although indirectly this influence may be considered as manifested through the Hubbert and Rubey papers). Perhaps Smoluchowski's quantitative capability as a physicist was insufficiently balanced by the

---

*However, Miall had reported in *Popular Science Review* (January 1872) on a series of ingenious experiments: plastic hinges were developed in thin limestone slabs subjected to bending moment over a period of months. Time was envisaged as a critical factor. His results were known to Reade (1886, p. 209).

general knowledge of time-dependent rock properties, boundary conditions, and geologic details required for theoretical modeling at the
turn of the century; perhaps he believed that he had elaborated sufficiently on that theme, and that a mathematical presentation of a simple
model, such as presented by Kehle, was not necessary. Hindsight
permits us to speculate, however, that the "mechanical paradox of large
overthrusts" was both recognized and quite likely reasonably resolved
at the turn of the century, in a few brief pages of the *Geological Magazine.*

Reprinted from *Geol. Mag.*, 5, 518 (1908)

THE MECHANICS OF OVERTHRUSTS

T. Mellard Reade

IN attempts to unravel some of the weightier problems of geology it has lately been assumed that certain discordances of stratification are due to the thrusting of old rocks over those of a later geological age. Without in any way suggesting that the geology has in any particular instance been misread, I should like to point out the difficulties in accepting the explanation looked at from a dynamical point of view when applied on a scale that seems to ignore mechanical probabilities. Some of the enormous overthrusts postulated are estimated at figures approaching 100 miles. Have the authors considered that this means the movement of a solid block of rock or rocks of unknown length and thickness 100 miles over the underlying complex of newer rocks? If such a movement has ever taken place, would it not require an incalculable force to thrust the upper block over the lower, even with a clean fractured bed to move upon? Assuming that the block to be moved is the same length as the overthrust, the fracture-plane would in area be $100 \times 100 = 10,000$ miles. I venture to think that no force applied in any of the mechanical ways known to us in Nature would move such a mass, be it ever so adjusted in thickness to the purpose, even if supplemented with a lubricant generously applied to the thrust-plane. These are the thoughts that naturally occur to me, but as my mind is quite open to receive new ideas I shall be glad to know in what way the reasoning can be met by other thinkers.

2

Reprinted from *Geol. Mag.,* **6**, 204–205 (1909)

SOME REMARKS ON THE MECHANICS OF
OVERTHRUSTS

M. S. Smoluchowski

University of Lemberg, Austria

MR. T. MELLARD READE evidently wished to elicit, by his note on the mechanics of overthrusts in the GEOLOGICAL MAGAZINE, 1908, p. 518, a discussion on these phenomena, as he also tells us in the February Number, 1909, p. 75. May I be allowed, therefore, to contribute some remarks on his paper?

It is easy enough to calculate the force required to put a block of stone in sliding motion on a plane bed, even if its length and breadth be 100 miles, and I do not think Mr. Mellard Reade meant to use the word 'incalculable' in a literal sense. However great this force may be, it certainly will be easy to mention instances of still greater terrestrial or cosmic forces. Still, I think Mr. Mellard Reade might dispute the analogy with the piling up of the Rev. O. Fisher's broken ice-sheets, and he could defend his statement "that no force applied in any of the mechanical ways known to us in Nature would move such a mass".[1]

Let us indicate the length, breadth, and height of the block by a, b, c, its weight per unit volume by w, the coefficient of sliding friction by e; then, according to well-known physical laws, a force a, b, c, w, e will be necessary to overcome the friction and to put the block into motion. Now, the pressure exerted by this force would be distributed over the cross-section a, c; hence the pressure on unit area will be equal to the weight of a column of a height b, e. Putting $e = 0\cdot15$ (friction of iron on iron), $b = 100$ miles, we get a height of 15 miles, while the breaking stress of granite corresponds to a height of only about 2 miles. Thus we may press the block with whatever force we like;

---

[1] T. Mellard Reade, GEOL. MAG., November, 1908, p. 518.

we may eventually crush it, but we cannot succeed in moving it. The conclusion is quite striking, and so far we cannot but agree with Mr. Reade's opinion.

But are we entitled, therefore, to condemn the theory of the Alpine overthrusts? I think the comparison is not quite fair. First, it may be remarked, the bed may not be horizontal but inclined; in this case the component of gravity is sufficient, at an inclination of 1 : 6·5, to put the block in sliding motion, and we need not apply any external pressure at all. And what seems still more important, nobody ever will explain Alpine overthrusts in any other way than as a phenomenon of rock-plasticity. Suppose a layer of plastic material, say pitch, interposed between the block and the underlying bed; or suppose the bed to be composed of such material: then the law of viscous liquid friction will come into play, instead of the friction of solids; therefore any force, however small, will succeed in moving the block. Its velocity may be small if the plasticity is small, but in geology we have plenty of time; there is no hurry.

Some features of these phenomena have been beautifully illustrated by Professor Sollas' pitch-experiments. Pitch is not the same as rock undoubtedly, a point which Professor Bonney lays much stress upon in the August Number, 1907, of the Quart. Journ. Geol. Soc., but, on the other side, let us realize the difference between two months (required for Professor Sollas' experiments) and hundreds of thousands or millions of years which must be allowed for the analogous process in the Alps. The analogy shows mountain-building to be a very slow, gradual process, in virtue of the smallness of plasticity, but it would be in the main a continuous process, and I do not think we are forced to assume its discontinuity, as Mr. Mellard Reade seems inclined to do.

The plasticity of rocks in greater depths is to be explained partly by elevation of temperature, partly by pressure. But whatever explanation we accept, there are too many evidences to deny the fact.

In conclusion, we must say Mr. Mellard Reade's paper is very instructive; indeed, it helps us to see, by contrast with the author's ideal example, what are the most essential features of the process as displayed in Nature.

12

Part II

INTUITIVE EXPERIMENTATION: "THE HIDDEN DÉCOLLEMENT"

Editor's Comments
on Paper 3

3   HALL

On the Vertical Position and Convolutions of Certain Strata, and Their Relation with Granite

It is to Horace Benedict de Saussure, the first of the distinguished band of Swiss Alpine explorers, that the experimental method can be traced. Saussure's experiments (1779–1796) involved the question of the fusion of granites; his results were negative. As Archibald Geikie explained: "He did not further advance along the path he had thus opened . . ."; thus the experimental approach was not applied by him to another field—*tectonics*—which he had anticipated through perceptive observations of natural phenomena. Proof of deformation was recognized in the tilted Valorsine conglomerates, which, Saussure had recognized in his "Voyage dans les Alpes," must have formed in a horizontal position. But on the whole it may be said that he dismissed subterranean distortion as an explanation for the typically irregular orientation of Alpine strata. It was, however, under Saussure's guidance that James Hutton "stood in imagination on the summit of the Alps, and watched from that high tower of observation the ceaseless decay of the mountains, the never-ending erosion of the valleys, and that majestic evolution of topography which he so clearly portrayed" (Geikie, 1905).

It is of some interest, therefore, that it is Hutton, with Playfair, who had joined Sir James Hall along the coast of Berwickshire in 1788, merely two years after the second of Saussure's quarto volumes had appeared. Here the strata exhibited "a succession of regular bendings, and powerful undulations, reaching from top to bottom of the cliffs, two or three hundred feet in height. These are occasionally interrupted, as might be expected, by the irregularities of the coast, by shifts and dislocations of the beds. . . ." Hall deduces (Paper 3, p. 82) that these strata, originally continuous and horizontal, have been compelled to assume their present contorted shape through "the exertions of some powerful mechanical force." In order to discover by what means this arrangement had evolved, he proposed to show (1) that this conformation could be given to a set of horizontal beds by a mechanical force of sufficient magnitude, and (2) that rational evidence existed for the exertion of such a force. The second point was approached by consideration of the possible consequences of observed volcanic phenom-

14

ena; the first point was approached by means of experiment. The first experiments in 1788, of magnificently crude design, were nonetheless effective; Hall's genius in the establishment of experimental methods in geology has been widely and rightfully celebrated. Subsequent experiments were conducted in the prototype "squeeze box." He concluded from his experiments that horizontal "thrust" (i.e., displacement) was required to produce the observed fold geometry; the remainder of the 1815 discourse was devoted to the geological implications of that assumption (cf. Paper 6, p. 42–43).

Several points are worthy of emphasis here. Hall's experimentally produced disharmonic buckle folds (Plate IV) do bear a close resemblance to natural phenomena, despite the fact that cloth is used as a model material. The use of cloth, of course, prohibits the development of local dislocations, which, in fact, had been observed by Hall in the field.

If we are to compare Hall's models with Buxtorf's (1915) Jura sections (e.g., for the area of Vellerat to Moutier) we are forcibly impressed by the geometric similarity. Perhaps the reason for this is that the *discontinuity* at the base of the folded Jura mass has been as accurately modeled as the folds themselves. In *both* instances we are dealing with a "thin-skinned" idealization (i.e., the deformation of sedimentary materials covering "rigid basement"); the décollement, or surface of "unsticking" or "ungluing", is as essential to the observed geometry as the deformability of the strata. Yet the mechanical significance of this décollement, invariably present in the models of Hall and subsequent workers by virtue (or defect) of a "rigid" model floor, was to go unemphasized—hence the phrase used herein—the *"hidden" décollement.* Indeed, the first field description of an overthrust fault was not to appear until 1826, and the recognition of décollement surfaces would not appear until much later.

Finally, it should be observed that the early vestiges of theoretical (as contrasted with experimental) tectonics may also be traced to Hall. In considering the cause of the observed displacements, Hall, who discovered the origin of dikes, pursued an approach wherein injected igneous material provides the motive force. Hall observes (Paper 3, p. 92) that "we should then have three forces more or less opposed to each other": (1) the force due to igneous liquid elevation (pressure head); (2) the superincumbent weight above the deforming strata; and (3) friction and inertia of strata propagating horizontally along their own beds. As a consequence of this force balance, "the strata, to a certain extent, would be thrust horizontally...." We, of course, proceed in a similar manner today in the application of the "laws of motion"; and for a consideration of the "obsoleteness" of Hall's views concerning the tectonic significance of fluid injections, the reader is referred to Paper 31.

3

Reprinted from pp. 85–86 of *Trans. Roy. Soc. Edinburgh*, 7, 79–108 (1815)

ON THE VERTICAL POSITION AND CONVOLUTIONS OF CERTAIN STRATA, AND THEIR RELATION WITH GRANITE

Sir James Hall

* * * * * *

This conjecture no sooner occurred, than I endeavoured to illustrate my idea by the following rude experiment, made with such materials as were at hand. Several pieces of cloth, some linen, some woollen, were spread upon a table, one above the other, each piece representing a single stratum ; a door (which happened to be off the hinges) was then laid above the mass, and being loaded with weights, confined it under a considerable pressure, (fig. 3. Plate IV.), two boards being next applied vertically to the two ends of the stratified mass, were forced towards each other by repeated blows of a mallet applied horizontally. The consequence was, that the extremities were brought nearer to each other, the heavy door was gradually raised, and the strata were constrained to assume folds, (fig. 4. Plate IV.), bent up and down, which very much resembled the convoluted beds of killas, as exhibited in the craggs of Fast Castle, and illustrated the theory of their formation.

I now exhibit to the Society a machine, by which a set of pliable beds of clay are pressed together, so as to produce the same effect, fig. 5. ; and I trust, that the forms thus obtained will be found, by gentlemen accustomed to see such rocks, to bear a tolerable resemblance to those of nature, as shewn in fig. 6., copied from the forms assumed in the machine, by an assemblage of pieces of cloth of different colours.

It still remains for us to consider how this *horizontal thrust* may have been produced.

* * * * * *

[*Editor's Note:* A row of asterisks indicates that material has been deleted.]

Figure 1 An ideal portion of a coast similar to that of Berwickshire.

Figure 2 The same, with continuity of strata given by dotted lines.

Figures 3–6 See text.

Editor's Comments
on Papers 4 Through 6

4   CADELL
    Experimental Researches in Mountain Building

5   JOHNSON
    Physical Processes in Geology

6   WILLIS
    The Mechanics of Appalachian Structure

By 1832 Dumont, the "father" of Belgian geology, had published on the geology of the Liège area and carefully defined the geometry of isoclinal folds. The referees who introduce his memoir for the Brussels Academy remark that this work suggests the "gliding of a section of the earth's crust down an inclined plane with resultant lateral pressure" against the ground obstructing the path. This is not the beginning of gravitational tectonics, which concept can be traced back at least to Gillet-Laumont in 1799 (Haarmann, 1930), but it may be the first in which fold geometry is employed to suggest that mechanism.

The first publication on the overthrusts of Canton Glarus by Escher von der Linth followed in 1841; the Rogers brothers' work on the physical structure of the Appalachians appears soon after (1843), as does Dana's proclamation on the "global cooling" (contraction) hypothesis for tectonic phenomena. The international recognition immediately given to these and similar studies may seem surprising, but the worldwide community of geologists was small, and their outlook was, at that time, anything but provincial.

At that happy time, wide geological traveling was commonplace, as illustrated by Sir Roderick Murchison's 1849 communication "On the geological structure of the Alps, Appenines, and Carpathians." He observed the order of Alpine structural operations to have been "first contortion, and then fracture," with the "nuclei of the folds" and the lines of dislocation being parallel to each other and to the great axis of the chain: "As far as mere outline goes, the undulations at one locality seem to conform to the wave-like progression so ably laid down by Henry Rogers and William Rogers" in the Appalachians, with the steeper sides of the anticlines most remote from the axis. Henry

Rogers has even expressed an opinion concerning the origin of Jura deformation (Murchison, 1849, footnote p. 253): judging from their observed form (i.e., with long slopes apparently parallel to the French side), he infers that the propelling force came from the Black Forest. [This false judgment is much later attributed by Bailey (1935, p. 31–32) to the misleading irregularity of the Jura folds observed by Rogers.] Murchison is escorted through the Glarus region by Escher, who "ingeniously urges me to try in every way to detect some error in his views, so fully was he aware of the monstrosity of the apparent inversion (p. 248)"; but the overthrust does exist, and Murchison, now nearly 60 years old, returns to the geology of the Highlands, rearmed with detailed knowledge of Alpine and Appalachian dislocations. It seems therefore ironic that Murchison, an excellent geologist who had mapped in the eastern Alps—who was responsible for founding the Silurian System and, most especially, who brought the Alpine discoveries of Escher and contemporaries to the attention of the English-speaking public—should not have recognized the existence of the major overthrust on his home grounds. Yet his role in the discovery was a major one.

This story is an important one in the history of tectonics. Following McIntyre (1954), the superposition of gneiss upon limestone and quartzite in the Eriboll district of Scotland was recorded first in 1819, with the first interpretation, appearing two decades later, that the gneisses somehow represented the younger rocks. Years later, in 1854, C. W. Peach discovered lower Paleozoic fossils in the limestone; its significance was realized by Murchison, who had worked in the area a quarter-century earlier in the company of Sedgwick, and in the following summer he returned there with his friend James Nicol. After further investigation, Nicol concluded (1857, p. 36): "whether we consider the gneiss resting on the latter (quartzite and limestone) as a newer metamorphic group, or merely as a portion of the lower gneiss forced up over it in some great convulsion, we have still views of very great interest opened up to us in the history of these Highland Mountains." In 1859 Murchison maintained, however, at the British Association Meeting at Aberdeen, that the stratigraphic passage from Lewisian to the schists of A'Mhoine peninsula was not only upward but conformable. Nicol, holding the Chair of Geology at Aberdeen University, dissented. In the following year Murchison published a supplement in which Nicol's dislocation hypothesis was denied; Nicol maintained, however, that the metamorphic Moine Schists could not have formed *in situ* above unmetamorphosed sediments, and pointed out the analogy with Alpine overthrusts on the basis of Murchison's own observations. However, several errors existed in Nicol's presentation, and these, coupled with the publication in the same year of a paper by Murchison and Archibald Geikie (1861), which again found authorative support, temporarily closed the controversy (McIntyre, 1954).

The Moine controversy reopened in 1878 with H. Hicks' publication on the region of Loch Maree. Hicks agreed more with Nicol than with "Murchison and the Survey," as he phrased it in 1897, for Murchison's successors as Director-General of the Survey, Ramsay and Geikie, chose to support Murchison's position. Nicol died in 1879, four years before Lapworth and Callaway were to publish their significant convictions concerning the geology of Durness, Eriboll, and Assynt. Lapworth, who later (1885) would coin the term "mylonite," was taken seriously ill with the excitement of his discovery. Indeed, his vivid imagination made him believe that the movements had not ceased! Only the first part of Lapworth's "secret of the Highlands" was published (in 1883); in it overfolds pass into overthrusts, and Lapworth acknowledges the inspiration of Heim's (1878) *Untersuchungen über Mechanismus der Gebirgsbildung.*

The same concept, it may be noted, had appeared in Murchison's (1849) report. Similarly, Callaway (1883) concluded that the Moine Schists had been thrust over the Cambrian; he called attention to Alpine analogs as cited in Geikie's text. Still, no opposition to these views were expressed at the Geological Society, for Geikie intended to place into operation his own plan to end the controversy; thus Survey geologists B. N. Peach and John Horne were sent in 1883 to the Northwest Highlands: "By the end of the following summer these careful observers had advanced so far as to be convinced that Murchison's interpretation of the ground could not be maintained. . . . Naturally I could not bring myself to abandon it until I had seen with my own eyes such evidence as would convince me of its error . . ." (Geikie, 1924, p. 214). Soon convinced by the "overwhelming" evidence, the Survey work was quickly summarized in a report and published in *Nature* one month later (Geike, 1884), wherein Geikie introduced the term "thrust-plane" in the literature. By 1888 the Survey was persuaded that the Alpine analogy of "overfolding preceding faulting" was inapplicable to the Highlands (Peach et al., 1888). That same year (1888) also witnessed publications of the first sketches of Törnebohm's *ofverskjutning* theory of Scandinavian deformation, with postulated overthrust displacements as great as 100 km, and Cadell's experimental production, excerpted as Paper 4. Henry Cadell, a British Survey geologist, was coauthor with Peach and other colleagues in the 1888 report previously referred to. After studying the Moine thrust phenomena in the field, it occurred to Cadell and his colleagues that further light on the causative deformational processes might be obtained by laboratory study. Accordingly, Cadell, with the approval of Geikie, instituted in 1887 a series of experiments.

Cadell was, naturally, familiar with the earlier experiments of Hall, Favre, Pfaff, and Mellard Reade; however, from observations of Highlands structure, he was chiefly concerned with brittle rather than ductile deformation. Hence, the model materials chosen for most experiments were chosen to effect dominantly brittle behavior. Cadell's *General Summary of Results* is quoted by Willis (Paper 6) and is therefore not reproduced here.

By 1896 the entire outcrop of the Moine Thrust had been mapped from Loch Eriboll to Skye; elsewhere, in the Scandinavian Caledonides, Törnebohm (1888) demonstrated 130 km of thrust displacement, enormous in comparison with the Peach and Horne estimate of 16 km for Scotland. The classic memoir on the geological structure of the Northwest Highlands of Scotland, published in 1907, was edited by Geikie after his Survey retirement. Thus closed an heroic era: the Highlands, in Suess's words, were rendered transparent.

In North America, meanwhile, the existence of overthrusts was becoming known. In the Canadian Appalachians, Sir William Logan (1861) postulated a zone of great thrust faults, and proceeded to trace the boundary thrust throughout its Canadian course; this boundary would later be recognized as "Logan's Line." Somewhat earlier, at the same time that James Hall and associates were recording the "New York System" as a standard for North American stratigraphy, the brothers Rogers were unraveling the succession of rocks in Pennsylvania and the Virginias, i.e., the central Appalachians (Fig. 1). Their synthesis reports on Valley and Ridge tectonics appeared at about the same time as Escher's Glarus work and were of comparable influence. In southwestern Virginia, W. B. Rogers recognized the great thrust faults typical of the southern Appalachians; he and his successors for a generation considered them as upthrusts along steep planes (Rodgers, 1970). His work was carried across into Alabama by J. M. Safford and E. A. Smith. There Safford (1869) discovered Valley and Ridge-type deformation on the Cumberland Plateau, and Smith (1893) first suggested the underthrust concept in order to account for observations of symmetry reversal. Toward the end of the century, the U.S. Geological Survey sent M. R. Campbell, C. W. Hayes, and Arthur Keith with adequate topographic base maps to the southern Appalachians; their discoveries included the clear recognition of overthrust faults, in contrast to the higher-angle Appalachian faults previously observed.

In the light of these new observations, Bailey Willis summarized the salient features of Appalachian tectonics and attempted to consider the question of mechanics of Appalachian deformation. His procedure involved the use of "squeeze-box" model experiments; he attempted to derive certain empirical "laws" from his experiments and then considered their significance to natural phenomena. His approach is

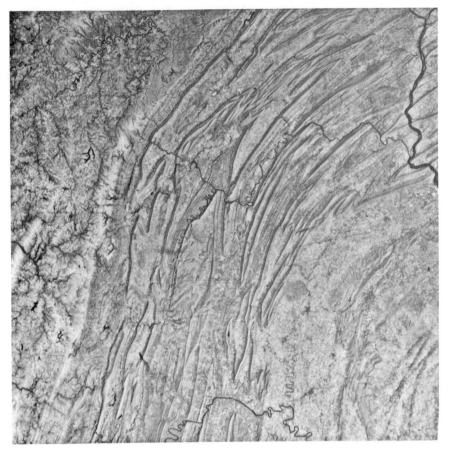

Figure 1 The Appalachian Mountains. The Valley and Ridge Province of Henry D. and William B. Rogers, and Bailey Willis, bordered on the west by the Appalachian Plateau Province and on the east by the Triassic Basin and Piedmont Province. The Susquehanna River is on the upper right, the Potomac River at the center base. (ERTS image, courtesy of NASA.)

clearly outlined in Paper 6 in an excellent section, "Problems of Structural Experiments." Willis admitted to the difficulty of comprehending the relative proportions of involved elements. However, his "bridge" analogy is instructive; he made use of available data on rock strength, he was aware of the brittle–ductile material transition, and he had a surprising understanding on the effect of stratification on structural flexibility and transmission of boundary forces. His use of lead shot as model overburden appears ingeneous. More important, his approach proved successful: he discovered that the conditions which determine the locations of buckling instability and break thrusting are inherent in the geometric attitudes of the layered material, and are not, in general, concerned with the nature of the applied force. This fundamental concept has not been accorded the recognition it deserves, perhaps because of criticism attached to Willis' work by other workers preoccupied with laws of similitude. An example of such criticism follows: "If one of Willis' models were to be regarded as representing an Appalachian section of about 100 km in length the overburden used would be equivalent to 150 to 300 km of rock. Even then the more rigid strata broke into discrete slabs and open cavities appeared. The Appalachian had no such overburden, but were held down solely by their own weight, yet the folding in the most rigid strata was continuous and no open cavities were formed. Hence, it must be concluded that the rocks of the Appalachians in respect to their environment were much weaker than the materials used by Willis, or conversely, that Willis' materials were much too strong to represent the Appalachian Mountains" (Hubbert, 1945, p. 1652).

Nonetheless, the question may be properly asked as to whether or not the scaling of rheologic properties is a necessary requirement for model experiments in which a geometric factor is the principal variable for the concept under investigation. The correct answer seems, emphatically, no. Because the employment of boundary conditions seems to be a more critical issue than material property idealization, I have prefaced Willis' report with a pertinent excerpt from Arvid Johnson's unique volume *Physical Processes in Geology*, a modern classic in its own right. Johnson has devoted an entire chapter to the question of fold trains in the Appalachian Mountains, a theoretical approach to some problems of folding as formulated by Willis; the selection reproduced as Paper 5 should help the reader to place Willis' work in modern perspective.

One final point requires discussion. In Willis' conceptual idealization, the thick crystalline basement underlying the superficial crust was assumed to be latently "plastic"; this basement was assumed to participate in the folding process. In the experiments, however, this "plastic basement" was idealized as a soft layer, only 2–6 cm thick, overlying the oak plank floor of his wooden box; this plank could be considered as rigid, as model dimensions were small (typical model length of 1 m; typical total thickness, excluding lead shot, of 5–15 cm). Willis' models therefore seem analogous to deformation over a décollement zone, rather than, as was intended, deformation over a quasi-infinite "plastic" basement. Ironically, with respect to Appalachian-type deformation, model accuracy seems to have been enhanced by this "error." The fact that the "hard beds" (for most models) transmitted most of the applied force to the far end of the model "vise" suggests that viscous and/or frictional transfer to the "rigid" model basement was small. The base of Willis' models acted as a décollement in a manner quite comparable to Buxtorf's later suggestion, which today seems as applicable to the Appalachians as to the Jura.

In retrospect it seems of interest that, most recently, workers have retreated from insistance on a rigid and entirely undeformed basement. Basement and Paleozoic irregularities exist but seem to have formed earlier than the buckle folds and associated faults and hence could not have "caused" them (Jacobeen and Kanes, 1974); however, in full accordance with Willis "initial dip" concept, such features could have controlled the location of subsequent instabilities. The same seems true in the Jura.

4

Reprinted from pp. 339–343 of *Trans. Roy. Soc. Edinburgh*, Pt. 1, 337–357 (1889)

EXPERIMENTAL RESEARCHES IN MOUNTAIN BUILDING

Henry M. Cadell

Geological Survey of Scotland

* * * * * *

The experiments were of three distinct kinds. The first series (A) was designed to explain the behaviour of different types and arrangements of strata when pushed horizontally over an immovable surface. The object of the second series (B) was to ascertain if possible how gently-inclined " thrust-planes" may have originated, and to trace their connection with " fan structure " and other phenomena observed in mountain systems of elevation. The third series (C) was conducted on principles suggested by the experiments of FAVRE, who placed layers of clay on a stretched india-rubber band, and on allowing it to contract, produced miniature mountain ridges by the wrinkling of the surface of the clay. I extended FAVRE'S experiments by removing the upper layers of the wrinkled clay, and observing the effect of the contraction on the deep-seated portions of the miniature mountain system.

A. *Thrusting over an Immovable Surface.*

The strata were formed in a rectangular box 6 or 8 inches broad and 3 to 5 feet long. One end of the box was movable, and could be pushed in so as to compress longitudinally the strata inside. At the beginning of the experiments, the sliding end piece, which may be called the pressure board, was pushed in, either by hand alone, or if the force required were considerable, with the help of a lever. In the last and most complete series of experiments, the pressure was applied by means of a strong screw running in bearings bolted to the prolonged sole of the box. The sides could be removed at pleasure, when it was desired to examine the section of the distorted strata inside. The figure on the following page gives a general idea of the size and nature of the whole apparatus.

In proceeding with the experiments, after the pressure board had been pushed in far enough to produce some marked change in the internal structure of the mass, the side of the box was removed, and the vertical section thus exposed was pared along the edge with a sharp knife, to reveal the beds clearly and remove all traces of friction with the wood. If the results were of interest the section was accurately sketched, traced, or in most cases photographed, a measuring tape having previously been attached alongside of the section to show the scale of the operations in feet and inches.

If it were desired to continue the experiment, the side was replaced, and the pressure board pushed further in, after which the new section was examined and recorded as before. In several of the experiments the process was repeated four or five times to indicate the successive steps in the formation of the ultimate structure of the mass.

The accompanying figures, selected from some sixty drawings and photographs of the sections obtained in the experiments, tell their own tale, and require but little description.

[*Editor's Note:* A row of asterisks indicates that material has been deleted.]

In some cases the section, although easily understood in the laboratory, is much less effective when seen in photographic form. In such cases, to make the meaning clear and bring out the structure effectively, a diagrammatic section from the photograph itself and from notes made at the time it was taken is exhibited alongside.

Experiments in Mountain Building.

In nearly every instance the pressure was exerted from the right, and the pressure board is seen in the figures supporting that end of the section. The thin white streaks are the edges of the plaster of Paris laminæ separating thicker layers of damp sand. These white beds were often so hard as to offer much resistance to the paring knife, and great care was required to avoid tearing the material out in cakes, and leaving cavities between the softer sand beds.

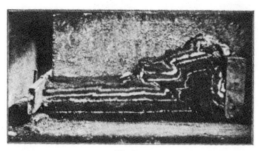

Fig. 1.

In this experiment (fig. 1) the strata had a thickness of about 4 inches, and consisted of damp sand with three thin bands of stucco. As soon as pressure was applied, the material immediately in front of the pressure board began to swell up into an anticline in exactly the same way as Pfaff's strata of loam and papier-maché pulp were observed to do. The right limb of the anticline being pressed in, gradually assumed a vertical position parallel to the face of the pressure board. At this stage the pressure from the right was uniformly distributed from the crown of the arch downwards. But the resistance on the left was only exerted as far up as the level of the surface of the undisturbed strata, so that the part of the arch above this level was free to travel forwards. It did so for a short space, and produced the monocline at the top of the section. Had the upper strata been more rigid, they would not have become bent into such a form, but would have snapped at once, and formed a reversed fault, as has indeed been done in the thin bed immediately at the surface. It was, however, impossible for the upper part of the anticline to move far forward with an increase of pressure from the right, since all particles in the same vertical plane were subjected to equal pressure from that quarter. In the lower part of the section, the horizontal pressure from the right was met by the horizontal longitudinal resistance of the strata combined with the vertical statical resistance of the sole, so that the resultant force tended to shear the strata obliquely along a series of planes inclined towards the right. The stiffness of the beds now came into play, and prevented this shearing strain from being distributed throughout the mass. Instead of this, the brittle strata snapped at one point, and all the movement was concentrated along the line of weakness thus produced. The whole mass above this thrust-plane moved obliquely upwards and forwards, and all interstitial movement ceased. The thrust-plane or reversed fault did not start directly from the bottom of the pressure board, but met the fixed sole a short distance in advance. As soon, however, as the front of the pressure board reached the point of the wedge of undisturbed

27

strata below the fault plane, the shearing ceased, and the forward motion was temporarily arrested. The photograph taken at this stage is reproduced in fig. 1.

Fig. 2.   Fig. 2a.

A new mass of strata was now brought under the influence of the pressure. In this case the beds were subjected to a certain amount of vertical pressure due to the heaping up of the strata on the slope of the wedge, in addition to the horizontal thrust from the right. The resultant shear plane might therefore be expected to meet the fixed sole at a slightly lower angle than before. An inspection of fig. 2 will show that this has been the case. The upper reversed fault is slightly steeper than the lower. Much importance is, however, not to be attached to this difference in hade, as the weight of the piled-up strata in such experiments is small in comparison with the horizontal force, and slight differences in the hade of successive reversed faults might be due to other causes. Fig. 2a is a diagrammatic representation of fig. 2, which is not very clear.

 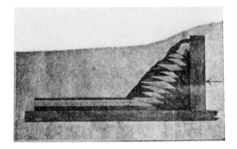

Fig. 3.   Fig. 3a.

In this case (fig. 3) the depth of strata was only $1\frac{3}{4}$ inches. The breadth was 8 inches, and the section was pressed in from an original length of 44 inches to a space of 15 inches. The same process went on here as in last experiment, but the cumulative effect is better displayed. A small overfold produced at the beginning of the experiment may be seen at the top of the section, but afterwards the strata underwent a process of piling up in separate slices by slightly-inclined reversed faults. The back part of the accumulating mass slipped vertically up the face of the pressure board as each new wedge was driven under the base of the slope in front. Fig. 3a is a diagrammatic representation of this structure.

These structures are almost identical in character with the structures found in advance of the great thrust-planes in Sutherland, where a similar process of heaping up of Silurian quartzites, shales, and limestones has at places given these strata an abnormally thick appearance, which, before the discovery of the thrust-planes, was quite inexplicable. For this arrangement of strata I would propose the name of "wedge structure."

*   *   *   *   *   *

28

5

Reprinted from pp. 130-131 of *Physical Processes in Geology,* Freeman, Cooper & Company, San Francisco, 1970

PHYSICAL PROCESSES IN GEOLOGY

Arvid M. Johnson

Stanford University

* * * * * *

G. K. Gilbert was instrumental in persuading Bailey Willis to apply the method of mechanical processes to a study of folds in the Appalachian Mountains. In the course of Willis' study, he introduced half a dozen fundamental concepts of folding, such as the effect of initial dips of sedimentary strata and of vertical forces on shapes and distributions of folds, as well as the concept of the strut member, which we will investigate in the next chapter.

Willis used very little mathematical theory during his study, but he was an unusually capable experimentalist; rather than use mathematics to check an idea, he performed a series of experiments. For example, in order to be sure that he correctly deduced the concept of initial dips, which states that a change in dip of strata can localize a fold, he performed tens of experiments with his squeezing machine. He built initial dips into the experimental strata at different distances from the driving piston. Each time a fold appeared in a predictable position, so that his concept was verified.

Willis was moderately concerned about producing a scale model of the folds in the Appalachian Mountains, and some critics claim that his study is not especially valuable because he did not produce a scale model of the natural folds. His use of

[*Editor's Note:* A row of asterisks indicates that material has been deleted.]

lead shot to load the experimental materials, for example, has been stated to be unrealistic because similar loading must have been absent in the Appalachian Mountains at the time the folds formed. The conclusion is that all of his results are suspect. These critics, however, do not understand what Willis was trying to do, nor do they understand experimentation.

Willis used the experiments to test and to illustrate his theories about the occurrence of certain anomalously large folds, such as Nittany arch, distant from the presumed source of the load applied in the southeast. Also, he wanted to test his theory about the effect of competence of strata on the transferral of loads across a wide area such as the Appalachians. It is unimportant whether or not he used a scale model of the Appalachians to check these theories experimentally. He might just as well have used materials such as corn syrup and a light oil, or hard and soft rubber strips, instead of the hard and soft wax layers he did use in his experiments. His theories relate to boundary conditions and mechanics, not to rheology, so that the rheological behavior of the experimental materials has no real bearing on the validity of experiments designed to test and develop his theories. An initial dip, for example, is a geometric feature, not a rheological property.

When we perform experiments and begin to worry about scaling factors and scale model theory, we would do well to ask ourselves this question: Does scale really have anything to do with the problem? Usually it does not. Also, contrary to much that has been written in the geological literature, the usual procedure of trying to guess what variables are important in a situation and of performing certain dimensional operations with the variables seems to be so much bunk for many if not most problems in geology. First, if we do not know enough about a problem to write the differential equations and the boundary conditions that describe it, we probably cannot *guess* which variables are important. Second, if we can write the differential equations and the boundary conditions, we need not bother ourselves with guesswork. All the variables are contained in the equations and they can be manipulated fairly easily to produce scale factors. For example, see the excellent book on dimensional analysis by Ipsen (5). Finally, if we can solve the differential equations for the conditions of both the model and the prototype, dimensional analysis will tell us nothing more than is already contained in the solution.

In my opinion, the value of experimentation in the mechanics of earth materials is not to produce scale models but, rather, to investigate certain *concepts*, such as Willis' concept of the effect of initial dips on folding, and to stimulate the imagination of the experimenter.

* * * * * *

REFERENCE

5. Ipsen, D.C., 1960, *Units, Dimensions and Dimensionless Numbers:* McGraw-Hill Book Co., Inc., New York.

6

Reprinted from pp. 217, 219, 222–223, 224, 227–228, 229–245, 247, 249, 250, 251, 268–269 of *U.S. Geol. Survey 13th Ann. Rept.*, 1891–1892, pp. 217–281

THE MECHANICS OF APPALACHIAN STRUCTURE.

By Bailey Willis.[1]

INTRODUCTION.

The facts of the following discussion are drawn from the belt of disturbed Paleozoic strata which extends from New York, through Pennsylvania, the Virginias, and Tennessee, to Georgia and Alabama.

This is an area of about 60,000 square miles—900 long and 50 to 125 miles wide. It is a geologic province distinguished by the age of its strata from the region on the east and by the facts of its structure from the horizontal rocks on the west. Toward the east extend crystalline rocks much older than the Paleozoic and part of that continent which yielded the materials for Paleozoic sediments. On the west is the area over which the mediterranean sea of North America prevailed during the periods from Cambrian to Carboniferous. Between the continental edge and the open sea was the narrow belt where mechanical and organic sediments accumulated in great bulk. This strip is the zone of strongly developed structural deformation.

The phase of deformation is that which follows from compression. Across this zone the arc once covered by the strata has shortened and the greater length of the beds has been taken up by folding and faulting. The folds and faults formed on a vast scale, with simple relations among themselves, and conditions of erosion have led to the development of a relief in close accordance with the occurrence of hard and soft rocks. Hence it follows that the general character of structure in this region is easily recognized, and through the great work of H. D. Rogers, followed by W. B. Rogers, Lesley, Safford, and many others, it has become widely known as a definite type. The term "Appalachian structure" conveys in geologic literature the idea of strata compressed into long narrow folds, generally parallel among themselves, and sometimes overturned, and overthrust.

* * * * * *

[1] To Mr. G. K. Gilbert and to my associates in the Appalachian province. Messrs. Hayes, Keith, and Campbell, I am indebted for many facts and for frank discussions of hypotheses, which have greatly aided in the preparation of this paper.—B. W.

[*Editor's Note:* A row of asterisks indicates that material has been deleted.]

TYPES OF STRUCTURE.

* * * * * *

CLASSIFICATION OF FAULTS.

The term fault is of general application to any dislocation of rocks involving movement of the separated masses past one another, and it therefore covers what has been called the normal or radial fault, which is a phase of deformation involving extension of an arc of the earth's crust. This type is precisely the opposite of the dislocation which arises from compression and which has been called reversed fault, compression fault, or thrust fault. Long usage has so identified the word fault with what is called the normal type, that clearness and precision are gained by employing a substitute for it in describing dislocations due to compression, and for this purpose there is none better than *thrust*.

Thrusts may arise from any one of four sets of conditions, and if classified genetically may be called—

First. The shear-thrust; this arises when either force or resistance is so concentrated as to produce a plane of easiest motion, along which shearing meets with a resistance less than that opposed by the strata to bending. The Scotch geologists first described this type, and it is illustrated by a figure from the article on the northwest Highlands.[1] (Pl. LIII.)

Second. The break-thrust; this develops when strata form first an anticline, so conditioned that in process of development folding soon becomes more difficult than breaking, followed by overthrust on the fracture plane. This is the characteristic type of faulting in the Appalachian province, and it is illustrated by drawings based on the interpretation of observed facts. (Pl. LIII.)

Third. The stretch-thrust; this is the result of extreme folding, with development of an overturned limb, which is stretched by the opposite pressures of the roof and floor. This type has been described by Heim, and is illustrated by diagrams from Mechanismus der Gebirgsbildung. (Pl. LIII.)

The shear-thrust is independent of flexure; the break-thrust follows moderate folding; the stretch-thrust is a final phase of a closed and overturned fold.

Fourth. The erosion-thrust; this may develop when a rigid stratum rises from a broad syncline to outcrop on an eroded anticline. Then, if compression follows, the stratum meets with no resistance and rides forward over the subaerial surface. Such a thrust, complicated indeed by a break-thrust, is shown by Hayes,[2] and a simpler form is suggested in Pl. LIII. The difference between the two illustrations lies mainly in the relative ages of the strata brought into contact.

[1] Recent work of the Geological Survey in the Northwest Highlands of Scotland. A. Geikie. Quart. Journ. Geol. Soc. for August, 1888.

[2] The Overthrust Faults of the Southern Appalachians. Bull. G. S. A., Vol. 2, pp. 141-154.

U. S. GEOLOGICAL SURVEY

THIRTEENTH ANNUAL REPORT   PL. LIII

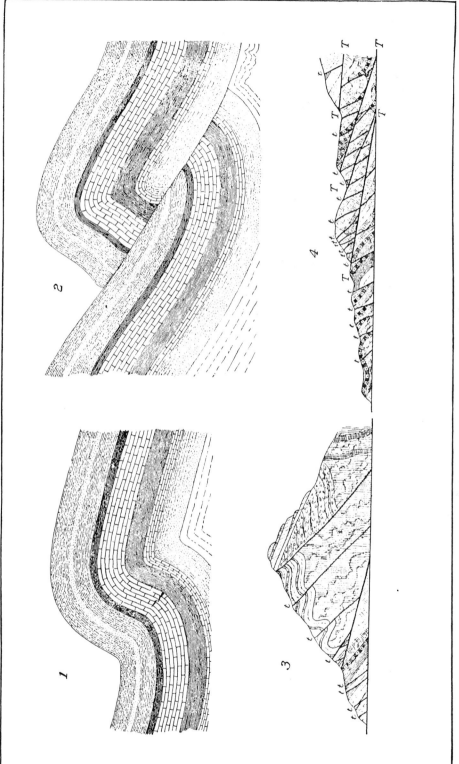

TYPES OF THRUSTS.

1. Step-fold, showing break in the massive limestone bed which determines the plane of the break-thrust (2), along which displacement results from further compression.
3 and 4. Examples of shear-thrusts from "Recent work in the Northwest Highlands of Scotland," by A. Geikie, 1888.
3. Horizontal section from Loch Assynt across the Silurian limestones to Cnoc an Droighinn (about three-quarters of a mile in length).
4. Horizontal section from Bealloch across Coinne-mheall to Corrie Mhadaidh (about half a mile in length).

U. 8. GEOLOGICAL SURVEY

THIRTEENTH ANNUAL REPORT   PL. LIV

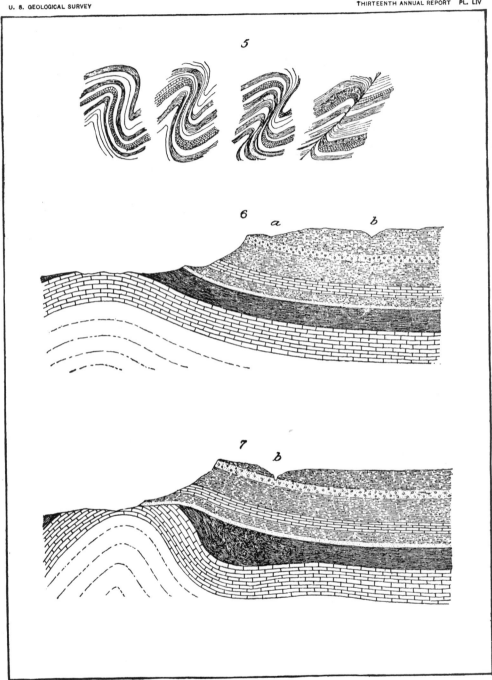

TYPES OF THRUSTS.

5. Stretch-thrust (Faltenverwerfung) developed from an overturned fold by stretching of the middle limb (after Heim).
6. Erosion-profile and section of a simple anticline.
7. Erosion-thrust developed from the condition shown in 6 by compression from the plateau side, accompanied by continued erosion.

STRUCTURAL DISTRICTS OF THE APPALACHIAN PROVINCE.

In the Appalachian province there are four districts, each of which is distinguished from the others by a prevailing structural type. These districts are as follows: *

(1) District of open folding: Alleghany region of Pennsylvania and West Virginia.

(2) District of close folding: Appalachian valley.

(3) District of folding and faulting: southern Appalachian region of Virginia, Tennessee, and Georgia.

(4) District of folding with schistosity: Smoky mountain region.

In regard to the relations of faults to folds we may quote from the writing of H. D. and W. B. Rogers of 1841. After describing faults transverse to the strike they say:

The other far more conspicuous class of dislocations connected with these crust undulations are the great longitudinal ones. These are of frequent occurrence in the more contorted portions of the Appalachian zone, especially in those where the chain is convex to the southeast, and in the straight sections of southwestern Virginia and eastern Tennessee. But I am persuaded from the descriptions of geologists and from my own observations that the fractures of this class are equally numerous in the Jura mountains, in the Alps, in the district of the Ardennes, in Belgium, and in the mountain chains of Scotland. A leading feature of these great fractures is their parallelism to the main anticlinal axes, or lines of folding of the chains to which they belong. They are, in fact, only flexures of the more compressed type, which have snapped and given way in the act of curving or during the pulsation of the crust. They coincide, in the great majority of instances, neither with the anticlinal nor the synclinal axis planes of the waves or folds, but with the steep or inverted sides of the flexures, and almost never occur on their gentler slopes. This curious and instructive fact may be well seen in the Appalachians of Pennsylvania and Virginia, and by tracing longitudinally any one of their great faults from its origin on the steep flank of an anticlinal wave along the base of its broken crest to where the anticlinal form is again resumed. The following brief description from our memoir on the physical structure of the Appalachians, taken from the transactions of the American Association, will show the general phases through which these fractures pass.

From a rapidly steepening northwest dip, the northwest branch of the arch (or flank of the wave) passes through the vertical position to an inverted or southeast dip, and at this stage of the folding the fault generally commences.

It begins with the disappearance of one of the groups of softer strata lying immediately to the northwest of the more massive beds, which form the irregular summit of the anticlinal belt or ridge. The dislocation increases as we follow it longitudinally, group after group of these overlying rocks disappearing from the surface, until in many of the more prolonged faults the lower limestone formation (Cambrian or Lower Silurian) is brought for a great distance, with a moderate southeast dip, directly upon the Carboniferous formations. In these stupendous fractures, of which several instances occur in southwestern Virginia, the thickness of the strata engulfed can not be less in some cases than 7,000 or 8,000 feet.

It does not appear that the Rogerses had accurate knowledge of the wonderful system of parallel thrusts extending through eastern Tennessee into Georgia, nor did Safford, who traced many of them, have a map adequate for the accurate delineation of their structural relations.

*See Plates LV–LVII (Ed.).

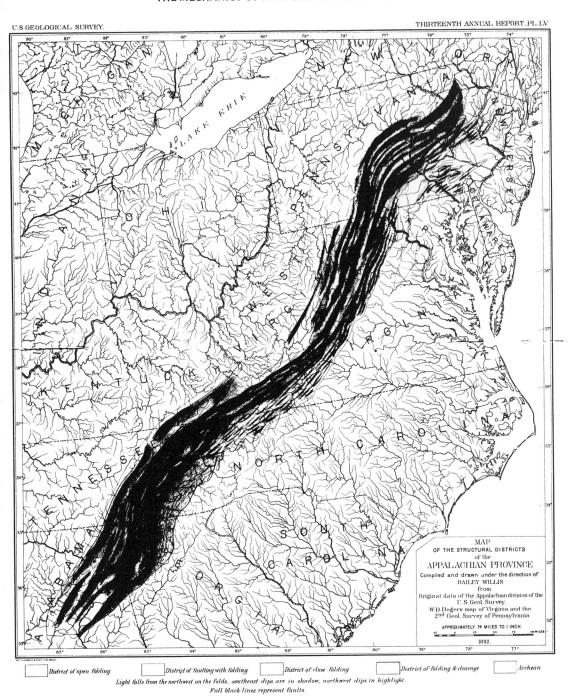

MAP
OF THE STRUCTURAL DISTRICTS
of the
APPALACHIAN PROVINCE
Compiled and drawn under the direction of
BAILEY WILLIS
from
Original data of the Appalachian division of the
U.S. Geol. Survey.
W.B.Rogers map of Virginia and the
2nd Geol. Survey of Pennsylvania

APPROXIMATELY 74 MILES TO 1 INCH

1892.

District of open folding   District of faulting with folding   District of close folding   District of folding & cleavage   Archean

Light falls from the northwest on the folds, southeast dips are in shadow, northwest dips in highlight.
Full black lines represent faults.

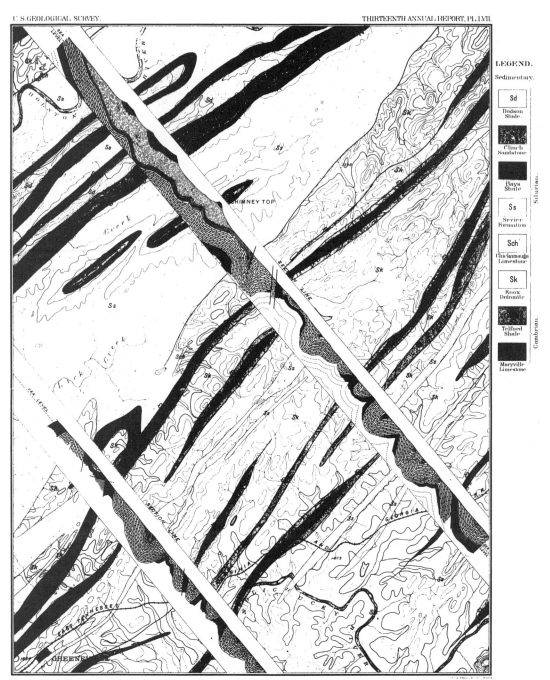

SPECIAL MAP OF DISTRICT OF CLOSE FOLDING.
(PART OF GREENVILLE, TENNESSEE, ATLAS SHEET.)

GEOLOGY BY ARTHUR KEITH.
SECTIONS BY BAILEY WILLIS.

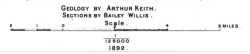

1892.

But the more recent work of the U. S. Geological Survey has mapped them from Georgia into Virginia, and the results are given on a small-scale map in Pl. LVIII. Within the area of this map there are 15 to 20 thrusts according to the distinctions made between thrusts and mere branches, with an aggregate length of about 4,500 miles. The longest single thrust extends northeast beyond the map and reaches a length of 375 miles. Inspection shows that they are intimately associated with folds, and whenever a fault fades out it is in the northwestern side of an anticline and in the direction of anticlinal pitch. This is in agreement with Rogers's observations, but his statement regarding the law of relation of thrusts to folds may be formulated with more general application, as follows: Appalachian thrusts arise in such relations to folds that the adjacent axis in the under thrust is always synclinal and the adjacent axis in the overthrust is anticlinal. But displacement may be so great as to override and bury the former, while erosion may remove all traces of the latter; hence a fault may appear between an anticline in the downthrust and a syncline in the upthrust, or between two synclinal or two anticlinal axes. When great displacement and erosion combine to destroy the evidence of folding the result is an isoclinal faulted mass.

These faults of great length, dividing the superficial crust into crowded scales, have provoked the wonder of the most experienced geologists. The mechanical effort is great beyond comprehension, but the effect upon the rocks is inappreciable. The strata beside a great fault are but rarely brecciated, squeezed or rendered schistose. The shearing planes are sharp and clean, the movement of overthrust was concentrated as by a knife cut, and the passing layers ground little grist one from another. Great vertical pressure and very slow movement probably conduced to this result, but however explained the fact is conspicuous that Appalachian thrusts are not associated with alteration of the faulted strata.

* * * * * *

THE STRUCTURAL PROBLEM.

The structural problems of the Appalachian province are indicated in the brief description of these four districts. It is observed that the strata have been tangentially compressed and the cause of that compression is the ultimate question. We can only know the force by its effect and we must clearly understand the determining conditions before we can approach the unknown cause. Let us reconsider the facts:

In order of development it has been usual to recognize the open fold, the closed fold, and the fault. How are these related to each other? Are they necessary stages of deformation from the flat strata to the last expression of the force? The open fold must precede the closed, but need one anticline close before the pressure can raise an adjacent one? Must faulting ensue when flexure reaches a definite phase? Over a large area many open folds lie side by side; clearly, conditions existed which permitted or required the growth of several arches simultaneously or the

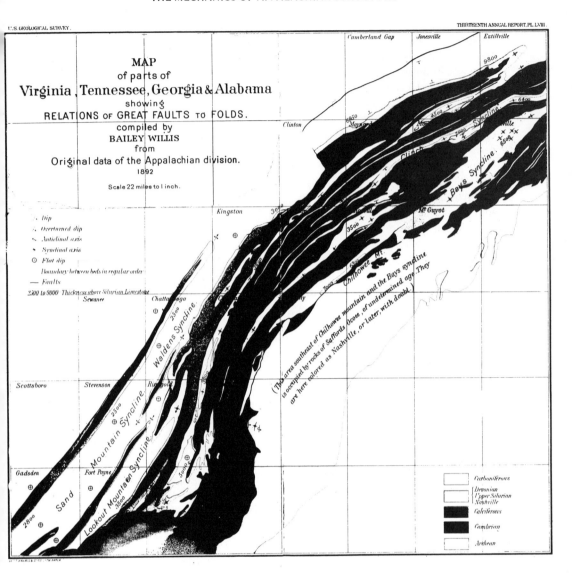

force failed to close one fold before it caused another; so continued compression may not close an open fold. Again over the valley district, where closed folds prove great compression, faults are few; yet they are numerous in the zone of comparatively moderate folding farther south. If the phase of folding in the latter area was alone appropriate for faulting, why did the strata in the valley district pass through it unfaulted? The fact stands as proof that faulting arises from some conditions more or less independent of the phase of folding; it is conceivable that faults may sometimes be independent results.

Folds lie here side by side, parallel, one far larger than its fellows, the others among themselves approximately of the same magnitude. Such are the relations of the Nittany arch to the group of minor folds southeast of it. And related to these is the Broad Top coal basin, a gently flexed syncline. What condition located and limited the Nittany fold? Why is it greater than its foot-folds, as a mountain range is higher than its foothills? Why did the Broad Top basin escape or resist compression that raised the anticlines which pitch into and die out in it? Or in the far southern field, what condition determined the parallel but widely separated anticlines of Alabama?

Over the entire province there is likeness of phenomena which argues unity of cause, but there is variety of effect, which suggests unlikeness of conditions. The conditions antecedent to deformation were the result of sedimentation. Does the distribution of strata afford any answer to the questions raised? The sedimentary deposits of the province are capable of threefold division: there is a laminated base, a massive middle, and a laminated top. The dominant fact of stratigraphy is the continuous limestone—the middle member—hard, resistent, relatively inflexible; it was the stratum which might best transmit a compressing force and would bend or break under increasing pressure or along a line of weakness. If a strut shall be strong it must be straight; crooked, it bends or breaks at the crook. Was the great limestone a plane or did it depart from horizontality—the direction of thrust? The uppermost member varies from one mile to four miles in thickness in the extent of the province. Were these variations rapid enough to cause deflections in the great limestone? Where the great Devonian and Carboniferous sediments give the upper members the maximum weight, folding is the type of deformation; where these strata are thin, faulting dominates. Was there relation between the load borne by the great limestone and the resulting type of deformation?

Questions like these suggest a general hypothesis that circumstances of sedimentation determined conditions which afterward controlled the place and type of deformation and influenced the size and relations of individual structures.

EXPERIMENTAL RESEARCHES.

The fact of compression is so patent in folded regions, the action of a force against the edges of the strata is so clearly suggested, that many geologists have been tempted to seek a ready solution of the problems of structure by imitative experiments. To put layers of sand, clay, plaster, or even cloth into a box and compress them endwise is a very simple operation, and the resulting plications often bear a likeness to the folds observed in rocks: but it is the lesson of experience in many directions that it is less difficult to imitate one of nature's processes than to understand either the imitation or, through it, the original. For this reason some have cast experiments aside as useless and others have been content to describe their unexplained results. Nevertheless two geologists have through experiments successfully attacked the problem of deformation by compression, Schardt and Cadell; and others whom I do not know of may have passed from the imitative to the explanatory stage of this study. The work of these two investigators became known to me only after my own experiments had led me to results in some cases in agreement with theirs. Thus, in so far as we have reached similar results, the conclusions carry the weight of independent corroboration. Some quotations may illustrate the methods and results of the more important experimental studies of which I have knowledge.

The first efforts to simulate the forms of folded strata by experiments with plastic materials were made as far back as 1812 by Sir James Hall, who presented a communication on the subject to the Royal Society of Edinburgh on February 3 of that year. His attention had been attracted by the folds exposed in the lofty cliffs on the coast of Berwickshire, England, of which he gives sketches with diagrams of their eroded and subterranean portions; the latter indicate that he interpreted the connections of the folds in the same manner that they have since been understood by later geologists.

He says [1]:

It occurred to me that this peculiar conformation might be accounted for by supposing that these strata, originally lying flat and in positions as nearly level as might be expected to result from the deposition of loose sand at the bottom of the sea, had been urged when in a soft but tough and ductile state by a powerful force acting horizontally; that this force had been opposed by an insurmountable resistance upon the opposite side of the beds, or that the same effect had been produced by two forces acting in opposite directions, at the same time that the whole was held down by a superincumbent weight, which, however, was capable of being heaved up by a sufficiently powerful exertion.

By either of these modes of action I conceived that, two opposite extremities of each bed being made to approach, the intervening substance could only dispose of itself in a succession of folds, which might assume considerable regularity and would consist of a set of parallel curves alternately convex and concave towards the center of the earth. At the same time, no other force being applied, any two particles which lay with respect to each other so that the straight line joining them was horizontal and at right angles to the direction of that active force, would retain their relative position, and of course that line would maintain its original straightness and horizontality; and thus, the forces exerted being simple, or, if compound, tending, as just stated, to produce a simple result, the beds would acquire the simple curvature * * * which belongs to them in the immediate neighborhood of

[1] Trans. of the Royal Society of Edinburgh, vol. vii, 1815, p. 84.

Fast Castle; whereas in Galloway and some parts of our coast, particularly near Gun's Green, to the eastward of Eyemouth, where the curvature deviates from that simple character and becomes in the utmost degree irregular, we must conceive the force to have been more complicated or most probably to have acted at successive periods.

This conjecture no sooner occurred than I endeavored to illustrate my idea by the following rude experiment, made with such materials as were at hand. Several pieces of cloth, some linen, some woollen, were spread upon a table, one above the other, each piece representing a single stratum; a door (which happened to be off the hinges) was then laid above the mass, and being loaded with weights, confined it under considerable pressure; two boards being next applied vertically to the ends of the stratified mass were forced towards each other by repeated blows of a mallet applied horizontally. The consequence was that the extremities were brought nearer to each other, the heavy door was gradually raised, and the strata were constrained to assume folds bent up and down, which very much resembled the convoluted beds of killas, as exhibited in the crags of Fast Castle, and illustrated the theory of their formation.

I now exhibit to the society a machine by which a set of pliable beds of clay are pressed together so as to produce the same effect, and I trust that the forms thus obtained will be found by gentlemen accustomed to see such rocks to bear a tolerable resemblance to those of nature, as shown in Fig. 6, copied from the forms assumed in the machine by an assemblage of pieces of cloth of different colors.

In 1878, M. Alphonse Favre, proceeding upon the hypothesis of a cooling nucleus which fails to support the hard outer crust of the earth, undertook some experiments with beds of clay subjected to contractive forces. Upon a stretched rubber band he placed a mass of clay from 25 to 26^{mm} thick, and allowed the rubber slowly to resume its proper length, carrying with it the mass of clay. In order that the clay should not slip upon the band of rubber, pieces of wood were attached to the ends, and thus the compression produced was in effect similar to that produced by Sir James Hall in his machine; but the materials of Sir James Hall's experiment were rigidly confined above, and the deformation of the upper surface was controlled by the cover. In M. Favre's experiments this upper surface was able to rise into ridges, and he obtained anticlinal and synclinal forms which bear a certain resemblance to those observed by geologists in the contorted strata of the Alps and other folded regions. The masses of clay employed by M. Favre were not divided by structure planes and acted simply as a homogeneous bed, yielding at the surface more completely than at the bottom simply because less confined. The description of the several experiments is published, with a discussion of the theories of mountain building, in the Bibliothèque Universelle.[1]

In 1884 Mr. Hans Schardt, having studied the geology of a portion of the Pays-D'Enhaut Vaudois, in the western Alps, was led to attempt the explanation of various structural facts by means of experiments with masses of clay and sand compressed in the same manner as in the experiments of M. Favre; but M. Schardt took a step beyond the theories of M. Favre, and attributed the character of individual structures and the differences observed between structures in different regions

---

[1] Archives des Sciences Physiques et Naturelles, No. 246, 1878.

to the nature of the strata in which they were produced. He was therefore not content to compress a single layer of plastic clay, but he made up the piles from various hard and soft layers of damp clay and of clay mixed with sand. I quote those statements[1] which I have had the pleasure of corroborating:

The earth's crust is not an homogeneous layer, but is a complex of layers of varied nature. It is therefore necessary (in experimenting for geologic structures) to multiply the beds of clay, and to vary their consistency, in such a manner as to represent in miniature the great strata of the mountains.   *   *   *

The first experiments were not crowned with the expected success. It is, indeed, very difficult to cause beds of clay of unequal hardness to adhere to each other. The bed of hard clay separates from the lower clay and forms hollow arches, in spite of the weight of soft clay which covers it. In nature the beds can scarcely separate, as they are subject to the action of weight, which does not act with the same importance in experiments on so small a scale. It was therefore necessary to replace this factor by the adherence of the beds among themselves. I accomplished this by placing between the layers of ordinary clay a small quantity of kneaded clay which was fine and tenacious. This method at the same time permitted the beds to slip one upon the other without separating. In nature the slipping of strata upon their bedding planes seems ordinarily to be produced when the beds are strongly bent. This kind of dislocation has certainly great importance, which has not always been sufficiently recognized.

In nature this adherence is replaced by the weight. It is unquestionable that the pressure of the upper beds upon the lower must be enormous at a certain depth. Now the moment that a compact calcareous bed begins to form an arch, the pressure which the upper beds exert upon the lower beds through this compact bed ceases exactly at the place of the anticlinal curve; it acts only at the two sides of the arch which suffices to force the lower soft beds to follow the fold of the compact bed and to conform exactly to its concave curve, in the same manner that a soft mass will pass between the fingers when it is pressed against the hand. Still more should this be true as the pressure, which acted previously equally over the entire surface, is localized and consequently increased toward the synclinal curves, where the upright legs of the arch, which must actively raise the superposed soft strata, find their points of support. All the experiments upon the action of compression show clearly this fact.

The effects of compression vary with the position of the beds. When they are horizontal and the pressure acts in the direction of the stratification, the resistance attains its maximum; but when they commence to form an arch, the pressure, which is transmitted always in the direction of a tangent,[2] acts obliquely to the stratification until the beds become vertical. From that time the pressure acts transversely to them; thence it follows that they are thinner on the legs of the folds than on the axes of curvature; they appear to have been laminated or flattened by the compression.

The opposite is produced, on the contrary, when there is a reaction of a hard bed upon a soft bed; then the latter is thinned around the convex curve of the hard bed.

I have thus far spoken of the case where a single hard bed was inclosed between two soft beds. But if experiments are made with a complex of beds alternately harder and softer, it will be found that all the hard beds are at the same time conductors of the compression proportionally to their thicknesses and consistencies. When there is folding their effect is combined and the plastic beds are simply carried with them in the rearrangement.

In February, 1888, the Royal Society of Edinburgh, before which

---

[1] Geological studies in the Pays-D'Enhaut Vaudois by Hans Schardt; Bull. de la Soc. Vaudoise des Sci. Nat., vol. xx, 1884, pp. 143-146.

[2] It will be seen later that I differ from M. Schardt in regard to the direction in which pressure is transmitted. B. W.

Sir James Hall had presented the first article on experiments of this nature, received a paper from Mr. Henry M. Cadell, on " Experimental Researches in Mountain Building." The purpose of Cadell's experiments was to simulate the " behavior of brittle rigid bodies, which, instead of undergoing plication when subjected to horizontal compression, had snapped across and been piled together in great flat slices like so many cards swept into a heap on a table." To this end he used plaster of Paris interstratified or mixed with layers of sand, and in some experiments black foundry loam and clay.

The experiments were of three distinct kinds. The first series (A) was designed to explain the behavior of different types and arrangements of strata when pushed horizontally over an immovable surface. The object of the second series (B) was to ascertain, if possible, how gently inclined thrust planes may have originated, and to trace their connection with " fan structure and other phenomena observed in mountain systems of elevation." The third series (C) was conducted on principles suggested by the experiments of Favre, and Favre's experiments were extended " by removing the upper layers of the wrinkled clay and observing the effect of the contraction on the deep-seated portions of the miniature mountain system."

The apparatus used by Cadell was a strong wooden box, in which pressure was applied to a removable end by means of a screw. The strata were subjected to no load but that of their own weight, and the conditions of the experiments simulated those of rocks at or near the earth's surface. The summary of his results is as follows:

(1) Horizontal pressure applied at one point is not propagated far forward into a mass of strata.

(2) The compressed mass tends to find relief along a series of gently inclined "thrust planes," which dip toward the side from which pressure is exerted.

(3) After a certain amount of heaping up along a series of minor thrust planes, the heaped-up mass tends to rise and ride forward bodily along major thrust planes.

(4) Thrust planes and reversed faults are not necessarily developed from split overfolds, but often originate at once on application of horizontal pressure.

(5) A thrust plane below may pass into an anticline above and never reach the surface.

(6) A major thrust plane above may, and probably always does, originate in a fold below.

(7) A thrust plane may branch into smaller thrust planes, or pass into an overfold along the strike.

(8) The front portion of a mass of rock being pushed along a thrust plane tends to bow forward and roll under the back portion.

(9) The more rigid the rock the better will the phenomena of thrusting be exhibited.

(10) Fan structure may be produced by the continued compression of a simple anticline.

(11) Thrust planes have a strong tendency to originate at the sides of the fan.

(12) The same movement which produces the fan renders its core schistose.

(13) The theory of a uniformly contracting substratum explains the cleavage often found in the deeper parts of a mountain system, the upper portion of which is simply plicated.

(14) This theory may also explain the origin of fan structure, thrusting, and its accompanying phenomena, including wedge structure.

The conclusion expressed in paragraph 9 was deduced from the experiments with rigid materials. By reference to Plates XCV and XCVI it may be seen that like effects are produced in butter-like substances under heavy load.

We may now turn to the theoretic considerations which governed the experiments of which this paper is partly a result.

PROBLEM OF STRUCTURAL EXPERIMENTS.

To bend, to break, to shear, these are purely mechanical operations. They require the application of a force external to the material bent, broken, or sheared, a force which overcomes the internal resistances. The processes of terrestrial folding and faulting involve these three operations and obey mechanical laws. The problem which the facts present is to ascertain: (1) what was the initial character and arrangement of the strata folded and faulted, and what consequently were the internal resistances; (2) under what conditions was the external force applied, and how was it transmitted; (3) what possible origin can be assigned for a force which is qualitatively and quantitatively sufficient to produce the observed results.

Mechanical laws do not vary with the magnitude of the active forces nor with that of the passive resistances; of a series of strata hundreds of feet thick and of a pile of layers only inches thick, the bending, breaking, or shearing will obey the same laws, if all the factors of pressure and resistance are proportionate in each case to the dimensions of the pile, and the similitude of results will be the closer the more exactly the conditions in the one case represent those in the other. If, then, we can make a reasonable analysis of the character and arrangement of the strata deformed and of the conditions governing deformation, we may be able experimentally to produce structures under conditions so similar to those of nature that the forms shall be of the same kind as are observed in strata, and with analysis thus confirmed by synthesis we may approach the problem of the origin of the sufficient force with more confidence.

The principal difficulty in this analysis is to comprehend the relative proportions of the elements of the problem. The masses involved are so extensive, the forces required are so utterly beyond expression in our foot-tons, that our usual conceptions of rock strength and of rock rigidity are worthless. In our constructions, opposed to our forces, to our tools, stones are hard, firm, unchanging, and the saying is "hard as a rock," but in resistance to forces of the earth's mass, this same rock may be relatively soft as wax. To arrive at a fair idea of conditions beyond our ordinary experience, we may consider the nature of the support of the earth's crust. Let us look upon it as the problem of a stone bridge. If an engineer wishes to span a culvert 5 feet wide, he may find a single flat stone to throw across. For a span of 50 feet he must build an arch; for 500 feet the arch must be so high and the masonry so massive that the structure is seriously weakened by its own weight. Increase the span and the construction ultimately becomes impossible; the weight of material required soon exceeds the crushing strength of the stone, and, however well proportioned, the structure must crumble as though built of sand. Now limit the engineer to an arch whose rise shall be 8 inches in a span of 1 mile—that is, limit him to the curvature of the earth. Is it conceivable that an arch, even of solid granite, a mile in span and 8 inches in rise, should be self supporting? Obviously not.[1] But the terrestrial crust, of which any arc is an arch of these proportions is composed of heterogeneous materials, some of them weaker than granite, and where granite falls short of self-support, the crust as a whole must fail. Hence, however thick we conceive the rigid outside shell to be, it rests with all its weight upon whatever lies within it.

As this statement is true for each layer of the earth's crust, at the surface and below it, it follows that the pressure due solely to weight increases from the surface downward; and as the attraction of gravity also increases in the same direction to a certain depth, the growth of this pressure is more than proportional to the depth below the surface. It is not necessary here to enter into the mathematical discussion of the relations of gravity, density, and pressure, but the following table gives the figures, according to the Laplacian hypothesis, as calculated by Mr. R. S. Woodward.

Variation of terrestrial density, gravity, and pressure according to the Laplacian law.

[By R. S. Woodward. 1890.]

Depth in miles.	Density.	Acceleration of gravity.	Pressure in atmospheres.	Pressure in pounds per square inch.
0	2.75	1.0000g	1	15
1	....	....	400	6,000
2	....	....	800	12,000
3	....	....	1,210	18,150
4	....	....	1,620	24,300
5	2.76	1.0006g	2,020	30,300
10	2.78	1.0012g	4,200	63,000
15	2.79	1.0018g	6,390	95,850
20	2.81	1.0024g	8,600	129,000
50	2.89	1.0060g	22,000	330,000
100	3.03	1.0116g	45,300	679,500
500	4.18	1.0379g	236,000	3,540,000
560	4.36	1.0389g	318,000	4,770,000
610	4.50	a 1.0392g	354,000	5,310,000
660	4.65	1.0389g	391,000	5,865,000
1,000	5.63	1.0225g	672,000	10,080,000
2,000	8.28	0.8312g	1,700,000	25,500,000
3,000	10.12	0.4567g	2,640,000	39,600,000
3,959	10.74	0.0000g	3,000,000	45,000,000

a This is the maximum value, and the corresponding depth; 610 miles is the depth at which a given mass would have the greatest weight.

[1] Physics of the Earth's Crust, Rev. Osmond Fisher, Chap. iv, 1st ed.

Clearly to comprehend the meaning of the figures in the last column of this table, consider the problem of support of the earth's crust as one of stability of a great structure. The engineer who would build to great height must have a secure foundation. If he build on yielding sands there is a narrow limit to the weight of the structure which can be sustained; if the foundation be granite there is also a limit beyond which the weight of the towering shaft will crush the support. Now the problem is not materially different if for height above the earth's surface we substitute depth below it. The crushing strengths of stones at the surface vary as follows:

	Pounds per square inch.
Granite	7,000 to 22,000
Limestone	11,000 to 25,000
Sandstone	6,000 to 14,000

These values probably increase with depth in the earth's crust and in an unknown ratio; but it is not likely that the increment of strength is as great as the increment of pressure. Mr. Woodward's table shows that at 5 miles below the surface the pressure exceeds the maximum resistance of rocks at the surface, and at 10 miles the pressure is more than double the resistance. This means that somewhere between 5 and 10 miles beneath the surface the weight of the superficial crust is sufficient to crush its support.

But crushing is not possible within the earth's mass in the way in which we see it at the surface. To crush is to separate into incoherent particles; and irresistible confinement, itself due to the pressures which are greater than coherence, holds any deep-seated rock mass to its coherent volume. In this condition, confined under pressures greater than its crushing strength, a substance may be said to be latently plastic. The cohesion between its particles is unimpaired, fracture or crushing into separated grains is impossible for want of space; but change of form may be induced by a sufficient disturbing force, and such change is plastic flow. The conception of this latent plasticity needs to be clearly understood. It is a mechanical condition, the result of external forces which are strong enough to overpower cohesion. It is not a plasticity due to internal tension like that of hot iron, for the temperature at a depth of 5 miles is probably not sufficiently elevated to modify greatly the firmness of rocks. The average rate of increase of temperature beyond the local unchanging mean of 51.3° Fahr. is 1° for every 75 feet, as recently determined in the well at Wheeling, West Virginia, to a depth of 4,500 feet. If we may assume that this rate continues to some depth, we should have at 5 miles below the surface a temperature of only 421°. It is not probable that the assumption is strictly valid, and the temperature may be considerably higher, but it can scarcely approach the melting point of rocks, which varies from 1,200 to several thousand degrees Centigrade. The independent evidence of stratified rocks, known to have been buried

20,000 to 30,000 feet in the crust and now exposed by erosion, bears on this point. Such strata are solidified by pressure, but have not suffered chemical metamorphism, as they must have done had they been heated to plasticity.

We may fairly conceive the earth's crust to consist of a superficial shell 5 to 7 miles thick, which rests upon and grades in substance and physical condition into a subjacent shell. The under is only differentiated from the upper by its relative position in consequence of which it supports a crushing load and forms a latently plastic foundation; and that immobility of the surface which is expressed in the phrase of "terra firma" depends upon the equality of the inert resistance to the downward pressure. Destroy that equality by increasing the pressure over one area beyond that at another until the strength of the rock is overcome, and there must result an adjustment of weights and supports in such wise that the latently plastic foundation flows from the greater toward the lesser load—that is to say, the earth's external mass is in a condition of hydrostatic balance. For this condition Dutton proposed the term isostatic, and he coupled the idea of isostatic adjustment with a theory of folding,[1] a theory to which we shall recur later. We have thus taken the first step in the analysis of our problem: The strata which have suffered folding and faulting floated upon and graded downward into a latently plastic mass.

In speaking of the earth's crusts resting upon a plastic support, it is easy to imply that the shell is homogeneous and distinct in character from the support. Neither implication is correct. That part of the earth's mass which it is convenient to call the crust can not be divided off from the spheroid within except by an imaginary boundary; and this same crust can not be regarded as homogeneous except by a disregard of plain facts. The consideration of the relations of its great rock types among themselves and of the resistances they respectively offer against earth-deforming forces forms the second step in the analysis of our problem. We need take account only of extensive bodies, and we may divide rocks simply into massive and stratified; the former may include great crystalline masses, either metamorphosed sediments or igneous rocks, and also closely folded stratified series; the latter consists simply of the flat-lying sediments. The distinction to be recognized between them is a difference of rigidity, and it is very like the difference between a heavy beam and the same wood sawed into boards. The beam resists a pressure which bends the pile of boards, and massive rocks are immovable in relation to a force which folds strata. To deform a massive rock requires that the cohesion of the particles in the mass shall be overcome and a rearrangement effected which results in schis-

---

[1] On some of the greater problems of physical geology. C. E. Dutton, Bull. Phil. Soc. of Washington, vol. XI, pp. 51, 64.

tosity. To deform stratified rocks demands that beds shall slip past one another and bend; the friction among beds and the interstitial resistances of different beds to folding are much less than the cohesive forces of a solid mass. It follows that strata are more easily deformed than masses, and if the two rock types sustain common compression the stratified series suffers the major deformation. Therefore when compression follows a period of deposition, and affects simultaneously a continental area of massive rocks and the adjacent area of sediments, it is in the sediments that we may most clearly observe the effects. The zone of folding and faulting may be miles in width and include anticlines of great height; the zone of schistosity may be but a few scores or hundreds of feet wide, and be masked by complex relations with the results of earlier actions of the same kind. The changes of form are precisely what would result from pressing a pile of sheet-iron irresistibly against a mass of soft but solid iron. The sheets may be bent while the mass is but bruised.

If the preceding statements are clearly grasped, we may proceed to consider the arrangement and characteristics of strata, with a view to understanding better the deformation of stratified rocks alone. In the Appalachian province strata have a maximum thickness of 30,000 feet near shore along a very narrow zone and thin away rapidly toward the west to less than 10,000 feet. These thicknesses are great, measured by our standards, but compared with the width of deposits they are but moderate. If the horizontal extent be represented by the width of this page, one hundred leaves will compare in thickness with the maximum of sediments; and it is obvious that a broad pile of strata, whose aggregate is relatively so thin, is rather flexible than rigid. As the beds would not sustain their own weight over any span of miles, so they would transmit a great force only while it coincided with their plane.

This idea of flexibility is strengthened by two considerations: such a mass of strata is not divided into a hundred but into thousands of layers, and great subdivision weakens it; and furthermore, in folding, the strata do not yield as parts of a simple mass, but resist individually and irregularly. Reference to the columnar sections of strata in the Appalachian province will show how heterogeneous is the pile in a vertical direction, and how varied are the deposits in adjacent parts of the same district. There is every class of sedimentary deposit: Conglomerate, sandstone, shale, and limestone, with gradations from one into another, giving an indefinitely varied series. Each bed of such deposits is, in relation to others, more or less flexible, more or less frangible, and the relative flexibility and frangibility of the principal members of a series have an important influence in determining the result of deformation. Flexibility is a direct function of lamination and toughness of the layers; its opposite, frangibility, is directly proportioned to the thickness and incoherence of the stratum. The following column read downward expresses the order of flexibility; read upward, that of frangibility of lithologic varieties:

Less frangible thick to thin bedded.		Less flexible thin to thick bedded.
	Argillaceous shales.	
	Calcareous shales.	
	Arenaceous shales.	
	Limestones.	
	Sandstones.	

Such a statement needs to be qualified by considerations of modifying conditions; of these, pressure and confinement are the most important, and, as we have already seen, they are effective in the earth's mass roughly in proportion to the depth below the surface. In discussing the support of the superficial crust we took account of depths at which pressure renders the rocks latently plastic; but pressure is an important condition far above that zone in the crust itself. The distinction between a frangible and a flexible stratum is that in process of deformation the particles of the former separate beyond the radius of cohesion; those of the latter do not. Pressure and confinement prevent this separation of the particles of an otherwise frangible mass and force it into a state of flexibility. Thus it is possible to explain that rocks which are brittle at the surface bend like iron within the crust; and thus we may comprehend that a thick stratum may fold without fracture in one district under great load and break in another district under less load. We may express this idea by saying that the flexibility of a layer is a function of its depth in the earth's crust, or of the load which the stratum bears;[1] and it follows that in an assumed homogeneous deposit of great depth the change from rigid beds at the surface to flexible beds at the base would be a gradual and continuous one. This assumption is never true for any depth. Deposits of strata are not homogeneous except in moderate thicknesses, for the alternation of shales, sandstones, and limestones in ever changing association is the rule, and with the lithologic changes go changes in rigidity. Near the top all are more frangible, toward the base all are more flexible, but from top to bottom each bed is different in frangibility or flexibility from its neighbor under like conditions.

The deforming force which folded Appalachian strata was one of compression, acting tangentially to the earth's circumference. The physical conditions necessary for such action are that an arc of the earth's mass shall shorten, and that this shortening shall take place in such manner as to restrain the superficial crust within the lessening length of the arc. We may conceive the strata confined between two crystalline masses as between two comparatively immovable buttresses, or as settling against one such buttress in consequence of movement

[1] G. K. Gilbert, Henry Mountains, p. 83.

51

of the stratified mass; but however we think of the force applied it is evident that it must be transmitted in the stratified rocks, and the mode of this transmission will affect the result of deformation. A push against one edge of a piece of bristol board reaches to the further edge; it does not to the same degree extend across a strip of tissue paper. So a thrust against a massive limestone may be effective at long distance from its origin, while the same force would be shortly expended in a thickness of shales. Again, a strut which is restrained from deflection by guides is stiffer than the same strut free to bend. So a stratum confined beneath a superincumbent load will more rigidly transmit a compression than the same bed near the surface. Thus, two conditions directly influence the transmission of a thrust tending to produce deformation; the one is lithologic character and massiveness of bedding; the other is the amount of load on the transmitting stratum.

The analysis of the conditions governing deformation of strata is thus carried theoretically as far as it safely can be. We have determined that: (1) the support of the superficial crust is latently plastic; (2) if massive and stratified rocks suffer like compression, the latter will exhibit the greater deformation; (3) the relation of thickness to extent of stratified rocks is such that the mass as a whole is flexible rather than rigid; (4) flexibility and frangibility, as applied to strata, are related in opposite ways to the thickness of the stratum and its toughness; and they may for one stratum be exchangeable according to the load it supports; (5) the transmission of a thrust tending to deform is a function of the firmness of any stratum and of the load upon it.

The synthetic study of the structural problem demands conditions similar to those of the earth's crust, and resistances which, in proportion to the force at command, are similar to those overcome by the terrestrial compressive strain. The conditions indicated by the preceding analysis, as necessary to successful experiment, are: (1) strata of such thinness in relation to their length as to fall far short of the rigidity required to support their own weight in a horizontal position; (2) a plastic support for these strata; (3) means of compressing the strata endwise. The forces at command for compression are necessarily of moderate power, and their capacity to deform must be greater than the resistance to change of form, hence the strata must consist of materials which are but moderately coherent, although firm. Furthermore, since strata in the crust pass through a wide range from frangibility to plasticity, the materials experimented with must be capable of like variations.

The substance chosen as most nearly possessed of the requisite qualities was beeswax, and its character was varied by adding other substances. Plaster of Paris to harden, and Venice turpentine to soften it, were adopted after trial of various materials, and with these added in different proportions, separately or together, the range of quality from brittle solid to semifluid may be covered. But when a mixture has been adopted as a standard by which to test any hypothetical condition, the temperature of the model must be kept approximately constant at successive stages of the experiment, since the plasticity of wax and turpentine is influenced by heat. If the wax be melted and the other substances be stirred in, the mixture can be cast into layers of any desired thickness, and these when cold can be arranged to simulate any given stratigraphic column.

The combination of weakness with reasonable firmness is fairly well obtained in strata so cast and piled; but the condition of plasticity in a high degree is not consistent with the stability of models which may be kept during days or weeks. Plasticity in the earth's crust is a result of pressure due to load, and if we can reproduce that condition during experiment, we may use materials which retain their form under ordinary circumstances. The load by which this is accomplished must be above the strata undergoing compression, and of such a nature that it will not interfere with the movement of the beds. In some respects mercury would be an ideal substance, but it is too difficult to handle and might cause buoyant strains of an undesirable character. A body of shot is at once heavy and yielding and has been found convenient to handle. Artificial conditions are introduced by such a load, as may be seen by reference to the illustrations of experiments, but they are easily observed, and no better means of representing vertical terrestrial pressures has yet been suggested. A maximum weight of 1,000 pounds has been used, evenly distributed over the models, giving a pressure of 5 pounds per square inch.

The machine used for compressing the piles of strata endwise is a massive box of oak provided with a piston which can be advanced by a screw. Several forms of this box have been tried, and that which is most convenient is represented in plate LXVI. The pressure chamber is 3 feet 3¾ inches (1 meter) long and 6 inches wide. The sides are removable, but are strongly bolted together during an experiment. The block which carries the screw and that against which the model is pressed are both bolted to a base which is stiffened by braces, and as other bolts which hold the sides in place pass through these blocks the distance between them is rigidly fixed. The piston is a massive box of oak, and the screw is so attached as to advance or withdraw it. The depth of this pressure box is only a foot, but when a model is in place additional height for the shot may be obtained by putting on frames that fit closely.

Given the materials, the load, and the pressure box, the making of an experiment involves the assumption of conditions of stratification, the casting and arrangement of strata in accordance with the assumption, and the compression of the resulting pile. The layers will usually be arranged with the expectation of producing some definite structural form, a fold or a fault, and the result of compression is the test of the

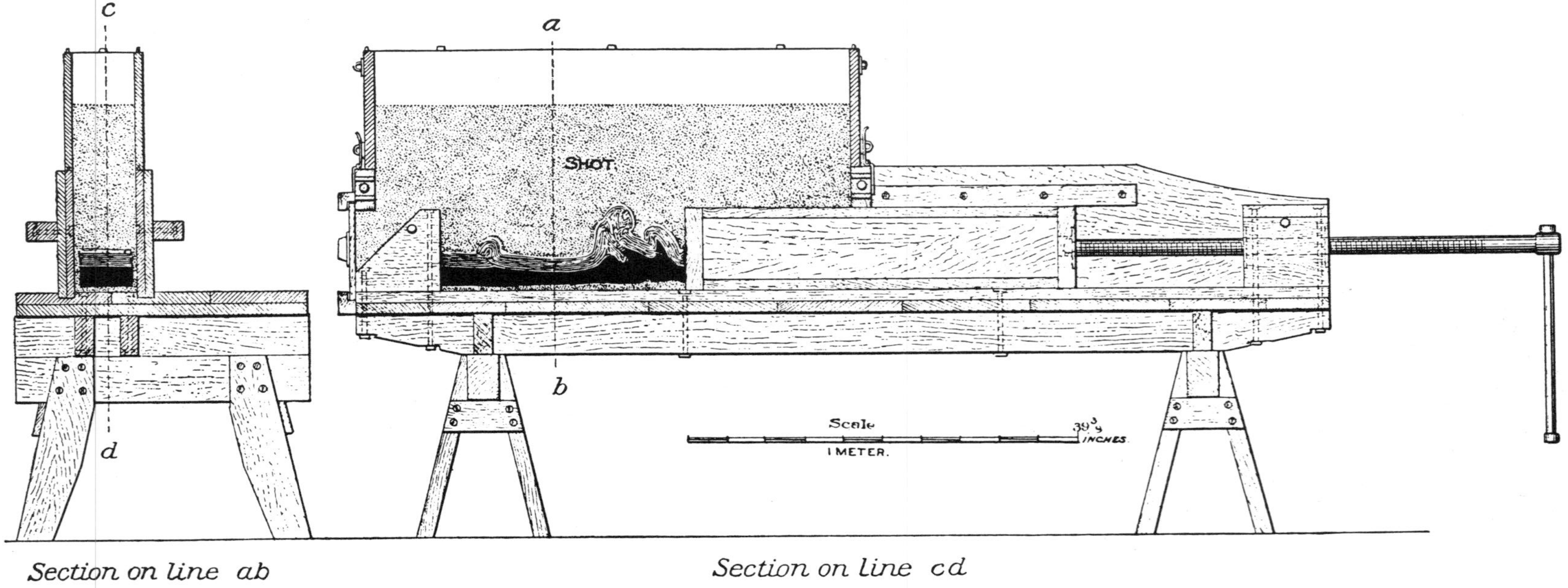

Section on line cd

Section on line ab

SHOT

Scale
1 METER.

COMPRESSION MACHINE FOR EXPERIMENTS.

hypothesis. Whether this be confirmatory or not it is desirable to know the progress of deformation, since this knowledge is an important aid toward improvement in assumptions and methods, and to secure this the compression is carried forward step by step, and the stages of shortening are successively photographed. These photographs furnish the accompanying illustrations. (Pls. LXXV to XCVI.)

The assumptions may be systematized and the experiments may be arranged accordingly. In beginning this research, in 1888, the general questions proposed were:

(1) What is the influence of stratigraphy?

(*a*) How do thin beds fold independently?

(*b*) How do thick beds fold independently?

(*c*) How do thick and thin beds fold combined in different vertical relations?

(*d*) How do thick and thin beds fold combined in different horizontal relations?

In order that the results of experiments based on these four queries may be comparable, the consistency of the materials used should be constant for a series from (*a*) to (*d*).

(2) What is the influence of load? To answer this the preceding tests should be repeated under different loads.

(3) Is the influence of plasticity the same as that of load? This may be tested by repeating the arrangement of strata in different materials and compressing under a constant load.

Putting these questions more concisely, we may say: Given three variables, stratification, load, and consistency, if any two be assumed constant, how will the variation of the third affect the result of deformation?

It was supposed that this was a complete statement of the structural problem; but difficulties soon arose in the mechanical management of the experiments and in the interpretation of results. These have led the inquiry from the direct course proposed, and the divergence has been found fruitful in hypotheses. The mechanical difficulties were two: The stresses developed in compressing the models proved to be unexpectedly great, and several boxes were burst in the early trials; and, again, friction of the plastic substances against the box sides was found to be a serious and artificial condition. The machine herewith illustrated has proved strong enough for experiments with models of firm substances, but even it has yielded so as to modify the amount of compression which a model was supposed to have suffered. Friction between the model and box has been practically abolished by introducing a layer of shot around the model; to accomplish this the layers are cast an inch narrower than the box, and the pile is placed upon shot, while the half inch space on either side is similarly filled.

After a number of experiments I began to be embarrassed to explain the constant occurrence of an anticline at the end of the model nearest the piston, and the question became prominent: How is the thrust transmitted through the model? The answer, when reached, suggested new hypotheses of conditions controlling deformation, and the discussion can be adequately treated only under a distinct heading.

THEORY OF STRAINS UNDER EXPERIMENTAL CONDITIONS.

Any block of material under the conditions of these experiments, that is, placed in a closed box under load and compressed from end to end, is subject to strains, to which it accommodates itself by that deformation which meets the least resistance. The active force is applied by the forward movement of the piston; the resistances are the firm walls of the box and the downward pressure of the load; only the latter can yield, and therefore the block tends to rise, lifting the weight. The first result of pressure is usually a reduction of volume; but when the block has a minimum volume under the load it must further yield to the sufficient compressing forces by change of form. This change may occur in one of three ways, according to the manner in which the pressure is transmitted through the block.

If the mass be semifluid or plastic the pressure will be transmitted equally in all directions; the block will shorten and correspondingly thicken. Under equally distributed load the form of equilibrium will present an even surface; under unequally distributed load the form of equilibrium will present an uneven surface, rising higher where the load is less. When pushed from one end the block assumes a form of equilibrium only after the lapse of an interval of time which is inversely proportioned to the degree of plasticity; the surface of a uniformly loaded plastic mass may therefore present temporary wave-like inequalities due to the compressing impulses and the rate of plastic flow in the mass. Whether stratified or massive, the sufficiently plastic block will adapt itself solely to the form determined by external forces, uninfluenced by its internal structure, but the strata may register the direction of flow. And this flow may take place by a more or less general but confused rearrangement of the particles of the mass, or by concentration of the movement along definite planes, which are then so related to each other that the mass is divided into bodies of the simplest forms and least number that will satisfy the conditions of the altered volume. This result is one phase of shear-thrusting (Pls. XCIV to XCVI.

If the mass be firm but flexible it will shorten by bending. The resistances involved by bending a free block are the internal resistances to compression on the concave and to extension on the convex side, and under the conditions of experiment the rising bend must also displace the load; this inert weight stiffens the block and modifies the result of flexure. Flexibility varies as the thinness of the mass; therefore a thin block, or one composed of thin layers, will bend more readily, than a massive one; and conversely a thick block, or one composed of thick layers, will transmit the thrust more persistently.

If the mass be rigid, uniform compression will tend to crush it, and if any condition direct the thrust into one plane, or if any plane of weakness exist in the mass in the line of thrust, the block will be sheared along that plane and shortened by the overthrust of one part upon another. It has been shown by Cadell's experiments that the resultant of the compressing force and the vertical direction of easiest movement is an inclined thrust which will shear a sufficiently rigid, flat mass.

The shear-thrusts obtained by Cadell in hard, brittle materials without load and those obtained in highly plastic material under load are very similar. In any process of deformation of strata there are three forces which influence the result; viscosity or internal friction, static pressure or load and the disturbing strain. In order that deformation shall take place, the last must be greater than the other two. Then three phases are determined by the relations of viscosity and load. If viscosity be relatively great fracture results, followed by thrusting on the planes of weakness. If viscosity and load are approximately balanced the form changes by flexure. If viscosity is relatively small, the result is flow, which may take the form of shearing.

In these experiments plastic and flexible masses have been combined; let us examine them separately in behavior with a view of ascertaining how they transmitted the thrust of compression.

DEVELOPMENT OF FOLDS AND LAW OF COMPETENT STRUCTURE.

If a strip of bristol board, lying on a table, be pressed from end to end, it will bow upward in a simple curve, the arc of longest possible chord and of least curvature. This is the form of least resistance, that which produces the minimum strains of compression and extension in the board. If a strip of tissue paper is pressed in the same way it will wrinkle irregularly. It differs from the bristol board in that it is not competent to sustain its own weight and is not so homogeneous in texture.

When these experiments were begun it was supposed that the hard strata would bow according to the law which governs the bristol board; it was thought that they would bend over the longest possible chord with the minimum curvature, and that the crown of the curve would be near the middle of the entire length; it was inferred that the distribution of load would modify this result; the hypothesis demanded that under uniform load a simple arch should rise in the middle of the block, and under unequal load the rise should occur where the weight was smallest. This hypothesis was not confirmed; in one experiment the entire absence of load over a central section determined the position of the rise, but the inequality of loading was in this case extreme, and in many other trials the uniformly distributed weight permitted the rise of an initial anticline near the force, an anticline which predominated until it was closed. Hence it was evident that some condition not foreseen exercised a controlling influence upon the locus of flexure. When in the earliest experiments this conclusion appeared, the control was attributed to external friction; but the effect continued after this supposed cause was removed. The importance of internal friction between the beds, forced in bending to move on one another, was next considered; but no adequate explanation could be founded on this, since the adjustment to bending is a local movement dependent upon the thrust from the concave toward the convex curve and is no more difficult in the middle than near the ends of the strata. The rate of compression was next studied. A very slow advance of the piston might create a pressure which, acting uniformly on the entire mass, would cause deformation at the weakest point; a more rapid advance might produce movement near the piston before the thrust could be transmitted to the further end. There is no doubt that these may be valid considerations, but they do not explain the local development of folds in these experiments, as the forward movement of the piston was always very steady at the slow rate of about one inch in five minutes. The fact that the models themselves, in early stages of compression, exhibit quite as much deformation at the farther end as at the piston, is the best proof that during the early stages the force was transmitted throughout their length; and the allied fact that in later stages deformation went on principally near the applied force, suggested that the deviation of the strata from the line of thrust was accompanied by deflection of the thrust itself from the direct line. This law has already been stated, but its influence in determining folds was not appreciated until many experiments had been made. Study of numerous models showed that slight dips arose in the supposed horizontal strata before compression, either through irregularities in the casting, or at an early stage of compression through lift of the piston with the ends of the beds in contact with it, or through swelling of the plastic base beneath the flexible layers. The conclusion reached through the experiments was that, when a firm but flexible stratum transmits pressure, it tends to yield by bending along any line, where there is a slight change of dip, and this deviation may be due to initial uplift or depression; the fold is further developed by that component of the thrust which is diverted by the inclined strata.

If we describe the sufficiently firm stratum by the word competent, we may formulate the law of anticlinal development, as deduced from these experiments, as follows: In strata under load an anticline arises along a line of initial dip, when a thrust, sufficiently powerful to raise the load, is transmitted by a competent stratum. The resulting anticline supports the load as an arch, and being adequate to that duty it may be called a competent structure. From the conditions of the case it follows that none other than a competent structure can develop by bending. If the thrust be not powerful enough to raise the load there will be no uplift; or if the layers be so plastic that they yield to the thrust by swelling, then the principal result of deformation is change of form other than by simple flexure, and it assumes some phase of flowing. This is incompetent structure.

CONSEQUENCES OF THE LAW OF COMPETENT STRUCTURE.

(a) *Folding redistributes load.*

(b) *The size of a competent anticline is directly as the competency of the effective stratum and inversely as the load.*

APPALACHIAN THRUSTS.

The district of Appalachian thrusts is 450 miles long, and within it the dominant structural facts are faults which (1) arise and die out in the northwest limb of anticlines characterized by gentle southeast and steep northwest dips; (2) have a fault-dip to the southeast, usually parallel to the gentler dipping limb; (3) are not marked by greatly thinned or schistose strata; (4) in spite of displacements, that sometimes must exceed 5 miles, never bring to the present surface any rock older than Cambrian strata; (6) are wonderfully persistent, the longest reaching 375 miles, and are remarkably parallel among themselves; (7) lie in a zone continuous with that of open folding, but occur in that part of it where the great Devonian sediments certainly, and most of the Carboniferous probably, never were deposited.

From these facts it has been inferred that (1) Appalachian thrusts are a result of peculiar anticlinal development and are produced by a force transmitted through the gentler dipping limb; (2) faulting checked flexure at a stage prior to excessive compression of the anticline; (3) the phenomena are confined to stratified beds and originate in them; (4) the condition which favored faulting rather than continued folding was general over the entire district and the antecedent folds were related to one another in a manner to produce parallelism; (5) the reason for faulting in the southern and folding in the northern half of the continuous zone is to be sought in the differences of stratigraphy between the two districts.

The stratigraphic contrasts are strikingly brought out by a simple statement of the fact that the thickness above the Cambro-Silurian limestone is 23,000 feet in the Pottsville basin, 10,000 feet in southwestern Virginia, and 4,000 feet in Alabama, including in each statement the highest Carboniferous strata known in each district. We know, as the few sections already cited prove, that strata vary greatly in thickness and within short distances. Thus the deposits of the Devonian period vary from 10,000 feet in eastern Pennsylvania to 7,000 in the central part of the State. This thickness they retain southwestward nearly through Virginia and then thin rapidly. Near Big Stone gap they are represented by the single formation, the black shale, 750 to 900 feet thick, and this extends through Tennessee to Alabama and Georgia with a thinness of 30 to 100 feet. Ten thousand feet of sediment represented by 30 feet! The statement is not strictly true since the 30 probably represents only the lower 2,000 of the 10,000 leaving 8,000 unrepresented, but none the less is the fact apparent that the Devonian record was never made in the South, while its bulk in the North is enormous.

*   *   *   *   *   *

PLATE LXXXIV.

Description of Model E 1:

    Original length, $39\frac{5}{8}$ inches = 1 metre. Fig. a.

    Width, 5 inches.

    Thickness, $1\frac{3}{4}$ to $2\frac{3}{4}$ inches.

Layers.	Composition (Parts by weight.)			Thickness.	Character.
	Wax.	Plaster.	V. turpentine.		
1				*Inches.*	
2 to 9 (inclusive).	1	1	0	$\frac{1}{4}$	Hard.
	1	1	0	$8 \times \frac{1}{16}$	Do.
10	1	0	$\frac{1}{2}$	1 to 2	Soft.

This model was made exactly like E, Fig. a, Plate LXXXIII, but the pressure was applied at the thinner end, remote from the principal assumed initial dip. A very slight initial dip limited the syncline on the right. Under a uniform load of 1,100 pounds the model was compressed nine times, as shown in Figs. b to k.

RESULTS.

Deformation went on during each compression at three places: at the applied force, at the minor initial dip, and at the sharper initial dip. In Fig. b the anticline at the sharper dip, furthest from the applied force, is just entering on the competent stage of development, and its growth from that on is continuous with the formation of an overthrust of typical Appalachian character. But the minor initial dip, exaggerated in Fig. b, has developed to a carinate anticlinal in Fig. c; at the next stage it is overthrust, and in Figs. c and f two consequent folds appear, one on each side of the original. These are caused by the resistance offered against the overthrust and by the weight which it must raise in developing. In Fig. c there is a broad swelling of the plastic base near the applied force, which caused a flat anticline that never rose above the inflowing soft material.

The relations of the two original anticlines in Fig. d are characteristic of the structure of northeastern Alabama.

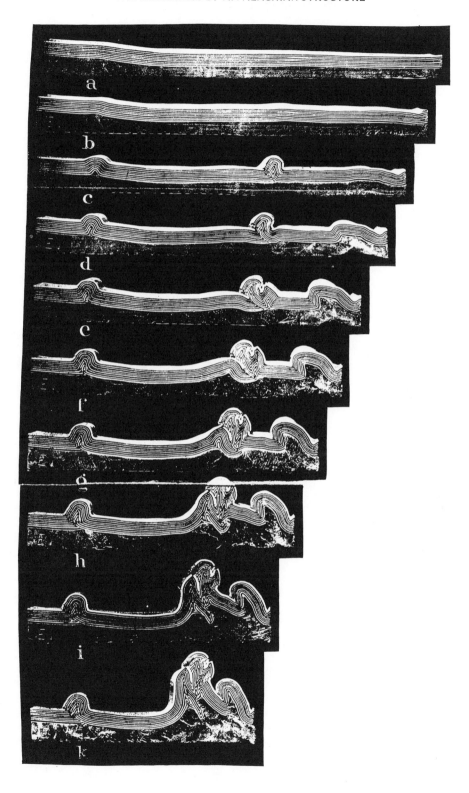

Editor's Comments
on Papers 7 and 8

7   SOLLAS
    Recumbent Folds Produced as a Result of Flow

8   KOENIGSBERGER and MORATH
    Theoretische Grundlagen der experimentellen Tektonik

A pioneering age in Appalachian tectonics came to a close with the publication of Willis' report; the same year, however, witnessed a renaissance of Alpine tectonics, with the publication of Hans Schardt's (1893) memoir on the origin of *des Préalpes romandes* shared by Switzerland and France. In the Prealps, Schardt overcame previous mistakes, which reflected an early tendency (one shared by the writer) to subordinate stratigraphy to structure. He demonstrated an imbricated thrust mass at structural levels well above those at Glarus, with enormous, consistently northward, displacements.

Bertrand, of course, had successfully reinterpreted the Glarus *Doppelfalte* in 1884, but he did it from such a distance as to be virtually unnoticed. (He worked in Paris, from the published literature.) Furthermore, there seems to be an instinctive distrust among geologists for regional syntheses written by authors personally unfamiliar with the ground under discussion, and more than one tectonic history could elaborate upon that theme. It was 1892 before Suess convinced Heim, in conversations at Zürich, of the "correct" interpretation, and thus Heim later credited both Bertrand (1884) and Suess (1892) in acknowledgment, even though Suess had never published on the subject. In the same paper Bertrand was also credited for anticipating Schardt's Prealpine discovery, but the superficial map resemblance does not withstand close examination (Bailey, 1935, p. 78–82); nonetheless, it could still be argued that Bertrand was the author of the concept subsequently nurtured by Schardt.

Although the scale of displacements was of the same order as that previously demonstrated by Törnebohm in Scandinavia, Schardt's Alpine discoveries were more appealing to the imagination of contemporary geologists (Bailey, 1935, p. 25); each Prealpine nappe seemed characterized by a unique stratigraphy, and the products of many widely separated districts had been stacked, one upon the other. The significance of horizons of "lubrication" was given emphasis, for Schardt recognized clearly the association of a Trias evaporite sequence at the sole of the overthrust horizons.

At first controversial, Schardt's view was soon to be commonly adopted. The French-Swiss Maurice Lugeon, for example, initially rejected the concept, but two years afterward announced his conversion; in subsequent work of a classic nature on the Alps, Carpathians, and Appenines (e.g., 1902) he may be said to have "unveiled the new tectonics to the outside world, strengthened and enriched by brilliant observations of his own" (Bailey, 1935, p. 88). Lugeon demonstrated, among other notions, the Prealpine connection with the High Calcareous Alps of the Rhône, elucidated the structure of the Helvetic and ultra-Helvetic nappes, and directed our attention (as Bailey puts it) to the magnificent Alpine game of *Sant de Mouton* (leapfrog).

In 1898, Schardt's conception of Alpine evolution was more fully elaborated. The Prealps were envisaged as a landslide on a Trias slip surface, gliding slowly down the front of an advancing Pennine nappe. In the same year the Simplon Tunnel excavations began; Schardt was Chief Geologist. According to Bailey its main service was to concentrate attention on the superb exposures in the vicinity of the tunnel line; thus Schardt, independent of Lugeon, developed similar conceptions, both in broad principle in agreement with modern views. The tunnel was completed in 1905, by which time most "modern" views on the Pennine roots had been developed (see, e.g., Fig. 2; cf. Milnes, 1973).

What of experimental investigations of Alpine tectonics? The name of Schardt reappears again. Having earlier studied a portion of the Pays-D'Enhaut Vaudois in the Western Alps, Schardt attempted to account for what he had observed by means of sand and clay compression models. He discovered the significance of lubricated stratigraphic horizons in his attempt to model "adherence" between individual clay layers, which seems an anticipation of his subsequent discovery of Triassic thrust soles. A selection from Schardt's paper is reproduced within Paper 6.

But what of the Alpine Renaissance? By now Schardt conceived simple *thrusts*, typically Prealpine, to be superficial structures, whereas *recumbent folds*, typically Helvetic, were taken to be deep-seated structures. Is there not considerable truth to this contrast? In the Helvetides, extraordinary complications have been encountered, but the general scheme seems uniform (Bailey, 1935, p. 121): "the Helvetides have the structure of a stream in which the upper currents have flowed more quickly than the lower. In plan one sees an object lesson, wherever water with streaky oily surface flows along a gutter on the side of a street. The edge of the gutter delays the adjacent current. It produced recumbent folds and involutions of the Helvetide type.

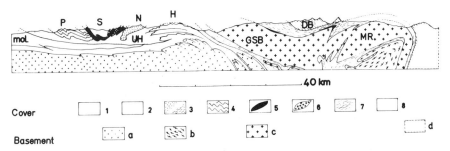

Figure 2 Schematic section of the Western Alps across the Pennine Alps and the Préalpes Romandes (after Bearth, Gagnebin, Lombard, Lugeon, Tercier, and others.) Cover units: (1) autochthonous and Helvetic (H, Helvetic nappes; mol, molasse); (2) Ultrahelvetic (UH); (3) Valais (Niesen nappe, N, in the Prealps; Valais-Simplon Schistes Lustrés below GSB basement nappe); (4) Subbrianconnais (P, Plastiques nappe); (5) Brianconnais Rigides nappe in the Prealps, and cover remnants on the GSB basement nappe); (6) Brèche nappe; (7) Piemont Schistes Lustrés; overthrust on the GSB nappe, they are tectonically substituted for the previous Subbrianconnais-Brianconnais cover, which lies now in the Prealps ("cover substitution" of Ellenberger); (8) Simme nappe (S). Basement units: (a) auto-chthonous and external crystalline massifs; (b) lower Penninic (Simplon-Ticino nappes); (c) upper Penninic (GSB, Grand Saint Bernard nappe; MR, Monte Rosa nappe); (d) Austroalpine (DB, Dent Blanche nappe). (After Lemoine, 1973.)

One realizes that much of the uniformity of Helvetide structure is due to drag in the depths over a resistant floor."

Thus we are led to the experiments of William J. Sollas, Professor of Geology at Oxford, whom we salute as editor of Suess' English-language edition of *The Face of the Earth.* Sollas had previously made experiments of glaciers using pitch as a model material (1895), the viscous nature of glaciers having been recognized over a century earlier by A. C. Bordier. He recognized that many of the features presented by recumbent folds were more suggestive of flowing than bending, which led to the (1895) comparison between some features presented by flow lines in pitch glaciers with Bertrand's Alpine reconstructions. In Paper 7 Sollas considers Lugeon's synthesis of the High Calcareous Alps (1902a, 1902b; Lugeon and Argand, 1905) and employs a "gravita-tional spreading" viscous model, an idea commonly attributed today to Carey (1954) or Bucher (Paper 44). In a discussion section, here omit-ted, M. S. Allorge pointed out that the experiment gave a good illustra-tion of the structure *en chapelet* of Alpine geologists: i.e., the tendency of a brittle bed to parcel out into a series of broken segments, recalling on a large scale Heim's dislocated belemnites. The term *boudinage* had not yet been coined by Lohest in the Ardennes.

In the Alps, Sollas indeed supposes the poets' words to be true

The hills are shadows, and they flow
from form to form, and nothing stands. . . .

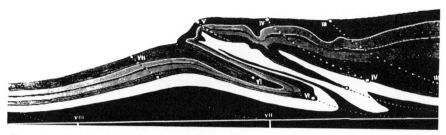

Figure 3 Result of gravitational gliding model experiment by E. Reyer (1892).

We recall here the criticism attached to the experimental models of Willis and his predecessors; our previous comments concentrated upon the significance of boundary conditions, as well as rheology, in model investigations. Nevertheless, the question of similitude cannot properly be ignored; its origins date, at the least, from Edward Reyer's (1888, 1892) qualitative appreciation of the subject; we quote here a translation from De Jong and Scholten (1973, p. xi footnote):

> When a deformation is produced in weak materials through the application of small forces and in a short time period, we may conclude that we should also be able to deform, as does Nature, materials with greater competence by the use of larger forces over longer periods of time. . . . With plastic materials and with material which ruptures easily we can study, at a small scale, the successive and final stages of the deformation process; such a material even permits us to make, with little force and in a short time, a typical mountain fold belt, and we may justifiably conclude that material of greater competence will behave analogously at a larger scale and over a longer time. . . .

Reyer, one of the early advocates of gravitational tectonics, performed experiments that still offer much by way of sophistication (see Fig. 3). As an example, in one series of experiments, gravitational gliding was induced by simply draining water from a model sedimentary basin into which fine-grained sediments had previously been gently deposited (Reyer, 1888, p. 484); Reyer's recognition of water level as a factor controlling sediment instability seems to be a direct anticipation of the "rapid drawdown condition" of slope analysis in modern soil mechanics theory.

By the turn of the century, the scale question was already grasped in more adequate terms; the surprising quotation here is freely translated from Otto Ampferer's 1906 classic paper, "On the Kinematics of Folded Mountains":

> In geology, scale reductions present a great danger for error because, as is especially ignored in tectonic questions, all other "constants" must also be reduced in accordance with the reduction in scale . . . Assume

65

we find in nature a rock 1 km thick, and originally 10 km wide, compressed now to a width of 5 km. We desire to reproduce these relations experimentally. We apply a reduction of 10^4 (i.e., 1 km in nature = 0.1 m in experiment). If we wish to introduce no unnatural conditions in our model we must reduce all other constants of the rock mass: stiffness, flexibility, tensional and compressional strength, in the same proportions. Unfortunately, we lack for many of these properties the appropriate measurements and applicable methods of comparison . . .

This is not the place to investigate such matters further: suffice it to say that the production of models requires comparable reduction in all constants as well as in dimensions. For our investigations the reduction in strength, a property easily overlooked, becomes especially important. We can consider cylinders of equal cross sectional area cut from various rocks: each, when of a specific height, will just crush its base. Thus we can express strength of rocks by maximum column height, and use these for convenient scale reduction. Following these considerations it is immediately clear that in a reduction which we must necessarily employ, even the strongest rock must be represented by very weak model materials. To return to the previous example, if the 1 km thick rock mass consisted of solid granite, the 10 cm model must then be composed of a material whose 1 mm^2 cross-sectional area would barely support a column 357 mm high. For most mountain-forming rocks a much weaker material would be required.

If we apply these considerations to the earth, we recognize that a somewhat thicker earth shell, insofar as it is conceived as being independent of its base, is characterized by its ready crumpling. An arbitrary wedge cut from it will not support itself but will crush its own base. Thus every larger arch of the earth shell, considered independently from its foundation, is too weak for self-support . . . !''

Shortly thereafter, Koenigsberger and Morath published their article, "Theoretical Foundations of Experimental Tectonics" (Paper 8); citing Helmholtz's 1882 theoretical work on hydrodynamic and aerodynamic modeling, they deal with the fundamental parameters of mass, length, and time. The models themselves seem comparatively crude, even though isoclinal folds and thrust faults are reproduced. But the approaches employed in regard to similitude are of benchmark significance, and this paper draws the final curtain on an era of primitive, if enormously clever, experimentation.

[*Editor's Note:* The glacial analogy advocated by Sollas continues to be attractive. Thus both David Elliott (Jour. Geophys. Res., v. 81, p. 949-963) and W. M. Chapple (Mechanics of thin-skinned fold and thrust sheets, *in press*) show that the basal shear stress along a sheet of thickness h is incremented by $\rho g h \alpha$, where ρ is density, g is gravitational acceleration, and α is topographic surface slope. This relation has been applied in glaciology for about a quarter-century, following the work of E. Orowan and J. F. Nye (cf. W. F. Budd, Jour. Glaciol. v. 9, p. 19-27). Elliott argues that because of (assumed) rock mass weakness, horizontal compression is of small importance for formation of thrust belts. Chapple's conclusion is nearly the opposite: rear compression of a thrust wedge is generally more important than gravity in overcoming the sliding resistance of a weak basal layer.]

Reprinted from pp. 716–719 of *Quart. Jour. Geol. Soc. London,* **62,**
716–721 (Nov. 1906)

RECUMBENT FOLDS PRODUCED AS A RESULT OF FLOW

William Johnson Sollas

University of Oxford

Our views as to the various kinds of deformation which affect the earth's crust received a remarkable extension with the memorable discoveries of Lapworth, Peach, and Horne in the north of Sutherland. These have now culminated in the long series of revelations which we owe to Bertrand, Rothpletz, Schardt, Kilian, Haug, Lugeon, Termier, Suess, and other workers in various regions of the Alps, who seem to have accomplished a veritable revolution in this branch of enquiry.

The long recumbent folds, which are perhaps the most surprising of the new forms of disturbance lately brought to light, can be more readily demonstrated than explained.

If we turn to one of the most complete and consistent accounts of these phenomena, Lugeon's description of the pre-Alps of Chablais, we perceive a series of recumbent folds, so greatly exceeding in horizontal extension their thickness vertically, that they are commonly spoken of as sheets rather than folds : they lie with remarkable flatness one on the other ; and as a rule those higher in the series extend farther to the front than those below, a feature referred to as 'déferlement' by the French, or 'leap-frog' as translated by my friend, M. Allorge.

The roots or origin of several of the lower of these folds are visible in the high Alps adjacent, but the roots of the higher folds, which form the pre-Alps, must be sought in the zone of Mont Blanc and the Briançonnais. Thus, some of the uppermost folds may have surmounted the obstacle presented by Mont Blanc on their way to the front in the pre-Alps.

It is no doubt true that overfolds may be traced into normal anticlines, and if long recumbent folds may be regarded as merely exaggerated overfolds, we may admit that their origin is not beyond our powers of comprehension ; but, even in this case, their subsequent history presents many difficulties to the imagination.

Many of the features presented by recumbent folds are more suggestive of flowing than bending, and long ago this led me to offer a comparison between some of the features presented by the flow-lines in pitch-glaciers with those made familiar to us by the sections of M. Marcel Bertrand.[1] An account of my first experiments with pitch-glaciers was brought before the Geological Society in 1895[2] ; but, as no detailed description has been published of others made subsequently, I take this opportunity to present the results of one or two of them, which recall in several striking peculiarities

---

[1] Rep. Brit. Assoc. 1895 (Ipswich) p. 689.
[2] Quart. Journ. Geol. Soc. vol. li, p. 361

some of the structures brought to light by Prof. Lugeon in the pre-Alps.

In the example that we may consider first (fig. 1, p. 718), the pitch-glacier was built up in the way already described in my previous paper. The three layers, of which the upper surface is indicated by the lines a, b, c, were placed in the experimental trough on April 3rd, 1895; it will be seen that they lie wholly behind the obstacle marked O. A fourth layer (d) was added on April 9th, its anterior termination lying just upon the summit of the obstacle. The experiment was brought to an end on June 12th, and, on cutting the 'poissier' or pitch-glacier longitudinally through the middle, the layers a, b, c were found to have assumed the forms shown by the lines a', b', c'. The general resemblance between these folded lines and some of the folded sheets in the Alps is sufficiently obvious; the second, marked c', with its 'carapace' of folds is not unlike the Morcles fold behind the Diablerets; and my friend Prof. Lapworth compares the third (b') with the Pilatus and Sentis, and the fourth (a') with the overslide of the Bavarian front.

The roots of the experimental folds lie on the other side of the obstacle O, which may be imagined to stand for Mont Blanc. In this respect they recall the views of Prof. Haug, who, to give one instance only, brings the zone of the Aiguilles d'Arve over the summit of Mont Blanc to form the recumbent fold of the Diablerets.

The whole of the four lines exhibit the phenomenon of déferlement, and may be compared in this respect with Prof. Lugeon's illustration of the three folds of Morcles, the Diablerets, and Mont Gond.[1] In the case of the pitch-glacier the déferlement is clearly a necessary consequence of the conditions of the experiment.

According to our present conceptions, there is one very marked difference which distinguishes the folds exhibited by this pitch-glacier from those of the Alps. In the former, the several folds originate from sheets which were superposed on each other at the commencement of the experiment, and in consequence the lower limb of each fold is adjacent to the similar limb of its neighbours, that is, the folds are 'emboîtés' one in the other. In the mountains, on the other hand, the lower limb of a superior fold reposes on the upper limb of the one immediately beneath it, that is, the folds are superposed, and not fitted one into the other.

In another experiment, however, made between March 24th and June 7th, 1895, even this difference has disappeared, or at least become greatly reduced. In this instance (fig. 2, p. 718), two obstacles O₁ and O₂ were placed in the path of the pitch: restricting our attention to the second layer, the original position of which is shown by the straight line a a', it will be seen to have formed three folds one behind the other,[2] all lying on the foreland beyond

---

[1] 'Les Grandes Nappes de Recouvrement des Alpes du Chablais & de la Suisse' Bull. Soc. Géol. France. ser. 4, vol. i (1901) fig. 3, p. 731.

[2] Indicated by the thickened lines, the front fold lies above the words '24 March' in fig. 2.

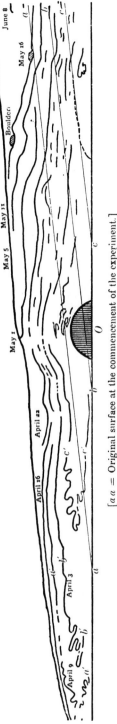

Fig. 1.—*Experiment commenced on April 3rd, and concluded on June 12th, 1895.*

[*a a* = Original surface at the commencement of the experiment.]

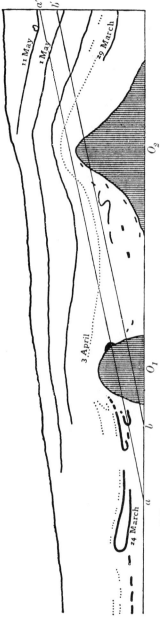

Fig. 2.—*Experiment commenced on March 24th, and concluded on June 7th, 1895.*

[*b b'* = Layer in its original position on March 24th, now seen overfolded immediately above *a b*.]

the obstacle O_1, while still farther on are a few scattered fragments, which may be compared with 'klippen.' In this instance, the several lobes of a complex fold would appear to have been sheared away from each other, passing through a series of stages such as are suggested by the folds in fig. 1.

If a flow has taken place in the Alpine regions at all comparable with that of these experiments, it is obvious that particular beds may have been very far from continuous at the conclusion of the movement. Thus, if we regard the layer just referred to as representing a limestone-series in the Alps, we might expect, on leaving the region of the 'klippen,' to enter another of recumbent folds; ascending the mountain-core O_1, we should find on its 'stoss'-side patches of the series, showing obvious signs of excessive pressure; and then we should have to traverse the wide interval between the two crystalline masses O_1 and O_2 before we again encountered insignificant remnants of the original sheet. The absence of the limestone-series between these several points might not, therefore, be the result of denudation, but the natural consequence of mountain-movement. If so, the removal of the stupendous masses of sediment by subaërial agencies, to which the Alps bear such striking witness, is brought closer to our powers of comprehension, since the material thus removed may have been the softer, more mobile argillaceous rock already deprived of its stiffening of comparatively-rigid strata.

Since, in the first experiment, it is only the first four layers that have assumed a complex folded character, and since these were the only layers which were compelled to make a somewhat abrupt ascent on passing over the obstacle O, it would appear probable that the folding stands in some relation to this fact. Such a suggestion is confirmed by reference to the illustration given in my previous paper,[1] where the layers marked 1 and 2, which commenced their existence wholly behind the barrier, have only taken the first step towards surmounting it, by forming comparatively-simple folds. It is in this region then, immediately behind an obstacle, that the production of folds originates, and the reason is clear: the anterior extremity of the layer is in contact, either with the floor of the experimental trough, or with the back of the obstacle, and in these regions the flow of the pitch is greatly retarded or even arrested by adhesion or excessive friction. The layer is held fast at this extremity, and as an upward and forward movement of the pitch takes place immediately behind, the layer is bulged upwards and forwards in the form of a fold. Slight inequalities of friction between the coloured layers of pitch and those to which pigment has not been added, may bring about the formation of secondary folds. In making these comparisons between flowing pitch and mountain-movements, I am anxious not to push the analogy too far; but observation has already so greatly outpaced explanation, that even remote resemblances may possess some value.

[1] Quart. Journ. Geol. Soc. vol. li (1895) fig. 3, p. 364.

[*Editor's Note:* The discussion has been omitted.]

Theoretische
Grundlagen der experimentellen Tektonik.

Von den Herren Joh. Koenigsberger und O. Morath.

(Mit 9 Textfiguren.)

Freiburg i. Br., im Oktober 1912.

Allgemein pflegt man im geologischen Unterricht die komplizierten Vorgänge bei der Gebirgsbildung durch einfache tektonische Modelle (Haut eines Apfels, Papier, Tücher) anschaulich zu machen. Manche Forscher haben dann das Ziel erstrebt, die Vorgänge in der Natur durch Modelle nachzuahmen und aus dem Verhalten des Modelles Schlüsse auf die Kräfte bei der Gebirgsbildung zu ziehen.

Wir erwähnen hier nur die Experimente von J. Hall, Favre, Daubrée, H. Schardt, H. Cadell, E. Reyer, B. Willis und namentlich von W. Paulcke[1]), der eine eingehende klare historische Übersicht der Arbeiten seiner Vorgänger gibt. Gerade in den neuesten Untersuchungen wird immer mehr Wert auf eine möglichst getreue Nachahmung der Natur gelegt, und es ist kein Zweifel, daß die Versuche von H. Schardt, von H. Cadell, von B. Willis, von W. Paulcke dem Ziel immer

---

[1]) W. Paulcke: Das Experiment in der Geologie. Karlsruhe 1912. Wir möchten z. S. 36, Anm 2 bemerken, daß die Wirkung des Wassers bei der Dynamometermorphose zuerst 1901 von dem einen von uns aus petrographisch-chemischen Gründen gefordert und dann durch Versuche von G. Spezia und solche von dem einen von uns gemeinsam mit W. Müller 1906 als wahrscheinlich nachgewiesen wurde. E. Riecke erwähnte nur beiläufig in einer theoretisch-physikalischen Arbeit die Rolle eines Lösungsmittels bei einseitigem Druck. — An die Verwandlung von Holz in Kohle nur durch Druck (S. 14) vermag der eine von uns nicht zu glauben. Möglicherweise war der Brückenpfeiler, wie das oft geschieht, schon vorher angekohlt worden. Andernfalls muß die Reibungswärme sehr groß gewesen sein.

näher gekommen sind. Diese Vervollkommnung der Mittel ist durch ein richtiges Gefühl der experimentierenden Geologen erreicht worden, ohne daß einer derselben hätte beweisen können, daß seine Anordnung wirklich besser als die früheren war. Die Ergebnisse können nicht als Beweis dienen; denn sie sollen gerade die Beobachtungen in der Natur kontrollieren. — Wir haben uns deshalb die Frage vorgelegt, wie ein Modell beschaffen sein muß, damit es möglichst genau die Vorgänge in der Natur wiedergibt. H. VON HELMHOLTZ[1]) hat zuerst das Problem des hydrodynamischen und aerodynamischen Modells theoretisch erschöpfend behandelt; das Studium an Modellen in der Praxis ist heute im Schiffsbau und Flugzeugbau allgemein üblich. — Auch in der Elastizitätslehre fester Körper und den damit zusammenhängenden tektonischen Problemen der Gebirgsbildung ist eine exakte Angabe der Beschaffenheit eines wirklich naturgetreuen Modells möglich. Das Problem ist mathematisch ziemlich einfach; wir wollen uns aber hier darauf beschränken, den Gedankengang der Ableitung darzulegen. Alle Eigenschaften oder physikalischen Konstanten einer Substanz, z. B. die von Granit, sind durch die drei Grundeinheiten, Länge, Masse und Zeit, gegeben[2]). Wenn wir also eine bestimmte Annahme über das Längenverhältnis der Natur zum Modell machen, z. B. daß 100 km = 1 m also das Verhältnis 100 000 : 1 sein sollen, ebenso bezüglich der Zeit und Masse, so sind theoretisch alle Eigenschaften der Modellsubstanzen eindeutig definiert; sie müssen in einem bestimmten Verhältnis zu denen der natürlichen Gesteine stehen. Praktisch entsteht dann nur die Frage, ob wir eine solche Modellsubstanz auch herstellen können.

Wir bezeichnen die Eigenschaften in der Natur mit dem Index O: l_0, ϱ_0 usw., die im Modell mit 1 : l_1, ϱ_1. Also das

$$\text{Längenverhältnis Modell : Natur} = \frac{1}{100\,000} = 1 . 10^{-5.}$$ Das Größenverhältnis bei dem Modell von W. PAULICKE dürfte wohl auch zwischen 10^{-4} und 10^{-5} liegen.

Hinsichtlich der Dichten oder spezifischen Gewichte der Modellsubstanzen haben wir nicht viel Auswahl; die verfügbaren, billigeren Substanzen haben ein spezifisches Gewicht zwischen 1 und 10, also von derselben Größenordnung

---

[1]) H. v. HELMHOLTZ: Wiss. Abhdlg., I, S. 158, 1882.
[2]) Man könnte auch die chemischen Vorgänge mit einbegreifen; doch sei hiervon abgesehen, da sie bei der Gebirgsbildung für die Tektonik nur von sekundärer Bedeutung sind.

wie das der Gesteine $(2,4-3,4)$. Wir wählen die Dichte der Modellsubstanzen etwa $= 3$ [1]), also $s_1 = s_0$ oder $[m_1 l_1^{-3}] = [m_0 l_0^{-3}]$, denn Dichte ist Masse m : Volumen l^3. Die Masse transformiert sich also im Verhältnis $\dfrac{m_1}{m_0} = \left(\dfrac{l_1}{l_0}\right)^3 = 1 . 10^{-5}$

Bezüglich der Z eit, mit der wir die Vorgänge am Modell sich abspielen lassen, haben wir keine willkürliche Wahl mehr; denn eine der Größen, welche die Zeit enthält, nämlich die Schwerkraft g, müssen wir so nehmen wie in der Natur. Wir können bis jetzt die Schwerkraft nicht beeinflussen oder eine andere Massenkraft ähnlicher Größe ohne viel Apparatur (Elektromagneten) hinzufügen [2]). Die Schwerkraftbeschleunigung g hat die Dimension $\dfrac{\text{Geschwindigkeit}}{\text{Zeit}} = \left[\dfrac{v}{t}\right] = [l \cdot t^{-2}]$.

Es ist also

$$l_0 t_0^{-2} = l_1 t_1^{-2} \quad \text{oder} \quad \frac{l_1}{l_0} = \left(\frac{t_1}{t_0}\right)^2$$

oder

$$\frac{t_1}{t_0} = \sqrt{\frac{l_1}{l_0}} . \quad \text{Da aber} \quad \frac{l_1}{l_0} = 1 \cdot 10^{-5}$$

angenommen wurde, so müßte sich die Zeit für den Modellvorgang zu dem in der Natur sich etwa wie 1 : 300 verhalten. Rein theoretisch müßten wir um ein vollkommen richtiges Modell herzustellen, die jetzigen Eigenschaften der Gesteine und von der geologischen Geschichte die Zeitdauer, den Anfangszustand und die wirkenden Druckkräfte kennen. Praktisch gestaltet sich die Sache hinsichtlich der Zeit einfacher. Auch wenn man, wie der eine von uns auf dem Standpunkt steht, daß einige tektonische Vorgänge bei der Gebirgsbildung sich relativ rasch in kurzen Perioden abgespielt haben, so wird man doch glauben dürfen, daß die Beschleunigungen [3]) äußerst gering und zu vernachlässigen sind. Sogar die Geschwindigkeiten werden recht klein gewesen sein. Deshalb ist es ziemlich gleichgültig, wie lange der Vorgang im Modell braucht; nur dürfen keine nennenswerten Geschwindigkeiten

---

[1]) Eine etwas andere Zahl wäre ohne wesentliche Bedeutung, wie aus dem folgenden zu ersehen ist.

[2]) W. PAULCKE hat diese Schwierigkeit bei seinen Versuchen umgangen, wie später erörtert wird.

[3]) Wir halten die Erdbeben nur für Anzeichen tektonischer Vorgänge, nicht für den Vorgang selbst. Doch weiß man hiervon noch fast nichts.

(mehr als 0,1 cm p. sec.) zustande kommen. Die Rücksicht auf die innere Reibung in den Gesteinen verlangt noch kleinere Werte der Geschwindigkeit, damit die Spannungen[1]) im Modell wie das in der Natur der Fall war, sich während des Vorgangs selbst ausgleichen und keinen nennenswerten Betrag erreichen. Wenn ein tektonischer Vorgang, z. B. im Tertiär, dreimal während 200 000 Jahren und innerhalb dieser Hauptperioden von vielleicht 1000 Jahren Dauer zehnmal in 6 Monaten vor sich gegangen wäre, so entspräche das einer wahren Zeitdauer in der Natur von etwa $30 \cdot 6$ Monaten $= 180$ Monaten; denn die Pausen sind ohne Belang. Im Modell müßte dann der Vorgang $\frac{180}{300} = 0,6$ Monate dauern. Wir haben auch bei unseren Versuchen gefunden, daß je langsamer und stetiger wir das Modell sich verändern ließen, um so ähnlicher die Ergebnisse der Natur werden.

Über den Anfangszustand vor der Bildung von Gebirgen ist man verschieden genau unterrichtet. In manchen Gegenden ist die geologische Geschichte vor der Hauptfaltung ziemlich gut, in anderen sehr wenig bekannt.

Bezüglich der wirkenden Kräfte bei einer Hauptfaltung steht es ähnlich. In einigen Fällen müssen Horizontaldrucke die Ursache gewesen sein, in andern sind noch Zusatzhypothesen möglich. Gerade diese Frage sollen die Modelle mitbeantworten und können es, wenn wir sie naturgetreu den theoretischen Forderungen entsprechend wählen. Die Brüche, Verwerfungen usw. soll unseres Erachtens das Modell automatisch wiedergeben. Man muß dazu im Modell die Erdkruste bis zur Tiefe der Druckausgleichung, der des „geschmolzenen" Gesteines, darstellen. Die Druckausgleichungsfläche für die Schwerkraft nach PRATT und die Schmelzfläche (Grenzfläche fest-flüssig) nach der geothermischen Tiefenstufe liegen übereinstimmend in etwa 100 km. Die Ausgleichsfläche für Spannungen möchten wir schon in etwa 50 km Tiefe suchen. Die Breite der darzustellenden Zone wird man nicht zu klein wählen dürfen. Wohl waren bei den meisten Gebirgsbildungen die Vorgänge einigermaßen auf kürzere Strecken parallel zu den Faltenachsen und senkrecht zu den wirkenden Kräften ähnlich. Doch sind überall erhebliche Wirkungen der seitlichen Massen bekannt.

[*Editor's Note:* Material has been omitted at this point.]

---

[1]) Es handelt sich hier um entsprechend große Spannungen; kleine Spannungen, wie sie sich im sog. Bergschlag usw. äußern, kommen für das Modell nicht in Betracht.

Part III

SQUEEZING INFORMATION FROM SEDIMENT AND ROCK

Editor's Comments
on Papers 9 Through 11

9   DAUBRÉE
    Expériences sur la déformation des fossiles, en relation avec la schistosité des roches

10 ADAMS and NICHOLSON
    Experiments on the Flow of Rocks Now Being Made at McGill University

11 KÁRMÁN
    Festigkeitsversuche unter allseitigem Druck

We arbitrarily distinguish between experimentalists who attempt to "scale" tectonic phenomena and who thus employ a variety of model materials, and those who in the laboratory attempt to determine the material properties of real geological materials. The contrast between these approaches has been generally well defined; for the most part, workers following the latter approach have (perhaps unfortunately) placed almost exclusive emphasis on cylindrical specimens subjected to triaxial pressurization, presumably because inhomogeneity renders interpretation difficult. In only a few instances has it seemed possible to combine rock or soil with boundary and environmental conditions appropriate to the solution of a specific geologic problem. The latter approach seems evident in some of the more recent work of Handin and his colleagues; but it was also present to a surprising extent in the work of A. Daubrée, whose experimental studies are a monument of meticulous and imaginative research. His torsion experiments in connection with jointing have been often cited; with unconfined experiments on prismatic specimens he demonstrated conjugate faults of acute dihedral angle bisected by the axis of compression. Perhaps less known are Daubrée's experimental studies of the action of pressurized gas at elevated temperatures on rocks, highly relevant to the origin of diatremes, his research on the origin of fault striations, and his simple but elegant models of buckling, which he compared to deformation in the Alps (citing Studer's *Geologie der Schweiz* and Rogers' Appalachian discoveries). He was aware of the existence of distorted and disrupted trilobites, brachiopods, and belemnites, as had been recently discovered

76

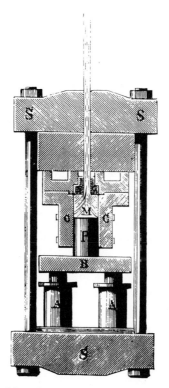

Figure 1 Production of schistosity in clay submitted to the action of the hydraulic press. P, compressor piston; M, clay to be compressed; N, steel plate; CC, cylinder; SSS, crossbars; A, A, pump body; B, plateau. An opening in the upper crossbar permits passage of the clay; its schistosity is parallel to the direction of movement. Scale: 1:10.

in Alpine nappes, and considered both fossil deformation and the associated question of schistosity within his experimental grasp. Using the hydraulic press designed by his colleague at the Ecole des Mines, Alfred Tresca (*see* Fig. 1), Daubrée beautifully reproduced schistose structure (Paper 9). His model belemnites were deformed by rupture, in accordance with Alpine observations; the reorientation phenomena of Sorby (1853), and the deformational phenomena of Phillips (1843), Sharpe (1847), and Tyndall (1856), were faithfully reproduced.

In contrast, as far as I have been able to determine, the first crushing tests on rock owed their inspiration not to mountain grandeur, but to the practical purpose of building construction. Extensive tests were carried out by the French engineers Gauthey, Soufflot, and Rondelet, at the time both of initial construction (1757) and subsequent remedial work of the church of St. Geneviève (the Pantheon) in Paris. In these tests, the compressional force required to crush the stone cubes was induced by means of weights attached to a lever system. A description is contained in Emiland Gauthey's *Traité de la construction des ponts* (1813), published after his death by his nephew Louis Navier, to whom we shall refer again. Robert Hooke had earlier employed stone-bending experiments associated with the proportionality principle named for him; but it was from Gauthey's time onward that crushing tests on stone were regularly conducted as a standard construction practice.

Experimental deformation under pressurized conditions appropriate to the geological environment was pursued only much later. Indeed, until 1892 plastic flow had not been experimentally developed in rocks*; in that year F. Kick embedded marble spheres in alum jackets and succeeded in flattening them. The closing decade of the past century saw the beginning of the classic work of Frank Dawson Adams at McGill University (Paper 10). Adams was a man of breadth, as attested to by his treatise *The Birth and Development of the Geological Sciences* (1838); his deformation experiments demonstrated beautifully that rocks can change their character and become ductile when confining pressure is applied. Because of the close-fitting thick-walled steel tube used to induce confinement, the values of the forces applied by Adams could only be approximately determined; nonetheless, he was able to determine the relation of strength to confinement. It seems ironic that, even so, his results seem more accurate than those produced by David Griggs' (1936) initial experiments, inasmuch as Griggs' endeavor was undertaken specifically to overcome the defects of Adam's procedure. Griggs, however, exposed most of his cylindrical specimens to the action of the pressure fluid, and his results were obscured by the effects of interstitial fluid pressure.

Adams believed that his experiments had a direct bearing on faulting, and his work was widely cited by his contemporaries concerned with faulting. Because rock strength appeared enhanced by pressure, Adams and Bancroft (1917, p. 635) suggested that only in the *upper* part of the zone of flow are the great Alpine "Decken" produced. They noted, however, that no account had yet been taken of the effect of temperature, "which would undoubtably tend to weaken geologic materials." Uniaxial experiments had been performed by Adams to 400°C.

Kármán's work represented a magnificent advance; using pressurized fluid as a confining medium, hydrostatic pressure of several kilobars could be applied independently of axial load, and both could be accurately measured. Kármán was a superb mechanicist, and his interest here concerned material behavior in general (i.e., failure theory, not geophysics); his well-known and highly regarded work includes studies of plastic buckling, turbulent fluid flow, and the development of

---

*However, in 1849 Thomson demonstrated ice flow in the laboratory, although the interpretation of his experiment subsequently created some confusion.

boundary-layer theory. His 1911 work, reproduced in its entirety as Paper 11, and that of his student, Böker (1915), form the basis for virtually all experimental rock-deformation studies to the present time. Triaxial compression, extension, and combined compression–torsion experiments were conducted; the brittle–ductile transition was noted. Mohr's theory was demonstrated to be more adequate than Coulomb's linear internal friction theory, (cf. Bouasse, 1901), fault angles were compared to predictions based on Mohr theory, and crystal-plasticity mechanisms were recognized.

Neglected for decades, Kármán's approach was followed in subsequent experiments of David Griggs, Jean Goguel, John Handin, Hugh Heard, Fred Donath, William Brace, and their coworkers and colleagues. A full elaboration of the developments in this field would more than fill a volume of this size, and the interested reader must be referred to other sources. The first phase of the post–World War II revival of interest in the subject was crowned by a Geological Society of America Memoir edited by Griggs and Handin (1960). Fairly up-to-date reviews of the state of knowledge have been adequately presented by Handin (1966) and Paterson (1970); important achievements include the discovery of temperature-enhanced ductility and decrease of yield stress, information on effects of anisotropy, and a crude appreciation of strength and "equivalent viscosity" reduction with increased duration of loading.

It seems necessary here to mention the quite different approach to rock deformation developed in the 1920s by the Austrians Bruno Sander and Walter Schmidt. These methods, in Goguel's words a veritable "corps de doctrine," aim at a statistical analysis of rock fabric. A major interpretive problem has been to decide whether the characteristics of a given distribution of fabric elements are significant or due, simply, to chance; in the east Alpine crystalline schists where the *Gefüge* methods developed, few independent controls existed. But as recognized by Eleanora Knopf (Knopf and Ingerson, 1938, p. 101), Griggs, and subsequently by Francis Turner and his associates, the guiding principles can be worked out by experimental deformation. Turner (1953) thus worked out a satisfactory statistical method to determine the orientation of the principal stresses producing twinning in calcite grains, and Melvin Friedman (1963, 1964; cf. Engelder, 1974) developed a similar relationship for microfractures in sand aggregate.

The application of the general approach toward an improved understanding of fault dynamics is exemplified by John Christie's work of the late 1950s and early 1960s. Christie had recently carried out a detailed field study of the extremely complicated Moine thrust zone; his 1958 and 1963 papers represent the extension of Turner's methods to dolomite fabric, and a renewed attack in modern terms on the remaining "secrets of the Highlands."

9

Reprinted from pp. 418–422 of *Etudes synthétiques de géologie expérimentale,*
Dunod, Paris, 1879

EXPÉRIENCES SUR LA DÉFORMATION DES FOSSILES, EN RELATION AVEC LA SCHISTOSITÉ DES ROCHES

A. Daubrée

Pour compléter la démonstration expérimentale des causes de la schistosité, il convenait de reproduire aussi les déformations de fossiles, qui en sont corrélatives et lui servent de témoins permanents. Quoique l'ensemble du phénomène ne puisse plus guère laisser de doute, il restait encore à en reconnaître les circonstances, par exemple, le degré de consistance que pouvait posséder la roche, lors de ces mouvements. C'était à l'expérience à nous éclairer sur ce sujet.

Les déformations considérables et variées que présentent les trilobites, les brachiopodes et en général les fossiles renfermés dans les roches schisteuses, sont de nature à guider

dans la recherche des forces, auxquelles les roches enve-
loppantes ont été soumises.

Un second type, non moins fréquent que les changements
de courbure, est représenté par les bélemnites de diverses lo-
calités des Alpes, qui ont été tronçonnées, et dont les segments
plus ou moins écartés laissaient primitivement entre eux
des vides, que des substances minérales sont ultérieurement
venues incruster ou remplir.

Lorsqu'un test n'a pas plus d'épaisseur que celui d'un
trilobite, il n'est pas difficile de le déformer, en l'empâ-
tant dans de l'argile, que l'on soumet ensuite à une pression.

Quant aux fossiles à test épais, comme une bélemnite ordinaire, leur résistance est trop grande pour qu'on pût la tron-
çonner au milieu de l'argile, au moins dans les conditions
de pression dont on pouvait disposer. Pour remédier à cette
difficulté et obtenir une rupture sous un moindre effort, on
a empâté, dans de l'argile, des cônes de craie très-allon-
gés, ayant la forme d'une bélemnite ordinaire, B (fig. 151).
Ce sont ces imitations de bélemnites qui ont été l'objet
d'une série d'expériences, dans lesquelles on a produit
l'écoulement', tantôt par écrasement, tantôt suivant le
sens de la pression; les fig. 152 et 153 représentent des
résultats ainsi obtenus.

La première de ces figures nous offre, pour ainsi dire,
une exagération du phénomène, les tronçons de la bélemnite étant beaucoup plus écartés qu'ils ne paraissent jamais
l'être dans les roches. Quant à l'échantillon de la figure
153, il a été obtenu par voie d'écrasement.

On arrive cependant à déformer les bélemnites naturelles,
mais il faut, pour cela, les enchâsser préalablement dans
une masse qui offre plus de cohérence que l'argile.

Plusieurs expériences, par voie d'écrasement, ont été faites sur des bélemnites (*Belemnites niger*), qui avaient été enchâssées très-exactement, au moyen du moulage, dans une masse

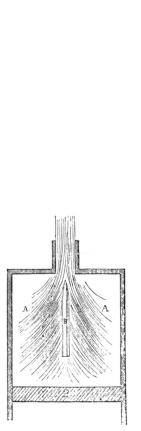

Fig. 151. — Imitation de bélemnite B en craie, placée dans une masse d'argile A A, que l'on force à s'écouler sous le piston P de la presse hydraulique, et destinée à subir un étirement et un tronçonnement. — Échelle de $\frac{1}{5}$.

Fig. 152. — Résultat de l'expérience précédente; B,B,B,B, B,B,B, tronçons fortement écartés les uns des autres, dans lesquels la bélemnite de craie a été réduite par l'écoulement de l'argile A. — Échelle de $\frac{1}{6}$.

Fig. 153. — Fragment de craie cylindro-conique, imitant une bélemnite naturelle, tronçonné par le laminage de l'argile où il était empâté. — Échelle de $\frac{1}{2}$.

de plomb, en forme de parallélipipède. La pièce de plomb était chaque fois soumise à une pression d'environ 50 000 kilogrammes. On a obtenu ainsi des bélemnites tronçonnées, dont les fragments sont plus ou moins espacés et qui, par conséquent, ont augmenté de longueur, exactement comme les types naturels que l'on avait en vue. L'échantillon (fig. 155)

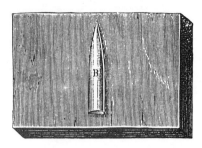

Fig. 154. —Belemnites niger B, exactement encastrée par moulage, au centre d'un prisme en plomb, formé de deux parties, dont une seule est représentée. Ce prisme est destiné à subir, perpendiculairement à ses plus grandes faces, l'action de la presse hydraulique. — Échelle de $\frac{1}{2}$.

comparé à l'état initial qui est représenté, par un moulage, (fig. 154), montre bien le changement qui s'est produit. Quelques-uns des tronçons se sont allongés, en s'écrasant.

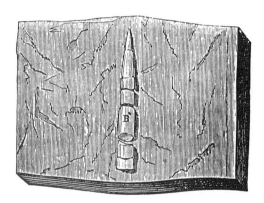

Fig. 155. — Étirement et tronçonnement de la bélemnite de la fig. précédente, par l'action de la presse hydraulique, sur le prisme de plomb où elle était encastrée. — Même échelle que pour la figure précédente.

Quant aux simples déformations de fossiles, leur imitation ne présente pas de difficulté. Ainsi, en enchâssant un test d'écrevisse dans une masse de plomb, que l'on comprime

ensuite, on le déforme, à la manière des trilobites des ardoi-
ses. Cette imitation trouve aussi son analogue, par exemple,

Fig. 156. — Déformation subie par une pièce de cuivre, à l'effigie de Georges III, sous l'action du laminoir, et rappelant certaines anamorphoses. — Grandeur naturelle.

dans le laminage par lequel on démonétise à Londres les
pièces hors de cours (fig. 156).

10

Reprinted from *Science*, n.s., 7(160), 82–83 (1898)

EXPERIMENTS ON THE FLOW OF ROCKS NOW BEING MADE AT MCGILL UNIVERSITY

Frank D. Adams and John T. Nicholson

The paper was presented by Dr. Adams and was illustrated by the lantern, by specimens of the results attained, and by a subsequent visit to the shops to see the machine. The authors have constructed a special crushing machine, much like the usual testing apparatus of engineering laboratories. Their object has been to subject cylinders of various rocks to pressures far above their crushing resistance, yet to confine them so that they could not shatter. After many unsuccessful trials of materials, strips of soft Swedish sheet iron were wrapped around a core of mild steel and welded together. The core was then bored out, the hole carefully polished and given a taper of one in a thousand. These cylinders were about $3\frac{1}{2}$–4 inches high and were turned down in the outer middle part so as to localize any bulging under pressure to this portion. They, therefore, looked like large spools, with thick ends. Cylinders of Carrara marble had meantime been prepared in Germany of the same taper as the holes and of such a size that, when the spools were heated and expanded, the cylinders dropped snugly in and were caught midway of the spool. The cylinders of marble were about two centimeters in diameter. Chrome-steel plungers were employed in the squeeze, and fitted perfectly in the spools. By using the city water mains, which give at the University a pressure of 135 pounds to the square inch, oil was forced in beneath the piston of the press, and cylinder pressure gauges and a recording curve-tracing mechanism were connected. The blocks were grad-

ually compressed until subjected to thirty tons' pressure. Under this squeeze the marble cylinder bulged at the middle, expanded its iron jacket and approximated a thick disc. When released it was found that it had flowed without losing its cohesion at all. When split down the vertical axis the cross section revealed two opposing paraboloids, or blunt cones of unchanged marble, filled in between with a dense, chalky variety, but all perfectly solid. Thin sections show a great abundance of twinning striations and gliding planes and evidences of strain. Cylinders of Baveno granite are now ready for experimentation, but have not yet been compressed. Peat has, however, been compressed into a black, shining and lustrous substance, very like high-grade lignite or coal, a result similar to that obtained abroad. Copper filings have been compacted also to solid metal. A further apparatus has been designed so that superheated steam can be introduced into the test, which can be kept at 500° F., for months at a stretch, while the compression is progressing, the gauges and recorder meantime registering the pressure at all times. Dr. Adams stated that two and a-half years had been spent in experimenting and six months in getting results.

The Fellows were outspoken in their praise of this work, and it was felt by all to be one of the most important contributions ever laid before the Society. It brings within the domain of experiment some of the obscure proceses of dynamic metamorphism and throws great light on the viscous flow of rocks.

11

Reprinted from *Z. Vereines Deut. Ing.,* 55(42), 1749-1757 (1911)

Festigkeitsversuche unter allseitigem Druck.

Von Dr. **Th. v. Kármán** in Göttingen.

(hierzu Textblatt 29)

Die Festigkeitsversuche unter allseitigem Druck, über die in den folgenden Zeilen berichtet werden soll[1]), haben den Zweck, zur Klärung der Frage beizutragen, durch welche Umstände die Elastizitäts- und Bruchgrenze beim allgemeinen Spannungszustande bestimmt wird. Die in der kontinentalen Praxis allgemein übliche, von St. Venant herrührende Berechnungsweise der auf zusammengesetzte Festigkeit beanspruchten Konstruktionsteile fußt bekanntlich auf der Annahme der größten Dehnung, während viele englische und amerikanische Ingenieure, die auf der Annahme der größten Spannung fußende Rankinesche Formel benutzen. Beide Annahmen sind jedoch durch neue Versuche als durchaus unzutreffend erkannt worden, und die Mohrsche Theorie, die sie ersetzen sollte, fand bisher auch keine ausreichende Bestätigung. Es sprechen zwar viele Versuche über die Elastizitätsgrenze bildsamer Stoffe (wie Kupfer, weiches Eisen, weicher Stahl) bei zusammengesetzter Beanspruchung zu ihren Gunsten, bezüglich spröder Körper stehen jedoch die Versuche, die sie bekräftigen, auch solche gegenüber, die ihr entschieden widersprechen. So erscheint es als eine durchaus wichtige Aufgabe, zur Klärung dieser Frage, die von jeher als eine der am meisten umstrittenen in der Festigkeitslehre galt, durch Versuche beizutragen und insbesondere die Grundlagen der Mohrschen Theorie planmäßig zu prüfen.

Außer der Prüfung der an die Mohrsche Theorie sich anknüpfenden Fragen bot die vorgenommenen Versuche Gelegenheit zur Untersuchung der Aenderungen im Kleingefüge des Versuchsmaterials bei Ueberschreitung der Elastizitätsgrenze. Die darauf bezüglichen mikroskopischen Untersuchungen gewinnen besonders dadurch an Interesse, daß ich durch Anwendung des allseitigen Druckes auch bei sonst als spröde geltenden Stoffen eine plastische Deformation zu erzwingen vermochte, so daß die die spröden Körper kennzeichnenden Bruchvorgänge und das plastische Fließen an einem und demselben Material verfolgt werden konnten.

In den folgenden Zeilen soll zunächst über die bisher

abgeschlossenen ersten Versuchsreihen berichtet werden: es sind dies Versuche unter allseitigem Druck an Marmor- und Sandsteinkörpern. Wenn auch einzelne grundsätzliche Fragen noch nicht erledigt worden sind, so glaube ich doch, daß auch die bisherigen Ergebnisse manches zur Klärung der Vorstellungen beitragen können.

Es ist mir eine angenehme Pflicht, Hrn. Prof. Dr. Prandtl für seine rege Anteilnahme an dem Zustandekommen dieser Arbeit meinen innigsten Dank auszusprechen.

I. Die Theorie von Mohr und die darauf bezüglichen Versuche[1]).

Während die erwähnten beiden Annahmen — jene von der größten Spannung und der größten Dehnung — eine einfache Größe suchen, die als Maß der Bruchgefahr gelten soll, liegt der Mohrschen Theorie mehr eine physikalische Vorstellung über Bruch und bleibende Formänderung zugrunde: die Vorstellung, daß man in beiden Fällen mit »Gleitvorgängen« zu tun hat. Einen ersten Ansatz in dieser Richtung lieferte bereits die Coulombsche Theorie der Druckfestigkeit. Coulomb stellte sich den Bruchvorgang beim Druckversuch in der Weise vor, daß die Schubspannung längs jeden Flächenelementes ein Gleiten hervorzubringen strebt, dies aber durch eine Art Reibungswiderstand verhindert wird. Setzt man diesen Reibungswiderstand aus einem konstanten Betrage s_0 und einem Betrage $f \sigma$, der dem auf das betreffende Flächenelement wirkenden Normalstück σ proportional ist, zusammen, so hat man als Gleichgewichtsbedingung für die Schubspannung die Ungleichung

$$\tau < s_0 + f \sigma.$$

Der Bruch erfolgt dann längs derjenigen Ebene, in der die Schubspannung τ mit zunehmender Belastung zuerst den Wert $s_0 + f \sigma$ erreicht. Man kann leicht einsehen, daß

a) nur die Ebenen in Betracht kommen können, die zu der Ebene der größten und der kleinsten Hauptspannung senkrecht stehen,

[1]) Der ausführliche Versuchsbericht wird in den Mitteilungen über Forschungsarbeiten erscheinen.

Die hier berichteten Versuche schließen sich an die durch die Jubiläumstiftung der deutschen Industrie veranlaßten Versuche von Prandtl und Rinne an; es ist mir eine angenehme Pflicht, zu erwähnen, daß für die Fortsetzung der Versuche einerseits die Jubiläumstiftung einen weiteren Beitrag gewährte, andererseits seitens der Kgl. ungarischen Akademie der Wissenschaften mir persönlich eine namhafte Unterstützung zuteil wurde.

[1]) Für Literaturangaben vergl. — außer der bevorstehenden ausführlichen Veröffentlichung in den Mitteilungen über Forschungsarbeiten — des Verfassers Bericht »Festigkeitsprobleme im Maschinenbau« in der Enzyklopädie der math. Wissenschaften Bd. IV, Art. 27; ferner P. Roth, »Die Festigkeitstheorien und die von ihnen abhängigen Formeln des Maschinenbaues«, Berlin 1902. Ueber die Problemstellung selbst vergl. Föppl, Mitteilungen aus dem mech.-techn. Laboratorium München 27 (1900) S. 1 u f.

b) daß die Elastizitätsgrenze (oder Bruchgrenze) durch eine Beziehung zwischen den äußersten Hauptspannungen σ_1 und σ_3 bestimmt wird. Diese lautet

$$\sigma_1 - \lambda\,\sigma_3 = C,$$

falls

$$\lambda = \mathrm{tg}^2\left(45^0 + \frac{\varrho}{2}\right),$$

$$C = 2\,s_0\,\mathrm{tg}\left(45^0 + \frac{\varrho}{2}\right)$$

und

$$f = \mathrm{tg}\,\varrho$$

gesetzt wird.

Die Größe der mittleren Hauptspannung spielt dabei keine Rolle.

Diese hier kurz angedeutete Coulombsche Betrachtung hat später zahlreiche Festigkeitstheorien veranlaßt[1]. Einen wichtigen Sonderfall der Coulombschen Theorie bildet der Fall $\lambda = 1$: die Annahme des größten Spannungsunterschiedes (maximum stress-difference theory[2]). Nach dieser Annahme würde die Elastizitätsgrenze durch einen konstanten Unterschied der äußersten Hauptspannungen oder mit andern Worten durch einen konstanten Wert der größten Schubspannung festgelegt.

Die Grundannahme der Mohrschen Theorie[3] kann in dem Satze zusammengefaßt werden, daß die Elastizitätsgrenze bezw. Bruchgrenze durch einen Grenzwert der Schubspannung festgelegt wird, welcher von der auf die Gleitfläche wirkenden Normalspannung abhängig ist.

Die Mohrsche Theorie ist somit allgemeiner als die Coulombsche, da sie keine »innere Reibung« voraussetzt, sondern nur soviel, daß der Bruch oder der Eintritt einer bleibenden Formänderung irgendwie von der Schub- und Normalspannung in der Gleitebene abhängen muß. Dementsprechend verlangt sie keine Proportionalität der zusammengehörenden Werte der Normal- und Schubspannung, sondern nur eine allgemeine Abhängigkeit, die für jeden Stoff besonders durch Versuche festgestellt werden muß.

Da nach der Mohrschen Annahme nur die Ebenen als Gleitebenen in Betracht kommen können, die senkrecht zu der Ebene der äußersten Hauptspannungen stehen, und die in diesen Ebenen wirkenden Schub- und Normalspannungen nur von den äußersten Hauptspannungen abhängen, so folgt daraus, daß die Elastizitätsgrenze von der mittleren Hauptspannung unabhängig ist. Man darf sich daher auf einen ebenen (zweidimensionalen) Spannungszustand beschränken, der durch die äußersten Hauptspannungen σ_1 und σ_3 bestimmt ist. Für einen solchen Spannungszustand hat Mohr eine sehr anschauliche Darstellung gegeben, indem er die auf dieselbe Ebene bezogene Normal- und Schubspannung zu Koordinaten wählte (vergl. Fig. 1). Jedem Spannungszustande mit den Hauptspannungen σ_1 und σ_3 entspricht alsdann ein Kreis, dessen Mittelpunkt in der Entfernung $\frac{\sigma_1 + \sigma_3}{2}$ auf der σ-Achse liegt und dessen Halbmesser gleich $\frac{\sigma_1 - \sigma_3}{2}$ ist. Den Koordinaten jedes Kreispunktes entsprechen je zwei zusammengehörige Werte der Normal- und Schubspannung, gegeben durch die Formeln

$$\sigma = \frac{\sigma_1 + \sigma_3}{2} + \frac{\sigma_1 - \sigma_3}{2}\cos 2\,\vartheta$$

$$\tau = \frac{\sigma_1 - \sigma_3}{2}\sin 2\,\vartheta$$

(ϑ = Neigung der Bezugsebene zu der Hauptspannung). Ist nun der Grenzwert der Schubspannung als Funktion der Normalspannung durch die »Grenzkurve« $\tau = f(\sigma)$ gegeben, so bildet diese Kurve die Umhüllende sämtlicher Spannungskreise, die den Belastungszuständen an der Elastizitätsgrenze

(»Grenzzustände«) entsprechen. Mohr zeigt auch, daß man die Neigung der Gleitebenen mit Hülfe dieser Darstellung unmittelbar ermitteln kann: der spitze Winkel zwischen zwei zu den Hauptspannungen symmetrisch gelegenen Gleitebenen ist durch den Neigungswinkel φ der Normalen zu der σ-Achse gegeben, s Fig. 1.

Der Coulombsche Fall ergibt sich nun als Sonderfall, indem man die Grenzkurve aus zwei Geraden bestehend wählt. Sind die beiden Geraden insbesondere parallel zu der σ-Achse, so haben wir den Fall der konstanten größten Schubspannung.

Fig. 1.

Darstellung der Grenzzustände nach Mohr.

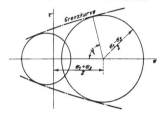

Wie schon erwähnt worden ist, sind Versuche sowohl für als gegen die hier dargelegte Mohrsche Theorie vorhanden.

Versuche zugunsten der Mohrschen Theorie.

Hier kommen in erster Linie die seit etwa 1900 in England und in den Vereinigten Staaten zahlreich unternommenen Untersuchungen über zusammengesetzte Beanspruchung in Betracht. J. Guest[1] hat als Versuchsgegenstand Rohre aus Flußeisen und Kupfer gewählt und bestimmte die Fließgrenze, indem er Zug mit Verdrehung, Zug mit Innendruck und Innendruck mit Verdrehung verband. Diese Versuche wiederholte in neuester Zeit W. Mason[2] und mit weichen Stahlrohren. E. L. Hancock[3] und W. A. Scoble[4] unterwarfen massive Stäbe aus Kupfer, Flußeisen und weichem Stahl gleichzeitig der Biegung und der Verdrehung. Schließlich ist eine ausgedehnte Versuchsreihe von Smith[5] zu erwähnen, der Zug und Druck mit Verdrehung verband und ebenfalls Probestäbe aus weichem Stahl benutzte. Alle diese Versuche ergaben sehr stark veränderliche Werte sowohl für die größte Spannung als für die größte Dehnung an der Elastizitätsgrenze (oder Fließgrenze), dagegen einen annähernd konstanten Wert für die größte Schubspannung oder — was auf dasselbe herauskommt — für den Unterschied der äußersten Hauptspannungen, und zwar unabhängig von der dritten Hauptspannung. Es sei bemerkt, daß nach diesen Versuchen das Verhältnis der zulässigen Schubspannung bei reiner Verdrehung zu der zulässigen Zugspannung bei reinem Zuge 0,5 beträgt, statt 0,7 nach St. Venantschen und 1,0 nach der Rankineschen Formel. Dieses Verhältnis 1 : 2 der zulässigen Schub- und Zugspannung wurde für plastische Stoffe durch zahlreiche ältere Versuche ebenfalls bestätigt.

Die soeben erwähnten Versuche sprechen entschieden für die Mohrsche Auffassung, namentlich für die Richtigkeit der Annahme, daß die mittlere Hauptspannung unwesentlich ist. Da aber bei den untersuchten ausgesprochen plastischen Stoffen augenscheinlich stets der allereinfachste Fall vorlag, der Fall der konstanten größten Schubspannung, so kann gerade der wesentlichste Punkt der Mohrschen Theorie: die Abhängigkeit der Schubspannung von der Normalspannung im Grenzzustand, offenbar erst bei spröden Körpern hervortreten. Nun liegen bedauerlicherweise vergleichende Versuche an spröden Körpern nur in sehr geringer Anzahl vor,

---

[1] Vergl. einen sehr interessanten Aufsatz von Bouasse (Annales de physique et chimie Band 23, 1902) über die Coulombsche Theorie, ferner A. Mesnagers Bericht für den internationalen Kongreß für Physik, Paris 1900.

[2] S. z. B. Thomson und Tait, Treatise on natural philosophy Bd. II S. 243. — Die Annahme der größten Schubspannung entspricht auch den Versuchsergebnissen von Tresca über Fließen fester Körper.

[3] O. Mohrs Arbeiten über den Gegenstand: Zivilingenieur Bd. 28 (1882) S. 113; Z. 1900 S. 1524; Abhandlungen aus dem Gebiete der technischen Mechanik, Berlin 1906, S. 167.

[1] Phil. Magazine Band 50 (1900) S. 690.

[2] Proceedings of the Institution of Mechanical Engineers 1909 S. 1205.

[3] Phil. Magazine Band 12 (1906) S. 418 und Band 15 (1908) S. 214.

[4] Phil. Magazine Band 12 (1906) S. 533.

[5] Proceedings of the Institution of Mechanical Engineers 1910 S. 1237.

und die vorhandenen Versuchsergebnisse sind auch infolge der Schwierigkeiten des Versuches weit unsicherer[1]).

Von grundsätzlicher Bedeutung sind die Versuche von Professor Föppl[2]) an Steinen und Zement über den Einfluß der Art des Spannungszustandes auf die Bruchgrenze. Der Vergleich der gewöhnlichen Druckfestigkeit ($\sigma_1 > 0, \sigma_2 = \sigma_3 = 0$) mit der von ihm untersuchten »Umschlingungsfestigkeit« ($\sigma_1 = 0, \sigma_2 = \sigma_3 > 0$) liefert unmittelbar Aufschluß über den Einfluß der mittleren Hauptspannung, da man die beiden Versuche so auffassen kann, daß die mittlere Hauptspannung einmal der kleinsten, dann der größten Hauptspannung gleich kommt.

Nun ergaben die Versuche, daß die Umschlingungsfestigkeit und die gewöhnliche Druckfestigkeit ungefähr gleich groß ausfallen, wenn man zwischen Versuchskörper und Druckplatte Schmiermittel einführt, um die Reibung zu vermeiden; bei ungeschmierten Druckflächen ist dagegen die Umschlingungsfestigkeit bis zweimal so groß wie die Druckfestigkeit. Ob nun tatsächlich mit Recht angenommen werden kann, daß man zu den richtigen Werten der Druck- bezw. Umschlingungsfestigkeit gelangt, wenn man zwischen Probekörper und Druckplatten Schmiermittel einführt, erscheint wenigstens zweifelhaft[3]). Wäre dies der Fall, so würde dies bedeuten, daß die mittlere Hauptspannung tatsächlich ohne Einfluß ist.

Versuche gegen die Mohrsche Theorie.

Als Versuche, die entschieden gegen die allgemeine Gültigkeit der Mohrschen Theorie sprechen, kommen lediglich die auf Anregung und unter Leitung des von Prof. W. Voigt im Physikalischen Institut der Universität Göttingen durchgeführten Versuchsreihen in Betracht[4]). Es wurden Zerreißversuche unter allseitigem Druck an spröden Körpern vorgenommen, und zwar in der Weise, daß der zwischen Zugköpfen eingefaßte Probekörper zuerst einem allseitig gleichen Druck unterworfen und dann die axiale Spannung durch Federkraft solange vermindert wurde, bis der Probekörper zerriß. Die Versuche ergaben für die Bruchgrenze stets denselben Unterschied der Hauptspannungen (d. h. der axialen Spannung und des auf die Mantelfläche wirkenden Druckes), und zwar war dieser Unterschied annähernd gleich der Zugfestigkeit bei gewöhnlichen Zugversuchen. Dieses Ergebnis würde an und für sich der Mohrschen Theorie noch nicht widersprechen; zieht man aber in Betracht, daß die Druckfestigkeit der untersuchten Körper die Zugfestigkeit derselben vier- bis fünffach übertrifft, so ist es klar, daß die Bruchgrenze nicht nur von den beiden äußersten Hauptspannungen abhängen kann.

Wenn die Voigtschen Versuche die Mohrsche Theorie auch nicht widerlegen, so erscheint es doch sehr wahrscheinlich, daß nicht alle Bruchvorgänge durch die Mohrsche Annahme erklärt werden können. Es ist überhaupt fraglich, ob so qualitativ verschiedene Vorgänge, wie der Bruch spröder Körper bei Zug und Verdrehung einerseits und bei Druck andererseits, sich einer einheitlichen Gesetzmäßigkeit fügen können. Professor Prandtl hat in einem vor der Versammlung der Naturforscher und Aerzte in Dresden 1908 gehaltenen Vortrage vorgeschlagen, zwei Arten des Bruches zu unter-

scheiden: den Gleitungs-(Verschiebungs-)bruch und den Trennungsbruch[1]). In der Tat lassen sich beide schon an der Verschiedenheit der Bruchflächen leicht erkennen: beim Trennungsbruch entstehen harte glänzende Bruchflächen, beim Gleitungsbruch sind dagegen die Bruchflächen mit feinem Mehl bedeckt. Es ist wohl möglich, daß die Mohrsche Theorie auf den Gleitungsbruch beschränkt werden muß, so daß Druckversuche und Zugversuche nicht unmittelbar verglichen werden können.

II. Strittige Punkte und Versuchsprogramm.

Aus den obigen Darlegungen geht hervor, daß die Frage nach der Richtigkeit der Mohrschen Theorie keineswegs hinreichend geklärt ist. Den heutigen Stand der Frage können wir etwa folgendermaßen zusammenfassen:

a) Erstens erscheint es fraglich, wie weit sämtliche Bruchvorgänge sich der Mohrschen Theorie fügen. Namentlich entsteht die Frage, unter welchen Umständen Gleitungsbruch und unter welchen Trennungsbruch erfolgt. Und zugegeben, daß für den Gleitungsbruch die Mohrschen Annahmen zutreffen, welche Gesetzmäßigkeit ist für Trennungsbruch maßgebend?

b) Zweitens ist es auch für diejenigen Fälle, in welchen zweifelsohne Gleitungsbruch oder bleibende Formänderung durch Gleitung entsteht, fraglich, wie weit die Annahme eines mit der Normalspannung wachsenden Grenzwertes für die Schubspannung zutreffend ist. Namentlich hatte bisher in allen Fällen, wo die Mohrsche Theorie unbestreitbar richtig dasteht, d. h. bei bildsamen Stoffen, stets den Grenzfall der konstanten größten Schubspannung beobachtet. Da sind also Versuche mit spröden Körpern notwendig.

c) Drittens ist es fraglich, wie es um den Einfluß der mittleren Hauptspannung steht? Für bildsame Stoffe ist gezeigt worden, daß die Elastizitätsgrenze von der mittleren Hauptspannung unabhängig ist. Für spröde Körper sprechen die Föpplschen Versuche dafür, die Voigtschen Zerreißversuche unter allseitigem Druck dagegen. Hier ist also ebenfalls weitere Klärung notwendig.

Bei den in den folgenden Zeilen zu berichtenden Versuchen wurden möglichst homogene Spannungszustände angestrebt, in der Weise, daß die zylindrischen Probekörper gleichzeitig einem bestimmten Flüssigkeitsdruck an der Mantelfläche und einer axialen Druckbeanspruchung unterworfen wurden. Namentlich läßt die Versuchseinrichtung, die weiter unten ausführlich beschrieben werden soll, zwei Reihen von Spannungszuständen verwirklichen, die in Fig. 2 schematisch dargestellt sind. Bei beiden Arten des Spannungszustandes

Fig. 2.

Schema der Spannungszustände bei Festigkeitsversuchen unter allseitigem Druck.

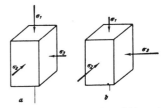

a   *b*

sind zwei Hauptspannungen σ_2 und σ_3 gleich, und zwar im Falle a) die beiden kleineren ($\sigma_1 > \sigma_2 = \sigma_3$), im Falle b) die beiden größeren ($\sigma_1 < \sigma_2 = \sigma_3$), falls man Druckspannungen mit positivem Vorzeichen rechnet. Die beiden Versuchsreihen sollen kurz als Druckversuche bezw. Zugversuche unter allseitigem Druck bezeichnet werden; bei der ersten Reihe ist

[1]) Man kann eine Reihe von Versuchen heranziehen, die allerdings nicht zu diesem Zweck unternommen wurden; so zunächst die vergleichenden Zug-, Druck- und Scherversuche Bauschingers an Zementwürfeln (Mitteilungen aus dem techn.-mech. Laboratorium München, Heft 8, 1879), ferner die bekannten Bachschen Zug-, Druck- und Verdrehungsversuche an Gußeisenkörpern (Berichte und Abhandlungen), die alle mit der Mohrschen Theorie in leidlicher Uebereinstimmung stehen. Vergl. auch das oben erwähnte Werkchen von P. Roth. Ich will jedoch betonen, daß diese Uebereinstimmung einzelner isolierter Festigkeitszahlen nicht viel beweist. Die Richtigkeit der Mohrschen Annahmen dürfte nur durch Vergleich ganzer Reihen von Spannungszuständen geprüft werden.

[2]) Mitteilungen aus dem techn.-mech. Laboratorium München. Heft 27, 1900.

[3]) Vergl. z. B. die Untersuchungen von F. Kick über diesen Gegenstand.

[4]) Die Versuche sind in den Annalen der Physik Band 53, 1894, S. 43; Bd. 67, 1899, S. 452; ferner W. Voigt, »Zur Festigkeitslehre«, Z. 1901 S. 1033 und Annalen der Physik Bd. 4, 1901, S. 567. — Die Versuche wurden mit ganz ähnlichem Ergebnis wiederholt von W. E. Williams, s. Phil. Mag. Bd. 15, 1908, S. 81.

[1]) Es ist bemerkenswert, daß schon Coulomb die beiden Bruchvorgänge gewissermaßen dadurch auseinanderhielt, daß er zu der oben dargelegten Bedingung gegen Gleitung die Bedingung $\sigma_{max} <$ konst. für Zugspannungen hinzufügte. Vergl. auch W. Thomson, Phil. Mag. Band 38 (1869).

der gewöhnliche Druckversuch ($\sigma_1 > 0$, $\sigma_2 = \sigma_3 = 0$), in der zweiten Reihe der Föpplsche Umschlingungsversuch ($\sigma_1 = 0$, $\sigma_2 = \sigma_3 > 0$) enthalten.

Die Versuche lassen eine unmittelbare Prüfung der Mohrschen Annahmen zu: Ist die Annahme richtig, daß an der Elastizitätsgrenze die Schubspannung mit dem Normaldruck auf die Gleitebene wächst, so müssen wir eine Erhöhung der Druckfestigkeit durch den allseitigen Druck beobachten. d. h. eine Zunahme der axialen Spannung um einen bedeutend größeren Betrag als der hinzugefügte Manteldruck. Was ferner den Einfluß der mittleren Hauptspannung anbelangt, so vertreten die Zug- und Druckversuche unter allseitigem Druck zwei Grenzfälle, in denen die dritte Hauptspannung einmal gleich der größten, dann aber gleich der kleinsten Hauptspannung wird, so daß der Vergleich der beiden Versuchsreihen erweisen muß, ob die mittlere Hauptspannung einen Einfluß hat oder nicht. Schließlich lassen die Zugversuche entscheiden, unter welchen Umständen der Bruch tatsächlich durch Gleiten erfolgt, entsprechend der Mohrschen Vorstellung, und unter welchen Umständen dagegen ein Trennungsbruch herbeigeführt wird.

III. Die Versuchseinrichtung.

Die Versuchseinrichtung mußte so beschaffen sein, daß der auf die Mantelfläche des Probekörpers wirkende allseitige Druck, den man kurz als Manteldruck bezeichnen kann, und die axiale Kraft voneinander unabhängig geändert und gemessen werden können. Fig. 3 gibt ein schematisches Bild der Versuchseinrichtung.

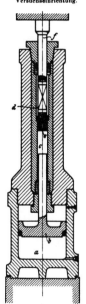

Fig. 3.

Versuchseinrichtung.

Der Manteldruck wird in der Weise ausgeübt, daß zunächst in dem Raum a durch eine Handpumpe Druck erzeugt und durch den Kolben b mit einer Uebersetzung von etwa 1 zu 25 auf den Raum c übertragen wird. Die beiden Räume c und d stehen durch die Bohrung des Gewindestückes e in Verbindung, so daß auf die Mantelfläche des Probekörpers der in dem Raume c erzeugte Druck wirkt. Als Druckflüssigkeit wurde für den Hochdruckraum Glyzerin gewählt, einerseits wegen seiner geringen Zusammendrückbarkeit, anderseits wegen seiner Dickflüssigkeit, die die Dichtung bedeutend erleichtert.

Der Längsdruck wird durch den Druckstempel f ausgeübt, und zwar in der Weise, daß der ganze Apparat auf dem Kolben einer sonst zu gewöhnlichen Druckversuchen dienenden hydraulischen Presse ruht, während der Druckstempel sich gegen die obere Druckplatte der Presse stützt. Wird nun das Probestück zwischen dem Druckstempel und dem mit Apparate fest verbundenen Gewindestück e gefaßt, so wird die durch den Kolben der hydraulischen Presse ausgeübte Kraft auf den Probekörper übertragen, der in dieser Weise eine axiale Druckbeanspruchung erfährt, die unabhängig ist von dem Manteldruck und die — abgesehen von der Reibung in den Dichtungen (s. weiter unten) — in derselben Weise wie bei gewöhnlichen Druckversuchen gemessen werden kann.

Die Versuchseinrichtung wurde von Fried. Krupp in Essen ausgeführt. Der auf einen inneren Druck von 6000 at berechnete Hochdruckzylinder besteht aus zwei warm aufeinander gezogenen Röhren von 24 bezw. 45 mm Wandstärke. Der Durchmesser des Hochdruckraumes beträgt 50 mm. Gewindestück und Druckstempel sind aus Spezial-Nickelstahl von 4 vH Nickelgehalt angefertigt; die Elastizitätsgrenze des

Nickelstahles liegt etwas über 10000 at, so daß die axiale Beanspruchung der Probekörper, die mit dem Druckstempel gleichen Durchmesser hatten, bis zu dieser Grenze gesteigert werden konnte.

Kraftmessung und Reibungsverhältnisse.

Als Kraftmesser dienten zwei Flüssigkeitsmanometer, von denen das eine (M_1) an den Raum a des Versuchsapparates, das andre (M_2) an den Zylinder der Presse angeschlossen wurde. Die Kraftmessung wird wesentlich beeinflußt durch die Reibungskräfte, die an den Dichtungen der beiden Kolben b und f auftreten. Die Reibungsverhältnisse sind jedoch bei näherer Untersuchung als so regelmäßig, daß die Reibungskräfte mit großer Genauigkeit in Betracht gezogen werden konnten und die Genauigkeit der Kraftmessung kaum verminderten.

Zur Bestimmung der Reibungskräfte wurde der Hochdruckraum völlig mit Glyzerin angefüllt und die Ablesungen an den beiden Manometern (M_1 und M_2) bei verschiedenen Bewegungsvorrichtungen der beiden Kolben verglichen, wie dies in Fig. 4 dargestellt ist. Es lassen sich insgesamt vier Bewegungszustände verwirklichen. Zunächst ergeben sich zwei Bewegungszustände, bei welchen sich Kolben b und Druckstempel f im Verhältnis zur Versuchseinrichtung nach derselben Richtung verschieben, und zwar einmal der Kolben den Druckstempel (I), dann der Druckstempel den Kolben (II) verdrängt. Außerdem hat man aber zwei weitere Bewegungszustände, wenn man Kolben und Stempel gegeneinander bewegen läßt (III) oder beide gleichzeitig entlastet (IV). Bezeichnet man die Reibung zwischen der Druckvorrichtung und dem Kolben b mit R_1, die Reibung zwischen der Druckvorrichtung und dem ruhenden Teile der Presse (einschließlich Stempel) mit R_2, so sieht man leicht ein, daß die Unterschiede der mit der Ueber-

Fig. 4.   Ermittlung der Reibkräfte.

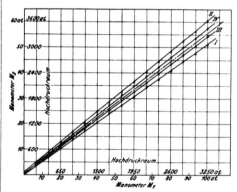

setzung multiplizierten Manometerablesungen in den Fällen I und II die doppelte Summe $2(R_1 + R_2)$, in den Fällen III und IV den doppelten Unterschied der Reibungskräfte $2(R_1 - R_2)$ liefern. In dieser Weise kann sowohl R_1 als R_2 als Funktion des im Hochdruckraum herrschenden Druckes bestimmt werden; beide ergaben sich mit dem Drucke sehr genau proportional. Die Reibung konnte daher bei der Auswertung der Versuche sehr genau berücksichtigt werden, falls man nur dafür sorgte, daß stets ein bestimmter Bewegungszustand vorhanden war, d. h. während der Belastung bezw. Entlastung kein Wechsel in dem Bewegungssinne der Kolben stattfand.

Messung der Formänderungen.

Die Formänderungen wurden durch Mikrometerschrauben gemessen, die Längenänderungen mit einer Genauigkeit von 0,01 mm abzulesen gestatten; da im ganzen Verschiebungen von 10 bis 12 mm zu messen waren, so schien mir die Genauigkeit von 0,01 mm reichlich genügend. Gemessen wurde die Entfernung der oberen Fläche der Druckvorrichtung von der oberen Druckplatte der Festigkeitmaschine an zwei gegenüberliegenden Stellen. Die elastische Deformation der Vorrichtung und insbesondere des Stempels wurde dadurch ausgeschaltet, daß Druckversuche nach derselben Anordnung mit einem Stahlzylinder von bekanntem Elastizitätsmodul vorgenommen wurden, der nur innerhalb der Elastizitätsgrenze beansprucht wurde.

Vorbereitung der Probestücke.

Die Probestücke wurden auf den Durchmesser des Stempels und der beiden Kugelplatten (40 mm) sorgfältig abgedreht und manche auch poliert. Die Länge schwankte zwischen 100 und 110 mm. Die sehr ausgedehnten Versuche von Prandtl und Rinne [1] ergaben für den Druckversuch als günstigstes Verhältnis zwischen Länge (*l*) und Durchmesser (*d*)

$$\frac{l}{d} = 2{,}5 \text{ bis } 3{,}5.$$

Bei so langen Probekörpern ist nämlich die Druckfestigkeit durch die Reibung an den Druckflächen schon sehr wenig beeinflußt, und die Körper sind noch durchaus knicksicher. Es zeigte sich auch bei meinen Versuchen, daß, wenn zur ungleichförmigen Verteilung der Deformation kein besonderer Grund vorlag, der Einfluß der Druckplatte sich nur auf ihre unmittelbare Nachbarschaft beschränkte. So erhält man bei hohem Manteldruck trotz der Reibung an den Druckflächen eine fast gleichmäßige Zunahme der Dicke.

Um das Eindringen der Druckflüssigkeit in die Poren zu verhindern, wurden die Probekörper mit einer Schutzhülse aus 0,1 mm dickem ausgeglühtem Messingblech versehen und deren überstehende Enden durch Lot an den Druckplatten abgedichtet. Um den Einfluß dieser dünnen Hülsen auf die Uebertragung des Mantel- bezw. Längsdruckes abschätzen zu können, habe ich das Messingblech Zugversuchen unterworfen und seine Fließgrenze bezw. Zugfestigkeit bestimmt. Die auf diese Versuche gegründete Rechnung zeigt, daß der durch die Hülse verursachte Fehler nur bei etwa 2 bis 3 Versuchen unter niedrigem Manteldruck 1 vH übersteigt. Das Messingblech erwies sich übrigens als äußerst bildsam und wurde durch den äußeren Druck so fest an die Oberfläche des Probekörpers gedrückt, daß sozusagen eine Prägung entstand und die Hülsen nach dem Versuch eine bis ins kleinste getreue Abbildung der Oberfläche zeigten.

IV. Versuchsergebnisse.

Wie unter II dargelegt wurde, sind besonders Versuche an spröden Stoffen wertvoll; so wurde zu den ersten Versuchsreihen, über die hier berichtet werden soll, als Versuchsmaterial Marmor (weißer Carraramarmor mit blauen Adern) und roter Sandstein (aus Mutenberg a. M.) gewählt. Marmor erschien hauptsächlich deshalb als Versuchsmaterial günstig, weil er bei den Versuchen von Prandtl und Rinne sehr gleichmäßige Werte für die Druckfestigkeit lieferte. Auch nach meinen Erfahrungen kommen bei den einzelnen Probekörpern, wenn sie aus demselben großen Block ausgeschnitten werden, größere Abweichungen als 3 bis 5 vH kaum vor.

Abhängigkeit des Formänderungsgesetzes vom allseitigen Druck.

Sämtliche Versuche wurden bei möglichst gleichbleibendem Manteldruck vorgenommen, so daß die Verkürzung als Funktion der wachsenden axialen Belastung gemessen wurde. Als »Belastung« kann singemäß der Unterschied der Hauptspannungen σ_1 (axialer Druck) und $\sigma_2 = \sigma_3$ (Manteldruck) betrachtet werden, und die Beziehung zwischen der so festgestellten Belastung und der spezifischen Längenänderung kann als Formänderungskurve bei gegebenem Manteldruck gelten. Die in diesem Sinne bestimmten Formänderungskurven des Marmors für verschiedene Werte des Manteldruckes zwischen $\sigma_2 = 0$ und $\sigma_2 = 3260$ at sind in Fig. 5 dargestellt. Die Formänderungskurve erscheint demnach als in

[1] Neues Jahrbuch für Mineralogie Band 12 1909 S. 121.

hohem Maße abhängig von dem allseitigen Druck, unter welchem der Versuch vorgenommen wird. Namentlich erscheinen in stetigem Uebergang die drei Arten der Formänderungskurve, die man sonst für spröde, plastische und zähe Stoffe als kennzeichnend ansieht. So liefert Marmor bei dem gewöhnlichen Druckversuch die für spröde Stoffe eigentümliche Kurve Nr. 1 und 2, bei der bis zur elastischen Formänderung, die von einer bleibenden Formänderung von höchstens derselben Größenordnung begleitet wird, eine Höchstlast erreicht wird, worauf unmittelbar der Bruch erfolgt. Der Einfluß des wachsenden Manteldruckes besteht zunächst in Verzögerung des Bruchvorganges. Etwa zwischen 700 und 800 at Manteldruck wird der Zustand der vollkommenen Plastizität erreicht, d. h. die Formänderung erfolgt unter annähernd konstanter Belastung. Wird der Manteldruck noch weiter gesteigert, so geht die Formänderung nunmehr nur unter zunehmender Belastung vor sich. Gleichzeitig bemerkt man aber, daß der

Fig. 5.

Formänderungskurve des Marmors beim Versuch unter allseitigem Druck.

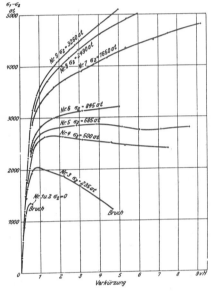

Einfluß der Steigerung des Manteldruckes allmählich abnimmt: so sind die Kurven für $\sigma_2 = 1650$ at und 2490 at nicht wesentlich verschieden, dagegen bringt eine weitere Steigerung des Manteldruckes um einen ungefähr ähnlichen Betrag auf 3260 at keine bemerkenswerte Aenderung mehr.

Zu ähnlichen Ergebnissen führten die Versuche mit Sandstein, Fig. 6.

Als erstes Ergebnis können wir daher feststellen, daß das Formänderungsgesetz von dem Manteldruck in der Weise abhängig ist, daß die Deformation bei niedrigem Manteldruck unter abnehmender, bei hohem Manteldruck unter zunehmender Belastung vor sich geht. Dazwischen liegt eine Grenze, wo die Belastung während der fortschreitenden Deformation gleich bleibt.

Die Abhängigkeit der Elastizitätsgrenze von dem allseitigen Druck.

Es muß zunächst bemerkt werden, daß man eine Elastizitätsgrenze in dem Sinne, daß innerhalb derselben alle Formänderungen vollkommen umkehrbar sind, überhaupt nicht bestimmen kann. Mit Hülfe von genügend genauen Meßinstrumenten kann man vielmehr neben jeder elastischen Formänderung eine bleibende Formänderung nachweisen. Zieht man den verwickelten Aufbau der für technische Zwecke in Betracht kommenden Stoffe in Erwägung, so ist dies kaum anders zu erwarten. Eine ausgeprägte Elastizitätsgrenze würde eben nur entweder bei einem völlig gleichwertigen, in seinen kleinsten Teilen isotropen Stoff oder bei einem vollkommen gleichartigen Kristall möglich sein. In der technischen Praxis ist man bekanntlich mehr oder weniger allgemein übereingekommen, als Elastizitätsgrenze die Belastung zu bezeichnen, bei welcher die bleibende Formänderung einen bestimmten Bruchteil der elastischen Formänderung erreicht. Eine in dieser Weise festgesetzte Grenze liefert zwar bei praktischen Versuchen, die stets in derselben Weise vorgenommen werden, sehr brauchbare Vergleichswerte; sie ist aber für unsern Zweck von geringer Bedeutung, weil sie

Fig. 6.

Formänderungskurven des Sandsteines beim Versuch unter allseitigem Druck.

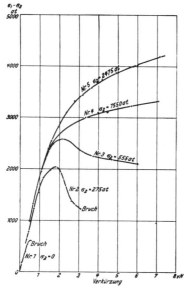

eher ein Maß für die Ungleichartigkeit der Probekörper als Aufschluß über den Einfluß des Manteldruckes liefern würde.

Bei Auswertung der hier berichteten Versuche habe ich im Einklang mit vielen andern Experimentatoren als Elastizitätsgrenze in weiterem Sinne, etwa als Grenze der vornehmlich elastischen Formänderungen, wenn beträchtliche bleibende Aenderungen unter gleichbleibender oder zunehmender Belastung auftraten, die Fließgrenze, in den Fällen, wo die Formänderungskurve einen Höchstwert der Belastung aufwies, den diesem Höchstwert entsprechenden Spannungszustand betrachtet. Die Figuren 5 und 6 zeigen, daß bei Aenderung des Manteldruckes Fließgrenze und Höchstwert der Belastung stetig ineinander übergehen. Ich glaube, man ist berechtigt anzunehmen, daß diese Grenzzustände für die Ueberschreitung der Elastizitätsgrenze in gewissem Sinne wirklich maßgebend sind, da die vorangehenden bleibenden Aenderungen höchstwahrscheinlich vornehmlich örtlichen Formänderungen infolge der Ungleichartigkeit des Materials zuzuschreiben sind und die bleibende Formänderung sich erst in diesen Grenzzuständen über den Probekörper ausbreitet.

Die Fließgrenze ist durch erhebliche Zunahme der Formänderungen unter gleichbleibender oder wenigstens unter schwach steigender Belastung bestimmt. In vielen Fällen — insbesondere bei Versuchen unter sehr hohem Manteldruck — findet man keine ausgeprägte Grenze dieser Art; in diesen Fällen können wir uns dadurch helfen, daß wir diejenigen Belastungszustände vergleichen, die denselben bleibenden Aenderungen entsprechen.

Beschränken wir uns zunächst auf die Versuche, bei welchen ein Höchstwert der Belastung oder eine ausgeprägte Fließgrenze beobachtet wurde, so gelangen wir zu den Zahlentafeln 1 und 2, die die Elastizitätsgrenze ($\sigma_1 - \sigma_2$) als Funktion des Manteldruckes ($\sigma_2 = \sigma_3$) angeben. Man kann auch die entsprechenden Grenzkurven nach der Mohrschen Darstellung zeichnen, wie dies z. B. in Fig. 7 für Marmor geschehen ist. Es zeigt sich — wie dies von Mohr angenom-

Nr. des Probekörpers	Manteldruck $\sigma_2 = \sigma_3$ at	Elastizitätsgrenze $\sigma_1 - \sigma_2$ at

Zahlentafel 1. Marmor.

1 u. 2	0	1360
3	235	2100
4	500	2650
5	685	2880
6	845	3210
7	1650	3900

Zahlentafel 2. Sandstein.

1	0	690
2	280	2040
3	555	2580
4	1550	3300
5	2475	4000

Fig. 7.

Elastizitätsgrenze des Marmors in der Mohrschen Darstellung.

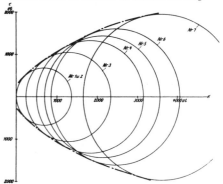

men wurde —, daß der Elastizitätsgrenze ein mit der Normalspannung wachsender Grenzwert der Schubspannung entspricht, der dann bei großen Werten der Normalspannung (bei großem allseitigem Druck) einem Höchstwerte zustrebt. Die Feststellung dieses Höchstwertes stößt allerdings auf die bereits erwähnte Schwierigkeit, daß unter hohem allseitigem Druck keine ausgeprägte Fließgrenze mehr beobachtet werden kann. Vergleicht man aber die Belastungszustände, die derselben bleibenden Längenänderung entsprechen, und zeichnet die Umhüllenden der entsprechenden Spannungskreise, so sieht man (vergl. Fig. 8), daß diese Umhüllenden in der Tat in wagerechte Geraden ($\tau = $ konst.) übergehen; d. h. unter sehr hohem allseitigem Druck hängen Auftreten und Maß von bleibenden Formänderungen nur von dem Unterschied der Hauptspannungen $\sigma_1 - \sigma_2$ ab; Belastungszustände, bei denen

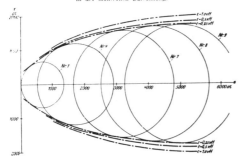

Fig. 8.

Kurven gleicher bleibender Dehnung bei Marmor
in der Mohrschen Darstellung.

dieser Unterschied derselbe bleibt, sind dann sozusagen gleichwertig. So fügen sich die von uns untersuchten spröden Stoffe unter hohem allseitigem Druck derselben Gesetzmäßigkeit, die für zähe Stoffe wie Metalle usw. bei gewöhnlichen Druckverhältnissen maßgebend ist.

Da bei sämtlichen Versuchen zwei Hauptspannungen, und zwar stets die kleineren, gleich groß waren, so lassen sich aus diesen Versuchen auf die Richtigkeit oder Unrichtigkeit der Grundhypothese der Mohrschen Theorie, ob die mittlere Hauptspannung tatsächlich belanglos ist und die Elastizitätsgrenze nur von den äußersten Hauptspannungen abhängt, keine Folgerungen ziehen; für den besonderen Fall zweier gleicher Hauptspannungen wurde aber die Mohrsche Theorie durch die Versuche sehr schön bestätigt.

Eine weitere Bekräftigung der Mohrschen Auffassung liefern die Risse und Faltungen, die an der Oberfläche der namentlich unter niedrigem Manteldruck beanspruchten Probekörper erscheinen und offenbar an die an Metallen beobachteten Fließfiguren erinnern. Diese sollen nach der Mohrschen Auffassung die Spuren derjenigen zur Oberfläche normalen Ebenen darstellen, längs deren die ersten bleibenden Gleitungen erfolgen. Auf dieser Grundlage kann man den Winkel, den die an den Probekörpern beobachteten und zu der Druckrichtung symmetrischen Linienscharen einschließen, mit den durch die Theorie geförderten Werten vergleichen. In der Mohrschen Darstellung wird nämlich der spitze Winkel zwischen zwei Gleitflächen durch den Winkel zwischen den Normalen der Grenzkurve und der σ-Achse gegeben (der Winkel φ in Fig. 1), so daß er aus den durch Versuche festgestellten Grenzkurven entnommen werden kann. Der Vergleich zwischen Theorie und Beobachtung führt zu den Zahlentafeln 3 und 4, und zwar sind dabei die beiden letzten Spalten einander gegenüberzustellen, da die an den Probekörpern unmittelbar gemessenen Winkel sich auf die Gestalt der Probekörper bei Ueberschreitung der Elastizitätsgrenze — die sich übrigens von der ursprünglichen Gestalt nur unwesentlich unterscheidet — reduziert werden müßten. Zu diesem Zwecke wurden die Probekörper vor dem Versuch mit Skaleneinteilung versehen. Die Uebereinstimmung zwischen

berechneten und gemessenen Werten darf mit Rücksicht auf die durch die Umrechnung bedingte Unsicherheit als recht befriedigend bezeichnet werden. Man ist dabei auf die Versuche unter niedrigem Manteldruck beschränkt, da diese Oberflächenerscheinungen überhaupt nur dann auftreten, wenn die Formänderung unter abnehmender Belastung vor sich geht (vergl. weiter unten).

Begleiterscheinungen der bleibenden Formänderung.

Die plastische Deformation des Marmors und mancher andrer spröden Körper unter hohem allseitigem Druck ist an und für sich keine neue Tatsache. Abgesehen davon, daß die Vorgänge im Innern der Erdkruste gewisser Gesteine, plastische Formänderungen zu erleiden, zweifellos erweisen, hat F. Kick[1]) den unmittelbaren Beweis durch seine bekannten Versuche für eine Reihe von spröden Körpern erbracht. Bei den Kickschen Versuchen blieb aber der Spannungszustand, unter welchem die plastische Formänderung vor sich ging, unbestimmt, da der seitliche Druck teilweise durch festen Umschluß, teilweise durch halbfeste Stoffe, wie Stearin, übertragen wurde und so die Art der Druckübertragung nicht bekannt war. So können diese an und für sich schönen Versuche zur Klärung der grundsätzlichen Frage der Abhängigkeit der Elastizitätsgrenze von der Art des Spannungszustandes kaum herangezogen werden.

Die plastische Formänderung unserer Probekörper ging durchaus ohne Veränderung der Dichte vor sich und unter genügend hohem Manteldruck — besonders bei Marmor — fast ohne Lockerung des Gefüges. Probekörper, die unter 2500 bis 3000 at Manteldruck 6 bis 9 vH Längenänderungen erfahren haben, zeigten, nochmals abgeschliffen und einem normalen Druckversuch unterworfen, eine Verminderung der ursprünglichen Druckfestigkeit um nur etwa 15 bis 20 vH. Beachtenswert ist es, daß Gestalt und Oberfläche der beanspruchten Probekörper wesentlich verschieden wurden, je nachdem die Formänderung unter hohem allseitigem Druck und dementsprechend unter steigender axialer Belastung, Fig. 9[2]), oder aber unter niedrigem allseitigem Druck und demzufolge unter abnehmender axialer Belastung vor sich ging, Fig. 10. Wird nämlich die Elastizitätsgrenze durch die Formänderung erniedrigt, wie dies der Fall ist, wenn letztere unter abnehmender Last erfolgt, so wird sich die Formänderung, sobald an einer Stelle infolge ungleichmäßiger Spannungsverteilung die Elastizitätsgrenze etwas überschritten wird, auf diese Stelle oder auf ihre Nachbarschaft beschränken, da die betreffende Stelle dadurch nur noch mehr an Widerstandsfähigkeit verliert, während die weiter entfernt liegenden Teile des Probekörpers infolge der Abnahme der Gesamtbelastung die Elastizitätsgrenze gar nicht erreichen. So genügt bei niedrigem Manteldruck die durch den Einfluß der Reibung an den Druckflächen hervorgerufene Ungleichförmigkeit der Spannungsverteilung, um die Formänderung wesentlich auf den mittleren Teil des Probekörpers zu beschränken. Bei steigender Formänderungskurve dagegen, d. h. in allen Fällen, wo die bleibende Formänderung eine Festigung des Materials zur Folge hat, wird eine örtliche Formänderung gerade ausgleichend auf die Spannungsverteilung wirken; dementsprechend erfahren die Probekörper unter hohem Manteldruck eine annähernd gleichmäßige Formänderung.

Durch die Art des Formänderungsgesetzes wird auch die Möglichkeit der Entstehung von Fließfiguren und sonstigen Unebenheiten der Oberfläche bedingt. Nimmt man die Mohrsche Vorstellung als richtig an, so erscheint es als sehr wahrscheinlich, und in einfachen Fällen kann es sogar durch weitergehende theoretische Untersuchungen nachgewiesen werden, daß eine örtliche Ueberschreitung der Elastizitätsgrenze sich zunächst längs der Gleitflächen fortpflanzt. Hat man daher etwa eine Fehlerstelle an oder in der Nähe der Oberfläche, so entstehen an der Oberfläche zwei Streifen, die die Richtung der Gleitflächen verfolgen und die stärker geändert werden als die übrigen Teile des Körpers.

[1]) Z. 1892 S. 278 u. 919; ferner Zeitschrift des österreichischen Ingenieur- und Architekten-Vereines 1891 Heft 1.
[2]) Fig. 9 bis 25 befinden sich auf Textblatt 29.

Manteldruck	φ gemessen		φ berechnet
at	ohne Reduktion	reduziert	
Zahlentafel 3. Marmor.			
0	54°	54°	53°
235	59°	58°	58°
500	72°	65°	63°
685	83°	70°	73°
Zahlentafel 4. Sandstein.			
0	38°	38°	40°
280	70°	69°	63°
555	82°	73°	70°

Wird nun durch die Formänderung das Material geschwächt, so werden diese einmal stärker beanspruchten Streifen an Widerstandsfähigkeit mehr und mehr verlieren, und die Verzerrung innerhalb dieser Streifen kann in solchem Maße zunehmen, daß Risse oder bei sanfter abnehmendem Formänderungsgesetz Faltungen entstehen, während die übrigen Teile des Probekörpers weit weniger beansprucht bleiben. Wird dagegen das Material durch die Formänderung gefestigt, so werden die durch einen Materialfehler zunächst in Mitleidenschaft gezogenen Streifen im Gegenteil widerstandsfähiger, und der Einfluß der Fehlerstelle wird, so wie dies innerhalb der Elastizitätsgrenze der Fall ist, auf die unmittelbare Nachbarschaft beschränkt.

Dementsprechend zeigen alle Probekörper, die bei steigender Belastung — d. h. unter hohem Manteldruck — beansprucht wurden, eine vollkommen glatte Oberfläche; dagegen erscheinen an der Oberfläche der bei abnehmender Belastung — unter niedrigem Manteldruck — beanspruchten Probekörper oft sehr ausgeprägte Oberflächenzeichnungen, die bei sanft abnehmendem Formänderungsgesetz durch Unebenheit der Oberfläche hervortreten, vergl. Fig. 13, bei rasch abnehmendem Formänderungsgesetz in feine Risse ausarten[1]), Fig. 11 und 12.

Bruchflächen.

Bei allen Probekörpern weist die primäre Bruchfläche durchaus die Merkmale eines Verschiebungsbruches auf.

Die Lage der Bruchflächen zur Druckrichtung, so weit die Flächen angenähert eben oder kegelförmig sind, entspricht ungefähr der Neigung der feineren Oberflächenrisse. Diese primären Bruchflächen sind durchaus mit feinem Mehl bedeckt, im Gegensatz zu den sekundären Bruchflächen, die sich durch ihre glatte und harte Oberfläche offenbar als Trennungsflächen erkenntlich machen. Eigenartige Bruchflächen zeigt der Sandsteinkörper Fig. 14; die primäre Bruchfläche bildet einen fast regelmäßigen Kreiskegel, der dann in das Gegenstück hineingedrückt wurde und dieses in mehrere Stücke zersprengte.

Mikroskopische Untersuchungen.

Beide untersuchten Gesteine — wie übrigens ein überwiegend großer Teil der technisch wichtigen Konstruktionsstoffe und insbesondere die meisten Metalle — gehören zur Klasse von festen Körpern, die man am besten nach W. Voigt[2]) als »quasiisotrop« bezeichnen kann. Sie bestehen aus kristallinischen Körnern, die man, um sie von den durch regelmäßige Flächen begrenzten eigentlichen Kristallen zu unterscheiden, als Kristalliten zu bezeichnen pflegt. Die Orientierung dieser Kristalliten ist in den quasiisotropen Körpern nach allen Richtungen gleichmäßig verteilt, so daß jeder Teil des Körpers, der gegen die einzelnen Kristalliten einigermaßen groß ist, sich in elastischer Hinsicht als isotrop erweist. Durch die Untersuchungen von Mügge[3]), Heyn[4]), ferner von Ewing und Rosenhain[5]) ist festgestellt worden, daß die kristallinische Struktur gerade bei der plastischen Formänderung der Metalle eine wesentliche Rolle spielt, da es der inneren Umlagerungsfähigkeit der Kristalliten zu verdanken ist, daß erhebliche bleibende Formänderungen ohne Lockerung des Zusammenhanges vor sich gehen können. Die einzelnen Kristalliten können nämlich durch Gleitungen (Translationen), ferner durch Umlagerungen zu Zwillings-

lamellen in ungemein vielfacher Weise umgeformt werden, so daß ein aus solchen Kristalliten bestehender Körper sehr erhebliche bleibende Formänderungen erfahren kann, ohne daß der Zusammenhang der Kristallite gestört werden müßte. Aehnliche Erscheinungen konnte ich bei Untersuchung der Dünnschliffe beobachten, die den Probekörpern entnommen waren, bei denen der Druckversuch unter gleichzeitigem hohem allseitigen Druck vorgenommen wurde; diese Beobachtungen gewinnen noch dadurch an Interesse, daß neben der plastischen Formänderung, die ähnlich wie bei den Metallen verläuft, an andern Probestücken aus demselben Material die Aenderungen des Kleingefüges beim spröden Bruche beobachtet werden konnten. Da von unsern beiden Versuchsmaterialien Marmor einen besonders einfachen Aufbau besitzt, so wollen wir uns in der vorliegenden Mitteilung auf dieses Versuchsmaterial beschränken.

Mikroskopische Untersuchung der Marmorkörper vor und nach der Formänderung.

Marmor ist ein Aggregat von ziemlich dicht gelagerten Kalkspatkristallen ohne oder mit ganz wenig Bindemittel. Da bei Kalkspat Zwillingsbildungen außerordentlich leicht entstehen[1]), so war zu erwarten, daß beim plastischen Fließen des Marmors die Zwillingsbildungen innerhalb der einzelnen Kristalliten eine wesentliche Rolle spielen. Dies wurde durch die mikroskopischen Bilder der Dünnschliffe, die den durch hohen allseitigen Druck beanspruchten Probestücken entnommen waren, sehr schön bestätigt. Zwar sind die Zwillingslamellen bei Marmor auch vor der Beanspruchung, Fig. 16, nur selten, was höchstwahrscheinlich auf ein Zeichen vorangehender Formänderung im Innern der Erdkruste anzusehen ist; doch finden wir ihre Anzahl bei den plastisch umgeformten Probekörpern ungemein vergrößert. Einzelne Kristalliten erscheinen von geradezu dichter Schraffierung (Zwillingsstreifung) durchzogen, so daß das mikroskopische Bild oft fast ein schachbrettartiges Aussehen gewinnt, Fig. 17 bis 20.

Ein ganz andres Bild liefern die Dünnschliffe, die denjenigen Probekörpern entnommen sind, die entweder dem gewöhnlichen Druckversuch unterworfen wurden oder bleibende Formänderung unter niedrigem Manteldruck erlitten haben, Fig. 21 und 22. Die Anzahl der Zwillingslamellen hat sich in diesem Falle nicht wesentlich vermehrt, dafür tritt aber die Begrenzung der Kristalliten stärker hervor, und zwar erscheinen nicht nur die Begrenzungslinien verstärkt, sondern oft sehen wir dicke Striche durch das ganze mikroskopische Bild hindurchgehend. Wir haben es offenbar mit einer ganz andern Art der Formänderung zu tun, die wesentlich in relativer Verschiebung der Kristalliten besteht, wobei die Begrenzungsflächen unvermeidlich etwas abgeschliffen und dadurch verdickt erscheinen. Die langen Striche, die zwischen den großen Anzahl von Kristalliten durchgehen, bilden dabei zweifellos die ersten Anfänge zu den später auch mit freiem Auge sichtbaren Rissen; die in der Nachbarschaft dieser Streifen liegenden Kristalliten sind augenscheinlich am stärksten in Anspruch genommen worden, manche scheinen sogar gespalten zu sein.

Die beiden Arten der Formänderung.

Man vermag daher auf Grund dieser mikroskopischen Bilder festzustellen, daß die Elastizitätsgrenze in zwei gänzlich verschiedenen Weisen überschritten werden kann: einmal durch Verschiebung der einzelnen an und für sich undeformiert bleibenden Kristallite gegeneinander, zum andern durch innere Deformation der Kristallite: sozusagen durch molekulare Umlagerung[2]). Diese beiden bezeichnenden Arten der bleibenden Formänderung springen besonders in den mikroskopischen Bildern mit stärkerer Vergrößerung ins Auge, Fig. 23 bis 25. Die beiden Beispiele bilden natürlich Grenzfälle,

---

[1]) Mit dieser Auffassung steht die bekannte Tatsache scheinbar in Widerspruch, daß bei zähen Stoffen, z. B. bei Eisen, oft sehr schöne Oberflächenzeichnungen entstehen. Dieser Widerspruch wird aber dadurch gelöst, daß gerade diejenigen Eisensorten Oberflächenzeichnungen zeigen, die ausgeprägte obere und untere Fließgrenze haben, d. h. bei welchen die Belastung nach Ueberschreitung der Elastizitätsgrenze zunächst abnimmt.

[2]) Annalen der Physik Bd. 38 (1889) S. 573, vergl. auch W. Thomson, Proc. Roy. Soc. Edinburgh Bd. 16 (1890) S. 693.

[3]) Neues Jahrbuch für Mineralogie Band 1 1898 S. 71 und Band 2 1899 S. 55.

[4]) Z. 1900 S. 433.

[5]) Phil. Trans. Band 193 (1899) S. 353 u. Band 195 (1900) S. 279, ferner W. Rosenhain, Journal of Iron and Steel Inst. Bd. 65 (1904) S. 335 und Bd. 70 (1906) S. 189.

[1]) Ueber das Wesen der Zwillingsbildung vergleiche die interessanten Ausführungen von E. Heyn, Z. 1903 S. 503 u. f.

[2]) Später (diese Versuche wurden bereits in meiner im Februar 1910 der philosophischen Fakultät der Universität Göttingen vorgelegten Habilitationsschrift beschrieben) haben G. Tammann und O. Faust in ähnlicher Weise zwei verschiedene Arten der Formänderung bei Metallen unterschieden, s. Z. für phys. Chemie 1910.

Fig. 9. Marmorkörper von ursprünglich gleichen
Abmessungen, rechts der Probekörper vor dem Versuch.
Links nach plastische Deformation unter 1650 at Manteldruck.

Fig. 10. Marmorkörper von ursprünglich gleichen Abmessungen,
links der Probekörper vor dem Versuch, in der Mitte nach dem Druckversuch
unter 685 at, rechts unter 500 at Manteldruck.

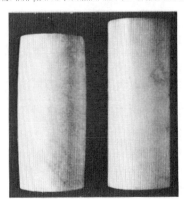

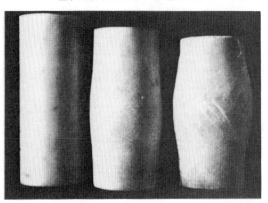

Fig. 15. Sandsteinkörper von ursprünglich gleichen
Abmessungen; links nach gewöhnlichem Druckversuch,
rechts nach Beanspruchung unter 2500 at Manteldruck.

Fig. 16. Marmor in natürlichem Zustand
40fach vergrößert.

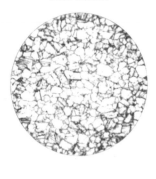

Fig. 20. Marmor nach 13 vH Längsänderung
unter 500 at allseitigem Druck.
40fach vergrößert.

Fig. 21 und 22.
Marmor nach gewöhnlichem Druckversuch.
40fach vergrößert.

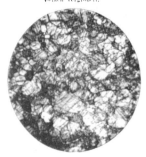

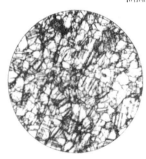

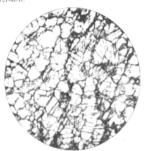

Fig. 11.
Marmorkörper nach gewöhnlichem Druckversuch.

Fig. 12.
Gleitlinien bei Marmor nach gewöhnlichem Druckversuch.

Fig. 13.
Gleitlinien bei Marmor nach plastischer Deformation.

Fig. 14.
Kegelförmige Bruchfläche bei Sandstein.

Fig. 17 und 18.
Marmor nach plastischer Längsänderung unter 2500 at allseitigem Druck. 10fach vergrößert.

Fig. 19. Zwillingslamellen bei einem großen Krystallit nach plastischer Deformation. 10fach vergrößert.

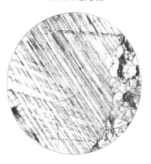

Fig. 23.
Marmor in natürlichem Zustande. 140fach vergrößert.

Fig. 24. Marmor nach plastischer Formänderung unter 2000 at allseitigem Druck. 140fach vergrößert.

Fig. 25.
Marmor nach gewöhnlichem Druckversuch. 140fach vergrößert.

Fig. 23 bis 25 sind mit Polarisator mit schiefer Belichtung unter 45° ohne Analysator aufgenommen.

die unter ganz niedrigem oder sehr hohem allseitigem Druck klar hervortreten. Dazwischen gibt es einen stetigen Uebergang. Ein Beispiel hierzu wird z. B. durch den Dünnschliff in Fig. 20 geliefert, der einem Probestab entnommen ist, der unter 500 at Manteldruck nahe bis zum Bruch beansprucht war. Einzelne Kristallite zeigen auch hier die für Zwillingsbilder charakteristische Schraffierung, im allgemeinen erscheint aber der Zusammenhang der Kristallite stark gelockert.

Vergleicht man diesen Befund der mikroskopischen Bilder mit den Beobachtungen über das Formänderungsgesetz und den mikroskopischen Beobachtungen über die Begleiterscheinungen des plastischen Fließens, so gewinnt man von den Vorgängen ungefähr folgendes Bild:

Die relative Verschiebung der Kristallite tritt besonders bei niedrigem allseitigem Druck auf. Je größer der Manteldruck, desto stärker werden die Kristallite aneinandergepreßt, und desto mehr wird eine Verschiebung verhindert. Dies erklärt offenbar die Zunahme der Elastizitätsgrenze mit wachsendem Manteldruck, oder in der Mohrschen Auffassung die Zunahme der »Grenzschubspannung« mit dem Normaldruck. Tritt aber die bleibende Verschiebung einmal ein, so werden dadurch die Begrenzungsflächen abgeschliffen, so daß eine weitere Formänderung erleichtert wird; daher die Abnahme der Elastizitätsgrenze bei zunehmender Formänderung, d. h. Verminderung der Belastung mit fortschreitender Formänderung.

Diese erste Art der Formänderung, die man als intergranulare Formänderung bezeichnen kann, ist daher durch folgende beiden Merkmale bezeichnet: Zunahme der Elastizitätsgrenze mit zunehmendem allseitigen Druck und Abnahme der Elastizitätsgrenze durch fortschreitende Formänderung.

Die zweite Art der Formänderung – die innere Formänderung der Kristallite – wird erst dann überwiegen, wenn der Manteldruck hinreichend groß ist, um die Verschiebung der Kristallite gänzlich zu verhindern. Der Einfluß des allseitigen Druckes auf die Elastizitätsgrenze hört damit auf. Die Möglichkeit von Zwillingsbildungen scheint eben von dem allseitigen Druck nicht abzuhängen, sondern nur von dem Unterschied der Kraftwirkungen nach verschiedenen Richtungen. So wird für die Elastizitätsgrenze nur der Unterschied der Hauptspannungen maßgebend sein, so daß eine Vermehrung sämtlicher Hauptspannungen um denselben Betrag für die Möglichkeit der Zwillingsbildungen gleichgültig ist. So ist es zu verstehen, daß bei sehr hohem allseitigem Druck die einfache Schubspannungstheorie oder die Annahme eines gleichbleibenden Unterschiedes der äußersten Hauptspannungen zutreffend sein kann. Besonders einfach wird die Erhöhung der Elastizitätsgrenze durch die bleibende Formänderung, d. h. die sogenannte Festigung des Materials, durch die Betrachtung des Kleingefüges erklärt. Die Möglichkeit der Zwillingsbildung ist offenbar von der Richtung der Kristallite gegen die Druckrichtung abhängig; so werden zuerst diejenigen Kristallite zur Formänderung herangezogen, die dafür am günstigsten liegen; dann kommen erst die daran, die weniger günstig gerichtet sind und deshalb einen größeren Spannungsunterschied beanspruchen. Da nun die verschieden gerichteten Kristallite gleichmäßig verteilt sind, sozusagen in gleicher Anzahl da sind, so folgt daraus, daß der gesamte Kraftbedarf bei fortschreitender Formänderung zunehmen muß. So gelangt man also zu einer sehr einfachen zwanglosen Erklärung der Festigung (der écrouissage der französischen Autoren), die in der Technologie der Metalle eine wesentliche Rolle spielt. Ich glaube, daß diese Erklärung auf diejenigen Metalle, bei denen das Fließen in plastischer Formänderung der Kristallite besteht, ohne weiteres übertragen werden kann, nur treten bei Metallen meistens Gleitungen statt Zwillingsbildung auf. Die Festigung setzt sich dann dadurch, daß stets ungünstiger liegende Kristallite an die Reihe kommen, so lange fort, bis in sämtlichen Kristalliten die Gleitflächen ausgebildet sind. Es muß aber hinzugefügt werden, daß bei den meisten Metallen diese Art der Formänderung, die bei Marmor erst unter hohem Manteldruck erfolgt, beim gewöhnlichen Zug- oder Druckversuch vor sich geht.

Als bezeichnende Merkmale der Formänderung innerhalb der Kristallite, die man entsprechend als intragranular bezeichnen könnte, gelten daher auf Grund der vorangehenden Ueberlegungen: die Unabhängigkeit der Elastizitätsgrenze von dem allseitigen Druck und die Verfestigung infolge bleibender Formänderung.

Die Grenzfälle der rein intergranularen bezw. rein intragranularen Deformation decken sich offenbar damit, daß sich der Stoff gegen Druckbeanspruchung als »spröde« bezw. »bildsam« erweist. Das spröde bezw. plastische Verhalten eines Stoffes beim Druckversuch hängt daher lediglich davon ab, ob die Verschiebung der Kristallite oder die innere Umlagerung leichter eintreten kann. Bei vollkommen plastischen Stoffen geschieht stets das letztere, bei sogenannten spröden Stoffen richtet sich die Aenderung nach der Art des Spannungszustandes.

Zusammenfassung und Schlußfolgerungen.

Durch die Versuche unter allseitigem Druck wurde die Mohrsche Vorstellung von der Elastizitätsgrenze sehr schön bestätigt. Die Versuche ergeben, so wie dies die Mohrsche Theorie erfordert, eine Zunahme der Elastizitätsgrenze mit wachsendem allseitigen Druck. Man darf die Annahme als bestätigt betrachten, daß die Elastizitätsgrenze durch einen Grenzwert der Schubspannung festgelegt ist, der mit der auf die Gleitfläche wirkenden Normalspannung zunimmt und bei sehr hohen Werten des Normaldruckes einem konstanten Höchstwerte zustrebt, so daß sich die Theorie der größten Schubspannung oder die Theorie der größten Unterschiede der Hauptspannungen (maximum stress-difference theory) als Grenzfall ergibt.

Die Grundannahme der Mohrschen Theorie, daß die mittlere Hauptspannung belanglos ist, konnte, wie schon oben erwähnt, durch die Druckversuche allein nicht entschieden werden. Allerdings zeigen die ersten Zugversuche, die ich in neuester Zeit durchführte, daß sie kaum auf unbegrenzte Gültigkeit Anspruch erheben kann. So ergab ein Zugversuch unter allseitigem Druck mit einem Marmorstabe die Elastizitätsgrenze für folgende Werte der drei Hauptspannungen:

$$\sigma_1 = 310 \text{ at}, \quad \sigma_2 = \sigma_3 = 4405 \text{ at}.$$

Wäre der Wert der mittleren Hauptspannung belanglos, so müßten beim Druckversuch die Werte

$$\sigma_1 = 4405 \text{ at}, \quad \sigma_2 = \sigma_3 = 310 \text{ at}$$

entsprechen. Entnimmt man aber aus Fig. 6 den entsprechenden Wert von σ_1 für $\sigma_2 = 310$ at, so gelangt man zu den Werten

$$\sigma_1 = \text{rd. } 2300 \text{ at}, \quad \sigma_2 = \sigma_3 = 310 \text{ at}.$$

Der Unterschied ist daher bei dem Zugversuch viel größer, als bei demjenigen Druckversuch, wo die kleinste Hauptspannung denselben Wert hat. Dieser Punkt wird hoffentlich durch die weiteren Zugversuche aufgeklärt werden.

Die mikroskopischen Untersuchungen zeigen, daß man mit zwei Möglichkeiten der bleibenden Formänderung zu tun hat, deren erste – die relative Verschiebung der Kristallite – hauptsächlich bei niedrigen Werten des allseitigen Druckes, deren zweite – die innere Formänderung der Kristallite – bei hohen Werten des allseitigen Druckes hervortritt. Mit der ersten Art der Ueberschreitung der Elastizitätsgrenze ist ein von der Normalspannung abhängiger Grenzwert der Schubspannung und eine Verschwächung des Materials durch fortschreitende Formänderung verbunden; diese Verschwächung hat dann ungleichförmige Verteilung der Formänderung und unter Umständen Auftreten von Fließfiguren zur Folge. Als Bedingung der Elastizitätsgrenze für die zweite Art der Formänderung ergibt sich dagegen ein gleichbleibender Wert der Schubspannung; die Formänderung ruft in diesem Fall eine Festigung des Materials hervor; dementsprechend erfahren die Probekörper eine gleichförmige Formänderung und bewahren die glatte Oberfläche. Die beiden Grenzfälle entsprechen einem — den gewöhnlichen Sprachgebrauche nach — spröden bezw. plastischen Verhalten des Stoffes gegen Druckbeanspruchung.

Dies ist ungefähr das Bild, welches man aus den Druckversuchen über die Vorgänge an und jenseits der Elastizitätsgrenze zu gewinnen vermag. Es ist zu erwarten, daß es durch die Zugversuche in mannigfacher Weise ergänzt werden wird.

Editor's Comments
on Papers 12 and 13

12 ANDERSON
The Dynamics of Faulting

13 NÁDAI
Plasticity: A Mechanics of the Plastic State of Matter

Of what use are pencil-and-paper operations to the tectonic geologist? The argument goes something like this: mathematics has the creative power of evolving hidden consequences from assumptions concerning observable portions of the earth, thus prompting predictions of previously unobserved phenomena. The earth is, of course, under no obligation to conform to these predictions! But if these consequences *appear* to be verified, then mathematics apparently has led to new discoveries; if they are not borne out by observation, then mathematics has demanded revision of the underlying general assumptions.

This seems to have been the approach of C. A. Coulomb. A physicist of distinction, a superior theoretician, and yet a practicing engineer, he stands in contrast to the "exclusive" scientists of the eighteenth century, who, like Euler and the Bernoullis, were mainly interested in finding applications for their new mathematical methods (Straub, 1949). Coulomb solved for the first time the beam-bending problem, among others; he can be regarded as the forerunner of the branch of mechanics known as structural analysis (building statics) later systematized by Navier. The way in which Coulomb approached problems of structural analysis is exemplified by his approach to the earth-pressure question in retaining-wall design. An earth prism tends to push the wall outward; the factors controlling the magnitude of the effect are, however, imprecisely known. Coulomb's method is to permit variation of the unknown factor (e.g., the inclination of slip planes) and to regard as significant the *limit values* which the pressure assumes at the point of instability. Thus, the concepts of active and passive earth pressure were developed. The material constants assumed are those indicating the *friction* and *cohesion* of the material (1776). Cohesion, "the resistance that solid bodies offer to the simple separation of their parts," is taken as either shear or tensile strength; interestingly, Coulomb had found little difference between them in rupture tests of rock (Heyman, 1972, p. 44–45).

Early in the nineteenth century Louis Navier joined the staff of the Ecole des Ponts et Chausées as Professor of Applied Mechanics. His brilliant lectures were published, for the first time in 1826, and it is chiefly through these lectures that the theories of structural analysis and strength of materials have become branches of modern engineering science. Navier integrated the isolated discoveries of his predecessors into a coherent whole, interspersed with methods either originated or reformulated by him.

In the solution of earth pressure and other problems he follows Coulomb. Navier's *Leçons* (1839) is subsequently cited by Anderson (1951, p. 157) as the source for the shear failure criterion which Anderson had earlier suggested in Paper 12; Jaeger (1962) followed Anderson's lead in this respect, and thus Navier's name has often been attached in the geological literature to what perhaps more properly should be termed the Coulomb criterion. Nonetheless, what we now refer to as the Coulomb equation was never written in that form in Coulomb's *Essai*.

Anderson, employing pencil-and-paper operations, developed a natural classification of faults based on consideration of causative principal stress orientations and the Coulomb criterion. His work is fundamentally important and is quoted in entirety as Paper 12.

In 1931, King Hubbert, then at Columbia University, was called upon by a Dr. Nádai, formerly a Professor of Engineering at Göttingen, but now a consulting mechanical engineer at the Westinghouse Research Laboratories in Pittsburgh. Nádai showed Hubbert a copy of his book *Plasticity* (Paper 13), just published as the first of a series of Engineering Societies Monographs. This was a revised English translation of an earlier German book, *Der Bildsame Zustand der Werkstoffe* (1927; a Russian translation of the American edition appeared later in 1936). "Dr. Nádai," Hubbert (1972, p. 6) writes, "expressed a great interest in the deformation of rocks, as he himself had seen them in the European Alps and Karpathian Mountains, as examples of the types of deformation with which he was concerned in engineering. He hoped that he might find a geologist with kindred interests."

Hubbert, not yet acquainted with Anderson's work, was forcibly impressed with the experimental demonstrations in Nádai's book of conjugate slip planes, in tests by Kármán, Prandtl, Rinne, and others, with the acute angle of intersection bisected by the axis of maximum compression. Nádai provided further demonstrations to Hubbert in his Pittsburgh laboratory by means of a sand-box experiment, in which a rigid blade, embedded vertically in loose sand, was moved horizontally in the direction of its normal. The experiment produced two slip surfaces, one in advance of the blade, the other behind it. Hubbert saw the immediate resemblance of these sand-box features to thrust and normal faults. Both groups of phenomena appeared to be satisfactorily explained by a theory due to Otto Mohr (1882; cf. 1914; the theory is fully derived in Nádai's book). Nádai's strong early influence in Hubbert's career seems clear (Hubbert and Willis, 1957; Papers 22 and 28).

Nádai's obvious fascination with geology had early roots. In our historical development we are able to trace a curious "feedback" mechanism, for it turns out that Nádai sat in Albert Heim's classroom in 1902, absorbing marvelous lectures and being much impressed with Heim's ability "to sketch from free hand on the blackboard pictures of mammoths, fossil fishes, or the tracings made by prehistoric man, discovered on the walls of caves, with the utmost perfection." Heim, of course, emphasized the plasticity of rocks long before the first high-pressure laboratory tests were made in demonstration of plastic behavior. Thus, in the excerpt reproduced herein, we see Nádai's emphasis of finite strain theory in connection with Alpine thrusts and recumbent folds. This approach contrasts with Anderson's emphasis on theory of brittle fracture, but the contrast seems completely explicable in view of Anderson's Scottish Highland experience. In terrain characterized by ductile deformation the strain approach seems most appropriate; for a superb modern example in which rolled garnets have been deciphered, the reader is referred to Rosenfeld's (1968) report on refolded nappes in the Green Mountains.

In his preface Nádai prophetically remarks: "The strange laws which seem to apply to the surfaces of slip have attracted recently the mathematicians and the engineers . . . . In the hands of the geologists, who in their faults have observed similar phenomena on a large scale for a long time, these surfaces might serve to decipher the riddles in the formation of high mountain chains . . . . They might possibly in the future still serve to study by mechanical means and to reconstruct the history of the crustal movements of the whole continents, which since the work of Alfred Wegener and F. B. Taylor we know are nothing but their shells drifting slowly over the earth globe and their plastic substratum." Are these not (merely!) large-scale décollement?

12

Reprinted from *Trans. Edinburgh Geol. Soc.*, **8**, Pt. 3, 387–402 (1905)

THE DYNAMICS OF FAULTING

Ernest M. Anderson

IT has been known for long that faults arrange themselves naturally into different classes, which have originated under different conditions of pressure in the rock mass. The object of the present paper is to show a little more clearly the connection between any system of faults and the system of forces which gave rise to it.

It can be shown mathematically that any system of forces, acting within a rock which for the time being is in equilibrium, resolves itself at any particular point into three pressures or tensions (or both combined), acting across three planes which are at right angles to one another.

Across these particular planes there is no tangential stress, but there will be tangential stress at that point across any other plane which may be drawn through it. There will evidently be positions of this hypothetical plane for which the tangential stress will be a maximum. It is evident that these maximum positions of the plane will have much to do with determining the directions of faults in the rock. We will therefore take the general case and investigate what the positions are. Suppose O to be any point in a rock, and let the three directions along which the pressures or tensions act (the directions perpendicular to the three planes mentioned above)

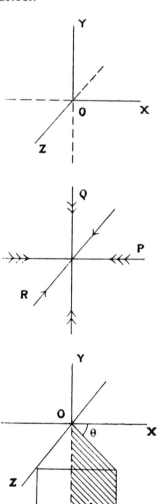

be OX, OY, OZ. Let the pressures, or tensions, acting along these three directions be P, Q, R, which we will suppose positive when they denote pressures, and negative when they denote tensions. Suppose further that P is the greatest pressure, or the least tension in the case where there are only tensions, and that R is the greatest tension, or the least pressure in the case where there are only pressures, so that P, Q, R are algebraically in descending order of magnitude. Then it can easily be shown that the planes of greatest tangential stress are parallel to the line OY, and inclined at an angle to the directions OX and OZ. That is to say, they are parallel to the direction of intermediate pressure, and inclined at certain angles to the directions of greatest and least pressure in the rock.

To determine what these angles are, suppose A to be a plane parallel to OY, and making an angle θ with the direction of OX. Then by a simple proof it can be shown that the tangential stress across A is $\frac{P - R}{2} \sin 2\theta$.

This proof is a well-known theorem, and can be found in any book which deals with the subject of stresses in solid bodies. The method, which I shall merely indicate, is to consider the forces acting on a right triangular prism, having its edges parallel to OY, one of its faces parallel to A, and two others parallel to the planes XOY and YOZ. This prism we suppose to exist in the rock, somewhat as the statue exists beforehand in the block of marble, and by considering the forces acting on it we are led to the above result, namely—

$$\text{Tangential stress} = \frac{P - R}{2} \sin 2\theta.$$

It is evident that this force will vanish when P = R. That is, there can be no tangential stress when the pressures in the two directions are equal. It will be large when P and R are of opposite sign; i.e. when one represents a pressure and the other a tension.

For any given system, however, the tangential stress is greatest when $\sin 2\theta = 1$, $2\theta = 90$, $\theta = 45°$. It is evident that it will be equally great when $\theta = -45°$, and thus we shall have two series of planes across which tangential stress is a maximum. The one set will be parallel to the plane which passes through OY and bisects the angle XOZ; the other set will be parallel to the plane passing through OY and bisecting the angle XOZ. Across any plane belonging to either of these series the tangential stress will be $\frac{P - R}{2}$.

Now, as we shall afterwards see, the planes of faulting in any ck do not follow exactly the directions of maximum tangential stress, but deviate from these positions in a more or less determinate manner. In endeavouring to explain this I have been led to suppose that the forces which hinder rupture from taking place in any rock are not the same in every direction. If we suppose that the resistance which any solid (otherwise isotropic) offers to being broken by shearing along any plane consists of two parts, one part being a constant quantity and the other part proportional to the pressure across that plane, we shall arrive at results which agree very well with the observed geological facts.

The second force will have an effect somewhat similar to friction, and I shall use the symbol μ in this connection, while by no means assuming that we are dealing with the same phenomenon. The effect of this force will be to make faulting more difficult along planes across which there is great pressure.

Supposing, as before, that P and R are the greatest and least pressures at any point. Then, as we found, the tangential stress across a plane parallel to the direction of Q, and inclined at an angle θ to the direction of P, is $\frac{P-R}{2} \sin 2\theta$, while the *pressure* across such a plane may easily be shown to be $P \sin^2\theta + R \cos^2\theta$.

Now, supposing a plane crack had actually formed in this direction, and that movement were just about to begin along it, the resistance to this movement due to friction would be $\mu(P \sin^2\theta + R \cos^2\theta)$, or $\mu \left(\frac{P+R}{2} - \frac{P-R}{2} \cos 2\theta \right)$.

If we assume the existence of the second force above referred to, then instead of considering the maxima of $\frac{P-R}{2} \sin 2\theta$, we must subtract from this quantity one of like form to that given above. We are supposing now that μ is the ratio which the variable part of the resistance to breakage bears to the pressure across the plane considered. We then get the following quantity, $\frac{P-R}{2} (\sin 2\theta + \mu \cos 2\theta) - \mu \frac{P+R}{2}$; and this will be a maximum in the directions in which faulting will be the most likely to occur.

For a maximum, it follows from the principles of the Differential Calculus that $\cos 2\theta - \mu \sin 2\theta = 0$, $\tan 2\theta = \frac{1}{\mu}$.

This gives us $\theta = 45°$ for $\mu = 0$

$\theta = 30°$ for $\mu = \frac{1}{\sqrt{3}}$ or ·577.

$\theta = 22\frac{1}{2}°$ for $\mu = 1$

102

It is difficult to form any estimate of what value must be assigned to μ for any particular rock, but we see what will be the general result. The planes of faulting, instead of bisecting the angles between the directions of greatest and least pressure, will deviate from these positions so as to form smaller angles with the direction of greatest pressure. This result agrees very well with the recorded facts.

I shall next consider a little more fully what takes place under (1) an increase and (2) a relief of lateral pressure in any rock. It is important to notice that it does not follow that because there is an increase of pressure in one horizontal direction, there will necessarily be so in all. On the other hand, it is quite possible that there may be an increase of pressure in one horizontal direction along with a relief of pressure in a horizontal direction at right angles to the first. This will form a third case to be treated separately.

(1) Suppose there is an increase of pressure in all horizontal

P *horizontal* $\Big\}$ *Thrust Planes.*
R *vertical*

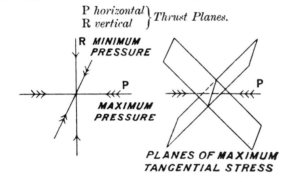

PLANES OF MAXIMUM
TANGENTIAL STRESS

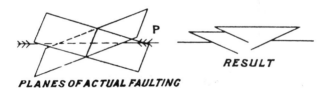

PLANES OF ACTUAL FAULTING   RESULT

directions. Then it may possibly happen that the pressure in all horizontal directions is equal. This will form a special case to be treated later on. For the meanwhile we shall assume, what is far more likely to happen in fact, that the different horizontal pressures are not exactly equal, but that there is one horizontal direction along which pressure is greatest. This maximum pressure we shall, as before, denote by P. Then R, the minimum pressure, will be vertical, and the intermediate principal pressure Q will be in a horizontal direction perpendicular to P.

Then there will be two sets of planes across which tangential stress will be a maximum. Both sets will have their "strike" parallel to Q and perpendicular to P. Both sets will "dip" at an angle of 45°, but they will dip in opposite directions.

Suppose now that the stresses are so great as to lead to actual rupture. Then the planes of faulting should strike in the same direction, but they should, as we have seen, be less inclined to the direction of greatest pressure, which in the present case is horizontal. Thus we should have a double series of fault-planes inclined to the horizontal at angles of less than 45° and striking perpendicularly to the direction of greatest pressure. Motion would take place along any of these planes in such a way as to relieve the pressure, that is, in the form of overthrust.

(2) Suppose next that there is a relief of pressure in all

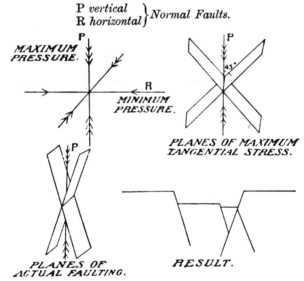

P *vertical* } *Normal Faults.*
R *horizontal* }

MAXIMUM PRESSURE.

P

R

MINIMUM PRESSURE.

PLANES OF MAXIMUM TANGENTIAL STRESS.

PLANES OF ACTUAL FAULTING.

RESULT.

horizontal directions, so that P, the greatest pressure, will be the vertical pressure due to gravity. Then it can only happen very rarely that the pressures, or tensions, in all horizontal directions, will be equal. In the general case there will be one horizontal direction for which the pressure is a minimum. Taking this as the direction of R, then Q, the intermediate principal pressure, will be in a horizontal direction perpendicular to R.

In this case the planes of maximum tangential stress will strike parallel to Q and perpendicular to R; while they will dip in opposite directions at angles of 45° as before.

The planes of actual faulting will deviate from these positions so as to form smaller angles with P, the vertical pressure. The result will be a double series of fault-planes dipping in opposite directions at angles of *more* than 45°, and striking perpendicularly to that direction in which the relief of pressure is the greatest. Motion will take place along these planes in the normal manner. I shall try to show later on how such faulting will tend to equalise the pressures.

(3) We have next to consider the case in which there is an

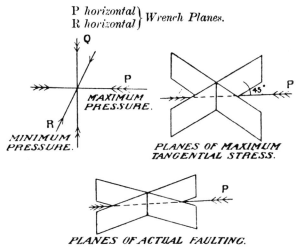

P *horizontal* ⎫
R *horizontal* ⎭ *Wrench Planes.*

MAXIMUM PRESSURE.

MINIMUM PRESSURE.

PLANES OF MAXIMUM TANGENTIAL STRESS.

PLANES OF ACTUAL FAULTING.

increase of pressure in one horizontal direction, together with a decrease of pressure in a horizontal direction at right angles to the first. In this case the maximum pressure P is horizontal; the intermediate pressure Q is vertical; while the third principal direction, which may correspond to a tension, or to the smallest pressure, is horizontal and at right angles to the direction of P.

Then the planes of maximum tangential stress are vertical, and inclined at angles of 45° to the directions of P and R. The planes of actual faulting will deviate from these positions so as to form smaller angles with the direction of P, the maximum pressure.

They might, in fact, form an arrangement not unlike that of the cleavage-planes of a hornblende crystal, supposing the crystal to be placed with its prism axis vertical. Motion would take place along any one of these planes in a horizontal manner. A plane of dislocation along which motion has actually taken place in a horizontal direction is sometimes called a "wrench-plane." Thus we see that under the system of forces last considered a network of such wrench-planes may possibly be developed. Each of these would hade vertically, and the two systems would cross each other at acute angles.

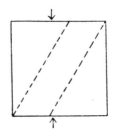

In this last case it is not particularly evident in what manner a single dislocation tends to relieve the stress. If, however, we take a number of such wrench-planes, crossing one another, it becomes optically evident how the stress will be relieved.

The accompanying diagram shows how a square area of land is deformed by a double system of wrench-planes; the particular case chosen being that in which the greatest pressure is from N. to S. and the least pressure from E. to W. The joint result is to decrease the N. to S. dimensions of the area, and to increase its dimensions from E. to W. Thus we see how the stress will tend to be relieved. The diagram also indicates in what direction motion will take place along a fault of either series.

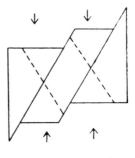

Regarded as a vertical section, the same diagram will do to illustrate the case of normal faults. As a matter of fact, however, it is almost impossible to tell, from surface indications, whether so complete a network ever does exist in this case. Two systems of faults like those in the diagram, originating under the

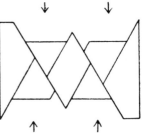

same system of pressures, might very well be called " complementary " systems.

There remain to be considered only the cases in which the stresses in two principal directions are equal ; while the stress in the third principal direction is not equal to the other two. In any such case shearing stress will reach its maximum across planes having an indefinite number of directions, but parallel to the tangential planes of a cone having its axis in the direction of unequal pressure. Thus faulting might take place in an indefinite number of directions. This is a case, however, which is very unlikely to arise in reality, as the odds are very great against the pressure in two principal directions being exactly equal.

I shall next consider in how far these theoretical conclusions are borne out by the recorded facts.

With regard to the first of the three cases, it is a well-known fact that reversed faults and thrust-planes do in general dip at a low angle; this is especially the case in connection with the great series of thrust-planes occurring in the N.W. Highlands. These are a series of dislocations dipping originally at a low angle to the east; motion has taken place along them by the country to the east being pushed over the country to the west. The pressure which produced them was obviously from east to west. There must at the same time have been a slightly more than normal pressure from north to south, while the least pressure would be in a nearly vertical direction.

In this case our theory would have led us to expect a complementary series of faults, dipping at a low angle to the west, and along which the country to the west had been pushed over the country to the east. In Scotland no such complementary series exists.

It is, however, very instructive to compare the Scottish system of thrust-planes with that which has been made out to exist in Scandinavia. In this peninsula rocks of Silurian age are overlain through more than 10° of latitude by rocks of what is called the Seve group; possibly Cambrian, and undoubtedly older than the rocks beneath them. The labours of many geologists, and notably Törnebohm (" Grundragen af det Centrala Skandinaviens Bergbyggnad "), have shown that this is due to thrusting on an enormous scale. In the case of the larger thrusts, the country to the west has been pushed over the country to the east. Törnebohm, however, states that in the case of some of the smaller thrusts this rule is reversed, and thrusting has taken place of east over west, as in Scotland.

In any case it seems probable that the thrust-planes in Scandinavia owe their origin to the same great series of pressures as produced those in Scotland, and that thus a pressure acting from east to west over a wide stretch of country may produce in one part thrusting of west over east, and in another of east over west.

When we come to consider angles, however, the results of observation are less in accordance with our theory. Thus in Scotland the prevailing dip of the thrust-planes to the east is too low to have been produced by any purely horizontal pressure. In Scandinavia the planes of dislocation have themselves been affected by subsequent movements, and they may dip west, east, or in any direction. If the original dip was to the west, however, it would seem to have been an extremely low one. The accompanying diagram is a rough attempt to give mathematical form to an idea of Dr Peach's with regard to the low angle of dip. What is intended to be expressed is as follows. Supposing a mountain chain was being produced by the same series of pressures that caused the thrusts, and suppose that it ran (as it naturally would) in a north and south direction. Suppose further that the west of Sutherland lay on the western margin of this mountain-chain, and that the district in Scandinavia where the thrusts have been produced lay on its eastern margin, or on the eastern margin of some parallel chain. Then the effect of the declivity of the ground

on either side would be to tilt the directions of greatest pressure, on either side, as indicated in the diagram. The planes of most likely thrusting would thus also be tilted, so as to become more nearly horizontal.

With regard to the second of our two cases, it is well known that normal faults have in general a dip much steeper than 45°. In the central lowlands of Scotland there is a great system of normal faults striking approximately east and west. Of these faults a certain number hade towards the north, and have a downthrow on that side; others have their hade and downthrow towards the south. We have thus here a beautiful example of a double series of faults produced by the same system of forces, and in this case we may regard it as proven—

(1) That at the same time when these faults were formed there was a relief of pressure in all horizontal directions.

(2) That this relief was greatest in the direction from north to south.

I have spoken everywhere above only of a relief of pressure, but it is possible that in some cases a relief of pressure may go so far as to amount to an actual tension. If this tension became so great as to produce actual rupture by pulling the rocks asunder, it seems likely that cracks would be formed perpendicular to the direction of tension. They would run parallel to the set of faults likely to be produced by a smaller tension, but their hade would most likely be vertical, and they would not necessarily produce any displacement of the strata on either side. In this connection it is extremely interesting to notice that sets of intrusive dykes are in some cases known to accompany systems of normal faults, and to run in the same direction.

Thus along with the east and westerly faults of the central valley of Scotland, occur, as is well known, a set of east and westerly dykes. Sets of parallel faults also accompany the north-westerly dykes of Tertiary age in some parts of the British Islands. Thus in the Cowal district Mr Clough has observed a series of north-westerly faults along with these dykes, and in Anglesea Mr Greenley has found a similar series accompanying dykes of the same age, which he knows to be normal faults. Although in a case like this the faults and the dykes may not be strictly contemporaneous, it seems almost certain that they owe their origin to the same great series of forces; and it may be possible that the dykes occupy fissures formed while the tension was at its strongest, by the actual pulling asunder of the strata.

With regard to the third class of faults, there is as yet much less recorded material with which to compare our conclusions. Planes of dislocation do occur, along which there has been much horizontal movement, but it is difficult to show in any particular case that this has not been accompanied by an equal or greater amount of vertical displacement.

In the Fassa Monzoni district of the Tyrol a double series of faults has been described by Mrs Gordon under the name Judicarian.[1] These form two sets running N.N.E. and N.N.W., and Mrs Gordon believes them to be strictly contemporaneous. On this assumption the only possible method of explaining their production is to suppose them to be wrench-planes which have formed under the influence of a great pressure in the north and south direction, accompanied by a relief of pressure in the east and west direction. If this explanation be the correct one, these faults form a very striking example of a double system originating under the same set of forces.

To come nearer home, a good example of faults accompanied by lateral wrench occurs in certain parts of the Highlands of Scotland. It is now well known that there is a great series of north-easterly or north-north-easterly faults traversing the Highlands, and giving rise to well-marked physical features. South of the Great Glen are four faults which have been named the Loch

[1] "The Geological Structure of Monzoni and Fassa." *Trans. Ed. Geol. Soc.*, 1903.

Tay fault, the Killin fault, the Glen Fine or Tyndrum fault, and the Loch Awe fault. North of the Glen occurs the Inbhir-Chorainn fault, which extends perhaps from the southern part of Skye to near Ben Wyvis in Ross-shire.

These are all faults with lateral displacement, and they run in the direction already mentioned. It is noticeable that the ground on the east side of these faults has in every case been shifted north-eastwards with regard to the ground on the west; the amount of displacement often amounts to two or three miles. From this fact we are justified in the following conclusion, that at the time when these faults were formed there was an increase of horizontal pressure in the north-south direction, accompanied by a relief of pressure in the direction from east to west.

It is just possible that the Loch Maree fault, which is a wrench running in the N.W. direction, may be complementary to the series above described, as the movement along it indicates a pressure which was greatest in a direction only a little W. of N. (and E. of S.). Otherwise we must suppose that we have a series like that of the thrust-planes which occur in another part of the Highland area, where for some reason only one set of a possible double set of faults has been developed.

According to the theory, a fault of the kind we are discussing should have a nearly vertical hade. I have seen this verified in the case of some small faults with horizontal slickensides, which occur in the valley of the Allt Coire Rainich in easter Ross-shire. Mr Clough has observed the Inbhir-Chorainn fault to hade in different directions in different parts of its course. It is possible that the *general* hade of this fault may be nearly vertical, and that what Mr Clough has noticed may be due to a sort of slickensiding or corrugating of the fault plane on a large scale.

In the above discussion I have everywhere assumed that one of the principal directions of pressure is vertical. If it were not, we might have systems of faults intermediate in character between the classes described above. Thus we might have a single fault, or a system of faults, each member of which was partly a wrench-plane and partly a normal fault. That such intermediate cases do occur seems certain, from the number of cases in which faults are accompanied by slickensides with a direction intermediate between the horizontal and vertical.

Or again, although, as we have seen, the majority of thrust-planes and normal faults do follow the rules above laid down, individual cases do occur in which a thrust is met with, with a steeper dip than 45°, or a normal fault which has a less dip than the above figure. These may very likely be explained in the same way, by a departure from the perpendicular and horizontal directions of the principal axes of pressure.

I have in the preceding part of this paper used the terms strike and dip, making them apply to fault-planes in the same sense as they do to planes of bedding. I have by this means avoided the use of the word " hade," which sometimes gives rise to ambiguity.

It is very difficult to estimate what amount of tangential force will be necessary in order to produce actual rupture and so lead to faulting. In Professor Ewing's book on " The Strength of Materials," figures are given for the amount of force necessary to produce crushing in prismatic blocks of various materials, the force being applied to the ends of the prisms. For prisms 1″ in section, the following are the results for a few common rocks

Granite .	. . .	6-10 tons.
Basalt .	. . .	8-10 tons.
Slate .	. . .	5-10 tons.
Sandstone	. . .	2-5 tons.

In § 9 of the same book occurs this sentence, which is of some significance in connection with the present subject :—

" When a bar is pulled asunder, or a block is crushed by pressure applied to two opposite faces, it frequently happens that yielding takes place wholly or in part by shearing on surfaces inclined to the direction of pull or thrust."

Now, in the case of a block crushed by pressure applied to two opposite faces, we are dealing with a single pressure, which we may denote by P; Q and R being nearly zero. Assuming, then, that yielding does take place by shearing to begin with (and it is difficult to imagine what else could happen), the amount of tangential stress necessary to produce this shearing cannot be greater than $\frac{P}{2}$

In granite this may be as much as 5 tons per square inch; in hard sandstone it would amount to $2\frac{1}{2}$ tons per square inch; in soft sandstone to only 1 ton; in shale, and in soft rocks of Tertiary formation, it would probably be even less.

If we suppose this tangential stress to be the result of a single pressure, the amount of such a pressure necessary to produce faulting is indicated by the figures already quoted from Professor Ewing.

For hard sandstone the pressure, if a single pressure, must amount to 5 tons per square inch. Now, supposing the S.G. of the rock we are dealing with to be 2·65 (the S.G. of quartz), a pressure of this amount will be caused by the weight of the superincumbent rock at a depth of $1\frac{5}{6}$ miles (1·844).

111

We are thus led to enquire what it is that prevents faulting taking place incessantly at this and all greater depths. We see from our formula that supposing P to be the vertical pressure, a lateral pressure of amount R diminishes the tendency to shear from P/2 to (P−R)/2. Thus the answer to the above question is that there must be lateral pressure in all directions. It is easily seen, too, that at the critical depth above mentioned, the lateral pressure cannot exceed twice the vertical, or else (P−R)/2 would become a negative quantity greater numerically than $2\frac{1}{2}$ tons per square inch, which we are taking for the critical amount of stress.

As we go further down in the substance of the earth's crust, the lateral pressures must increase along with the vertical, as the difference between the vertical pressure and the pressure in any horizontal direction, even for a rock as hard as basalt, can never exceed 10 tons per square inch. At a depth of, say, 25 miles, the vertical pressure will be something very much greater than 10 tons per square inch, and so at this depth the differences between the pressures must be small quantities when compared to the pressures themselves. Thus there must be a condition of things, at great depths, similar in one respect to fluid pressure.

The question next arises, what is to produce this lateral pressure, which we see is necessary to preserve equilibrium, altogether apart from the production of anticlines or thrust-planes. The question may be answered by considering the case of an arch consisting of a single layer of bricks. If we take the brick at the summit of the arch, it is easy to show by drawing a triangle of forces that the horizontal forces acting on the brick, and due to the pressure between it and its next neighbours, are great in comparison with the vertical force acting on it due to gravity. This will be the case even when the arch is loaded by the weight of further material resting on it. For a brick in this position, then, the horizontal pressure is not equal, but much greater than the vertical.

The same would be the case if all the extraneous forces, instead of being directed in parallel lines, were directed towards the centre of the arch ; and it is easy to see that the statement also applies to the case of a hollow globe, acted on by a system of forces tending towards its centre. Under a force as great as that of gravity at the earth's surface, however, a hollow globe of the size of the earth could not exist, at least if composed of any known rock. The horizontal pressure would necessarily be so great as to cause shearing, being uncompensated by any vertical pressure of corresponding magnitude.

Thus it is impossible to look on the earth as being a series of independent, self-supporting, concentric shells.   At the same time, if, by a mathematical fiction, we suppose the earth divided into a series of concentric shells, it is not the case that each shell will have to bear the whole weight of those above it.

Each of the latter acts *to a small extent* as an arch, and so, as it were, bears part of its own weight.   This part will only be a small fraction of the whole   In fact, if A denote the lateral pressure which would exist in any such shell, supposing it to be entirely unsupported, and to act as an arch, and if H denote the actual mean horizontal pressure, then, roughly, H/A will denote the fraction of its own weight borne by such a shell.

The same will hold in the case of a liquid globe, in equilibrium under its own attraction ; only in this case the problem will be far more definite, as the pressure must be the same in all directions at any point.

I have brought in this discussion to show how we might account for a lateral pressure even much greater than that which actually exists.   We have seen that the actual lateral pressure existing in each layer of the earth's crust is fixed within certain limits, so long as equilibrium is to be maintained. I shall try to show how it is that adjustment takes place whenever the lateral pressure transgresses these limits.

Suppose that during any period the outer surface of the earth is not contracting so quickly as the earth's interior.   The result will be that the outer crust ceases to have so much support from the underlying layers (whether liquid or solid) as it would normally have.   It has thus to support a greater than usual fraction of its own weight, and the result is greater lateral pressure.

If the lateral pressure only exceeds the vertical by so small a quantity as to lie within the above mentioned limits, it may result in the gradual formation of anticlines or isoclinal folds. If, however, the lateral pressure exceeds the vertical by a greater quantity than the rigidity of the rocks concerned will allow, a series of reversed faults or thrust-planes will result. In either case the result is the same, the circumference of the surface layer is diminished ; it thus settles down on the layers below it, and ceases to bear a more than normal fraction of its weight, and so lateral pressure is diminished to within the before mentioned limits.

Suppose, on the other hand, that during some geological period the surface layer is contracting more rapidly than the more central portions.   The result will be that the latter are compressed, and that there is an increase of pressure in the central portions which tends to set up a tension in the surface layer. In this way lateral pressure in the surface layer will be partially, or it may be wholly, relieved.

As soon as the lateral pressure falls short of the vertical at any point, by more than a certain amount, a series of normal faults will result. These increase the earth's circumference so that the arch-like condition of the surface layer is restored, and lateral pressure is increased to within the required limits.

When considering the formation of faults, we have generally to deal with systems of forces which existed in past geological time. In a few cases the contrary may be the case. Thus in Japan and other countries which are liable to earthquakes, there is no doubt that faults are being produced or are growing in magnitude even at the present day. To a less extent this may be the case in our own country ; at least we know that movement is going on in some districts along faults which have already been formed. This brings us to the interesting question whether it may be possible, by studying the direction of fault-lines along which motion is taking place, to arrive at some conclusion with regard to the system of forces at present at work in any given area.

From a consideration of the facts published by Mr Davison, we see that in Scotland there is a tendency for earthquakes to occur along faults which run in a north-easterly direction. This may perhaps be an accident due to the fact that the largest faults in the country do happen to run in that direction. At the same time the fact is worth noticing ; it might be connected with a slight increase of horizontal pressure in a direction from south-east to north-west; it might, however, be due to other forces, and we are very far from being in a position to form any definite conclusion on the subject. In England, earthquakes seem to take place along faults running in every direction of the compass ; perhaps there may be a very slight tendency in some parts to select faults which run north-eastwards, but it would be very unsafe to base any hypothesis on this supposition. In a case like this it is impossible to say whether we are dealing with a simple system of pressures or not. On the other hand the extent of country over which roughly parallel folding may often be found is a proof that at certain past geological epochs the same sets of forces must have extended over wide areas. The length which certain mountain chains extend in what are practically straight lines may be taken as additional evidence to the same effect.

It has been observed that fault breccias occur more frequently in connection with normal faults and wrench-planes than they do in connection with thrusts. This may be easily explained as follows. The normal pressure due to gravity in a rock mass is a fairly constant quantity at a given depth. It may therefore be taken as a standard with which to compare our other pressures.

Now, in the case of normal faults both horizontal principal pressures are less than the vertical; in the case of wrench-planes one is greater and one is less; while in the case of thrusts the pressure in all horizontal directions is greater than the vertical. If we take the formula $P \sin^2\theta + R \cos^2\theta$, which for a certain value of θ denotes the pressure *across* a fault plane, we see that in the first case the pressure must be less than the vertical. In the second it may be greater or it may be less, while in the third case, that of thrusts, it is bound to be greater. The explanation seems therefore to be that fault-breccias form more readily along faults across which the pressure is not very intense.

Some interesting conclusions are suggested by what has already been brought forward as to the ultimate strength of rocks. Thus it is impossible to have a sheer cliff face of sandstone more than $1\frac{5}{8}$ miles in height, this figure expressing what we have before denoted as the "critical depth" for hard sandstone. If the cliff exceeded this height, fracture would immediately be set up at the base of the cliff.

Before I conclude, I may as well briefly summarise the results to which I have been led in this paper. They are as follows :—

Faults may be grouped roughly into the three classes, known as reversed faults, normal faults, and wrench-planes, but varieties intermediate in character between these three types also occur.

(*a*) Reversed faults and thrust-planes originate when the greatest pressure in the rock mass is horizontal, and the least pressure vertical. They "strike" in a direction perpendicular to that of greatest pressure, and dip in either direction at angles of less than 45°.

(*b*) Normal faults originate when the greatest pressure is vertical, and the least pressure in some horizontal direction. They "strike" in a direction perpendicular to that of least pressure, and dip in either direction at angles of more than 45°.

(*c*) The third type of faults, to which the name of wrench-planes has been applied, originate when the greatest pressure is in one horizontal direction, and the least pressure in another horizontal direction, necessarily at right angles to the first. They "strike" in two possible directions, forming acute angles which are bisected by the direction of greatest pressure; their hade is theoretically vertical.

13

Reprinted by permission of the publisher from pp. 298–302 of *Plasticity: A Mechanics of the Plastic State of Matter*, McGraw-Hill, Inc., New York, 1931

PLASTICITY

A. Nádai

* * * * * *

The Strain Ellipse.—Recapitulating briefly the preceding, we see that *for simple shear* a circle of radius a is distorted into an ellipse (the "strain ellipse") with semi-axes A and B

$$A = a \cdot \sqrt{\lambda_1}, \quad B = a \cdot \sqrt{\lambda_2}, \tag{20}$$

in which the values of λ_1 and λ_2 are determined by

$$\lambda_1 = 1 + \frac{\gamma}{2}(\gamma + \sqrt{4 + \gamma^2})$$
$$\lambda_2 = 1 + \frac{\gamma}{2}(\gamma - \sqrt{4 + \gamma^2}). \tag{21}$$

Using these equations, the two principal strains ϵ_1 and ϵ_2, in a rock where the verticals are all displaced obliquely through an angle β or in which there is a unit shear of value $\gamma = \tan \beta$ may be calculated from

$$\epsilon_1 = \frac{A}{a} - 1 = \sqrt{\lambda_1} - 1$$
$$\epsilon_2 = \frac{B}{a} - 1 = \sqrt{\lambda_2} - 1. \tag{22}$$

Obviously, since the volume is assumed constant, we have

$$(1 + \epsilon_1)(1 + \epsilon_2) = \sqrt{\lambda_1 \lambda_2} = 1. \tag{23}$$

These formulæ serve to make it possible to estimate quantitatively the actual strains occurring during relative slippage or shear of rock layers in nature. The angles α_1, α_2 for the principal directions, are given before the distortion (Fig. 360) by

$$\tan \alpha_1 = \frac{\lambda_1 - 1}{\gamma}, \ \tan \alpha_2 = \frac{\lambda_2 - 1}{\gamma} \tag{24}$$

and after the displacement (Fig. 361) by

$$\tan \alpha_1' = \frac{\lambda_1 - 1}{\gamma \lambda_1}, \ \tan \alpha_2' = \frac{\lambda_2 - 1}{\gamma \lambda_2} \tag{25}$$

A simple finite shear is therefore connected with a finite rotation through an angle

$$\alpha_1 - \alpha_1'. \tag{26}$$

There are two series of parallel straight lines which in the case of simple shear have the property that their length remains unchanged. One series is obviously parallel to the x axis. The lengths parallel to the x axis remain in each intermediate

[*Editor's Note:* A row of asterisks indicates that material has been deleted.]

position unchanged. There is yet a second series of parallel lines which retain their original length, namely, the series of straight lines $y = -\dfrac{2x}{\gamma} + \text{const.}$ These become after the displacement the straight lines $y = \dfrac{2x}{\gamma} + \text{const.}$ In the case

of this second system, however, the straight lines do not retain their initial length in all intermediate positions during the process of deformation. It will be recognized that a simple shear or the above-considered homogeneous plane strain may be described as follows:

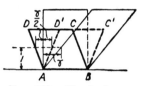

Fig. 362.—Simple shear.

There exists in the stressed body a rhombus $ABCD$ (Fig. 362), which becomes after the distortion a congruent rhombus $ABC'D'$. The angles of the rhombus are interchanged so that the obtuse angle before distortion becomes the acute angle and *vice versa*.

e. Simple Shear in Rock Layers.—Traces of homogeneous strains such as described above, which indicate that the rock layers have undergone severe permanent deformations in the plastic state, are sometimes disclosed by fossils. An example observed by A. Heim can be seen in Fig. 362a. The remains of the fossil fish clearly indicate that the rock was deformed by plastic flow in the direction of the arrow.

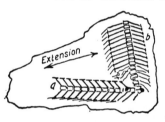

Fig. 362a.—Fossil fish remains deformed together with rock in which they were imbedded. The two pieces *a* and *b* of the broken spine of a fossil fish belonged to the same skeleton of Lepidopus. Being buried at a different angle to the main direction of the plastic extension (marked with arrows) they appear distorted by different amounts. Tertiary rocks; Alps. (*According to Albert Heim, Geologie der Schweiz, part 3, p. 88, 1920, Tauchnitz, Leipzig.*)

The observed displacements of the rock layers which occur in nature, often have as their fundamental characteristic a simple shear such as has been described above. Geologists call a sudden discontinuity in the rock layers, as shown in Fig. 363, a *fault*. In other cases, the distortion is in the form of a flat wave or undulation in the position of the rock layers, with crests (anticlines) and troughs (synclines) (Fig. 364). In the case of large distortions, geologists speak of "folds," vaults of folding, or arching, etc. The axis of the fold appears to be considerably inclined to the vertical, and in the case of large inclinations there result the so-called "recumbent" folds (Fig. 365). Relative to these, geologists have long ago observed that the *thickness of the layers* (measured perpendicular to the boundary of the layers) is often quite variable; for example, in the "middle limb" of the fold (Fig. 365) the thickness is much smaller, while in the "vault or trough," it is much larger than in the "upper" or "lower

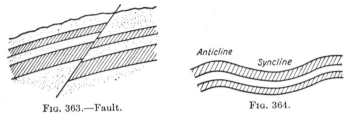

FIG. 363.—Fault.   FIG. 364.

limbs," where the distortion is small. As W. Schmidt has recently pointed out, similar deformation phenomena which at first glance are hard to understand *may be very simply explained* if the observed distortions in the rocks are considered, not as a "bending" or "arching action," but as a *consequence of simple shears of variable value parallel to some given direction, more or less inclined to the undisturbed position of the strata.* A more exact

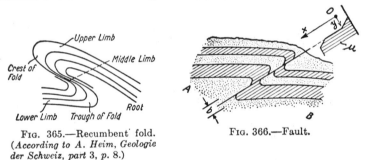

FIG. 365.—Recumbent fold.   FIG. 366.—Fault.
(*According to A. Heim, Geologie der Schweiz, part 3, p. 8.*)

observation of the profile of a fault surface shows that the boundary layers are often distorted somewhat as shown in Fig. 366. Obviously, the fault is not confined to a single plane as is theoretically assumed. The rock is distorted in a layer of thickness b and the value of the probable relative displacement of a part A with respect to a part B of the mountain follows a curve somewhat as shown in the upper right part in Fig. 366 by the displacements u. In this layer the rock structure is sometimes greatly changed and in certain particles or grains of the rock definite traces of large distortions may be found.[1]

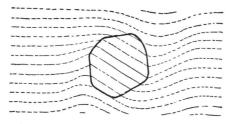

Fig. 367.—Rotation of laminar markings in rock around and within garnet inclusion. (*According to W. Schmidt.*)

If the amount of the displacement u is measured in the direction in which the two large masses are displaced relative to each other, we have in general $u = f(y)$, *i.e.*, u is a function of the coordinate y perpendicular to the direction of displacement. Since the unit shear γ is defined as the value of the dis-

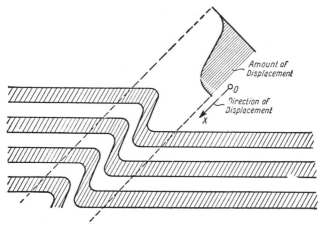

Fig. 368.—Simple shearing displacements in a strip crossing strata at an angle.

placement per unit of length (in the perpendicular direction) in such a case, the unit shear

$$\gamma = \frac{\partial u}{\partial y} = \frac{df}{dy}$$

[1] An instructive example is given by W. Schmidt: In certain rock layers in the eastern Alps, the garnets in the severely distorted rock show a rotation of the previous stratification (Fig. 367); this phenomenon is literally as if rigid balls or rolls had been rolled along the fault surface, surrounded by the plastic rock.

is given by differentiating the displacement curve u in Fig. 366 or 368 with respect to y. Not the displacement component u but rather the unit shear γ is the precise measure of the degree of specific deformation of the rock.

In Figs. 368 to 370 additional examples of simple shear are given. In Fig. 369 is given an example of a group of strata distorted in two mutually

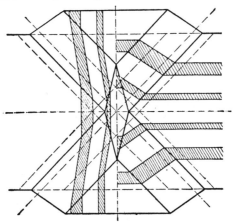

FIG. 369.—Sketch showing result of two simultaneous shearing displacements of equal amount in two perpendicular strips.

perpendicular directions by the same amount of unit shear γ = const. In this it was assumed that the two slip layers were formed *at the same time*. Figure 370 shows a case in which the assumption was made that the slip layer I was first formed, and the slip layer II later. As seen from Fig. 370 it is noticeable that a portion of one layer is almost completely enclosed by

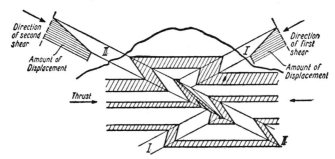

FIG. 370.—Sketch showing distortion of horizontal strata as result of two consecutive shearing displacements acting in two different directions I and II.

the distorted portion of the other. Similar distortions which seem so mysterious at first glance may, however, be simply and easily explained. It is noticeable that in the so-called "Deckentheorie" of the Alps a great use was made of similar explanations.

120

Editor's Comments
on Papers 14 and 15

Few men in a lifetime have exerted as strong an influence on their profession as that of Karl Terzaghi. Terzaghi's earliest papers were concerned with essentially geological phenomena, but he soon thereafter began to lay foundations for the field of *Erdbaumechanik* (i.e., soil mechanics). The key principle concerned the recognition of a two-phase model for material behavior (i.e., the role of fluid pressure in the mechanics of geological materials); once discovered, rational design of building foundations became at once possible, and landslides became understood and susceptible to control. The discovery of the *effective stress principle,* for thus the paradigm was called, is presented in Paper 14 by Alec Skempton of Imperial College, himself a celebrated engineer-scientist; the selection is taken from a memorable *Festschrift* volume honoring Terzaghi on his 75th birthday (Bjerrum and others, 1960). Terzaghi was also the first, apparently, to appreciate the significance of effective stress in experimental rock deformation, as demonstrated in an excerpt from his 1945 paper, *Stress Conditions for the Failure of Saturated Concrete and Rock* (Paper 15).

14

TERZAGHI'S DISCOVERY OF EFFECTIVE STRESS

by A. W. SKEMPTON

The principle of effective stress has been stated by Terzaghi in the following terms.*

The stresses in any point of a section through a mass of soil can be computed from the *total principal stresses* σ_1, σ_2, σ_3 which act in this point. If the voids of the soil are filled with water under a stress u, the total principal stresses consist of two parts. One part, u, acts in the water *and* in the solid in every direction with equal intensity. It is called the *neutral stress* (or the porewater pressure). The balance $\sigma_1' = \sigma_1 - u$, $\sigma_2' = \sigma_2 - u$ and $\sigma_3' = \sigma_3 - u$ represents an excess over the neutral stress u and it has its seat exclusively in the solid phase of the soil.

This fraction of the total principal stresses will be called the *effective principal stresses*. . . . A change in the neutral stress u produces practically no volume change and has practically no influence on the stress conditions for failure. . . . Porous materials (such as sand, clay and concrete) react to a change of u as if they were incompressible and as if their internal friction were equal to zero. All the measurable effects of a change of stress, such as compression, distortion and a change of shearing resistance are *exclusively* due to changes in the effective stresses σ_1', σ_2', and σ_3'. Hence every investigation of the stability of a saturated body of soil requires the knowledge of both the total and the neutral stresses.

This principle is of primary importance in soil mechanics. Its realization is entirely due to Terzaghi, and his earliest

* K. Terzaghi, "The shearing resistance of saturated soils," *Proc. First Int. Conf. Soil Mech.*, Vol. 1, pp. 54–56, 1936.
The notation in the above quotation has been altered to conform with modern standards.

use, in 1923, of the equation $\sigma' = \sigma - u$ marks the beginning of the modern phase of our subject. Preceding work, even by such great engineers as Coulomb, Collin, Rankine, Résal, Bell, and Forchheimer was of limited validity, owing to the absence of this fundamental unifying principle.

Like all truly basic ideas, however, the concept of effective stress is deceptively simple, and its full significance has only become apparent quite slowly; but, as would also be expected, there were several occasions in the past when it seems almost to have been appreciated, at least in a restricted sense. In 1871, for example, the celebrated English geologist, Sir Charles Lyell, wrote: †

When sand and mud sink to the bottom of a deep sea, the particles are not pressed down by the enormous weight of the incumbent ocean; for the water which becomes mingled with the sand and mud resists pressure with a force equal to that of the column of fluid above. The same happens in regard to organic remains which are filled with water under great pressure as they sink, otherwise they would be immediately crushed to pieces and flattened. Nevertheless, if the materials of a stratum remain in a yielding state, and do not set or solidify, they will gradually be squeezed down by the weight of other material successively heaped upon them, just as soft clay or loose sand on which a house is built may give way. By such downward pressure particles of clay, sand and marl may be packed into a smaller space, and be made to cohere together permanently.

† Charles Lyell, *Student's Elements of Geology*, pp. 41–42, London, 1871.

Here, apart from a clear description of the process of consolidation or compaction of sediments, Lyell shows that the pressure in the water in the clay or sand is not contributing in any way to the initial density of packing of the particles. At about this time, also, Boussinesq came very close to a more general statement of the principle of effective stress. In a memoir on the mechanical properties of dry sand he writes: *

One can ignore the atmospheric pressure . . . for this acts in all directions within the sand and all around each grain. . . . It therefore has no influence on their mutual action and consequently will not modify the supplementary pressure that acts at the contact between the grains. . . . Only this supplementary pressure has to be considered.

Ten years later, in 1886, Osborne Reynolds described his famous experiments on dilatancy.† Having demonstrated that, if free to change its volume, a dense sand expands on being subjected to uniaxial compression, he then showed that large negative pressures were developed in the pore water of the sand if loaded in the same manner, but without allowing any volume change to take place. Moreover, in this condition, the sand was very resistant to deformation, whereas on releasing the negative pore pressure the sand lost most of its strength.

The principle of effective stress is clearly involved in Reynolds' tests, but it could not be deduced quantitatively from them. By the end of the century, however, the influence of pore pressure on the strength of concrete was becoming an important question for the designers of dams, requiring precise data for its evaluation. Some rather inconclusive tests on this problem were made by Föppl in 1900 and by Rudeloff in 1912. But more definite evidence was provided by Fillunger in 1915 when he published the results of tension tests on unjacketed specimens of Portland and slag cements, carried out under water in an apparatus in which the water pressure could be varied.‡ Care was taken fully to saturate the voids in the cement; and the average results of the two series of tests are summarized in the following table, where the strength under zero hydrostatic pressure (about 35 kg./sq. cm.) is taken as unity.

FILLUNGER'S TESTS ON CEMENT MORTARS, 1915

Each result is the mean of 24 tests

Cell pressure (kg./cm.²)	0	100	200
Tensile strength	1.00	0.99	0.97

Fillunger naturally concluded that the tensile strength does not vary with water pressure, at least within the pressure range and limits of accuracy of his experiments. This

amounts to a corollary of the principle of effective stress, in the special case under consideration, yet neither he nor anyone else at the time realized the significance of these results. The same remark applies in the field of soil mechanics at this period for, as will be mentioned later, tests approximating to the undrained condition had been carried out by Bell (1915) and Westerberg (1921) which showed that the gain in strength in saturated clays under increasing external pressure was practically zero. This result is similarly a direct consequence of the principle of effective stress. Nevertheless it is clear that the physical meaning of these tests was in no way understood, and it required the genius of Terzaghi to clarify and enunciate this basic law of the mechanical properties of porous materials.

We may conveniently note here that it was also Terzaghi who first showed that a variation in pore pressure alone has a negligible influence on the *compression* strength of concrete.§ Using three types of concrete, with compression strengths ranging from 70 to 580 kg./sq. cm., he tested unjacketed specimens in a triaxial apparatus under hydrostatic pressures up to 400 kg./sq. cm. The results are summarized below, and they show this phenomenon in a more convincing manner than is possible with tension tests.

TERZAGHI'S TESTS ON CONCRETE, 1934

Each result is the mean of 7 tests

Cell pressure (kg./cm.²)	0	100	200	300	400
Compression strength	1.00	1.03	1.02	0.99	0.99

Consolidation of Clays

The concept of effective stress was first explicitly stated by Terzaghi in relation to the consolidation of clays. Geologists and civil engineers had long recognized that clay under load gradually consolidates as water escapes from the voids. We have already seen Lyell's remarks on the subject, and as early as 1809 Telford preloaded a 55 ft. thick bed of soft clay, on which the eastern sea lock of the Caledonian Canal was to be founded, by building an embankment and allowing it to settle for about nine months "for the purpose of squeezing out the water and consolidating the mud" before constructing the lock.‖ Similarly, from his experience with buildings founded on the Chicago clay, Sooy Smith ¶ knew that "the slow progressive settlements result from the squeezing out of water from the earth"; whereas Shankland in 1896 not only expressed the same opinion, but also published time-settlement curves for the foundations of the Masonic Temple covering a period of four and one-half years.**

* J. Boussinesq, "Essai théorique sur l'equilibre d'elasticité des massifs pulvérulents." *Mem. savants étrangers, Acad. Belgique,* Vol. 40, pp. 1–180, 1876 (see p. 28).
† O. Reynolds, "Experiments showing dilatancy, a property of granular material." *Proc. Roy. Inst.,* Vol. 11, pp. 354–63, 1886.
‡ P. Fillunger, "Versuche über die Zugfestigkeit bei allseitigen Wasserdruck." *Osterr. Wochenschr. offentlich Baudienst.,* Vienna, pp. 443–48, 1915.

§ K. Terzaghi and L. Rendulic, "Die wirksame Flächenporosität des Betons," *Zeit. österr. Ing. Arch. Ver.,* Vol. 86, pp. 1–9, 1934.
‖ The quotation is from Telford's article on "Inland navigation," *Edinburgh Encyclopaedia,* Vol. 15, pp. 209–315, 1830
¶ W. Sooy Smith, "The building problem in Chicago from an engineering standpoint." University of Illinois, *The Technograph,* No. 6, pp. 9–19, 1892.
** E. C. Shankland, "Steel skeleton construction in Chicago." *Min. Proc. Inst. C.E.,* Vol. 128, pp. 1–27, 1896.

An early attempt to assess quantitatively the volume changes of clays under pressure was published by Sorby in 1908, from a consideration of field data on the porosity of argillaceous sediments at various depths.* But he made no allowance for hydrostatic uplift below groundwater level, and thus incorrectly translated depth into pressure by using the full density of the overburden.

The first experimental work seems to have been that carried out by Frontard in 1910. A sample 2 in. thick and 14 in. in diameter was placed in a metal container with a perforated base and loaded through a piston; the test was made in a room at a very high humidity to prevent drying out of the clay. Each increment of pressure was left in position until equilibrium was attained, and the results were plotted as a graph relating water content and pressure. Frontard then remarks that "one of the most interesting facts which have been revealed is the great length of time required for the escape of the excess water" and, in spite of the small thickness of the sample, five days was the minimum period for complete consolidation.† He did not investigate this phenomenon; but in 1914 Forchheimer published a theoretical treatment in which the mathematics were based on the fact that the settlement of a clay layer is equal to the volume of water expelled during consolidation, and he derived an expression for the time required for a settlement of given magnitude to take place. The solution was oversimplified, however, and in particular no account was taken of compressibility or pore pressure dissipation.‡

This was the state of the subject when Terzaghi com-

* H. C. Sorby, "On the application of quantitative methods to the study of the structure and history of rocks." *Quart. Journ. Geol. Soc.*, Vol. 64, pp. 171–231, 1908.
† J. Frontard, "Notice sur l'accident de la digue de Charmes." *Ann. Ponts et Chaussées*, 9th Series, Vol. 23, pp. 173–280, 1914.
‡ Ph. Forchheimer, *Hydraulik*, pp. 26, 494–495, Leipzig, 1914.

menced his research in 1919 at Robert College, Istanbul. During the next six years he carried out a series of masterly experiments, both on consolidation and shear strength, the results of which were given in the following series of publications:

1921   "Die physikalischen Grundlagen der technischgeologischen Gutachtens." *Zeit. österr. Ing. Archit. Ver.*, Vol. 73, pp. 237–241.

1923(a)   "Die Beziehungen zwischen Elastizität und Innendruck." *Sitz. Akad. Wissen. Wien Math-naturw. Kl.*, Part IIa, Vol. 132, pp. 105–124.

1923(b)   "Die Berechnung der Durchlässigkeitsziffer des Tones aus dem Verlauf der hydrodynamischen Spannungserscheinungen." *Ibid.*, pp. 125–138.

1924   "Die Theorie der hydrodynamischen Spannungserscheinungen und ihr erdbautechnisches Anwendungsgebiet." *Proc. Int. Cong. App. Mech. (Delft)*, pp. 288–294.

1925(a)   *Erdbaumechanik auf bodenphysikalischer Grundlage.* Deuticke (Vienna).

1925(b)   "Principles of soil mechanics." A series of eight articles in *Engineering News-Record*, Vol. 95.

References to the first four papers will be by date. Page references will be given to *Erdbaumechanik* and the *Engineering News-Record* articles.

In consolidation tests, starting with clay at the liquid limit, increments of pressure up to about 1 kg./sq. cm. were applied in the odometer shown in Fig. 1(a). The apparatus was then dismantled and the 8 cm. diameter bronze ring with its contained clay was removed and set up in the manner shown in Fig. 1(b), under a pressure of 2 kg./sq. cm. After equilibrium had been attained, pressures up to about 20 kg./sq. cm. were applied, each increment remaining constant for two days to allow consolidation to be completed before the next load was added. The pressures were then progressively decreased to zero, when the clay was

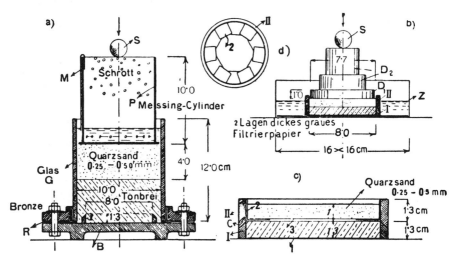

FIG. 1

again loaded to a pressure rather greater than the first maximum. The data were expressed as a relationship between void ratio e and pressure p; the first set of curves were published in 1921.

The principle of effective stress is implicit throughout the 1921 paper, and Terzaghi writes:

A mass of clay is in hydraulic equilibrium when the water content has been everywhere reduced to the value corresponding to the pressure acting at any point (as given by the p-e curve). . . . The change in water content due to a change in pressure, during the transition from one state of hydraulic equilibrium to another . . . takes place very slowly in clays owing to their low permeability.

In 1923(b) sets of p-e curves are shown for three more clays, and the first statement of the classic theory of consolidation is given, in which the concepts of pore pressure and effective stress are fundamental. The process of consolidation is now described as follows:

A change in pressure at any point in a clay causes an alteration in water content. But for a change in water content to occur some water must flow out, and this flow must be the consequence of a gradient in the pore water pressure. Due to the low permeability of clays the rate of flow will be correspondingly small, but finally the pore pressures will disappear.

In deriving the theory of consolidation Terzaghi considers a clay layer originally in equilibrium under p_0. A pressure increment p_1 is then applied, and, in an accompanying diagram, it is shown that the effective pressure increment at any depth, at a time t after the load application, is p and the pore pressure is

$$u = p_1 - p$$

This equation is given in the text; and the physical significance of p is made evident from the expression for the change in void ratio

$$\Delta e = -ap$$

where a is the coefficient of compressibility as deduced from the relation between pressure and void ratio for the condition of "hydraulic equilibrium."

We thus see that the principle of effective stress is fully comprehended in this 1923 paper; but in 1924 Terzaghi again deals with the same problem and gives the following even clearer account:

Before the application of the pressure increment the pressure acting in the solid phase of the clay at each point in the layer is p_0 and the excess porewater pressure is zero. Following an increase in pressure on the upper surface of the layer from p_0 by p_1 the hydrostatic excess pressure at any point has a positive value u and at the same point the pressure acting in the solid phase of the clay has a value p, where $p < p_1$ and, from the equilibrium of the layer,

$$p_1 = p + u$$

. . . If the hydrostatic pressure u decreases by du in a time interval dt, this involves an increase in the pressure acting in the solid phase of the clay by $dp = -du$ and a corresponding decrease in water content equal to $-a\,dp$ where a can be taken as approximately constant for a small increment of pressure.

For the term "pressure acting in the solid phase of the clay," we now use the more convenient expression "effective stress," but otherwise the statement falls little short of the complete formal definition quoted at the beginning of this essay.

It may be noted that the apparatus in Fig. 1 and the method of carrying out the consolidation tests were first described in 1924. The well-known time-consolidation curves were not published until 1927,[*] however, and in 1923(b) the rate of consolidation was studied by means of an experiment in which observations were made on the decrease in pressure in a loaded clay specimen held at constant strain, rather than measuring the compression under constant pressure. In spite of the rather considerable difficulties inherent in this test Terzaghi was able to show that the permeability as deduced from the theory of consolidation was practically identical with the results obtained in 1919 (published in 1921) from direct measurements on the same clay in the combined falling-head permeameter and odometer shown in Fig. 2 and in Plate I. The basic assumptions of the theory were therefore confirmed.

Shear Strength

The principle of effective stress was first clearly stated in 1924, as we have seen, but it was understood by Terzaghi already in 1920. It was used in the theory of consolidation and also checked experimentally by 1923. Subsequent developments in the consolidation of clays involved essentially only the working out of details. The application of the concept of effective stress and pore pressure in the more complex problems of shear strength, however, was a more lengthy process.

The shearing resistance of dry sand was fairly well understood towards the end of the nineteenth century. The expression

$$s = \sigma \tan \phi$$

was widely accepted, and whereas most engineers had not progressed beyond the point of equating ϕ with the angle of repose, the values of which were usually given as 30 to 40°, Darwin's earth pressure tests had shown in 1883 that the value of ϕ operative in a mass of sand depended on the manner of packing, and for a dense sand could be appreciably greater than the angle of repose.[†]

As for saturated sand, there were the remarkable experiments of Osborne Reynolds, which have already been mentioned; and they could have led to a major advance in soil mechanics had they been followed up. For Reynolds demonstrated not only the volume change during shear, but also the interdependence of this volume change with pore pressures in the case where no drainage of the porewater was allowed to take place. But this work was so far ahead of its day that it had no direct influence.

In the case of clays Coulomb had suggested the expression

$$s = c + \sigma \tan \phi$$

[*] K. Terzaghi, "Principles of final soil classification." *Public Roads*, Vol. 8, pp. 41–53, 1927.

[†] G. H. Darwin, "On the horizontal thrust of a mass of sand." *Min. Proc. Inst. C.E.*, Vol. 71, pp. 350–78, 1883.

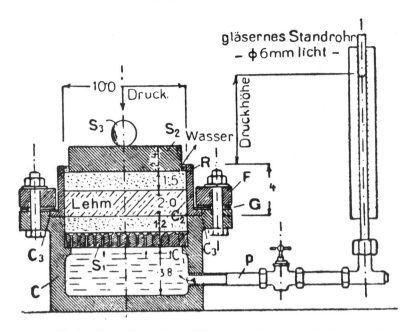

FIG. 2

PLATE I. PERMEABILITY AND CONSOLIDATION APPARATUS, ROBERT COLLEGE, 1919.

as long ago as 1773, and various attempts were made to determine the values of c and ϕ from field observations, by Navier,[*] for example, and from laboratory tests by Leygue [†] in 1885. But the earliest shear tests of any practical value in this connection were those by Frontard [‡] in 1910 on the same clay that he used for the consolidation experiments. In his tests the clay was compacted (i.e., not fully saturated) and for a sample of thickness 4 in., taken to failure in 15 minutes, he obtained the values:

$$c = 400 \text{ lb./sq. ft.} \qquad \phi = 8°$$

the pressure normal to the shear surface ranging, in different tests, up to 10,000 lb./sq. ft.

The first shear box tests covering a wide variety of clays were carried out by Langtry Bell.[§] Little opportunity was allowed for drainage or consolidation and the whole test, including the application of vertical load and the shear force, was completed in about 30 minutes. Some typical results are:

A. L. BELL'S BOX SHEAR TESTS, 1911-1912

Soil	c, lb./sq. ft.	ϕ, degrees
Soft clay	650	½
Firm clay *	1000	2½
Stiff clay	1400	2
Firm sandy clay *	1300	2½
Stiff boulder clay	1600	6½

* Undisturbed samples.

Further tests of a rather similar nature were reported during the next decade by Chatley (Shanghai clay, $c = 230$ lb./sq. ft., $\phi = 9°$), Fellenius (Järna clay, $c = 65$ lb./sq. ft., $\phi = 4\frac{1}{2}°$) and others.[||] But, as we can now see, the technique was not altogether satisfactory; for the tests, although approximating to the undrained condition, were not rigorously controlled, and some drainage undoubtedly was allowed to occur.

[*] C. L. Navier, *Resumé des leçons . . . sur l'application de la mécanique.* Paris, 1833.
[†] L. Leygue, "Nouvelle recherche sur la poussée des terres." *Anns. Ponts et Chaussées*, 6th Series, Vol. 10, pp. 788–998, 1885.
[‡] J. Frontard (1914), *op. cit.*
[§] A. L. Bell. "Lateral pressure and resistance of clay." *Min. Proc. Inst. C.E.*, Vol. 199, pp. 233–72, 1915.
[||] Summarized in K. Terzaghi, "The mechanics of shear failures on clay slopes." *Public Roads*, Vol. 10, pp. 177–92, 1929.

In fact, the importance of controlling the drainage conditions was not appreciated by these investigators, since they were not aware of the existence or significance of pore pressures. In 1920, however, concurrently with his work on consolidation, Terzaghi began a fundamental research into friction and, so far as clays were concerned, he clearly understood that the test specimens must be in a state of "hydraulic equilibrium" before their shear strength was measured. He therefore allowed the clay to consolidate fully before applying the shear loads; and the first results obtained in this way were summarized in 1921 as items 1 to 3 in the following table. Item 4 is quoted in 1923(a),

TERZAGHI'S SHEAR TESTS ON CLAY, 1920-1923

Soil	Coefficient of Internal Friction f	ϕ'
1. Yellow pottery clay I	0.450–0.473	24° –25½°
2. Sandy clay from a landslip	0.357–0.373	19½°–20½°
3. Colloidal fraction of No. 2	0.243–0.250	13½°–14°
4. Blue marine clay IV	0.250–0.300	14° –17°

and in *Erdbaumechanik* values of f and ϕ' are given for clays I and IV (page 195) together with their index properties (page 70):

	Type	LL	PL	PI	Clay Fraction
Clay I	Residual	58	24	34	28%
Clay IV	Sedimentary	58	26	32	42%

It should be remarked that the symbol ϕ' corresponds to the use in Figs. 5 and 7, not to the present-day definition.

Details of the shear apparatus used in these tests (Fig. 3) were first illustrated in 1925(b), page 1027. The specimen of soft clay was placed between a wooden block and the base of a shallow zinc tray, and then submerged under water to prevent drying. Double layers of filter paper covered with No. 20 brass wire mesh were fixed to the tray and the block to provide drainage surfaces and to eliminate the possibility of the clay sliding at the contacts. A vertical pressure was applied and maintained for about two days, at which time the shear strength was measured. The rate of application of the shear force was fairly rapid,

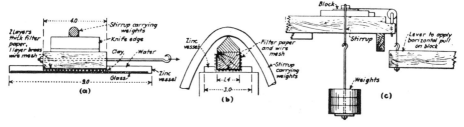

FIG. 3

however, and the test approximated to what would now be called the consolidated-undrained condition; and the clay was normally consolidated.

In the papers of 1921 and 1923 Terzaghi says little about these tests, but in 1925(*b*), pages 1027, 1029, he gives the following highly significant explanation:

Suppose the plane surface of a mass of air-free wet clay is under a definite external pressure. . . . As the voids of the clay are filled with water an increase in pressure involves escape of the excess water. But as the permeability of the clay is very small, the escape of the water proceeds very slowly. During the drainage process part of the excess pressure is compensated for by a corresponding hydrostatic pressure of the water in the voids; the surcharge partly floats on the water, so to say, and only the surplus is carried by the solid parts of

the mass. Immediately after the surcharge is applied, the compression of the clay is practically equal to zero, hence the hydrostatic pressure at this time is almost equal to the surcharge. While the excess water gradually escapes, the hydrostatic pressure becomes smaller and approaches the value zero. . . . A hydrostatic pressure does not produce any static friction. Hence it is merely the remainder of the surcharge which counts. The frictional resistance increases with decreasing hydrostatic pressure, and does not assume its normal value until the excess water has completely escaped from the layer of clay. . . . In a test [in which] the clay is not yet in hydrostatic equilibrium any friction value between zero and the normal value may be furnished. Such values are not the coefficient of static friction, but coefficients of what may be called the momentary hydrodynamic friction; they depend on the pressure which acted on the clay prior to the time

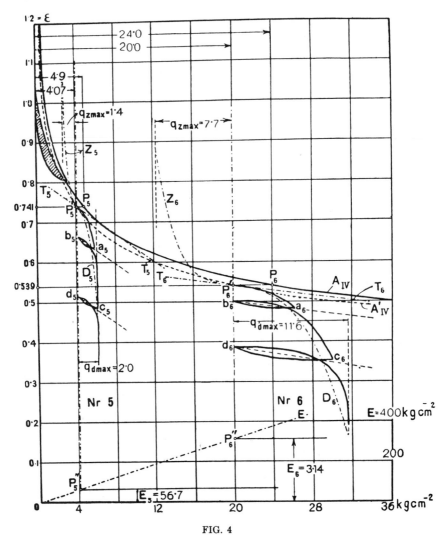

FIG. 4

when the surcharge was applied, on the time during which the surcharge was allowed to act, on the thickness of the layer, and on various other factors.

This may well be considered one of the most important statements in the whole history of soil mechanics, and it undoubtedly provided the basis for most subsequent research on the shear strength of clays.

Before discussing this later work, however, it is worth examining in some detail the thorough way in which already by 1925 Terzaghi had applied the idea of effective stress to an analysis of the unconfined compression test. As shown in Fig. 4, a clay was remolded at various water contents, such as that corresponding to the point P_6', for example, and then a compression test was carried out on a cubical sample cut from the clay, the sample being surrounded at a short distance away by soaked cotton wool held in a wire mesh to prevent drying. The stress-strain relation for such a test is shown by the curve $P_6'a_6b_6 \cdots$; the compression strength is $q_{d\,max}$ and the modulus is given by the slope of the initial tangent and the mean slope of the hysteresis loops. The procedure was then repeated for another water content.

Now, in order to analyze the results, it is necessary to know the capillary pressure in the clay at the start of the unconfined compression test. An odometer test was therefore made, as shown by the curve A in Fig. 4. In this test the clay was under vertical and horizontal pressures p_s and $K_0 p_s$ where K_0 is the coefficient of earth pressure at rest. But the cubes were under an equal all-around capillary pressure p_k and, in $1923(a)$, Terzaghi estimated this pressure from the relationship

$$p_k = \tfrac{1}{3}(1 + 2K_0)p_s$$

The value of p_k is shown by the point P_6'.

The determination of K_0 required a separate test. Two superimposed samples were consolidated in a special odometer equipped with steel tapes arranged horizontally in one and vertically in the other sample, and K_0 was expressed as the ratio between the horizontal and vertical pressures acting within the clay as measured by the forces required to overcome the friction between the clay and the two tapes. The results for the yellow pottery clay I and the blue marine clay IV were 0.70 and 0.75 respectively ($1923a$).

Both the strength and the modulus were found to increase in direct proportion to p_k and an interesting comparison was made between the capillary tension in soils and the intrinsic pressure in metals. But even more interesting is the further investigation of these tests published in 1925. For Terzaghi then points out that

in the case of a solid (such as metal) the intrinsic pressure remains constant during a compression test, while the volume decreases. . . . But in the case of a clay cube the volume cannot possibly change as long as the moisture content is unchanged, and therefore the capillary pressure decreases with increasing load ($1925b$, p. 798).

Here we see a point of great consequence, namely that the pore pressure must change during an undrained test on saturated clay. To arrive at some quantitative assessment of this effect Terzaghi noted the decrease in Young's modulus as the failure stress was approached and assumed that the value of p_k dropped accordingly.* Moreover, he then went on to reason that if the capillary pressure at failure is p_k' the angle of friction ϕ' could be deduced from the equation

$$\tan^2 (45 - \phi'/2) = \frac{p_k'}{p_k' + q_{d\,max}}$$

In this way the values of ϕ' were found to be of the same order as those measured in the shear tests (*Erdbaumechanik*, pp. 188–89, 195).

It is, of course, quite obvious today that much of the foregoing analysis can be criticized in detail. Yet, for its basic clarity, and considered in relation to its period, this is an outstanding achievement.

Later Developments

A full inquiry into shear strength research after 1925 would far exceed the limits of the present study. But a few matters relevant to the principle of effective stress must be considered briefly, in order to complete our survey of Terzaghi's work in this subject.

We have seen that he realized there would be no appreciable increase in strength immediately after the application of a pressure increment on a saturated clay. And by direct implication, no change in strength of such a clay can occur unless the water content changes. It will be recalled, however, that Bell and others found values of ϕ typically between about 2 and 8° in tests carried out with little change in water content; and during the 1920's it seems to have been rather widely assumed that even with no water content change a value of ϕ greater than zero would be obtained. Thus in 1927 Krey combined, as it were, the tests of Terzaghi and Bell in a diagram shown in Fig. 5.† The line AB, inclined at ϕ', represents the strength of a clay which has been consolidated and tested in shear under a pressure such as n_3. But if, after consolidation, the pressure is varied without any further water content change, the strength also varies with pressure as indicated by the line W_3 inclined at an angle ϕ''.

For a partially saturated clay this statement is acceptable, but not for a fully saturated clay. Nevertheless there was as yet no convincing experimental evidence to prove the latter point, and in his 1929 paper on the stability of clay slopes Terzaghi felt bound to follow Krey's view as one of practical validity.

In fact, shortly before this paper was published, Streck, working in Franzius' laboratory at Hanover, had been able to show that ϕ'' was as small as 1° for a clay initially consolidated under a range of pressures ($\phi' = 24°$) and then tested as soon as possible after a rapid increase in load. This result was obtained on samples 2 in. thick in torsion shear tests made for the express purpose of finding if ϕ''

* We may note that the corresponding increase in pore pressure ($p_k - p_k'$) is equivalent to values of the pore pressure coefficient $A = 0.5$ (Clay I) and 0.4 (Clay IV).

† H. Krey, "Rutschgefährliche und fliessende Bodenarten." *Bautech.*, Vol. 35, pp. 485–89, 1927.

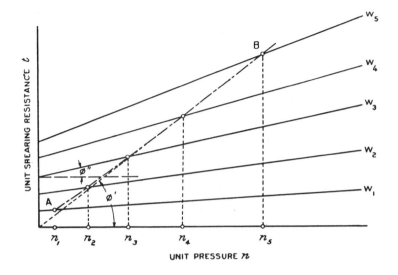

FIG. 5

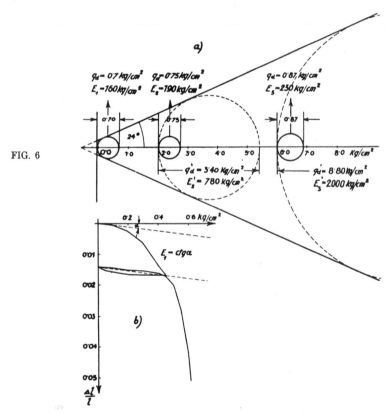

FIG. 6

was zero, as would be expected on theoretical grounds from the principle of effective stress, and taking into account the insignificant compressibility of the pore water compared with that of the clay structure.* And to confirm this, Terzaghi built a triaxial apparatus in Vienna at the end of 1930, in which undrained triaxial tests (as well as tests approximating to the consolidated-undrained condition) were carried out by Gottstein in 1931 on a saturated clay with a plasticity index of 23. The results are given in Fig. 6 and they show hardly any increase in strength or elasticity with increasing total pressure in the undrained state.†

Mention should be made here of some early triaxial tests on clay which were briefly described (unfortunately without the actual results being quoted) in 1921 in a paper by Westerberg.‡ In these tests, which were strictly plane-strain rather than triaxial, the stress difference $\sigma_1 - \sigma_3$ for a given clay at failure was found to be constant and independent of the magnitude of the stress. Westerberg rightly deduced from his experiments that the clay was behaving as an $\phi = 0$ material, and that the shear strength equaled exactly one-half the compression strength. But he took these conclusions to be of general validity and did not appreciate that they were relevant only to saturated clays in the undrained conditions; nor was he aware of pore pressures and effective stress.

Terzaghi's work on shear strength which has so far been described is related chiefly to the question of pore pressures.

* A. Streck, "Fortschritte auf dem Gebiete der Baugrundforschung." *Zentralblatt Bauverwaltung*, Vol. 48, pp. 306–12, 1928.
† K. Terzaghi, "Trägfahigkeit der Flachgründungen." *Int. Assoc. Bridge Struct. Eng.*, Prelim. Publ., pp. 659–83, 1932.
‡ N. Westerberg, "Jordtryck i Kohesionara jordarter." *Tek. Tidsk. Vag. o. Vatten.*, Vol. 51, pp. 25–29, 1921. The apparatus is also illustrated in Krey (1926), *op. cit.*

But research of a more general nature was started by him at M.I.T. in 1927, in conjunction with Janiczek.§ The tests were carried out with a clay initially at the liquid limit and consolidated under different pressures in the odometer. The samples were then removed and specimens prepared for unconfined compression tests. The results are represented by points P_1 and P_2 in Fig. 7 and are similar to those obtained in the 1923 tests. In another series the clay was overconsolidated before being tested, when the strength was found to lie on a curve such as P_2C_2; whereas in a third series the clay was cyclically consolidated, leading to curves of the type C_2P_3. In this way the effects of overconsolidation were demonstrated and one began to see the shear strength of clays in relation to their geological history.

The meaning of the "cohesion" term in Coulomb's equation also became clearer; and this aspect of the work was followed up by plotting the results of Fig. 7 in the form shown in Fig. 8.‖ The compression strength q for the normally consolidated samples is plotted against the pressure under which the clay had been consolidated, giving a straight line through the origin as before. For the overconsolidated samples, however, the strength was plotted against the "equivalent consolidation pressure," defined as the pressure on the normal consolidation curve corresponding to the actual void ratio of the sample. The strength at point m for example is shown at m_1' and the strength at point h at h_1'. Now these samples were under zero consolidation pressure; therefore their strength was considered to be made up of "true cohesion" only and, as will be seen, this cohesion is directly proportional to the equivalent consolidation pres-

§ K. Terzaghi (1929), *loc. cit.*
‖ K. Terzaghi, "Festigkeitseigenschaften der Schüttungen, Sedimente und Gele." *Handbuch physik. Tech. Mechanik.*, Vol. 4, pp. 513–78, Leipzig, 1931.

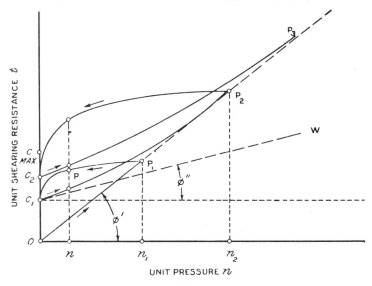

FIG. 7

sure and a function of the void ratio. Thus were the foundations laid for the important research on the basic components of shear strength, carried out by Hvorslev during the years 1933 to 1936 at Vienna under Terzaghi's direction.[*]

A further aspect of the shear strength of clays remains to be mentioned. We have seen that Terzaghi realized in

* M. J. Hvorslev, "Conditions of failure for remolded cohesive soils." *Proc. First Int. Conf. Soil Mech.*, Vol. 3, pp. 51-53, 1936.

1925 that pore-pressure changes are bound to occur during a compression test on saturated clay. But no real significance was given to this conclusion for several years and, indeed, the tacit assumption was made that, in tests where volume changes were permitted to occur, the application of a shearing force had a negligible influence on the water content of a clay. In other words, no distinction had been made in practice between a consolidated-undrained and a drained test. But in tests carried out at M.I.T. during

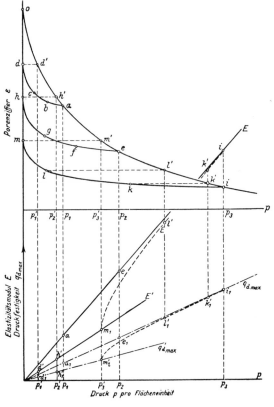

FIG. 8

the years 1930 to 1932 Casagrande, using an improved form of shear box, observed that significant volume changes occurred during shear and, in normally consolidated clays, the slower the shear stresses were applied the greater the consolidation and the greater the strength, up to a limiting value which we now recognize as the *drained strength*.[*] These tests were carried further by Jürgenson, and the results published in 1934.[†]

Finally another classic research, which may be said to be the culmination of all the early work based on effective

* A. Casagrande and S. G. Albert, "Research on the shearing resistance of soils." Unpublished Report M.I.T., 1932.
† L. Jürgenson, "The shearing resistance of soils." *Journ. Boston Soc. C.E.*, Vol. 21, pp. 242-75, 1934.

stress, was started by Rendulic under Terzaghi's direction in Vienna, in the spring of 1933. Using an improved triaxial apparatus, in which the pore pressures inside a clay specimen could for the first time be measured, Rendulic carried out both drained and undrained tests, and was able to demonstrate the validity of the principle of effective stress by direct observation.[‡]

The experimental proof of this principle as a fundamental physical law makes a fitting conclusion to our study of one of Terzaghi's great contributions to the science of soil mechanics.

‡ L. Rendulic, "Ein Grundgesetz der Tonmechanik und sein experimenteller Beweis." *Bauing.*, Vol. 18, pp. 459-467. 1937.

Reprinted from pp. 784–786, 791, 792 of *Amer. Soc. Testing Materials Proc.*,
45, 777–792; discussion, 793–801 (1945)

STRESS CONDITIONS FOR THE FAILURE OF SATURATED CONCRETE AND ROCK

K. Terzaghi

Harvard University

* * * * * *

EFFECT OF PORE WATER PRESSURE ON STRENGTH

The preceding discussions have left no doubt that the increase in strength due to the increase of the confining pressure p_c on specimens with empty voids is chiefly due to the fact that the grains themselves are very much stronger than the bond between the grains. Otherwise the failure would take place across the grains, and, as a consequence, the strength would be practically independent of the confining pressure. As a matter of fact, the strength of steel, which fails only across the grains, decreases slightly with increasing confining pressure instead of increasing (Roš (11)).

It has also become evident that the strength of the specimens depends on the pressure which is carried by the bond between grains along the potential surfaces of failure. The state of stress in the grains is irrelevant. If the voids of the specimen are empty, the normal stress due to a confining pressure p_c is entirely carried by the solid material which constitutes the bond. An invasion of the voids by a liquid under pressure p_c does not alter the stress, but it reduces that part of the stress p_c which is carried by the bond. Therefore it reduces the strength. The magnitude of the decrease depends on the degree of continuity of the intergranular bond. This degree can be expressed by the ratio of that part of the area of the potential surface of failure which is in contact with

the interstitial liquid and the total area of this surface. It will be referred to as *boundary porosity*, n_b.

The data required for estimating the boundary porosity can be obtained by means of compression tests on saturated specimens with unprotected surface, immersed in a liquid under pressure p_c. Figure 7(a) represents a section through a completely saturated specimen surrounded with and permeated by water under a pressure p_c. While the water pressure rises from zero to p_c, the density of the water increases more than the density of the solid. Hence, water

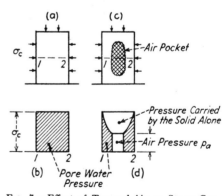

FIG. 7.—Effect of Trapped Air on Stress Conditions in Uncovered Specimen Immersed in Water Under Pressure.

flows into the voids. However, since the total volume decrease of the water is very small, the water pressure in the interior of the specimen becomes equal to p_c a short time after the external pressure has assumed this value (Fig. 7(b)). In a

[*Editor's Note:* A row of asterisks indicates that material has been deleted.]

specimen saturated with oil the stress adjustment takes place much more slowly than in a specimen saturated with water. Therefore water is preferable to oil.

If the specimen contains air (Fig. 7(c)), a large quantity of water must enter the voids of the specimen until the pressure in the air becomes equal to the external fluid pressure. Figure 7(d) shows the state of stress while the volume of the air is still decreasing. Since the pressure p_a in the air is not known, the state of stress illustrated by Fig. 7(d) precludes a correct interpretation of the test results. The transition from the state illustrated by Fig. 7(d) to that illustrated

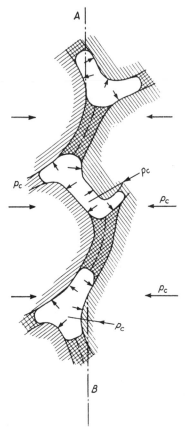

FIG. 8.—Magnified Section Through Surface of Splitting in Uncovered Specimen of Granular Material Immersed in Liquid Under Pressure p_c.

by Fig. 7(b) may require weeks or months.

To make sure that the test specimen does not contain more than a trace of air,

the specimen should be dried at 105 C. prior to the test and then saturated under vacuum. The success of this operation can be checked by keeping the specimen under liquid pressure p_c for one or two days while all valves are closed. If the specimen contains an appreciable volume of air under a pressure of less than p_c, the pressure in the liquid will drop because the volume occupied by the air is still decreasing. Any departure from the procedure recommended in this paragraph may preclude the possibility of correct interpretation of the test results.

The effect of the presence of water under pressure in the voids on the strength of porous materials which fail only by splitting is illustrated by Fig. 8. If the voids of the specimen were empty, a confining pressure p_c would increase the strength of the specimen from the standard strength q_u to the confined compressive strength q_c, Eq. 4. As explained before, the increase in strength is due to the fact that the confining pressure forces the two halves of the specimen together with a force p_c per unit of area of the total surface. If the voids of the specimen are filled with a liquid under pressure p_c, the pressure which is transmitted from right to left from solid to solid is not p_c as in jacketed specimens but only $p_c(1 - n_b)$. Hence, in order to compute the compressive strength q'_c of specimens without a jacket, one must replace p_c in Eq. 4 by $p_c(1 - n_b)$ whence

$$q'_c = q_u\left[1 + \frac{p_c(1 - n_b)}{q_t}\right] \dots (7)$$

Solving for n_b we obtain

$$n_b = 1 - \frac{q_t}{p_c} \times \frac{q'_c - q_u}{q_u} \dots (8)$$

This equation would make it possible to compute the boundary porosity of materials, such as porcelain, which fail only by splitting, but so far only jacketed specimens of such materials have been tested.

Most materials fail by pseudo-shear along surfaces of sliding. If the interstices along the surface of sliding represented by Fig. 6(a) are filled with a liquid under a pressure p_c, the normal pressure which presses the sliding part of the specimen onto its uneven seat is equal to $p_c (1 - n_b)$ per unit of area. Hence Eq. 6 must be replaced by

$$s' = \frac{c}{\cos \delta}$$
$$+ [p_c (1 - n_b) + \sigma_q] \tan \delta \dots (9)$$

Triaxial compression tests on specimens without jacket have been made only on concrete (Terzaghi (13)) and on Solenhofen limestone and marble (Griggs (3)). Both series of tests have shown that fluid pressures up to several hundred atmospheres have no measurable effect on the compressive strength of these materials, whereas equal pressures acting on jacketed specimens increase q_c very considerably. To account for the test results, it is necessary to assume that the value of n_b in Eq. 9 is almost equal to unity. In other words, the area of actual contact or merging between the individual grains could not have exceeded a small fraction of the total intergranular boundaries. Yet, since the volume porosity of both materials is very low, the voids must consist of very narrow, but continuous slits.

The presence of a continuous network of slits in crystalline and in clastic rocks in general is also suggested by the striking influence of drying on the compressive strength of rocks. Complete statistics regarding this influence are not yet available, but it has repeatedly been observed that the immersion of dry rock specimens, including specimens of granite and other igneous rocks with a very low volume porosity, reduces their compressive strength by amounts up to 10 per cent, in exceptional cases even by 20 per cent.[6] There are no known rock constituents which decrease in strength

[6] The effect of immersion on strength is commonly expressed by the ratio between the compressive strength of the saturated and the dry specimens. It is called the *coefficient of softening* η. A. Hanisch obtained for 11 different granites from Austria (cube strength in a dry state q_d = 1070 to 1923, average 1482 kg. per sq. cm.) values of η between 0.74 and 0.98, average 0.88. For nine crystalline limestones from Austria, q_d = 618 to 1417, average 981 kg. per sq. cm., he found η = 0.84 to 0.96, average 0.90. M. Gary tested 25 different granites from Sweden and Germany, q_d = 1163 to 3519, average 2440 kg. per sq. cm. and found η = 0.82 to 0.99, average 0.94. (More complete data can be found in a book by Hirschwald (4)).

upon contact with water. Therefore the loss of strength can only be due to one of the two following conditions. Part of the intercrystalline matrix consists of soluble material which goes into solution when the dry rock is immersed, whereby the intercrystalline connections are weakened, or else part of the strength of the rock is due to the surface tension in that part of the interstitial liquid which does not evaporate at room temperature. Immersion eliminates this source of strength as it eliminates the entire strength of a dry clay. If the boundary porosity n_b were equal to or only slightly greater than the volume porosity n (tubular or pocket-shaped voids), neither one of these two conditions could possibly account for a measurable loss of strength due to immersion. On the other hand, if the boundary porosity is close to unity, either one of the two processes could account for the observed phenomena. Hence the effect of drying on the compressive strength also indicates that the value of n_b is close to unity.

* * * * * *

REFERENCES

(3) D. T. Griggs, "Deformation of Rocks under High Confining Pressures," *Journal of Geology*, Vol. XLIV, pp. 541–577 (1936).

(11) M. Roš, and A. Eichinger, "Weitere Versuche zur Klärung der Frage der Bruchgefahr," *Proceedings* 3rd Intern. Congress Applied Mechanics, Vol. II, pp. 254–262, 1930 (Stockholm).

(13) K. Terzaghi, "Die Wirksame Flächenporosität des Betons," *Zeitschrift österr. Ing. u. Arch.-Ver.*, Heft ½, 1934.

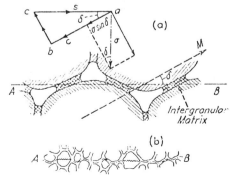

FIG. 6.—Magnified Section Through Surface of Shear in Granular Material.

Part IV

RESOLUTION OF THE MECHANICAL PARADOX

Editor's Comments
on Papers 16 Through 20

By the turn of the century, Bailey Willis had followed Horace Greeley's dictum and was mapping in the Northern Rockies. His report of the Lewis overthrust discovery followed soon thereafter (1902), fifteen years after McConnell's initial discoveries of overthrusts across the Canadian border. In subsequent years major overthrusts were discovered along the geosynclinal boundary of western Wyoming (Veatch, 1907; Schultz, 1914) and adjacent parts of Utah (Blackwelder, 1910; Billingsley and Locke, 1933, 1939; Nolan, 1935), Idaho (Richards and Mansfield, 1912; Mansfield, 1927), and Montana (Scholten et al., 1955). The incredible Heart Mountain and South Fork diverticulation faults along the Absaroka Mountain eastern boundary were identified within this period by Dake (1918); first thought to be part of the western Wyoming thrust belt, numerous puzzling characteristics were recognized by Dake, Hewett (1920), and subsequently by Walter Bucher (1933a). Geometric aspects of the problem required four decades of meticulous mapping by W. G. Pierce (1941, 1957, 1960, 1973), as well as a recently completed remapping of the Yellowstone Park area by Survey volcanologists, in order to approach a solution. The Heart

138

Mountain structure is thus revealed to consist of discrete, widely separated but often enormous rockslide fragments lying on a slope that probably never, on the average, exceeded 2°! Analogous landslide-like features were discovered by Reeves (1925, Paper 19) in the Bearpaw Mountains in Montana. Meanwhile, in Nevada, Longwell (1922, Paper 20, 1949; Longwell et al., 1965) had discovered the Muddy Mountain overthrust, and comparable structures were found by Merriam and Anderson (1942; cf. Gilluly, 1957), Misch (1957, 1960) and others. Elsewhere in North America, major thrusts were reported in the Ouachita Mountains of Arkansas and Oklahoma (Miser, 1924), but above all in the Appalachians, where research activity was intense. The Cumberland–Pine Mountain overthrust block, for example, was described by Wentworth (Paper 16), Charles Butts (1927), Rich (Paper 18), and with the aid of oil and gas well data, by Miller et al. (1947), Miller and Brosgé (1954), Stearns (1954), and Young (1957); other major thrusts were discovered throughout the Valley and Ridge Province. In the northern Appalachians and Newfoundland the concept of the Taconic overthrust was developing on the intuition and detailed mapping of Ruedemann (1909), Keith (1912, 1932), Swinnerton (1922), and Cady (1945); that concept has, of course, been more recently revived and extended, in particular by E-an Zen (1961, 1967, 1972; cf. Cady, 1968, 1969; Williams, 1975; St. Julien and Hubert, 1975; Voight and Cady, in press).

In the central Appalachian Piedmont, Ernst Cloos discovered imbricate thrust sheets and large-scale flowage (Cloos and Heitanen, 1941; Cloos, 1947), while Bailey and Mackin (1937) discovered nappes (recumbent folds) subsequently refolded in domes (cf. Mackin, 1950, 1962); similar structures were established by King et al., (1944) and Bryant and Reed (1962) in the Blue Ridge and Smokies, and by Thompson (1956; Thompson et al., 1968) and coworkers in the Green Mountains of Vermont.

In several of these studies, prior to 1960, some attention is devoted to mechanical aspects; the few excerpts presented here are helpful for perspective regarding the state of geological awareness prior to the Hubbert and Rubey papers. In a few cases (e.g., Paper 16) the mechanical issue seems to have been included as an afterthought. But the "paradox of overthrust faulting," alluded to previously by Smoluchowski, was gradually being recognized, as illustrated by the article by R. D. Oldham (Paper 17). Oldham, who had first discovered Himalayan overthrusts (1893), appealed qualitatively to a kind of progressive failure mechanism which later would be compared by Hubbert and Rubey (Paper 28, p. 129) to the locomotion of a caterpillar. Subsequently, Rich, in a critically significant article (Paper 18), showed how the whole Cumberland fault block structure was the natural consequence of thrust movement predominantly parallel to weak (shale) layers for long distances; local high-angle-fault segments were demonstrated to occur only where the major overthrust jumps to a higher-level stratigraphic horizon (cf. Raleigh and Griggs, Paper 30). He recognized the frictional problem of overthrust faulting, and suggested simply that both the material (e.g., clay shale vs. granite) and water content (wet vs. dry) may have a great deal to do with the frictional coefficient. Reeves appealed to a similar mechanism associated with bentonite layers (Paper 19); the montmorillonite-rich bentonite clay shales were particularly involved with stability problems in the nearby Fort Peck Dam construction operations, which Reeves does not cite but could certainly have been aware of. His study may represent the first statical consideration of an overthrust toe (cf. Raleigh and Griggs, Paper 30).

The frictional coefficient required by Reeves' analysis is extremely low, 0.02, which value stands in strong contrast with the assumption considered reasonable by Longwell in an important article published in the same period (Paper 20). Longwell emphasized friction coefficients as published in an engineer's handbook, and demonstrated the implications of the crustal "tilt" required if these friction coefficients were to be accepted.

Longwell considers, for his specific case, important concepts raised previously by Haarmann (1930) and Van Bemmelen (1936). Geodetic measurements had demonstrated geologically significant rates of vertical velocity in many areas of the world, and under certain circumstances such vertical movements could provide appropriate slopes for "low-angle" tectonic transport. Subsequently, possibly in association with a reversal in heat-flow patterns, these same "uplifts" could "revert to base," leaving few traces of their formerly enhanced position. The existing attitude of décollement are, clearly enough, not a *necessary* indication of slopes at the time of tectonic movement.

Longwell also called attention to the relatively rigid behavior of the Nevada thrust plate in contrast to supposedly analogous ductile landslides, but his citation of the 1925 Gros Ventre slide may prove to be ironic inasmuch as the Amsden formation, which forms the base of the Gros Ventre slide, is, in fact, a dominant thrust sole horizon (with, however, some possibly significant facies changes) in the Idaho–Western Wyoming Belt that was subsequently referred to by Longwell (cf. Heard and Rubey, Paper 35, p. 751).

Reprinted from pp. 351, 368–369 of *Jour. Geol.,* **29**, 351–369 (1921)

RUSSELL FORK FAULT OF SOUTHWEST VIRGINIA

CHESTER K. WENTWORTH

University of Iowa

[*Editor's Note:* In the original, material precedes this excerpt.]

In the course of his meditation on this study the writer has made very briefly a few computations, based on extremely general and only very approximate assumptions, which are given below. Their value is solely to indicate orders of magnitude, and it is hoped that they may serve, as they did in the case of the writer, to visualize the immensity of forces involved.

FORCE TO SHEAR AND FORCE TO THRUST

ASSUMPTIONS

Block 125 miles $\times$ 25 miles $\times \frac{1}{2}$ mile
Density 170 lbs. per cubic foot
Coefficient of friction, mean between rough and smooth granite, 0.60
Shearing strength 200 pounds per square inch
Average extent of overthrust, 6 miles

RESULTS

Force to shear block loose over entire area $= 25 \times 10^{14}$ pounds
Force to move block against friction on horizontal plane $= 23 \times 10^{15}$ pounds
Work done in moving block 6 miles at angle of 5 degrees $= 85 \times 10^{19}$ foot pounds
Equivalent to 420,000 horse-power working for 100,000 years
Estimated coal in block $= 50 \times 10^9$ tons

Burning of this coal would produce power enough to move the block 2.2 feet, assuming the usual engine efficiency. It has actually moved an average of at least six miles.

It is especially interesting to note that the force required to shear the block loose over the whole area is only about one-tenth of that required to produce motion against the resistance of friction. Since both forces are proportional to area and only one—that of motion against friction—proportional to thickness, we find that for a block of any area and of a thickness of 287 feet, according to the conditions assumed, the shearing force is just equaled by the force to overcome friction, and as thickness is greater than this amount the latter force is greater in proportion. It is evident, then, that in the case of most overthrust faults the motion of the rock involved against the resistance of friction is more impressive than the production of the break which separated it.

17

Reprinted from pp. 77, 88–90 of *Quart. Jour. Geol. Soc. London,* 77, Pt. 1, 77–92 (1921)

KNOW YOUR FAULTS

ΓΝΩΘΙ ΣΕΑΥΤΟΝ

R. D. Oldham

Custom has decreed that on these occasions your President shall deliver an address, which is usually devoted to a review of the past history, of the present condition, or of the future needs of some department of Geological Science. To-day I propose to follow neither of these courses, but to make a digression into the philosophy of our science, to examine the meaning of some of the words which we use, and to take for my text that motto which, blazoned in letters of gold from the ancient temple of Delphi, may be translated by geologists as ' know your faults.'

[*Editor's Note:* A row of asterisks indicates that material has been deleted.]

From the consideration of these two matters which have given rise to controversy, concerned almost entirely with words, by which things that really matter may be described, I now come to one which is a vital one, for it may involve a modification, and in some respects a radical change, in some of the fundamental principles, which have rather been tacitly accepted than definitely proved. In discussions, as in descriptions, of the phenomena or of the origin of these overthrusts, the masses involved have generally been regarded as passive, moving under the influence of external forces in the production of which they took no part. The notion is a natural one, it is the simplest and easiest way of interpreting the facts of observation; but its general acceptance must be very largely attributed to the influence of experiments on a small scale, which have themselves been suggested and directed by the hypothesis which they were intended to illustrate and investigate. In these we have an inert mass, variously composed to imitate, more or less, the rocks of the Earth's crust, and this mass is subjected to deformation by the application of external forces. In this way many of the structures which have been worked out by geological observations in the field were imitated on a small scale in the experiment, and the resemblance was accepted as evidence that the large-scale structures, met with in Nature, were produced, like the small-scale structures of the experiment, by the application of external forces. Difficulties, however, arise when we consider the conditions which are introduced by an increase of dimensions to the scale of Nature; and, when the mechanics of overthrusts are investigated, these difficulties become insuperable.

When one body is pressed against another by any force at right angles to the surface of contact, it may be caused to move by another force acting at right angles to the first, and the magnitude of the second force needful to produce movement bears a definite ratio to the first, a ratio which depends on the nature of the material and the character of the surface of separation.   This ratio is known as the 'coefficient of friction,' and is, numerically, the same as the tangent of the angle of repose.   For a flat-dressed surface of stone the coefficient is about three-fifths of the weight of the stone, for a surface such as that of a so-called 'thrust-plane' it would not be less: consequently, to move a block of rocks 5 miles wide would need a pressure equal to that due to the weight of a column, of the same rocks and of the same cross-section, having a height of at least 3 miles, or just about the limit of height of column which average hard rock can bear without crushing.

From these figures it appears that the maximum possible width of the overthrust must be somewhere about 5 miles, if it moved as an inert mass under the influence of some external impulse : for, if the width exceeded this limit, the stresses would be greater than those which rock could bear or transmit, and relief would be found in some other way than by a general displacement along the whole width of the overthrust ; but 5 miles is less than half the width of the mass moved in the Highland over-thrusts, it is not more than a tenth of that of the Scandinavian, and a still smaller fraction of those which have been deduced in the region of the Alps.   From this it might seem to be established that none of these overthrusts could possibly have been produced, and that there must be some error in the obser-vations, or the inferences which have been drawn from them as to structure.

This reasoning, however, is not justifiable.   We have again a case very like that which has been mentioned in connexion with what are ordinarily understood as reversed faults, and once more we have to face the alternative that the hypothesis of origin needs correction, not the facts of observation ; but, before examining this, it is necessary to refer to one possible means of getting over the difficulty which has been encountered.   If we might believe that the coefficient of friction along the surface of the thrust was less than that adopted in the calculation, the width of the blocks which could be moved would be correspondingly increased ; but not in this way can sufficient increase be obtained, for even with the most perfectly formed and lubricated surfaces in mechanism the coefficient is not materially less than one-tenth, and the maximum width of block which could be moved would not be increased beyond about 30 miles. The actual surface along which movement took place being, to say the least, much less perfect than those which give so small a coefficient of friction, the maximum width that could be moved would in any case be less than has, in some instances, been shown by observation in the field. Resistance to movement might, however, be reduced if the downward pressure due to the weight of the upper block were, in some way or other, temporarily relieved, and if this relief were complete there would be no limit to the width of block that could be moved. It is not easy to conceive the means by which this could be brought about, nor is it necessary to consider the possibility, for the existence of mylonites, and other indications, of resistance to movement, given by the deformation and fracture of rock, are eloquent of the resistances which had to be overcome when the existing displacements were brought about. Taking these into consideration, it is evident that the frictional resistance must have been at least as great as is represented by the coefficient made use of, so that the width of 5 miles must be regarded as a maximum rather than a minimum limit of the width of the overthrust which could be moved by pressure from without.

From this we are led to the conclusion that the thrusts did not move simultaneously over the whole of their extent, but partially, first in one part then in another, each separate movement involving an area limited by the strength of the rocks and their power to transmit, or resist the effect of, pressure. Some years ago it might have been said that any supposition of this kind was physically impossible; but at the present day the change of volume which results from an alteration of the molecular grouping of the same chemical elements, expressed geologically as a different mineralogical constitution of rocks having the same chemical composition, or more briefly as a change of mode of the same norm, has opened up at least one means by which the desired effect might be produced. Doubtless the advance of knowledge will open up other possibilities, some of which might be indicated, though I shall not refer to them, as my present purpose is not to deal with things themselves, but with the words in which they are expressed.

* * * * * *

18

Reprinted from pp. 1584, 1587, 1589, 1595–1596 of *Bull. Amer. Assoc. Petrol. Geol.*, **18**(12), 1584–1596 (1934)

MECHANICS OF LOW-ANGLE OVERTHRUST FAULTING AS ILLUSTRATED BY CUMBERLAND THRUST BLOCK, VIRGINIA, KENTUCKY, AND TENNESSEE[1]

JOHN L. RICH[2]
Cincinnati, Ohio

ABSTRACT

The structural relations of the Cumberland overthrust block are such as would occur if gliding on the thrust plane took place parallel with the bedding along certain shale beds in such a way that the thrust plane followed a lower shale bed for some distance, then sheared diagonally up across the intervening beds to a higher shale, followed that for several miles, and again sheared across the bedding to the surface.

Reasons are given for the belief that subsidiary faults and folds within the block are superficial and do not extend below the thrust plane. This possibility should be borne in mind when exploration of such structures for oil or gas is contemplated.

Study of the Cumberland block throws new light on the broader problems of the nature of folding and faulting in the sedimentary rocks bordering great mountain ranges and on the function of friction in setting limits to the distance through which overthrust blocks can be moved.

[*Editor's Note:* Material has been omitted at this point.]

EXPLANATION OF CUMBERLAND THRUST

The peculiar structural features of the Cumberland block, namely, the flat-topped Powell Valley anticline, the sharp monoclinal structure at Cumberland Mountain, the flat-bottomed Middlesboro syncline between Cumberland Mountain and Pine Mountain, and the

Fig. 4.—Diagram representing course of incipient thrust plane in series of sedimentary rocks. Plane follows beds of easy gliding, such as shales, and breaks diagonally across more brittle beds from one shale to another, and, finally, up to surface.

apparent warping of the thrust plane, all fit into a consistent picture when the thrusting is considered as having taken place along and across the bedding of the sedimentary rocks along some such path as that represented by the dotted line, *CD*, of Figure 4, and the whole region as having been subsequently tilted slightly toward the north.

* * * * * *

[1] Presented before the Association at the Houston meeting, March 24, 1933. Manuscript received, May 4, 1934.

[2] Department of geology, University of Cincinnati.

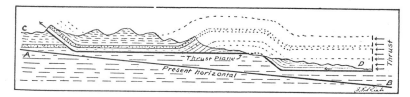

FIG. 5.—Diagram showing result of movement on thrust plane such as that shown in Figure 4. Tilting so that horizontal is represented by line AB and erosion down to line CD gives structure and topography to be compared with present cross section of Cumberland block (Fig. 3).

BROADER BEARING ON GENERAL PROBLEM OF OVERTHRUST FAULTING

Revelation of the way in which thrust faults may follow for long distances along the bedding of strata, such as shales, on which movement is easy, completely changes the basis of calculations of the possible distance the rocks may move on overthrust faults. In the past such calculations have been based on the coefficient of friction of dry granite. The friction involved in movement along the bedding of clay shales, probably wet with water at the time of movement, is something of an entirely different order than that for granite.

Wentworth, for example, made a calculation of the force required to move the Cumberland block,[15] but he used the coefficient of friction of dry granite. His results would have been radically different had they been based on movement along the bedding of shales.

Recognition of the importance of thrusting along bedding planes throws a new light on many problems in thrust faulting. For example, the striking fact that in crossing the Appalachian valley southeastward from the Cumberland block toward the great up- and overthrust mass of the crystalline Appalachians one passes half a dozen or more large thrust faults which bring up repeatedly, shingle-fashion, almost the whole of the unmetamorphosed Paleozoic sedimentary series without once bringing up the crystalline basement, makes it appear likely that the thrusting in that part of the Appalachian valley is entirely confined to the sediments, which have been sheared off from the underlying crystalline basement, pushed forward, and piled up in shingle-fashion by a great plunger moving from the southeast—presumably the pre-Cambrian mass of the crystalline Appalachians—which came up diagonally along a shear plane to the base of the sediments, then rode forward horizontally, pushing and piling up the sediments before it. It is not unlikely that the pre-Cambrian basement beneath the Appalachian valley would be found to be undisturbed.

[15] Chester K. Wentworth, "Russell Fork Fault of Southern Virginia," *Jour. Geol.*, Vol. 29 (1921), pp. 351–69.

19

Reprinted from pp. 1033–1034, 1042, 1043–1045 of *Geol. Soc. America Bull.,*
57(11), 1033–1048 (1946)

ORIGIN AND MECHANICS OF THE THRUST FAULTS ADJACENT TO THE BEARPAW MOUNTAINS, MONTANA.

BY FRANK REEVES

ABSTRACT

The hypothesis is presented that the thrust faults in the plains adjacent to the Bearpaw Mountains in north-central Montana were produced by a plainsward sliding of the volcanic rocks and underlying strata down the flanks of the low sedimentary arch on which the volcanics rest. It is predicated that this slipping took place on a bed of bentonite in the Colorado shale and was aided by earth tremors that accompanied the last stages of the volcanic outbursts. The flat-lying Upper Cretaceous and lower Tertiary strata in the adjacent plains which, if more rigid, would have obstructed the plainsward sliding of the tilted mountain mass or been overridden by it were shoved plainsward and thrust-faulted for a distance of 20 to 30 miles.

The faults show no progressive plainsward decrease in amount of thrust across the broad area in which they occur and are separated by wide belts of undeformed flat-lying strata. These anomalous features are attributable to slight frictional resistance to movement offered by a slippery bentonite bed and the probability that the sedimentary rocks over part of the faulted belt were overlain and strengthened by a cover of volcanic rocks that thinned toward the outer margin of the faulted belt. The deformed strata consist chiefly of shales that were too weak to support the load involved in lifting the strata in anticlinal folds. Therefore, thrust faulting, instead of anticlinal folding, took place.

$$* \quad * \quad * \quad * \quad * \quad *$$

Quantitative estimate of forces.—If slipping occurred on $ABCD$ of Figure 6, it was caused by the component of gravity of M acting parallel to AB. If this force produced the faulting in N, at its maximum, it must have been equal to, and not materially greater than, the frictional resistance encountered in M sliding on AB and N sliding on BD, plus the force necessary to produce the outermost fault at D. If $G =$ weight of M, or 55,000,000 tons; $G' =$ weight of N, or 58,000,000 tons; $C =$ coefficient of statical friction; $a =$ inclination of M to the horizontal, or 3°; $R =$ force necessary to produce the fault at D, or 500,000 tons[2], then the component of gravity causing the slipping $= G \sin a$ and the frictional resistance on AB and $BD = CG$ cos a and CG', respectively, and $G \sin a = CG \cos a + CG' + R$ (*1*) and

$$C = \frac{G \sin a - R}{G \cos a + G'} \ (2).$$

Substituting in (2):

$$C = \frac{2,878,700 - 500,000}{54,924,650 + 58,000,000} = .0216 \qquad (3)$$

---

[2] The value of R is assumed to be equal to the crushing strength of the rocks in the forepart of N. The shales in the section are regarded as having practically no crushing strength. The sandstones have a total thickness of approximately 1200 feet but are not well indurated and are assumed to have a crushing strength of around 6000 pounds per square inch, or 432 tons per square foot. On these assumptions the total crushing strength of the forepart of N is 1200 times 432 tons, or approximately 500,000 tons.

[*Editor's Note:* A row of asterisks indicates that material has been deleted.]

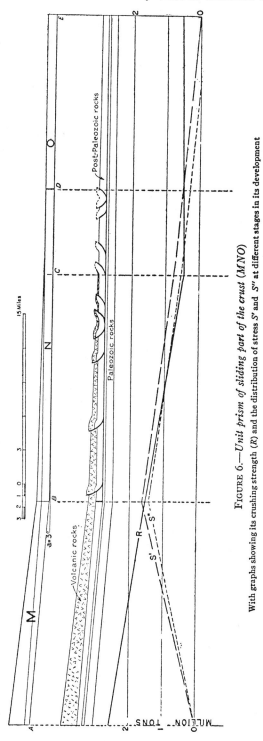

FIGURE 6.—*Unit prism of sliding part of the crust (MNO)*

With graphs showing its crushing strength (R) and the distribution of stress S' and S'' at different stages in its development

Assuming a value of 0.0216 for the coefficient of friction, M, if unopposed by N, would have slipped plainsward on a surface whose angle of slope had a tangent value of 0.0216 or an angle of approximately 1°.   Consequently, the amount that M was tilted above this angle gave it the force to shove N forward on the surface BD and produce the thrust faults.   This was evidently equal to the total force exerted by the component of gravity of M acting parallel to the surface of slip minus the frictional resistance encountered by movement on that surface. $G \sin a - CG \cos a = 1{,}692{,}-000$ tons.   The horizontal component of this force, effective in shoving N forward, is equal to $\cos a$ times 1,692,000 tons, or approximately 1,600,000 tons.

A graphic representation of stresses in a unit prism of the sliding mass and their relation to the crushing strength of the rocks is presented in Figure 6.   Graph R of this figure represents the crushing strength of the rocks in any vertical section of the unit prism.   It is based upon calculation of the total crushing strength at points A, C, and D.   The total crushing strength at D, as previously estimated, is 500,000 tons.   Beyond the area formerly covered by volcanic rocks the thickness of the sliding part of the crust probably varied little, and the value of R is assuméd to be constant between C and E.   The total crushing strength of the rocks increases regularly from the former outer margin of the volcanic rocks at C to their thickest part, in the center of the mountains, where it is roughly estimated to be five times its value at D, or 2,500,000 tons.

The theoretical distribution of the stress when it attained its maximum intensity is shown in graph S'.   It is based upon a determination of the values of S', at A, B, C, D, and E.   Its value at A is zero.   Calculations made above give its value at B as 1,600,000 tons and at D as 500,000 tons.   The graph between B and E is not a straight line because more force is required per unit distance in shoving forward the part of N laden with volcanic rocks than the part not so laden.   The value of S' at C equals its value at B minus the frictional resistance between B and C, which is roughly 900,000 tons.   Its value at C therefore equals 1,600,000 tons minus 900,000 tons, or 700,000 tons.

The value of S' in O, the part of the crust that is only elastically compressed, decreases from 500,000 tons at D to zero at some distant point E.   The distance DE or the length of O is governed by the frictional resistance encountered in the slipping that accompanies the elastic shortening.   It may be determined as follows:

Let F be the maximum force applied to O in cross section, P the weight of the rocks elastically shortened by this force, and C the coefficient of friction then, by definition,

$$C = \frac{F}{P} \qquad (1)$$

$$F = PC \qquad (2)$$

If d be the length and h the thickness of the rocks in O and w their weight per cubic foot (0.08 tons) then

$$P = dhw \qquad (3)$$

Substituting in (2),

$$F = dhwC \qquad (4)$$

$$d = \frac{F}{hwC} \qquad (5)$$

$F = 500{,}000$ tons; $h = 3{,}500$ feet; $w = 0.08$ tons; $C = 0.0216$.   Substituting these values in (5), $d = 82{,}600$ feet or approximately 15 miles.

*   *   *   *   *   *

150

20

Reprinted from pp. 426–432 of *Amer. Jour. Sci.,* 243A (Daly Volume), 417–447 (1945)

THE MECHANICS OF OROGENY

C. R. Longwell

* * * * * *

Some geologists consider sliding under the urge of gravity an attractive mechanism as applied to specific tectonic masses. More than one student has suggested that the structure of the Swiss Alps would present less difficulty if certain far-travelled "rootless" nappes, such as those of the Prealps, could be accepted as gigantic slides. There is no question that the component of gravity acting laterally on a tilted stratified section is adequate for the work of deformation, provided the angle of tilt is large enough on a sufficiently long slope. Quantitative evaluation of these elements requires assumed values for

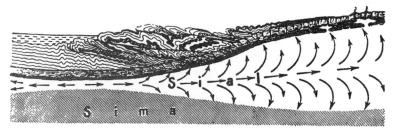

Fig. 7. Schematic representation of "free gliding" on flank of a "geo-tumor." (After Haarmann, Ferdinand Enke, 1930.)

several variables, and proponents of the Haarmann scheme who discuss the problem in other than vaguely qualitative terms usually assume special conditions under which abnormal values obtain. Without doubt submarine slides occur on low slopes, involving strata that are weak and so saturated with water that the coefficient of friction is almost negligible. Deformation of this kind can ordinarily be distinguished from results of orogeny. The Paleozoic formations in the Cordilleran geosyncline were not weak, partially lithified deposits when orogenic forces belatedly overtook them. In Cretaceous time the thick quartzites and carbonate rocks in southern Nevada were hundreds of millions of years old, and surely were as strong as they are at present. The Beltian rocks in the Lewis thrust plate offer a similar illustration, among many that can be cited from mountain zones the world over.

To present the problem graphically, Fig. 8 is drawn repre-

[*Editor's Note:* A row of asterisks indicates that material has been deleted.]

senting a tilted sedimentary section 4 miles thick and 20 miles long. A shale formation lies near the base of the section. The force available for sliding is the component *t* of gravity, or *w sin a* where *w* is the weight of the block and *a* the angle of tilt. If *f* is the coefficient of friction on the shale, the frictional resistance is *fn*, or *fw cos a*. Assuming that the section farther down the dip has been removed, say by erosion or by downfaulting, silding on the shale may occur whenever *sin a* slightly exceeds *f cos a*. The value of *f* therefore is critical. Since we are considering a section of old indurated deposits,

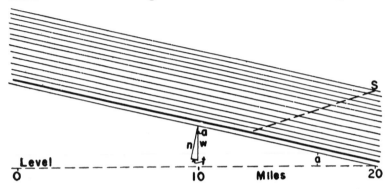

Fig. 8. Inclined stratigraphic section 4 miles thick, with shale (black) near base. Sliding can occur only if angle *a* is steep enough to overcome frictional resistance and to cause folding, or shearing along some such plane as *S*. w=weight of block, *t*=component parallel to bedding, *n*=component normal to bedding.

and the shale is far below the zone of abundant ground water, there is no apparent reason for an abnormally low coefficient of friction. Development of isoclinal drag folds and slaty cleavage below the Johnnie thrust (**Fig. 4**) suggests large frictional resistance. Engineers' handbooks give the following experimental values of the coefficient for materials that may apply in the present problem:[2]

Kind of Surface	Coefficient of Friction
Hard limestone on hard limestone, both well-dressed ..	0.38
Granite, soft-dressed upon soft-dressed	0.70
Granite, hard-dressed upon hard-dressed	0.55
Common brick on well-dressed hard limestone	0.60
Masonry on dry clay	0.51
Masonry on wet clay	0.38

[2] Troutwine, John C.: 1922, The Civil Engineer's Pocket-Book, p. 411; Merriman, T., and Wiggin, T. H.: 1930, American Civil Engineers' Handbook, p. 894.

Since the laboratory determinations involve plane surfaces that are almost ideally smooth, the resulting values probably are lower than for corresponding materials on a fault surface. Nothing in the experimental data justifies a value as low as 0.3, and yet with this low value the downslope component of gravity will barely overcome frictional resistance when the angle of tilt is 16°42′. However, we are considering the most favorable case, with no resistance other than that caused by friction on a surface of sliding. For the Nevada thrusts enormous energy was required to buckle and shear across the thick section, before movement became possible. Moreover, each plate had to be forced up a "ramp" inclined against the movement; for the Wheeler Pass thrust (Fig. 4) the "ramp" must have measured 10 miles from bottom to top. In addition, after the "ramp" was surmounted the Muddy Mountain plate was propelled at least 15 miles across uneven topography underlain by sandstone, with a coefficient of friction which, on the basis of laboratory data, must have been at least 0.6.

With an assumed inclination of 16°42′, any horizon at the left end of the section in Fig. 8 stands 30,000 feet above the same horizon at the right end. If the reasonable value 0.6 is assumed as the coefficient of friction for limestone on standstone in the Muddy Mountain thrust, inclination of 31° would be required to overcome frictional resistance alone. In other words, the western edge of the 15-mile-wide remnant now exposed must have stood nearly 8 miles higher than the eastern edge; and since the plate that now lies on younger rocks formerly lay in place altogether west of its present site, the full uplift to produce an inclined plane for gravitational sliding would have had fantastic height.

Arguments for gravitational movement of thrust plates commonly cite the migration of landslide masses on low slopes. The supposed analogy calls for critical analysis. Subsurface water plays an essential rôle in all landslides. Most sliding and creeping masses contain abundant earthy material, which in combination with water acts as a lubricant to facilitate sliding on definite surfaces and in many cases also produces earthflow in the body of the mass. Even masses that consist chiefly of bedrock, such as the Gros Ventre slide of 1925, are in the zone of abundant ground water and rest on clayey strata which when saturated offer exceptionally low frictional resistance. Nothing analogous to earth-flow has occurred in thrust masses made of strong rock. Except for localized imbrication, the strata in each of the great Nevada plates are almost undisturbed, indicating that the plate moved essentially as an intact mass. Moreover, even if the base of a plate that is many thousands of feet thick were subject to saturation by ground water, frictional resistance to movement over such a formation as the Jurassic sandstone of southern Nevada would remain high. The general problem is not to be solved by assuming favorable conditions that do not apply to numerous known field examples.

Engineers recognize that repeated jarring of an inclined mass will induce slow movement by small increments, at angles appreciably less than the "angle of friction." To provide a margin of safety, the theoretical angle is decreased by 5° in designing structures whose stability depends on frictional resistance under jarring disturbances. Sharp jarring which doubtless occurs in all active orogenic zones might by countless repetitions become a factor of some importance in the movement of sliding masses. However, a decrease of the theoretical angle by 5° or even 10° does not reduce to reasonable limits the measure of uplift required to explain the Muddy Mountain thrust as the product of gravitational sliding, even if all resistance other than that due to friction is ignored.

It has been suggested that a thrust moves not merely by sliding on a basal shear-surface but in large part by plastic creep within a zone of limited thickness directly above this surface. According to this view the zone of mylonite and breccia commonly found at the base of a thrust mass behaved like tractor treads equipped with slow-motion roller bearings for transport of the rigid mass above. The suggestion may have merit; but the supposed mechanism demands adequate motive power. At the bottom of a stratified section 4 miles thick the confining pressure is about 21,000 pounds per square inch; and with a tilt of 10° the differential pressure parallel to the base would be about 3600 pounds per square inch. Only an extremely weak rock would yield by plastic creep under a shearing stress so low. Many types of rock, including strong limestone, have an elastic limit several times greater than 3600 pounds per square inch. As Griggs (1936) states, "Below the elastic limit it appears that the rock will not fail, no matter how long the force is applied. Also, no permanent set occurs within the limits of sensitivity of the present measurements." Since the

Muddy Mountain plate consists of strong dolomite and lime-stone, and the Johnnie plate of thick-bedded quartzite, the angle of inclination sufficient to cause plastic creep in either mass would be fully as extreme as that ·required for sliding on a basal surface.

There are still those who argue, in effect, that work in a geologic sense must be defined with a factor of time which, if large enough, compensates for any deficiency of the force involved. Everything known about mechanical principles indicates that this concept is fallacious. The point is discussed further on a later page.

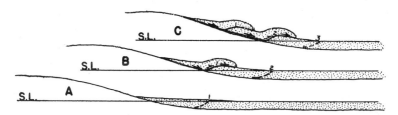

Fig. 9. Schematic representation of successive bedding thrusts developed in a geosynclinal prism (stippled) as a "geotumor" or "undation" migrates from left to right. Horizontal and vertical scales approximately equal. S. L., sea level. *1, 2, 3,* successive thrusts in the stages of development *A, B, C.*

In an effort to save something of value in the sliding concept, let us assume that a "geotumor" advances steadily across the position of a geosyncline, providing opportunity for a succession of slides and for repeated movement of any one plate, as shown in Fig. 9. In *A* the angle of tilt is not steep enough for sliding, but the front of the first potential thrust plate is shown with broken line. In *B* this plate has moved forward, and the front of a second thrust mass is indicated. In *C* the second mass has moved, and the first plate has kept pace by "riding on the back" of the first. Thus eventually the over-lapping thrust sheets may affect the entire width of the sedimentary prism. This mechanism in vague, qualitative outline may appear to have merit, but it does not survive a quantitative test. If the first two plates formed have the dimensions of the Bannock and Absaroka plates in Idaho and Wyoming, together they measure at least 40 miles from front to rear. The incline on which the two plates slide together in

Fig. 9, *C* has a slope of 19°—surely a minimum for the required energy, in view of the analysis given above. Such a slope of the necessary length would demand a "geotumor" at least 13 miles high. This height would have to increase as other thrust plates develop, unless each plate is to slide away from the next older, leaving a wide tensional gap between. There are no such gaps, which would reveal belts of older rocks alternating with the thrust masses, in the Idaho-Wyoming area and in similar structural units that have escaped partial concealment by subsequent block-faulting and sedimentation.

In a variation of the general mechanism assumed above, the "geotumor" might migrate continuously across the sedimentary prism from east to west. If the steep slope were toward the east, each sliding mass in its turn would be torn from the prism and would leave behind it a tensional gap, which would be wholly or partially healed as the next succeeding mass slid forward. With ideally uniform action of the travelling welt, the final result might be several overlapping major plates, with no visible evidence of the tensional scars. It is hardly conceivable, however, that irregularities in a long orogenic belt would not betray the operation of such a mechanism. Moreover, since considerable time would be required for the full succession of slides, and since the assumed conditions would favor rapid erosion and sedimentation, each plate following the first should override and deform a series of young beds laid down in the tensional gap in front of it. However, the primary difficulty encountered by either version of the "rolling geotumor" hypothesis lies in the concept of an incline adequate for the movement of a single intact plate as wide as either the Muddy Mountain or the Johnnie remnant. If each of these plates consisted of several imbricate slices, which might have slid in succession down the front of a travelling welt, the hypothesis could be considered as reasonable. However, each of the plates was propelled as a unit.

Thus the Haarmann theory of gravitational sliding, as a self-sufficient explanation of orogeny, seems to fail under an elementary test. However, we are not therefore justified in discarding the concept as of no possible value. A tendency to slide on the long flanks of geanticlinal uplifts may have reduced appreciably the primary force required for thrusting and folding in thick stratified sections. The mechanism of thrusting is difficult to understand at best, and the explanation may lie in some combination of forces and conditions that are invoked separately by current hypotheses.

* * * * * *

Editor's Comments
on Paper 21

21   **GOGUEL**
Introduction à l'étude mécanique des déformations de l'écore terrestre

The publication of Jean Goguel's *Memoire* in 1943 (Paper 21) ought to be regarded as a milestone in the history of tectonic theory. A systematic overview, solely concerned with the application of mechanical theory to tectonic problems, was formally attempted. Goguel had already considered theory and conducted extensive experimental deformation experiments in the pre-World War II era. In this work he distinguished carefully between plastic and viscous deformation, and considered the significance of similitude to such tectonic problems as *tectonique d'écoulement* (gliding tectonics) and salt domes. Goguel is no physicist blindly tossing around equations appropriate only in highly oversimplified and often irrelevant models. His mechanical approaches seem modern, his appreciation of details of field relationships are refined, and his awareness of idealization error is acute: the result is an uncommonly sophisticated treatise.

In the excerpt presented as Paper 21, Goguel shows that the limit angle associated with a gliding nappe is a function of its size (i.e., height). Choosing a yield point of 40 bars for the gypsum-bearing sole horizon (this is not a wholly arbitrary choice; *see* Goguel, 1960) and assuming ideal plasticity (i.e., yield strength independent of mean pressure), Goguel demonstrates that whereas a 1-km-thick slab requires an 8.5° slope to achieve instability, a 3-km-thick slab slides on a 3° slope. Following the same logic, a 6-km slab can slide down a slight incline of 1.5°; indeed, at this stage one may wonder if the overthrust paradox has not disappeared.

21

Reprinted from pp. 406–408 of *Introduction à l'étude mécanique des déforma-tions de l'écorce terrestre* (Mémoires pour servir à l'explication de la carte géologique détaillée de la France), 2nd ed., Imprimerie Nationale, Paris, 1948

INTRODUCTION À L'ÉTUDE MÉCANIQUE DES DÉFORMATIONS DE L'ÉCORCE TERRESTRE

J. Goguel

EXEMPLES NUMÉRIQUES.

Pour étudier une déformation de la couverture, on devra, comme nous venons de l'indiquer, tenir compte de trois causes de résistance, dont chacune absorbe un travail déterminé, pour une déformation donnée : la pesanteur, la rigidité des roches, le frottement à la base. De ces trois facteurs, le premier seul se prête à un calcul absolu; pour les deux autres, le résultat dépend de la valeur d'un seuil de plasticité, dont les déterminations au laboratoire, outre qu'elles sont peu nombreuses, prêtent le flanc à de graves critiques de principe. Aussi doit-on profiter de toutes les occasions de comparer directement le travail correspondant à la pesanteur, et celui absorbé par la déformation ou par le glissement à la base.

Considérons, par exemple, une série qui glisse dans son ensemble sur un socle horizontal, et qui forme plusieurs plis successifs. La résistance au glissement, dans l'intervalle entre les anticlinaux, a pour conséquence que l'effort de compression n'a pas la même valeur pour les anticlinaux successifs. Si le

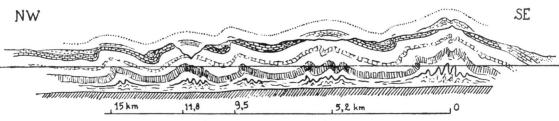

NW — SE

15 km — 11,8 — 9,5 — 5,2 km — 0

Fig. 150. — Coupe du Jura, au Sud de Gex (d'après de Margerie, Le Jura, *Mém. Carte Géol. Fr.*)

glissement se fait toujours dans le même sens, les anticlinaux successifs auront subi une compression de moins en moins importante, et il y a des chances pour qu'ils soient de moins en moins élevés.

Tel est précisément le cas dans le Jura, où le premier anticlinal, en partant de la Plaine Suisse, est généralement le plus important, les chainons suivants décroissant régulièrement. La figure 150 donne une coupe du Jura, prise un peu au Sud de Gex et de Saint-Claude, et qui montre nettement cette décroissance. Nous avons rétabli, sur cette coupe, les couches enlevées par l'érosion, pour permettre le calcul du travail absorbé. Les cinq anticlinaux que montre cette coupe ne sont, ni très serrés, ni réduits à des simples amorces. Nous admettrons donc que le travail absorbé par la déformation des roches, pour une accentuation de ces différents plis, est proportionnel à la contraction Δx.

Le travail absorbé par la pesanteur (travail non compensé) est, pour chaque anticlinal, égal au produit de la contraction par l'épaisseur de la série, que nous évaluons à 2.500 mètres, et par la pression due au poids d'une épaisseur de terrain égale à la hauteur, au-dessus de la surface non plissée, du centre de gravité de l'intumescence supplémentaire due à l'accentuation du pli. Assimilant la section du pli à une parabole, de base constante, on trouve que la hauteur du centre de gravité de l'augmentation de section est les quatre cinquièmes de la hauteur du pli.

Enfin, la différence du travail absorbé par le glissement à la base, suivant que l'un ou l'autre de ces anticlinaux s'accentue, est égale au produit de Δx par la distance et par la valeur du seuil de plasticité des couches de base.

Si les plis se sont terminés dans une même phase tectonique, sous l'action d'une même compression d'ensemble, ils doivent être en équilibre, c'est-à-dire que l'accentuation de n'importe lequel d'entre eux absorberait le même travail. De ce travail, nous ne calculerons que la partie qui diffère de l'un à l'autre des plis. En évaluant leurs hauteurs à 1.500, 1.200, 800, 600 et 500 mètres respectivement et les distances d'axes en axes à partir du premier à 5 km. 2, 9 km. 5, 11 km. 8 et 15 kilomètres, on trouve que le travail absorbé par la pesanteur, en kilogrammètre par mètre de contraction, pour une longueur de 1 centimètre, suivant la direction de la chaîne, est respectivement 810.10^5, 590.10^5, 430.10^5, 320.10^5 et 270.10^5. D'autre part, le supplément du travail de glissement à la base absorbé par les anticlinaux autres que le premier, est donné par $5,2.10^5$ S, $9,5.10^5$ S, $11,8.10^5$ S et 15.10^5 S. On trouve immédiatement que pour S = 40 kg./cm² la somme de ces deux expressions a la même valeur pour tous les plis.

Cette valeur du seuil de plasticité d'une assise plastique de base — vraisemblablement les couches salifères du Trias moyen — présente le gros intérêt d'être indépendante de toute mesure de laboratoire, si bien que son ordre de grandeur au moins — car les calculs ci-dessus ne peuvent prétendre à une grande précision — est parfaitement établi. Cet ordre de grandeur n'est d'ailleurs pas très différent de celui que l'on aurait pu déduire des expériences de laboratoire.

Partant de cette valeur $S = 4o$ kg/cm² on trouve immédiatement que l'inclinaison de glissement limite est telle que $\sin \alpha_0 = \frac{15o}{Em}$. Pour une série épaisse de 1.000 mètres, on a $\alpha_0 = 8°\ 3o'$, pour 2.000 mètres, $\alpha_0 = 4°\ 2o'$ et pour 3.000 mètres, $\alpha_0 = 2°5o'$.

Pour nous rendre compte de la force que pourrait produire ce glissement, considérons une série épaisse de 2.000 mètres, inclinée de $\alpha = 8°$, et admettons que la plus grande partie de la résistance de cette série soit due à un étage calcaire, épais de 3oo mètres, et caractérisé par un seuil $S = 1.000$ kg/cm² (calcaire très rigide). Admettons qu'il existe déjà des ondulations de hauteur $h = 3oo$ mètres. L'accentuation de ces ondulations absorbe un travail $2 \times 1000 \times 3.10^4 \times 0,5$ kgm. par mètre de contraction (et par centimètre de longueur en direction). Le travail produit par le glissement d'un mètre d'un panneau de largeur L étant

$$\mathfrak{C} = 2.1o^3 \times 0,27 \times L \times (\sin 8° - \sin 4°2o') = 4o\ L,$$

on doit avoir : $L = 3.10^7 : 4o = 7,5.10^5$ cm. $= 7$ km. $5oo$, ce qui correspond à une dénivellation de 1.05o mètres seulement. Comme nous l'avions annoncé à la fin du chapitre précédent, le glissement d'une série épaisse, sur une assise salifère de base, est susceptible de produire des dislocations à sa partie aval, même pour une dénivellation très modérée.

[*Editor's Note:* Material has been omitted at this point.]

Editor's Comments
on Papers 22 Through 24

22   **HUBBERT**
      Mechanical Basis for Certain Familiar Geologic Structures

23   **HAFNER**
      Stress Distributions and Faulting

24   **ODÉ**
      Faulting as a Velocity Discontinuity in Plastic Deformation

In 1946 King Hubbert, then with Shell Development, constructed a squeeze box with a glass front with which to conduct sand-box experiments of the Nádai type for the benefit of members of the Shell staff concerned with Gulf Coast normal faulting (Hubbert, 1972, p. 10). In addition, an internal report was prepared describing the application of "Mohr theory" (actually Coulomb theory employed in association with Mohr's stress circle) to geological deformation. At about the same time the 1942 edition of Anderson's *The Dynamics of Faulting* came to Hubbert's attention; that Anderson's analysis was equivalent to Hubbert's became immediately obvious, and the book was passed on to Hubbert's Swiss-born colleague, Willy Hafner, who reviewed it in an internal report subsequently used along with Hubbert's in Shell training courses. Publication of both reports was contemplated inasmuch as the subjects discussed were still apparently not common knowledge among geologists, despite previous publications in the English language by Anderson (Paper 12, 1942) and Nádai (Paper 13); however, because Hafner's was in essence only a review, he chose to greatly expand his work. Hafner employed the theory of elasticity for his work. Elasticity is a well-known field, of course, older indeed than Escher's Glarus discoveries, dating back to the work of Cauchy (a contemporary of Navier); yet it had been infrequently applied to tectonic problems. The powerful stress-function method of Airy was employed by Hafner, and he solved several boundary-value problems of geologic interest. The two articles by Hubbert and Hafner (Papers 22 and 23) were jointly submitted and published in 1951, and together they represented an important advance. Boundary conditions associated with complex geologic deformations had been evaluated, stress distributions associated with the applied body and surface forces had been considered, and families of faults were predicted on the basis of stress trajectories.

The fact that predictions based on the theoretical elastic models seemed to be consistent with the results of the sand-model (Coulomb-material) experiments has since led some workers to seriously question the model approach. Instead, it may be more reasonable to conclude that boundary conditions may often be more significant in terms of effected results than the details of model rheology; it will be recalled that a similar suggestion was postulated in regard to Willis' experiments (Paper 6). This statement appears qualitative and indefinite, however, and thus seems unsatisfactory.

Fortunately, we need *not* invariably assume elastic materials in stress boundary-value problems of interest; indeed, we have already examined the beginnings of a principal alternative, for it seems that plasticity theory can be traced to extrusion experiments of ductile metals by Tresca, circa 1868, in the *same device* employed by Daubrée (Paper 9) for his schistosity and ruptured-belemnite experiments. The first mathematical theory for plastic solids was developed by Barre de Saint Venant in 1871 in order to quantitatively interpret Tresca's flow experiments; subsequent advances of significance were obtained by Prandtl, Mises, Hill, and many others. Thus Helmer Odé, also at the Houston Shell laboratory, considered the phenomenon of *ductile faulting* and suggested its relevance both to natural faulting and to Hubbert's sand-box experiments (Paper 24).

22

Reprinted from pp. 355, 367–372 of *Geol. Soc. America Bull.*, 62(4), 355–372 (1951)

MECHANICAL BASIS FOR CERTAIN FAMILIAR GEOLOGIC STRUCTURES

By M. King Hubbert

Abstract

A simple experiment with loose sand shows that this material exhibits faulting under deformational stresses in a manner remarkably similar to rocks. Moreover, the sand experiment is amenable to theoretical analysis with good agreement between predicted and observed behavior. The same theoretical treatment, with slight modification, is alsoa pplicable to the behavior of rocks, and appears to afford a basis of understanding for a variety of empirically well-known geologic structures.

[*Editor's Note:* Material has been omitted at this point.]

Tension in Normal Faulting

We have already seen that, in the case of cohesionless materials, normal faults are produced only under compression. For cohesive materials (Fig. 12), it will be seen that, for any given material, there must exist at the time of faulting a certain critical value of σ_1, say σ_1^* for which σ_3 will be zero. Then, for all values of $\sigma_1 > \sigma_1^*$, the horizontal stress σ_3 in normal faulting will be compressive, and for all values of $\sigma_1 < \sigma_1^*$ the horizontal stress σ_3 will be one of tension.

Since, in the case of normal faulting, the vertical stress at a given depth is equal to the pressure of the overburden, there will be some critical depth for a given material below which the horizontal stress producing normal faults will be compressive and the existence of a tensile stress will be impossible. At less than this depth, the horizontal stress will be one of tension.

The approximate depth at which this transition will occur can be obtained from crushing-strength data of rocks. The conditions

$$\sigma_3 = 0,$$

$$\sigma_1 = \sigma_1^*,$$

are those which are satisfied approximately in the crushing-strength tests for rocks at atmospheric pressure. Consequently, for a given rock the critical stress σ_1^* is equal approximately to its crushing strength. For dry rocks the depth at which this vertical stress occurs would be

$$z^* = \sigma_1^*/\rho g$$

where ρ is the bulk density of the material. In the earth, however, the rocks ordinarily are saturated with ground water, which exerts an upward force of buoyancy. In this case, the effective stress is that transmitted by the solid system. Hence this effective stress $\bar{\sigma}_1$ at a given depth is

$$\bar{\sigma}_1 = (\rho_s - \rho_w) \, g \, (1 - \epsilon) \, z,$$

or for the critical depth

$$z^* = \bar{\sigma}_1^*/[(\rho_s - \rho_w) \, g \, (1 - \epsilon)], \qquad (16)$$

where ρ_s is the grain or mineral density ρ_w the density of water, and ϵ the porosity of the rock.

The crushing strengths of rocks range from values near zero, for recent sediments, to as high as 3×10^9 dynes/cm² for granites. Such tests, however, are made upon nearly flawless specimens, so that, for large volumes of rock broken by numerous joints, the effective values will no doubt be considerably smaller. The Beaumont clay underlying Houston for example has a bearing strength of only about 10^6 dynes/cm². Recorded crushing strengths of sedimentary rocks are commonly of the order of 5—15×10^8 dynes/cm².

Introducing these values of σ_1^* into equation (16) and solving for z^*, we find the maximum depths at which tension can exist in rocks of various types. In the Gulf Coast sediments, for example, this depth is found to be of the order of 10^3 cm., or 10 meters (about 30 feet). In the more consolidated sediments, it may increase to a few hundred meters; and, in the strongest crystalline rocks, it may reach values of several kilometers.

Yet the Gulf Coast sediments are cut by numerous normal faults extending from the surface to depths of thousands of meters. It is clear, therefore, that aside from a surface veneer, these sediments have been broken by a series of normal faults in response exclusively to compressive stresses.

PROBLEM OF ASYMMETRICALLY FOLDED AND FAULTED BELTS

One-sided Thrust Hypothesis

With this background, let us consider an equally familiar but more complex problem. One of the commoner types of orogenic structure is the belt of asymmetrically folded and faulted sediments as exemplified in North America by the Appalachian Mountains, the Ouachitas, the foothills of the Rocky Mountains, and some of the West Coast ranges, and in Europe by the Alps, the Juras, and the Scottish Highlands. In these cases, the folds are unidirectionally overturned and accompanied by reverse faulting, with the direction of thrusting —that is, the direction of the relative displacement of the upper block with respect to the lower—in the same sense as the overturning of the folds. Moreover, where data are available, as in the foothills region of Alberta where seismic surveys have been made and numerous oil wells drilled, the fault surfaces have been found to be concave upward, the dip decreasing with depth.

Considerable attention has been given to the causes of these structures, and the hypothesis most often favored has been that formulated by J. D. Dana (1847a; 1847b)—that such structures are formed by a one-sided "active thrust" from the direction opposite to that of the overturning. Thus, the Appalachian Moun-

tains would have been formed by an active thrust from the southeast, which presumably died out with distance toward the northwest. An auxiliary question often considered is how far rocks of a given strength are capable of transmitting such a thrust.

The state of stress thus postulated and its associated orogeny is illustrated in Figure 13, where a large "active thrust" σ_{x_1} acts on the left-hand end of the section and dies out to the much smaller stress σ_{x_2} on the right-hand end. Not only does this appeal to the intuition, but it is supported by the deformational evidence as well. Since the rocks on the left are intensely deformed whereas those on the right are undistorted, the inference is justified that σ_{x_1} is very much greater than σ_{x_2}.

Application of the Newtonian Laws of Motion

That this picture is somehow deficient becomes evident when considered in the light of the Newtonian laws of motion, according to which the sum of all the forces acting upon the block in a given direction must equal the product of the mass of the block and the

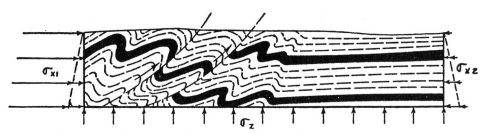

FIGURE 13.—COMPRESSIVE STRESSES ON ENDS OF SECTION OF ASYMETRICALLY FOLDED AND FAULTED SEDIMENTS

component of its acceleration in that direction. Taking the x-axis as horizontal, this requires that in the x-direction

$$\Sigma F_x = \int_0^{z_1} (\sigma_{x_1} - \sigma_{x_2})dz = ma_x \quad (17)$$

As an order of magnitude, $(\sigma_{x_1} - \sigma_{x_2})$ may be taken to be about 10^{+9} dynes/cm². Then if the block of Figure 13 is assumed to be 100 kilometers wide, 10 kilometers deep, and 1 centimeter thick, ΣF_x will be about 10^{16} dynes.

For the ma-term, the mass of the block will be about 3×10^{13} grams. Velocities in known orogenies are rarely more than a few centimeters per year, so that secular accelerations much greater than a centimeter per year per year or about 10^{-15} cm/sec² are unlikely. Hence, the *ma*-term would have a magnitude only of the order of 10^{-2} gm cm/sec², which is infinitesimal as compared with the forces acting, and the sum of all forces must effectively be zero throughout the orogenic process.

Since, with the stress distribution postulated in Figure 13, the sum of the horizontal forces acting upon the block is manifestly not zero, some essential element must be lacking. Since body forces are ineffective horizontally, the only remaining possibility is shear stresses upon the vertical faces parallel to the plane of the drawing and along the bottom of the block. In arcuate structures, shear stresses on the vertical faces normal to the strike exist, but they vanish as the curvature approaches zero and the structure becomes rectilinear.

Assuming the structure to be rectilinear,

value near the lower left-hand corner of the block and would decline to near zero at the lower right-hand corner. The complete equation of forces in the x-direction must therefore be

$$\Sigma F_x = \int_0^{z_1} (\sigma_{x1} - \sigma_{x2})dz - \int_{x_1}^{x_2} \tau_{zs} dx = 0. \quad (19)$$

As noted, the components of shear stress parallel to a plane normal to both are the same on any two mutually perpendicular planes, but of opposite rotational senses. Hence there must also be a downward-directed shear stress on the left-hand vertical face of the block equal in magnitude at the bottom corner to the stress on the bottom, and approaching zero toward the top. A similar shear stress, but of magnitude near zero and directed upward, must exist on the right-hand end.

In addition, all surface forces and body

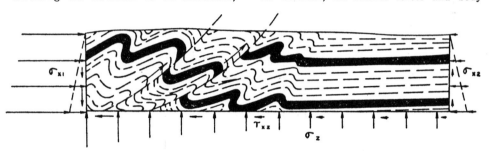

FIGURE 14.—COMPLETE SYSTEM OF TWO-DIMENSIONAL STRESSES ACTING UPON BOUNDARIES OF BLOCK

we are then left only with the bottom shear stress τ_{zs}, whose magnitude at any given distance x along the bottom must just be sufficient to balance the force due to the stress difference $d\sigma_x$ in the horizontal distance dx. Thus, at distance x

$$\tau_{zs} dx = + \int_0^{z_1} \left[\left(\sigma_x + \frac{\partial \sigma_x}{\partial x} dx \right) - \sigma_s \right] dz$$

or

$$\tau_{zs} = + \int_0^{z_1} \frac{\partial \sigma_x}{\partial x} dz; \quad (18)$$

that is the shear stress on the bottom of the block at a given point is proportional to the gradient of the normal stress in that direction. The gradient of the normal stress would be approximately the same as the gradient of the intensity of the deformation. Hence, in Figure 13 the shear stress would have a maximum

forces must be so related that all turning moments acting on the block are zero. When these several conditions are satisfied the approximate set of boundary stresses shown in Figure 14 is obtained as being necessary to produce the generalized type of orogeny shown.

Interior Stress Distribution

From the boundary stresses shown in Figure 14, the approximate pattern of the stress distribution in the interior of the block can be inferred.[1] Trajectories of principal stress are curves which at every point are tangent to a given principal stress. Since the principal

---

[1] The reasoning here is intentionally only qualitative, the objective being to convey an intuitive sense of the principles involved. A more formal quantitative analysis is given in a companion paper by the author's colleague, W. Hafner (1951).

stresses are mutually perpendicular, then three perpendicular stress trajectories, one each for σ_1, σ_2, and σ_3, must pass through each point. In a two-dimensional stress system on a plane parallel to σ_1 and σ_3, two families of orthogonal stress trajectories, one everywhere parallel to σ_1 and the other to σ_2, will exist.

Recalling that, when the principal stresses are unequal, shear stresses exist upon all surfaces except those perpendicular to the stress trajectories, the approximate pattern of the stress trajectories in the interior of the block may be determined from the combination of normal and shear stresses on its boundaries.

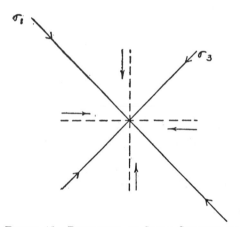

FIGURE 15.—DIRECTIONS OF SHEAR STRESSES IN VARIOUS QUADRANTS DEFINED BY PRINCIPAL STRESSES

downward. Without knowing the actual magnitudes of both the normal and shear-stress components, the angles of incidence of the stress trajectories can be determined only qualitatively, but the quadrant in which each falls can be definitely known. Across any plane on which a compressive normal stress σ and shear stress τ exists, the axis of greatest principal stress, σ_1, will lie in the quadrant between the plane normal and the direction *from* which the shear stress occurs (Fig. 15).

The boundary stresses of Figure 14 will produce the approximate pattern of principal-stress trajectories in the interior shown by the solid curves of Figure 16. The family of greatest-stress trajectories, σ_1, is tangential to the upper surface of the block, plunging downward and divergent to the right, and convex upward. The family of least-stress trajectories, σ_3, is orthogonal with this.

Surfaces of Potential Faulting

If we now assume that faulting is most likely to occur along surfaces tangent to the intermediate stress, σ_2, and at an angle of about $45° - \dfrac{\phi}{2}$ to the greatest stress, where ϕ may be taken to be about 30°, we are able to sketch in on Figure 16 the traces of the surfaces of potential faults, and indicate their directions of relative displacement. These comprise two families of curves (shown by dashed

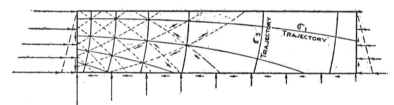

FIGURE 16.—TRAJECTORIES OF PRINCIPAL STRESSES (SOLID LINES), AND OF POTENTIAL REVERSE-FAULT SURFACES (BROKEN LINES) COMPATIBLE WITH THE BOUNDARY STRESSES OF FIGURE 14
Arrows indicate directions of potential slip

On the upper surface of the block, the normal stress is the atmospheric pressure and the shear stresses are zero. Therefore, at any point upon this surface, one of the stress trajectories must terminate perpendicularly, and the other must be tangential to the surface.

Across the left-hand end of the block, and the bottom, the stress trajectories must cross obliquely, with the angle of obliquity approaching zero near the upper surface, and increasing

lines) which everywhere intersect the σ_1-trajectories at an angle of about 30°. In virtue of the divergence and curvature of the stress trajectories, these surfaces are also curved, one set convex and the other concave upward.

At any given point, the stresses on each of these conjugate surfaces are the same; so that if they are at the critical value for fracturing on one surface, the same is true for the other. However, slippage along a finite surface involves an integral of conditions along the entire surface. Whatever this integral should be, in homogeneous materials with symmetrical stress distributions, its value over each of a pair of conjugate surfaces of potential slippage will be the same, so that equal slip should occur on each family of surfaces.

In asymmetrical cases, the integral of any stress quantity over one surface or family of surfaces will in general be different from that over the conjugate surface or family. Because of this inequality, slippage should occur on one set before on the other, or on only one of the two sets. In the asymmetrical geological system here considered, faulting develops almost exclusively on the concave-upward set of surfaces. The same was true for the reverse faulting in the sandbox experiment.

In experiments with symmetrical systems, slippage on both sets of surfaces is observed.

Returning to the sandbox experiment, it will be recalled that the average angle of hade for the normal faults was about 29°, and of dip for the reverse faults was 25°, whereas our theory indicated that the two should be the same. Perhaps we can now explain this discrepancy. We assumed the principal-stress trajectories to be strictly horizontal and vertical. However, in the reverse-fault case, the sand was pushed along the bottom which exerted a frictional reaction in the form of a bottom shear stress. The effect of this would be to deflect the horizontal stress trajectories downward and thus to lower the dip angle of the reverse faults.

Other shear stresses along the vertical faces of the partition must also have existed which would likewise tend to produce some deviation in observed angles from the idealized conditions assumed.

Conclusion

It thus appears that some of the more common large-scale geological structures are in satisfactory agreement with theoretical deductions based upon the stress patterns inferred from the observed deformation and Newton's laws of motion, and the empirically determined properties of rocks under known stress conditions.

References Cited

Anderson, E. M. (1942) *Dynamics of faulting and dyke formation*, Oliver and Boyd, Edinburgh and London, p. 191.

Balmer, Glenn G. (1946) *A revised method of interpretation of triaxial compression tests for the determination of shearing strength*, U. S. Bur. Reclamation Basic Structural Lab. Rep. No. SP-9.

Billings, Marland P. (1946) *Structural geology*, Prentice-Hall, Inc., New York, p. 206.

Blanks, R. F., and McHenry, Douglas (1945) *Large triaxial testing machine built by Bureau of Reclamation*, Eng. News-Record, vol. 135, p. 171–173.

Chamberlin, R. T., and Shepard, F. P. (1923) *Some experiments in folding*, Jour. Geol., vol. 31, p. 490–512.

Coulomb, C. A. (1776) *Essai sur une application des regles des maximis et minimis à quelques problèmes de statique relatifs a l'architecture*, Acad. Sci., Paris, Mem. pres. divers savants, vol. 7.

Dana, J. D. (1847a) *Geological results of the earth's contraction*, Am. Jour. Sci. (2), vol. 3, p. 176–188; vol. 4, p. 88–92.

—— (1847b) *On the origin of continents*, Am. Jour. Sci. (2), vol. 3, p. 381–398.

Hafner, W. (1951) *Stress distributions and faulting*, Geol. Soc. Am., Bull., vol. 62, p. 373–398.

Jones, Valens (1946) *Tensile and triaxial compression tests of rock cores from the passageway to penstock tunnel N-4 at Boulder Dam*, U. S. Bur. Reclamation, Basic Structural Res. Rept. No. SP-6.

von Kármán, Th. (1912) *Festigkeits Versuche unter allseitigen Druck*, Mitteilungen über Forschungzarbeiten auf dem Gebiete des Ingenieurswesens, H. 118, V.d.I.

Krynine, D. P.. (1941) *Soil mechanics*, McGraw-Hill, Inc., New York, p. 451.

Leith, C. K. (1913) *Structural geology*, 1st Ed., Henry Holt & Co., p. 55.

McHenry, Douglas (1948) *The effect of uplift pressure on the shearing strength of concrete*, International Congress on Large Dams, June 1948.

Mead, W. J. (1925) *Geologic role of dilatancy*, Jour. Geol., vol. 33, p. 685–698.

Mohr, Otto (1871; 1872) *Beiträge zur Theorie des Erddrucks*, Zeitschr. Architekten und Ingenieur—Ver. Hannover, vol. 17, p. 344; vol. 18, p. 67 and 245.

—— (1882) *Über die Darstellung des Spannungzustandes eines Körpelementes*, Zivil Ingenieure, p. 113.

—— (1900) *Welche Umstände bedingen die Elastizitätsgrenze und den Bruch eines Materials*, Zeitschr. Vereins deutsches Ing., p. 1524.

Nádai, A. (1928) *Plasticität und Erddruck*, Handbuch der Physik, VI, J. Springer, Berlin, p. 428–500.

—— (1931a) *Plasticity*, McGraw-Hill, Inc., New York, p. 349.

—— (1931b) *Phenomenon of slip in plastic materials*, Edgar Marburg Lecture, Am. Soc. Test. Mat., Pr., vol 31, pt. II, p. 11–46.

Nijboer, L. W. (1942) *Onderzoek naar den weerstand van bitumen-mineraalaggregaat mengsels tegen plastische deformatie*, N. V. Noord-Hollandsche Uitgevers Maatschappij, Amsterdam, p. 232.

Rankine, W. J. M. (1857) *On the stability of loose earth*, Royal Soc. London, Philos. Tr., vol. 147.

Sax, H. G. J. (1946) *De tectoniek van het Carboon in het Zuid-Limburgsche mijngebied*, Mededeelingen van de Geologische Stichting, Ser. C-I-I no. 3, p. 1–77.

Ros, H. C. M., and Eichinger, A. (1928) *Experimental attempt to solve the problem of failure in materials—nonmetallic materials*, Federal Mat. Testing Lab., E. T. H., Zürich, Rept. No. 28. Translation by F. Stenger, U. S. Bur. Reclamation, Denver.

23

Reprinted from pp. 373, 380-390, 398 of *Geol. Soc. America Bull.*, 62(4), 373-398 (1951)

STRESS DISTRIBUTIONS AND FAULTING

By W. Hafner

Abstract

Tectonic deformations result from a condition of internal stress caused, in turn, by primary and secondary forces. In the geological literature, a great deal of discussion is based on a direct connection between forces and deformation, completely by-passing the concept of stress. This paper is a contribution in the intermediate field of stress relations. It presents the complete solutions of certain stress systems caused by various forms of boundary forces. Furthermore, the location and attitude of the fault surfaces likely to be associated with them is determined.

The basic concept of stress is briefly reviewed and some of the fundamental differences between the force-vector and the stress-tensor are pointed out. The fallacy of applying the familiar methods of vector-addition of forces to problems in stress is demonstrated.

For certain systems of external boundary forces acting on a portion of the earth's crust, the internal stress distribution can be calculated by means of the familiar equations of elasticity. Appropriate calculation methods for two-dimensional cases are shown and the basic equations applicable to a series of important boundary conditions are derived. The examples here presented include: (1) superposed horizontal compression with constant lateral and vertical gradients; (2) horizontal compression with exponential attenuation; and (3) sinusoidal vertical and shearing forces acting on the bottom of a block. The latter equations provide solutions for differential vertical uplift and for the important case of drag exerted on the bottom of the crust by convection currents in the substratum. Diagrams show configuration of the stress trajectories and distribution of the maximum shearing stress for the resulting stress systems.

A parallel series of diagrams shows the disposition between the relatively stable and unstable segments of the blocks and the probable attitude of the fault surfaces likely to be associated with the individual stress systems. The construction of the fault surfaces is based on the original stress distributions alone, the influence of local stress alterations due to the occurrence of fracture being disregarded. The full effect of this inter-action is not known, due to the extreme complexity of the problem. The fault patterns shown are strictly applicable only to the initial stages of fracture, but may also represent fair approximations during the more advanced stages, since the original stress remains the dominating influence and stress-alterations due to faulting diminish rapidly with distance.

CONTENTS

* * * * * *

[*Editor's Note:* A row of asterisks indicates that material has been deleted.]

The Standard State

E. M. Anderson (1942) has introduced the term "standard state," defined (p. 137) as

"a condition of pressure which is the same in all directions at any point, and equal to that which would be caused by the weight of the superincumbent material, across a horizontal plane, at the particular level in the rock. It is assumed in this definition that the surface is flat, and that the strata are of uniform specific gravity."

It will be shown that the standard state—according to this definition—represents a stress system composed of two parts: (1) the effect of gravity, and (2) a superposed horizontal stress which is constant in any horizontal plane but increasing uniformly with depth. The manner in which gravity affects our investigations must be clearly understood, since it represents one of the major causes of stress. Being a body force, it plays a somewhat different role in the analytical treatment than do the surface forces. For this reason—and also to illustrate the discussed calculation methods—the standard state is briefly derived as follows:

We select the Airy stress function in the form of a polynomial of third degree

$$\Phi = k_1 x^3 + k_2 x^2 + k_3 x y^2 y + k_4 y^3,$$

which satisfies equation (2). The stress components (1)′ then are:

$$\sigma_x = \frac{\partial^2 \phi}{\partial y^2} = 2k_3 x + 6k_4 y;$$

$$\sigma_y = \frac{\partial^2 \phi}{\partial x^2} - \rho g y = 6k_1 x + 2k_2 y - \rho g y;$$

$$\tau_{xy} = \frac{\partial^2 \phi}{\partial x \partial y} = -2k_2 x - 2k_3 y.$$

In view of the boundary conditions at the surface, k_1 and k_2 must be set zero; k_3 and k_4 can be chosen arbitrarily. We select $k_3 = 0$; $k_4 = -\frac{1}{6}\rho g$ and obtain:

$$\sigma_x = \sigma_y = -\rho g y; \qquad \tau_{xy} = 0 \qquad (6)$$

These are the stress components for the standard state. The normal stresses are equal in all directions and for all points on a horizontal plane ($y = $ constant) and the shear stress is

zero throughout the body. This stress system is, therefore, equivalent to the hydrostatic pressure in a liquid. Its physical significance for solid bodies is apparent from the following consideration. In a homogeneous solid body of infinite horizontal extent, uniform lateral extension is prevented by lack of space. Therefore, if the weight of the body is the only source of stress, the strain components (elongations) in the horizontal directions are zero. From this the relation

$$\sigma_{hor} = \sigma_{vert} \frac{\nu}{1 - \nu}$$

is derived, where σ_{hor} and σ_{vert} represent the horizontal and vertical stress components, respectively, and ν is Poisson's ratio. It is seen that $\sigma_{hor} = \sigma_{vert}$ only if $\nu = 0.5$. Hence, the hydrostatic state is produced by gravity alone only in liquids. In solid bodies the horizontal stress components due to weight are substantially smaller than the vertical component; they are only a third of the latter if Poisson's ratio is taken to be 0.25, which is a good approximation in most cases. The term $6k_4y$ in the above derivation of the standard state thus includes two effects: roughly a third of it is due to gravity, while the remaining portion represents a superposed horizontal stress, constant in any horizontal plane, but directly proportional to depth.

The important result derived from these considerations is that the effect of gravity can be incorporated into a stress system which is essentially hydrostatic in character. This is a consequence of the fact that a stress system of the form:

$$\sigma_x = ky; \qquad \sigma_y = \tau_{xy} = 0$$

is likewise a valid one—as shown by the above derivation—and thus can be combined with the weight of the body to produce a hydrostatic system. It does not matter whether the complementary horizontal stress ky actually exists in any particular case; the standard state represents an idealization which is probably only seldom realized in nature. Its principal usefulness—aside from offering a convenient "standard of reference—"arises from an application of the already discussed principle of superposition of stresses, which can now be formulated more precisely. Let Φ_1 and Φ_2 be two stress functions, each of which represents a correct solution. Then the sum of the two is

$$\Phi = \Phi_1 + \Phi_2 \qquad (7)$$

also a valid solution. Now let Φ_1 be the standard state which includes the effect of gravity. Since no other body force is involved, equation (7) states that we may superimpose upon it any other valid stress system Φ_2 caused by surface forces alone. Anderson has named such systems "supplementary stresses." Their mathematical treatment is now divorced from the effects of gravity, and the stress components are given by the simpler expressions of equations (1). The physical implication of this procedure is, of course, merely a statement of the fact that the standard state contributes nothing to the shearing stress and has no influence on the configuration of the stress trajectories. Conversely, it forms a major component of the total stress and, therefore, must be taken into account in phenomena depending upon the magnitude of the confining pressure. One of these phenomena is faulting.

Faulting Associated with Stress Distribution

While the determination of the stress properties is amenable to exact mathematical analysis, the occurrence of faulting presents an intricate problem for which no complete theory has yet been worked out. It is, therefore, important to discuss its critical aspects and to point out the shortcomings of the present treatment.

The principal object of the fault analysis is two-fold: (1) to know the attitude of the fault surfaces and their variations in the calculated stress systems, and (2) to determine the location of faulting in the stressed medium, i.e., the distribution of the stable and unstable portions.

The first problem is readily solved if we accept a widely held theory on the relation between the fault surfaces and the directions of the principal stresses. It states that fracture occurs along two planes which have a specific angle θ with the direction of the maximum principal pressure. This angle is a material constant and, therefore, variable. It is, how-

ever, always less than 45°, and for most types of rock falls between 20° and 40°. A value of 30° appears to form a good overall approximation. In three-dimensional stress systems the direction of the two fault surfaces is parallel to that of the intermediate principal stress, the angles θ being measured in the plane containing the maximum and least principal stresses. In the two-dimensional cases here analyzed, we shall presume that this is the plane containing the variable stress components—that the third dimension is parallel to the intermediate principal stress. Then the attitude of the fault surfaces is obtained by drawing two sets of curves having everywhere a selected constant angle with the direction of maximum principal pressure. This is evidently a simple procedure, once the configuration of the stress trajectories has been completely determined. While the choice of θ is arbitrary within certain limits, a value in the neighborhood of 30° appears most appropriate. Corresponding to the plus and minus sign of θ, the method always yields two sets of potential fault surfaces, which intersect each other at an oblique angle. While there is no theoretical basis for a discrimination between the two systems in a homogeneous medium, geological observations indicate that in most areas only one of them has been utilized, or that it strongly predominates. Nevertheless, examples of the co-existence of both sets within the same structural belt are known for all classes of faults.

The validity of the above theory on the attitude of the fault surfaces has been questioned by some authors (Jeffreys, 1936), but it finds much support both from the results of experiments and from geological observations. It is based on the assumption that fracture in a stressed medium takes place along that plane which offers the least resistance to it. To produce faulting, the available energy represented by the condition of strain has to overcome not only the strength of the material but also the internal friction opposing the motion. The latter is proportional to the normal pressure across the fault surface. The shear stress is always greatest in a direction at 45° to those of the maximum and minimum principal stresses. Now if we rotate a plane from this position towards the axis of maximum principal pressure, both the shear stress—which tends to produce faulting—and the normal pressure—which opposes motion—become smaller. Depending upon the specific properties of the material, there will be one particular angle for which the combined effect of the two opposing factors becomes most favorable for the occurrence of faulting; this is the angle θ above referred to. A detailed mathematical formulation of this principle is found in Anderson (1942, p. 9–10).

The second problem of faulting accorded attention here deals with the determination of the boundaries between the stable and unstable portions of the stress systems. We will first examine the problem solely upon the basis of the statical theory as applied to the original stress distribution.

The basic principle is easily stated: faulting occurs where the shearing stress exceeds the strength of the material. The magnitude of strength depends, first, upon the material, falling in the range of approximately 500–3000 kg/cm^2 under atmospheric pressure for the most common rock types. Second, it varies widely for a given material as a function of other factors, such as the confining pressure, temperature, presence or absence of solutions, and even the time of application of the shearing stress. Only the first of these has been adequately investigated in laboratory experiments. It was generally found that the breaking pressure increases with the confining pressure and that in many cases the rate of increase is nearly linear over a wide pressure range. Since little information is available on the influence of the other factors, the boundary equation for the separation of the stable and unstable portions adopted in the present paper is:

$$\sigma_{min} \leqq n\sigma_{max} + \sigma_0 \qquad (8)$$

where:

σ_{min} = maximum compressive stress (compressive stresses being given the negative sign, the maximum compressive stress is, algebraically, the smallest),

σ_{max} = minimum compressive stress,

σ_0 = breaking strength under atmospheric confining pressure,

n = material constant.

As pointed out, while the standard state contributes nothing to the shearing stress in a homogeneous medium, it adds a vital component to the total confining pressure; hence σ_{min} and σ_{max} in (8) represent total principal stress, including that due to the standard state. They are obtained by adding to the components from equation (4) the hydrostatic pressure defined by (6):

$$\begin{aligned} \sigma_{min} &= \tfrac{1}{2}(\sigma_x + \sigma_y) - \tau_{max} - \rho gy \\ \sigma_{max} &= \tfrac{1}{2}(\sigma_x + \sigma_y) + \tau_{max} - \rho gy \end{aligned} \Big\} \qquad (9)$$

For the calculation of numerical examples it will be necessary to assign the constants ρg, σ_0 and n specific values. The selections for the present paper are:

ρg = 250 kg/cm^2 per kilometer or 400 kg/cm^2 per mile,

σ_0 = 1000 kg/cm^2,

n = 4.

While in nature each of these quantities is variable, the figures chosen represent the right order of magnitude. The value for n is based mostly on geological evidence, but it is also substantially in agreement with the results of experimental data.

In the practical application, a simple method was used for the solution of equation (8) which is illustrated in Figure 5. The curve σ_{st} represents the standard state, to which the principal supplementary stresses derived from equation (4) are added algebraically (compressive components being negative). Then the curve: $(n\sigma_{max} + \sigma_0)$ is drawn (dashed line). Faulting occurs where the latter is to the left of σ_{min}. The intersections between the two curves mark the boundaries between the areas of stability and those in which the shear stress exceeds the strength. In the specific examples, the boundary lines were determined for not only one, but several, values of the significant constants appearing in the formulae for the external boundary stresses.

In the preceding discussion, the distribution and mode of faulting were examined solely in relation to the original stress system. With the first occurrence of faulting, an entirely new problem arises. The process of fracturing causes an instantaneous and drastic alteration in the pre-existing stress system, accompanied by the introduction of a new internal boundary surface. It is thus evident that any subsequent faulting must take place in accordance with the changed conditions, and the value of the original stress system as a guide to further deformation may rightfully be questioned. This interaction between repeated changes in the stress distribution and the successive occurrences of faulting introduces a complication which presents great difficulties to an adequate mathematical analysis. No attempt is made here to deal with this problem. However, some qualitative arguments are advanced which may shed some light on the degree of control which the original stress distribution still exercises during the more advanced stages of faulting.

A mathematical solution of the alteration of stress due to the occurrence of a fault is given by Anderson for a simple case (1942, p. 144–156); this is the only example known to the writer. The results demonstrate that pronounced changes in both direction and intensity of the stresses occur in the immediate neighborhood of the fault surface, but that they diminish rapidly with distance. This is especially true in a direction at right angle to the fault where the alterations soon become insignificant. Near the central portion of the fracture the shear stress is greatly reduced, which has the effect of preventing further faulting in this vicinity. Conversely, there is a substantial increase in stress at both ends of the fracture, indicating a tendency for lateral extension, once the process has been initiated. While the solution applies to a specific case only, it is probable that the basic results are similar in other, more complicated situations.

In the light of these data we may visualize the following sequence of events. Under the action of external forces, a condition of stress is produced in a block of the earth's crust. As long as the shearing stress is everywhere below the strength of the material, the stress distribution is static and can be determined by means of the discussed methods. The various examples given below are rigorously correct during this stage only. With further intensification of the stresses, a point is reached where the strength is exceeded at some particular place, and faulting will occur there.

The general location of this place is determinable from the initial conditions. How far the fracture extends itself laterally into regions where the original shear stress is still well below the strength is an unsolved problem. Theoretical results suggest the possibility of

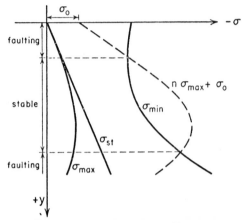

FIGURE 5.—DETERMINATION OF BOUNDARIES OF STABLE PORTIONS OF BLOCK

pronounced extension, whereas geological observations demonstrate that the majority of faults are of rather limited extent. Regardless of this apparent conflict, the principal result will be a drastic diminution of stress in the neighborhood of the origin of the fracture, while farther away the initial stress distribution remains substantially unaltered. The first fault thus provides local relief in the most intensively stressed portion of the block. After its occurrence, the further building-up of the stress system is presumed to go on, and additional portions of the block will gradually reach and exceed the strength limit. Renewed faulting then occurs some distance away from the first fracture, at a place where the stress-relief is of negligible amount. Hence, while no longer exactly the same, the conditions producing the new fracture should still conform closely to the original stresses. A further implication of this process of reasoning is the probability that the later faulting belongs to the same set as the initial one, since a secondary fracture utilizing the alternative complementary direction would lead straight into the protective zone of the preceding one. This, perhaps, is the explanation of the relative scarcity of complementary faults.

In this manner the process may continue until the maximum development of the system has been reached. At this stage the extent of the unstable areas has likewise reached its maximum expansion which, depending upon the ultimate intensities of the causal forces, will include a smaller or greater portion of the entire block. Due to the limited extent of the influence of previous fractures in adjacent areas, it appears probable that the initial faulting in each portion of the unstable segments is primarily controlled by the original stresses. The local relief of stress provided by the existing fractures will then gradually disappear again due to the tendency of the original stress system to re-establish itself. This can lead to renewed faulting in already fractured segments. The process thus becomes more and more complicated, on account of the introduction of a new internal boundary surface along each additional fault and the adjacent zone of pronounced stress alterations. In extreme cases, the stress system may be substantially changed during the final stages of deformation and the further course of events is then beyond the scope of theoretical analysis.

From this line of reasoning it appears justified to assume that the original stress system constitutes the controlling influence on the location and type of faulting at least during the initial stages of deformation in each faulted segment. The author, therefore, believes that illustrations showing the attitude of fault surfaces and the boundaries between stable and unstable segments based purely on the original stress system serve a useful purpose in studies of this nature.

Examples of Two-Dimensional Stress Systems

Introduction

Using the methods outlined in the preceding sections, we will now present a series of examples of two-dimensional stress systems. One or several numerical examples for each group are illustrated graphically in Figures 6–8 and on Plate 1. In each figure, the upper diagram shows the complete solution of the internal stress distribution in the form of stress trajectories and lines of equal maximum shearing stress; the lower diagram shows the attitude

of the fault surfaces and the distribution of the stable and unstable portions of the block. All graphs show the respective systems of boundary stresses.

Supplementary Horizontal Stress without Superposed Vertical Stress

In the first group to be analyzed, we assume the presence of a supplementary horizontal stress but absence of an associated vertical stress component, i.e., there is no pressure or tension across any horizontal plane in the body in addition to the normal hydrostatic component. Mathematically this is expressed by:

$$\sigma_y = \frac{\partial^2 \Phi}{\partial x^2} = 0 \quad \text{for all values of } y.$$

Integrating we obtain the stress function:

$$\Phi = cf(y)x + ax + bf_2(y) + d.$$

To satisfy equation (2)

$$\frac{\partial^4 \Phi}{\partial x^4} + 2\frac{\partial^4 \Phi}{\partial x^2 \partial y^2} + \frac{\partial^4 \Phi}{\partial y^4} = cxf_1{}^{IV}(y) + bf_2{}^{IV}(y) = 0$$

the fourth order derivatives of f_1 and f_2 must be zero. Hence the second order derivatives may be either linear functions of y, constants, or zero. The stress components then are (equations (1))

$$\left. \begin{array}{l} \sigma_x = \dfrac{\partial^2 \Phi}{\partial y^2} = cf_1{}^{II}(y)x + bf_2{}^{II}(y); \\[2mm] \sigma_y = \dfrac{\partial^2 \Phi}{\partial x^2} = 0; \\[2mm] \tau_{xy} = -\dfrac{\partial^2 \Phi}{\partial x \partial y} = -cf_1{}^{I}(y). \end{array} \right\} \quad (10)$$

The boundary conditions at the surface require that $f_1{}^{I}(y) = 0$ for $y = 0$. Keeping within the limits of the above restrictions we can set up the following three subgroups:

a)
$$f_1{}^{I}(y) = 0; \quad f_2{}^{II}(y) = y + d$$
$$\left. \begin{array}{l} \sigma_x = by + d; \\ \sigma_y = 0; \\ \tau_{xy} = 0. \end{array} \right\} \quad (10a)$$

b)
$$f_1{}^{I}(y) = y; \quad f_2{}^{II}(y) = 0$$
$$\left. \begin{array}{l} \sigma_x = cx; \\ \sigma_y = 0; \\ \tau_{xy} = -cy. \end{array} \right\} \quad (10b)$$

c)
$$f_1{}^{I}(y) = \tfrac{1}{2}y^2; \quad f_2{}^{II}(y) = 0$$
$$\left. \begin{array}{l} \sigma_x = cxy; \\ \sigma_y = 0; \\ \tau_{xy} = -\dfrac{c}{2}y^2. \end{array} \right\} \quad (10c)$$

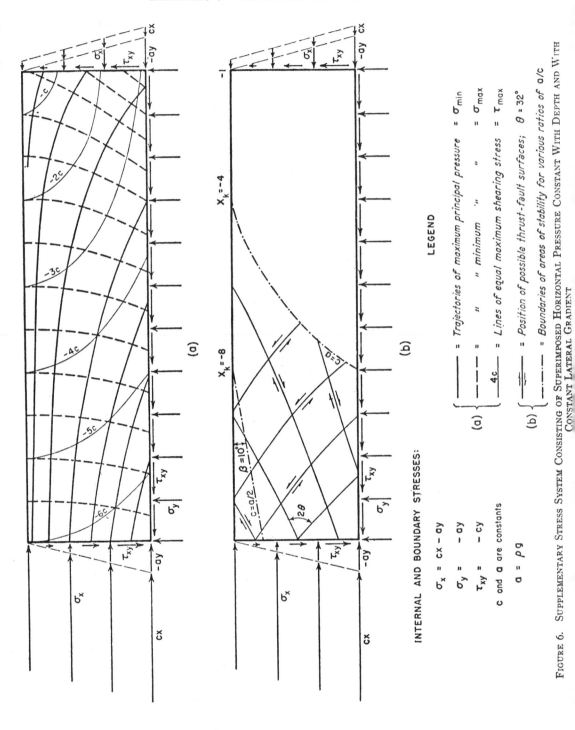

INTERNAL AND BOUNDARY STRESSES:

$$\sigma_x = cx - ay$$
$$\sigma_y = -ay$$
$$\tau_{xy} = -cy$$

c and a are constants

$$a = \rho g$$

LEGEND

(a) $\left\{\begin{array}{l} \rule{2cm}{0.4pt} \\ \rule{2cm}{0.4pt} \\ \rule{2cm}{0.4pt}\ 4c \end{array}\right.$ = Trajectories of maximum principal pressure = σ_{min}
= " " minimum " = σ_{max}
= Lines of equal maximum shearing stress = τ_{max}

(b) $\left\{\begin{array}{l} \rule{2cm}{0.4pt} \\ \rule{2cm}{0.4pt} \end{array}\right.$ = Position of possible thrust-fault surfaces; $\theta = 32°$
= Boundaries of areas of stability for various ratios of a/c

FIGURE 6. SUPPLEMENTARY STRESS SYSTEM CONSISTING OF SUPERIMPOSED HORIZONTAL PRESSURE CONSTANT WITH DEPTH AND WITH CONSTANT LATERAL GRADIENT

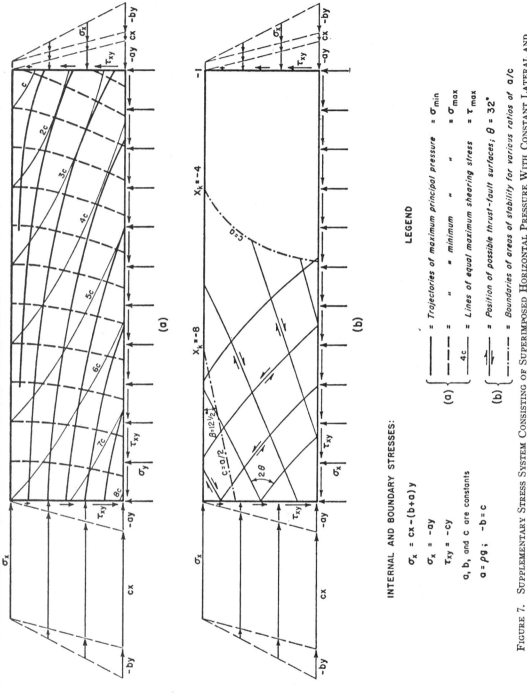

INTERNAL AND BOUNDARY STRESSES:

$$\sigma_x = cx - (b+a)\,y$$
$$\sigma_x = -ay$$
$$\tau_{xy} = -cy$$

a, b, and c are constants

$$a = \rho g \; ; \quad -b = c$$

LEGEND

(a) $\left\{\begin{array}{l} \text{———} = \textit{Trajectories of maximum principal pressure} = \sigma_{min} \\ \text{– – –} = \text{" } \text{" } \textit{minimum } \text{" } \text{" } = \sigma_{max} \\ \underline{4c} = \textit{Lines of equal maximum shearing stress} = \tau_{max} \end{array}\right.$

(b) $\left\{\begin{array}{l} \rightleftarrows = \textit{Position of possible thrust-fault surfaces; } \theta = 32° \\ \text{–·–·–} = \textit{Boundaries of areas of stability for various ratios of } a/c \end{array}\right.$

FIGURE 7. SUPPLEMENTARY STRESS SYSTEM CONSISTING OF SUPERIMPOSED HORIZONTAL PRESSURE WITH CONSTANT LATERAL AND VERTICAL GRADIENT

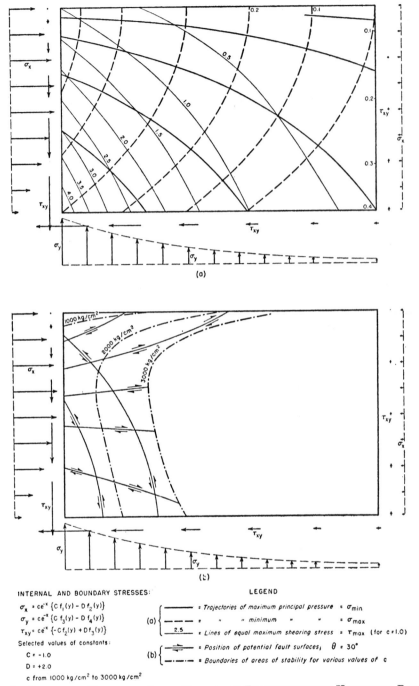

INTERNAL AND BOUNDARY STRESSES:

$\sigma_x = ce^{-x} \{C f_1(y) - D f_2(y)\}$

$\sigma_y = ce^{-x} \{C f_3(y) - D f_4(y)\}$

$\tau_{xy} = ce^{-x} \{-C f_2(y) + D f_3(y)\}$

Selected values of constants:

C = -1.0

D = +2.0

c from 1000 kg/cm² to 3000 kg/cm²

LEGEND

(a) {
= Trajectories of maximum principal pressure = σ_{min}

------ = " " minimum " " = σ_{max}

2.5 = Lines of equal maximum shearing stress = τ_{max} (for c=1.0)
}

(b) {
= Position of potential fault surfaces; $\theta = 30°$

—·—·— = Boundaries of areas of stability for various values of c
}

FIGURE 8. SUPERPOSED HORIZONTAL STRESS DECREASING EXPONENTIALLY IN HORIZONTAL DIRECTION

In these equations, b, c, and d are arbitrary constants and may be assigned any value, including zero. Furthermore, any linear combination of equations (10a) to (10c) represents another allowable system.

The simplest case of a supplementary stress

system is given by (10a) if we let b equal zero. Then the superposed stress is restricted to a constant horizontal component $\sigma_x = d$, the other two components σ_y and τ_{xy} being zero everywhere. From (3) and (5) it follows that the maximum shearing stress is also constant—equal to half the horizontal stress—and that the stress trajectories are everywhere horizontal and vertical, respectively. The associated fault planes will be inclined at angles of about $\pm 30°$ to the horizontal (thrust faults).

We will next examine the stress systems of the first two sub-groups. The constant d is now selected to be zero since its contribution has already been discussed. The combination of (10a), (10b), and the standard state gives the total stress components:

$$\sigma_x = cx + by - ay,$$
$$\sigma_y = -ay,$$
$$\tau_{xy} = -cy;$$

where $a = \rho g$. The term $-ay$ represents the hydrostatic pressure and is the only one appearing in the vertical stress component. The horizontal component σ_x contains two additional terms, the first a linear function of x only and the second a linear function of y only. Hence σ_x has constant gradients in both the horizontal and vertical directions. The shear stress τ_{xy} is seen to be constant in any horizontal plane; further, it increases vertically at a constant rate which is equal to the lateral gradient of σ_x. The term ky in σ_x appeared previously as a component of the standard state; in the present formulae, it denotes the difference (excess or defect) between the actual magnitude of this stress and that portion of it which is absorbed in the standard state.

Figure 6[1] illustrates the case where the supplementary horizontal pressure is constant with depth ($b = 0$) and Figure 7 where it has a vertical gradient equal to the horizontal one ($b = c$). The external boundary stresses are shown on the end-surfaces and along the bottom of a rectangular block. On the two sides, the horizontal stress σ_x is broken down into its individual components. It is seen that addition of the term by does not change the equilibrium of the block in the lateral direction, a fact which

[1] Similar to Figure 25 in the companion paper by M. King Hubbert (1951).

explains that the shearing stress along the bottom is the same in both cases.

The trajectories of maximum principal pressure are curved lines, dipping downward away from the area of maximum compression. The curvature is stronger if the vertical gradient of σ_x is small. Since the trajectories are curved, so are the potential fault surfaces. The latter obviously belong to the class of thrust faults. The set dipping towards the area of maximum pressure is slightly concave upwards, the complementary set concave downwards. Thrust faults of the former type are very common in nature and the theoretically deduced curvature is frequently observed. The latter type appears to occur only rarely and little is known regarding the curvature.

The lines of equal maximum shearing stress are expressed in terms of multiples of the constant c. The shearing stress naturally increases towards the area of greater horizontal pressure but it also increases with depth, the latter effect being more pronounced if σ_x has a vertical gradient. To determine the boundary of the stable portion of the block, the constant c has to be fixed numerically. This is done best by selecting certain values of the ratio c/a, thus expressing the lateral gradient of the superposed horizontal stress in terms of a fraction of the vertical pressure gradient due to weight. In nature this ratio is evidently widely variable with place and time and, in fact, may assume any value from zero up to a magnitude large enough to cause thrusting. The most useful procedure, therefore, consists of assuming several ratios and calculating their respective boundary lines. In both Figures 6 and 7, two such lines are shown corresponding to the ratios $c = a$ and $c = a/2$.

These boundary lines are seen to be dipping towards the area of greater horizontal pressure. The rate of dip is primarily a function of the ratio c/a, being greater the larger this fraction is. It is steep only if the latter is near unity, when the horizontal pressure gradient approaches the magnitude of the vertical pressure gradient due to weight. For smaller values of c, the boundary lines are nearly straight and their inclination diminishes rapidly. Numerical examples on the dip β are: for $c = a/2$, β is about 10° in Figure 6 and $12\frac{1}{2}°$ in Figure 7, for $c = a/10$, β is less than $2°$ in the former case.

These results suggest the following conclusions. If the supplementary horizontal pressure has only a small lateral gradient, say of the order of half the vertical pressure gradient or less, the potential area of thrusting is confined to a shallow, gently dipping wedge. The resulting deformation will consist of a series of slice-thrusts covering a broad belt but extending only to shallow depth. The presence of a vertical gradient slightly steepens the wedge. However, the boundary becomes very steep or nearly vertical only if the horizontal gradient approaches the magnitude of the vertical increase of the hydrostatic pressure. In that event, thrusting can take place throughout a thick, but probably only narrow, zone of the crust. Such a condition appears to have occurred, for instance, in the marginal belts of several ranges in the Rocky Mountain province.

The boundary stresses of the third sub-group (formulae 10c) are also characterized by the fact that the shearing stress is constant in all horizontal planes. Its vertical gradient, however, is now a function of the second order of y, instead of the first order. The shearing stress is balanced by a horizontal pressure, again having constant lateral and vertical gradients but increasing more rapidly in the diagonal direction. This case has not been further analyzed.

The most general expression for the stress systems satisfying the assumption of absence of a vertical stress component is given by the superposition of equations (10a), (10b) and (10c), as follows:

$$\left.\begin{aligned}\sigma_x &= c_1xy + c_2x + by + d; \\ \sigma_y &= 0; \\ \tau_{xy} &= -\frac{c_1}{2}y^2 - c_2y\end{aligned}\right\} \quad (11)$$

It is seen that the stipulation $\sigma_y = 0$ is associated with two additional general properties of the internal stress system: (1) that the shearing stress is a function of y only, i.e., constant in all horizontal planes, and (2) that σ_x has linear gradients in both the horizontal and vertical directions. That the first two properties are reversible can readily be demonstrated by deriving the stress system based on the assumption:

$$\tau_{xy} = -\frac{\partial^2 \Phi}{\partial x \partial y} = -k_1 f(y).$$

Using the same direct integration method as before and satisfying equation (2) and the boundary conditions at the surface, this leads again to equations (11). Hence, the reversed statement: "constancy of the shearing stress in all horizontal planes is associated with absence of a vertical supplementary stress throughout the body," is also true. Thus the heading of this Section could equally well read "Supplementary shearing stress constant in all horizontal planes."

Supplementary Horizontal Stress Decreasing Exponentially in the Horizontal Direction

We have investigated the condition of the presence of a supplementary horizontal stress coupled with absence of a superposed vertical component. In this section we again take a supplementary horizontal stress, but drop the assumption that σ_y be zero throughout the block. Instead, we now impose the condition that σ_x decrease exponentially in the horizontal direction.

For this purpose we take the Airy stress function in the form:

$$\Phi = ce^x f(y). \quad (12)$$

Applying equation (2) to this function yields the differential equation

$$f(y) + 2f''(y) + f^{IV}(y) = 0$$

the general solution of which is

$$f(y) = A \sin y + B \cos y + Cy \sin y + Dy \cos y. \quad (13)$$

Substituting (13) in (12) and differentiating we obtain the stress components (1). The compulsory boundary conditions $\sigma_y = \tau_{xy} = 0$ for $y = 0$ can be taken care of by a proper disposition of the two constants A and B. After eliminating these constants, the stress components become:

$$\begin{aligned}\sigma_x &= ce^x\{C(2\cos y - y\sin y) - D(\sin y + y\cos y)\}; \\ \sigma_y &= ce^x\{Cy\sin y \qquad\qquad - D(\sin y - y\cos y)\}; \\ \tau_{xy} &= ce^x\{-C(\sin y + y\cos y) + Dy\sin y\}.\end{aligned}$$

Let:

$$\begin{aligned}f_1(y) &= 2\cos y - y\sin y; & f_2(y) &= \sin y + y\cos y; \\ f_3(y) &= y\sin y; & f_4(y) &= \sin y - y\cos y;\end{aligned}$$

then the stress components can be written in the form:

$$\begin{aligned}
\sigma_x &= ce^x\{Cf_1(y) - Df_2(y)\} = ce^x F_1(y); \\
\sigma_y &= ce^x\{Cf_3(y) - Df_4(y)\} = ce^x F_2(y); \\
\tau_{xy} &= ce^x\{-Cf_2(y) + Df_3(y)\} = ce^x F_3(y).
\end{aligned} \quad (14)$$

The direction of the stress trajectories is given by:

$$\tan 2\beta = \frac{2\,F_3(y)}{F_1(y) - F_2(y)}. \quad (15)$$

The functions F_1 to F_3 are constant in the horizontal direction since they are functions of y only. Therefore, equation (15) shows that all stress trajectories are parallel curves and the same applies to the fault surfaces under the assumption of constancy of angle between the latter and the directions of maximum principal pressure.

When writing equations (14) in the form:

$$\sigma = c\,\epsilon(x, y),$$

it is evident that the constant c has the dimension of stress. Its value can be selected arbitrarily and determines the absolute magnitude of the stress components, whereas the function $\epsilon(x, y)$ determines their areal variations. The two remaining integration constants C and D are dimensionless and, being likewise fully independent, permit the selection of numerous variations. The degree of variability is further enhanced by the fact that the position of the bottom of the block can be chosen at any desired lower limit of the y functions, the unit of length also being arbitrary. The range of opportunity thus provided can easily be exploited by drawing graphs of the four functions f_1 to f_4 and making visual estimates of the combinations obtainable after multiplication with various constants.

Figure 8 illustrates the case where $C = -1$ and $D = +2$. In this and all succeeding illustrations, only the supplementary boundary stresses are shown; the normal hydrostatic components are omitted to simplify the drawings. The shearing stress acting on the bottom of the block is no longer constant but, like the superposed horizontal stress, increases exponentially towards the pressure area. The same is true for the vertical component which now forms a part of the stress system. All three boundary stresses diminish rapidly away from the area of compression. The zone of potential faulting consists of a narrow, nearly vertical, belt and a shallow outwardly protruding wedge near the surface. Figure 8b indicates three boundary lines limiting the area of stability; they correspond to values of the constant c of 1000, 2000, and 3000 kg/cm^2, respectively.

The general fault system in the shallow portion of the unstable segment is similar to that obtained in the preceding stress systems, except for a slight change in attitude. The thrust faults dipping towards the pressure area are less steeply inclined. This divergence increases with depth where, because of a gradual clockwise rotation, the fault surfaces of this set become practically flat and finally even overturned. It appears doubtful, however, that the extreme stage is ever realized, because the sudden decrease of σ_x near the lower left-hand corner of the block does not represent a boundary condition which one would expect in nature. As already stated, we are at liberty to match the bottom of the block with any desired point of the (σ_x, y) curve, for instance, that for which σ_x reaches a maximum. In that case the lateral pressure increases throughout the entire thickness of the block, thus providing a stress system which fully satisfies our intuitive concept of geological probability. Figure 8 demonstrates that even within these limitations the thrust planes reach a practically horizontal attitude (in the vicinity of approximately half of the total depth to which calculations were carried out).

It is thus evident that stress systems of this type produce thrust faults which are only gently inclined at shallow depth and become nearly horizontal at greater depth. Such attitudes are frequently observed in nature and are usually interpreted to indicate subsequent tilting of the fault surfaces. It has been shown here that the gently inclined attitudes may equally well be explained as original features of the stress system.

*   *   *

References Cited

Anderson, E. M. (1942) *The dynamics of faulting*, Oliver and Boyd, London, 183 pages.

Hubbert, M. King (1951) *Mechanical basis for certain familiar geologic structures*, Geol. Soc. Am., Bull., vol. 62, p. 355

Jeffreys, Dr. H. (1936) *Note on fracture*, Royal Soc. Edinburgh, Pr., vol. 56, pt. 2, p. 158–163.

24

Reprinted from pp. 293-294, 311-312, 313, 366, 369, 371 of *Rock Deformation* (Geol. Soc. America Mem. 79), D. Griggs and J. Handin, eds., 1960, pp. 293-321

Faulting as a Velocity Discontinuity in Plastic Deformation

Helmer Odé

Shell Development Co., Houston, Texas

ABSTRACT

Triaxial loading of certain rocks sometimes results in faulting without fracture in the ordinary sense, or sudden displacements. A brief discussion of the theory of brittle fracture of Griffith is given to indicate why it cannot apply to this kind of ductile faulting. Such ductile faulting can be explained by the theory of plasticity for plane strain. This theory is presented for more general yield conditions than that of von Mises, and it is assumed that the stress–strain rate relations can be derived from the yield condition by a process of differentiation. This formalism is by no means original, but it is little known among geologists. Across certain planes in the plastic mass, discontinuities in the velocity are possible, whereas the stresses remain continuous across these planes. These "characteristic planes" of the velocity equations are identified with the planes of ductile faulting. The results obtained in the theory of plane strain for the von Mises, Coulomb, and Torre yield conditions do not hold for the more general three-dimensional theory, unless very stringent conditions on the strain rates are satisfied. Smooth characteristic surfaces in the plastic domain are possible only if these conditions are satisfied. Possibly, however, there are yield conditions which lead to smooth characteristic surfaces under weaker restrictions on the strain rates.

INTRODUCTION

It is well known that fault patterns closely resembling structures observed in the field can be initiated by deforming masses of dry sand or wet clay. It is not our intention to describe or to enumerate here the many experiments performed by various investigators. [*See* for example some of the experiments of Cloos (1929) and Hubbert (1951).] The purpose of this paper is rather to investigate the mechanism of faulting in materials like sand and clay and to see whether it is possible to predict the systems of fault planes arising in a mass of sand or clay when forces are applied in a prescribed manner to the outer boundaries of this mass.

It will be assumed that the geometry of fault systems in materials like sand and clay depends solely on the state of stress within the body. This is a fundamental assumption which seems justified by the experimental evidence mentioned above.

In experiments with earthy materials the deformations in the stage where faulting becomes visible are almost exclusively outside the elastic range. Therefore, an analysis of fault patterns, based entirely on an assumed relation between the various components of the stress tensor, cannot be made using the theory of elasticity. Faults in sand and clay grow slowly as long as sufficient external force is applied and can be stopped at any time by decreasing these forces. The regions between the faults remain permanently deformed. It is apparent that a "ductile" process is involved in contrast to the phenomenon of sudden rupture of glass wires under tension or the shattering of a glass plate under a sudden impact.

[*Editor's Note:* A row of asterisks indicates that material has been deleted.]

* * * * * *

It follows from this that velocity discontinuities can arise only along characteristics of the velocity equations.

If such a velocity discontinuity is present, it will appear that the plastic medium, after flowing for some time, has been faulted along the discontinuity. In the case of plane strain the fault planes contain the z direction. If discontinuous behavior is possible in the general three-dimensional case of plasticity, then it may occur along

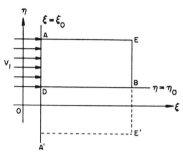

FIGURE 12.—*Velocities prescribed on characteristic arcs in the ξη plane (V₂ = 0 in the figure)*

curved planes. It is the purpose of this paper to point out that faults in materials like sand, clay, and possibly rocks can be interpreted as velocity discontinuities arising along one or more members of the characteristic surfaces of the velocity equations.

SOME EXPERIMENTS WITH DRY SAND

The theory can be verified by some experiments. One series of experiments with dry sand has been described by Hubbert (1951). In a box filled with dry sand, a wooden partition was moved forward by a screw mechanism, and the deformation of the sand, marked with thin bands of dry plaster of Paris, could be observed through a plate-glass wall. In one half of the box the sand was compressed, and thrusting occurred; in the other half the sand was extended, and a normal fault formed. The conditions of plane strain were well simulated except for some friction at the front and back walls of the box. It will be assumed that in the thrust compartment of the box the following conditions were satisfied: (1) the sand deformed plastically according to von Mises' yield condition, and (2) the shear along the bottom of the box, close to the moving partition, was constant and equal to k. It is then not difficult to show that a theoretical solution of the stress field within the sand can be constructed from three parts: (1) a constant-state area, (2) a centered fan, and (3) a curvilinear part. A constant-state area is by definition an area in which the functions α and J_1 are constant throughout. This implies that both families of characteristics are straight lines (Prager and Hodge, 1951). A center fan is an area in which one family of characteristics are straight lines through one point. The stress fields for the constant-state area and the centered fan can be written down explicitly. In the curvilinear part the stress field can be computed by numerical methods. The characteristics of this solution are shown in Figure 13. This solution can be compared with the results of Hubbert's experiments. A number of drawings, prepared after photographs of his experiments are shown in Figure 14 A–D. Especially in the initial stages of the experiment the correspondence between theory and observation is excellent. It should be noted that Hubbert (1951, Fig. 16) has proposed an alternative solution which also agrees well with the experimental observations.

* * * * * *

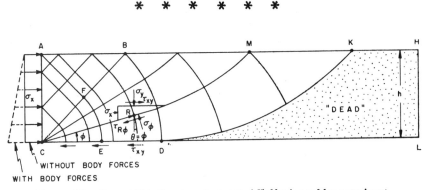

FIGURE 13.—*Stress field in thrust compartment of Hubbert's sand-box experiment*

* * * * * *

REFERENCES

Cloos, H., 1929, Künstliche Gebirge: Natur und Museum, Frankfurt, Senckenbergische Naturforschende Ges., v. 5, p. 225–243.

Hubbert, M. K., 1951, Mechanical basis for certain familiar geologic structures: Geol. Soc. America Bull., v. 62, p. 355–372.

Prager, W., and Hodge, P. G., 1951, Theory of perfectly plastic solids: New York, John Wiley and Sons, 264 p.

[*Editor's Note:* Figure 14 appears on the next page.]

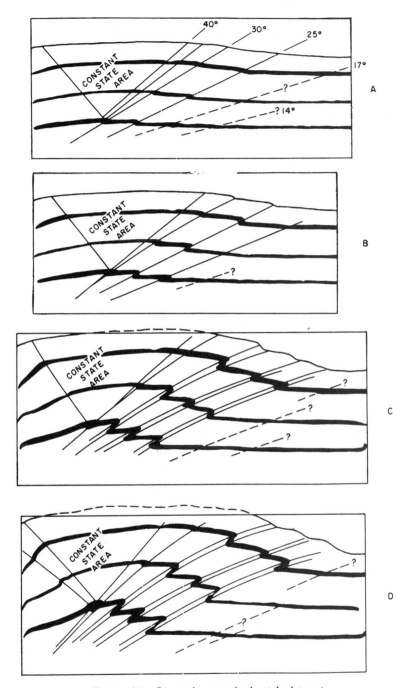

FIGURE 14.—*Consecutive stages in thrust development*
After Hubbert (1948)

Editor's Comments
on Papers 25 Through 27

The transition from the classical theories of civil engineers in the pre–World War I era to what is commonly termed "soil mechanics" occurred in a relatively brief period of professional adventure. The transition occurred simultaneously in several countries, caused by the widening gap between the requirements of canal and foundation design and an adequate mental grasp of the essentials involved (Terzaghi, 1936). In the United States, the catastrophic descent of the Culebra cut on the Panama Canal issued a warning that led to quantification of earth properties; at the same time (1913), a Geotechnical Commission was appointed by the Swedish State Railway to eliminate, if possible, the recurrence of catastrophic landslides in post-glacial sediments, and their labors resulted in development of methods still employed in stability investigations; and in Germany, failures of the quay wall along the Kiel Canal construction led to the development of new computational methods of retaining wall and bulkhead design.

Meanwhile, Karl Terzaghi was in the United States searching for the "philosopher's stone" by collecting and coordinating geological data in dam-site construction camps of the U.S. Reclamation Service. When funds ran out, he worked temporarily as a driller at the Celilo lock construction of the Columbia River. When he returned disheartened to Austria, World War I had broken out; Terzaghi then served as a first lieutenant in the design and research unit of which Kármán and Mises were important members. In 1916 he was sent to the Imperial School of Engineers in Istanbul, and after the war accepted a lectureship at Robert College. The rest is well known, for in the discovery of the effective-stress principle, Terzaghi had indeed found the "philosopher's stone."

The effective-stress principle played a major role in Terzaghi's studies, as exemplified by an excerpt from his outstanding 1950 review of slide mechanisms (Paper 26). The principle applies, in essence, to all geological materials; thus the Coulomb–Terzaghi strength equation is applied by Terzaghi as successfully to stratified rocks as to sediments, as shown by his discussion of the classic Goldau rockslide.

Terzaghi's 1950 paper is preceded by a generally unrecognized paper, *A Phenomenal Landslide,* which nevertheless seems as remarkable as its title suggests. Its background is as follows. In 1893 and 1894, two reservoirs constructed by the City of Portland, Oregon became involved in a massive, deep-seated, reactivated slide over 500 meters in length. Extensive investigations showed that ground water was the prime factor producing the slide; remedial drainage was suggested, which later proved effective as a cure. In 1898 the owners of the ground instituted a suit against the city for property damage, and a trial began the following year, which is referred to in Paper 26. I know of no slide which, up to that time, had received such exhaustive study, and few since which have so well documented the role of water in the sliding process.

Indeed, the conclusion expressed by George Dillman, in a discussion appended to Clarke's work (Paper 25), strikes directly to the heart of the matter: water pressure is compared to a "myriad of jackscrews" lifting the slide mass. This, of course, is comparable to Terzaghi's analogy of a "hydraulic jack" (Paper 26, p. 92); yet it preceded the publication of *Erdbaumechanik* by two decades.

Finally, we present in Paper 27 a selection from an important article by L. F. Harza, a consulting engineer concerned with the problem of the sliding of concrete dams over their foundations. Harza attempted to show by deductive reasoning that under and within any concrete structure (e.g., a gravity dam) founded on soil or rock, hydrostatic uplift acts over the entire basal area. This, in hydraulic engineer's jargon, is the doctrine of "100 percent effective uplift area" and is equivalent to application of the Archimedes principle of buoyancy. We have seen that Terzaghi had already shown that the "surface porosity" associated with his microstructural idealization is required to be approximately unity (0.97–1.03) to account for experimental results. Accordingly, to Harza this microstructural approach is unnecessary; buoyancy is complete and requires only that the material contain interconnected pores. Thus, the effective-stress concept for concrete or rock is wholly accepted; the simplest equations, requiring no "surface-porosity" term, can be employed.

Subsequent critics of Harza's interpretations "subpoenied as witnesses the great men of the past back to Galileo and . . . even invoked the German language," but in an artful 20-page closure following extensive discussion, Harza appears to have maintained his ground.

25

Reprinted from pp. 331, 375, 379–381, 398–399 of *Amer. Soc. Civil Eng. Trans.*, 53 (Paper 984), 322–397; Discussion, 398–412 (1904)

A PHENOMENAL LAND SLIDE.

By D. D. CLARKE, M. AM. SOC. C. E.

* * * * * *

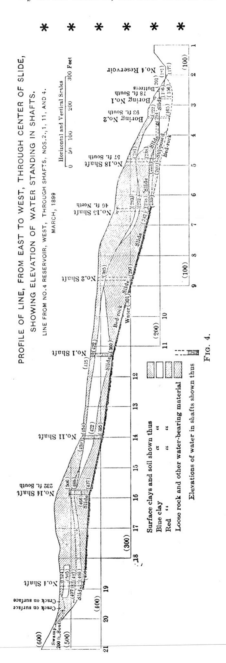

FIG. 4.

[*Editor's Note:* A row of asterisks indicates that material has been deleted.]

*   *   *   *   *   *

Another interesting exhibit used upon the trial was a diagram showing a comparison of the volume of the rainfall and the movement of the slide for corresponding months, from January, 1895, to the date of the trial. This diagram has since been extended so as to cover the entire period during which the surveys have been continued. This diagram is shown in Plate XXV.

The measurement of rainfall was taken from the reports of the United States Weather Bureau in Portland. It is to be noted that while the amount of precipitation is extremely large for some months, notably 13.12 ins. for November, 1896, the average for any one season rarely exceeds 44 ins.

The movement, as recorded on the diagram, is the average of fourteen points observed at intervals along the central portion of the sliding ground, where the maximum movement was supposed to have taken place, from September, 1895, to December, 1899. Since the latter date the average of a large number of points has been taken, two hundred and sixty points having been established and observed regularly. The average of the readings at fifty-one of these points, along a central belt 100 ft. wide, has been used in plotting the diagram of movement for 1900–1903. Merely casual inspection and comparison will serve to show the apparently close relation which exists between the volume of the rainfall during what may be called the wet and dry seasons of the year—say, from December to May, inclusive, for the former, and from June to November, inclusive, for the latter —and the movement of the slide during the same period. It will be seen from the diagram that with each recurring dry season there was a corresponding cessation of the movement, and that with the beginning of the winter rains the movement increased. It is to be

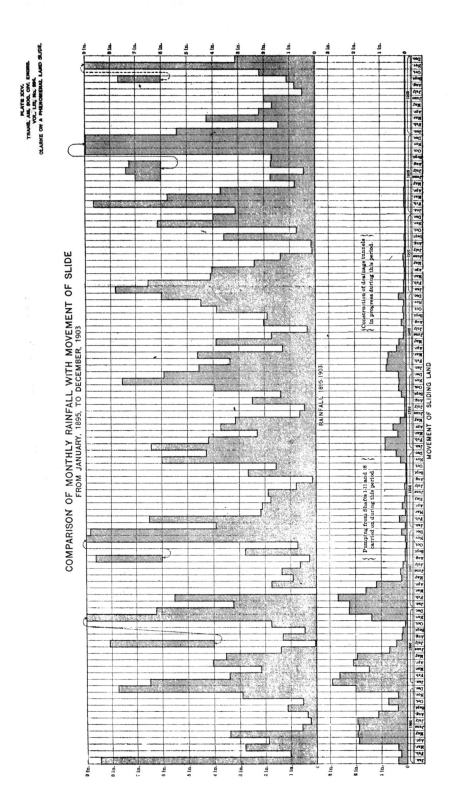

COMPARISON OF MONTHLY RAINFALL WITH MOVEMENT OF SLIDE
FROM JANUARY, 1895, TO DECEMBER, 1903

PLATE XXV.
TRANS. AM. SOC. CIV. ENGRS.
VOL. LII, NO. 994.
CLARKE ON A PHENOMENAL LAND SLIDE.

RAINFALL 1895-1903

MOVEMENT OF SLIDING LAND

{ Pumping from Shafts 1-11 and 18 }
{ carried on during this period. }

{ Construction of drainage tunnels }
{ in progress during this period. }

noted, also, that, in each instance, an interval of about one month elapsed between a change in the volume of rainfall and the corresponding change in the rate of movement.

An inspection of the diagram will show at once one irregularity in the yearly movement, namely, a marked increase in the monthly movement during the winter months is followed usually by a corresponding diminution with the return of dry weather. The explanation of this irregularity in the movement, which occurred during the years 1897 and 1898, forms one of the most interesting of the developments which have occurred since the study of this problem was begun.

It will be recalled that while the excavation of Shaft No. 1 was in progress considerable water was found, the draining of which was accomplished by deep-well pumps. The draining of this well was in progress from August 19th, 1897, until January 31st, 1898, the total volume of water removed during that period being more than 3 900 000 galls. Even then, the underground reservoir had not been drained thoroughly, for, when pumping ceased, the water rose in the shaft to a height of several feet.

By observing the diagram it will be noted that when pumping from this shaft was in progress, the movement did not increase as rapidly during the winter months as it had during former seasons, although there was a slight increase from August to February when the maximum for that year was reached. The movement for February, 1898, was only 0.30 in., as compared with 2.69 ins. for February, 1897, and 0.85 in. for February, 1899. Perhaps this irregularity can be shown best by comparing the total movement for the entire season preceding with the total movement for the season following the drainage operations. From May, 1896, to May, 1897, the total movement was 13.10 ins. From May, 1897, to May, 1898, during a portion of which period pumping was in progress, the total movement amounted to only 1.57 ins.; and this was followed by an increase to 3.69 ins. during the year beginning May, 1898, and ending May, 1899.

It is probable that the decreased movement noted during the year 1898 was due in part to the pumping of water from Shaft No. 18, in July and August, amounting to 650 000 galls., and from Shaft No. 11, in October and November, amounting to 540 000 galls., but it is believed to have been due principally to the thorough drainage of the underground reservoir connected with Shaft No. 1. As has been stated, the drainage of this body of water was practically completed in January, 1898, pumping having been suspended on the 31st of that month, and the water was thereafter allowed to accumulate, the rise in the shaft indicating the rate at which the reservoir was being filled. This rate was noted carefully, and it was observed that the underground reservoir was not filled to its original height until the end of December of that year.

During the period when this underground reservoir was being drained, and for some months thereafter, the movement was at a greatly diminished rate. When the underground reservoir was again filled the rate of movement began to increase until it was more than double that of the preceding year, and this uniformity of rise and fall was continued for the two seasons following.

It was this seeming coincidence of the cessation of the movement with the draining of the underground reservoir at Shaft No. 1 which first changed conviction into certainty that the cause of the slide was to be found in the underground water stored in various portions of the sliding ground, and hence it followed that a remedy for the difficulty was to inaugurate a thorough system of drainage.

This diagram was one of the most important and convincing "exhibits" introduced by the Water Committee in defending the suit, as it showed, clearly, that the movement depended more, and chiefly, upon the volume of water falling upon the surface and stored in underground reservoirs in the interstices of the broken rock forming a large portion of the mass of the slide, than upon the removal of a small fragment of earth from the toe of the slope.

DISCUSSION.

Mr. Dillman. GEORGE L. DILLMAN, M. AM. Soc. C. E. (by letter).

The investigation of this slide is by far the most complete of which the writer has any knowledge. It seems to make plain that the water does the damage, not so much by its volume or weight as its pressure, and, at the surface of motion, acts like a myriad of jack-screws to lift, and at the same time lubricate it, the same pressure forcing the water into and through otherwise impervious strata.

The writer knows of many slides which have been cured by drainage, and knows of no failures to do so when the remedy was applied properly, and in the right places.

26

Reprinted from pp. 83, 91–94, 121, 122, 123 of *Application of Geology to Engineering Practice: Berkey Volume,* Sidney Paige, Chairman, Geological Society of America, 1950, pp. 83–123

MECHANISM OF LANDSLIDES

By Karl Terzaghi

Harvard University, Cambridge, Mass.

* * * * * *

LUBRICATING EFFECT OF WATER

If a slide takes place during a rainstorm at unaltered external stability conditions, most geologists and many engineers are inclined to ascribe it to a decrease of the shearing resistance of the ground due to the "lubricating action" of the water which seeped into the ground. This explanation is unacceptable for two reasons.

First of all, water in contact with many common minerals, such as quartz, acts as an anti-lubricant and not as a lubricant. Thus, for instance, the coefficient of static friction between smooth, dry quartz surfaces is 0.17 to 0.20 against 0.36 to 0.41 for wet ones (Terzaghi, 1925, p. 42–64).

Second, only an extremely thin film of any lubricant is required to produce the full static lubricating effect characteristic of the lubricant. Any further amount of lubricant has no additional effect on the coefficient of static friction between them (Hardy, 1919). In humid regions such as the eastern United States and within less than 1–2 feet from the sloping surface every sediment—sand included—permanently contains far more than the quantity of water needed for "lubricating" the surfaces of the grains (Terzaghi, 1942, Fig. 15, p. 356). Yet in humid regions rainstorms start landslides as often as they do in arid ones. In other words, since practically all the sediments located beneath slopes are permanently "lubricated" with water, a rainstorm cannot possibly start a slide by lubricating the soil or boundaries between soil strata.

However, the rain water which seeps into a slope affects the stability of the slope in various other ways. If the voids of the ground are partly filled with air, the water eliminates the surface tension which imparts to fine-grained, cohesionless soils a considerable amount of apparent cohesion (Terzaghi and Peck, 1948, p. 114–128). The water which enters the voids also increases the unit weight of the soil though, as a rule, this increase is commonly unimportant.

Some soils, such as typical loess, owe their cohesion to a soluble binder. If a slope on such a soil is submerged for the first time, or if the soil becomes saturated by seepage from a newly created, artificial source of water, the binder is removed by solution and the soil loses its cohesion.

Last—but not least—water which enters the ground beneath a slope always causes a rise of the piezometric surface[3], which, in turn, involves an increase of the pore-water pressure and a decrease of the shearing resistance of the soil. Since water can affect the stability of slopes in several radically different ways, its actions will be discussed under different subheadings.

---

[3] The piezometric surface is the locus of the points to which the water would rise in piezometric tubes. If the permeability of a soil, such as a soft clay, is too low to permit locating the position of the piezometric surface by means of observation wells, pressure gages must be used.

[*Editor's Note:* A row of asterisks indicates that material has been deleted.]

RISE OF PIEZOMETRIC SURFACE

Throughout a saturated mass of jointed rock, soil, or sediment, the water which occupies the voids is under pressure. Let

p = pressure per unit of area at a given point P of a potential surface of sliding, due to the weight of the solids and the water located above the surface,
h = the piezometric head at that point,
w = the unit weight of the water, and
ϕ = the angle of sliding friction for the surface of sliding.

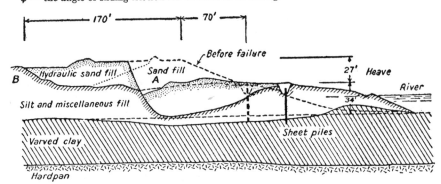

FIGURE 3.—*Section through a slide which was caused by an excess hydrostatic pressure in the silt layers of a stratum of varved clay*

The row of sheetpiles advanced in a few minutes over a maximum distance of up to 60 feet toward the river.

Regarding the relation between these four quantities, soil mechanics has led to the following conclusions (Terzaghi and Peck, 1948, p. 51–55). If the potential surface of sliding is located in a layer of sand or silt, the shearing resistance s per unit of area at the observation point is equal to

$$s = (p - hw) \tan \phi \qquad (3)$$

Hence, if the piezometric surface rises, h increases, and the shearing resistance s decreases. It can even become equal to zero. The action of the water pressure hw can be compared to that of a hydraulic jack. The greater hw, the greater is the part of the total weight of the overburden which is carried by the water, and as soon as hw becomes equal to p the overburden "floats." If a material has cohesion, c per unit of area, its shearing resistance is equal to the sum of s, equation (3), and the cohesion value c, whence

$$s = c + (p - hw) \tan \phi \qquad (4)$$

The effect of a decrease of the shearing resistance s on the stability of slopes on stratified sediments is illustrated by Figure 3. The figure shows a vertical section through sand dikes which were constructed along a river for flood-protection purposes. The dikes rest on a layer of soft silt and miscellaneous fill which covers the surface of a horizontal stratum of varved clay with a thickness of about 50 feet. The dash line shows the position of the dike and of the boundaries between the underlying strata prior to the slide.

Some time after the dike A was completed by depositing and compacting moist sand in layers, the space between the landward slope of this dike and the outer slope of an older dike, *B*, was filled with sand. The sand was excavated by means of a hydraulic dredge and deposited in a semiliquid condition. The sluicing operations raised the piezometric level in the pore water of the silt layers in the varved clay to a considerable height *h* above the original water table. As a consequence the resistance *s*, equation (3), against sliding decreased.

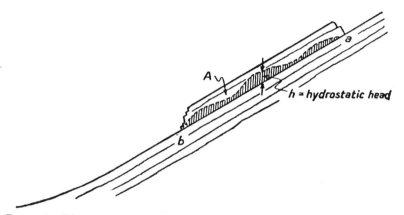

FIGURE 4.—*Diagrammatic section through site of rock slide of Goldau (1806) prior to slide*

Slab *A* was separated from its base by a thin layer of weathered rock. The dashed line represents the piezometric surface in this layer during a heavy rainstorm.

During the construction of the hydraulic fill the dike *A* suddenly subsided, and the row of sheet piles, together with the foreland, moved over a distance up to 60 feet and over a length of about 1200 feet toward the river. The row of sheet piles remained perfectly intact. This fact showed that the failure had occurred by sliding along one or more horizontal surfaces of sliding located in the varved clay. If the shearing resistance along these surfaces had not been extremely low, the wedge-shaped body of silt, located on the river side of the sheet piles, could not possibly have advanced over a distance up to 60 feet without undergoing intense compression and shortening in the direction of the movement.

After a slide has occurred the excess pore-water pressure in the zone of shear always decreases, on account of progressive consolidation, and approaches a value zero. In order to get information on the rate of decrease of the pressure which prevailed in the varved clay, a great number of pressure gages were installed. The first readings were made more than 3 months after the slide occurred. Yet, even beneath the banks of the river, the piezometric elevation *h*, equation (3), still amounted to more than 10 feet, with reference to the river level.

The relation expressed by equation (3) also applies to stratified or jointed rocks. To illustrate its bearings on rockslides, the classical slide of Goldau in Switzerland will be discussed. This slide has always been ascribed to the "lubricating action" of the rain- and meltwater. Figure 4 is a diagrammatic section through the slide area. It shows a slope oriented parallel to the bedding planes of a stratified mass of Tertiary Nagelflue (conglomerate with calcareous binder) which rises at an angle of 30° to the horizontal. On this slope rested a slab of Nagelflue 5000 feet long, 1000 feet wide, and about 100 feet thick. It was separated from its base by a porous layer of weathered rock.

The fact that the slab had occupied its position since prehistoric times indicates that the shearing force, which tended to displace the slab, never exceeded the shearing strength, in spite of the effects of whatever hydrostatic pressures, hw, in equation (3), may have temporarily acted on the base of the slab in the course of its existence.

On September 2, 1806, during heavy rainstroms, the slab moved down the slope, wiped out a village located in its path, and killed 457 people (Heim, 1882). This catastrophe can be explained in at least three ways. One explanation is that the angle of inclination of the slope had gradually increased on account of tectonic movements, until the driving force which acted on the slab became equal to the resistance against sliding. A second explanation is based on the assumption that the resistance of the slab against sliding was due not only to friction, but also to a cohesive bond between the mineral constituents of the contact layer. The total shearing resistance due to the bond was gradually reduced by progressive weathering, or by the gradual removal of cementing material, either in solution or by the erosive action of water veins. The third explanation is that h in equation (3) or (4) assumed an unprecedented value during the rainstorm, whereas the cohesion c, in equation (4), remained unchanged, provided cohesion existed. In Figure 4 the value h is equal to the vertical distance be-between the potential surface of sliding, ab, and the dash line interconnecting ab which represents the piezometric line. During dry spells h is equal to zero. In other words, the piezometric surface is located at the slope. During rainstorms the rain water enters the porous layer located between slab and slope at a and leaves it at b. Since the permeability of this layer is variable, the piezometric line descends from a to b in steps, and the value h in equations (3) and (4) is equal to the average vertical distance between the piezometric line and the slope.

The maximum value of h changes from year to year, and if the exits of the water veins at b are temporarily closed by ice formation while rain- or melt-water enters at a, h assumes exceptionally high values. However, the seasonal variations of h, the corresponding variations of s, equations (3) and (4), and the occasional obstruction of the exits at b have occurred in rhythmic sequence for thousands of years, without catastrophic effects. It is very unlikely that h assumed a record value in 1806, in spite of unaltered external conditions. Therefore it is more plausible to assume that the slide was caused by a process which worked only in one direction, such as a gradual increase of the slope angle or the gradual decrease of the strength of the bond between slab and base. In no event can the slide be explained by the "lubricating effect" of the rain water. One might as well ascribe a theft to some mysterious effects of the presence of the thief in the house instead of inquiring about his physical actions.

*   *   *   *   *   *

REFERENCES

Hardy, W. B. and T. V. (1919) *Note on static friction and on the lubricating properties of certain chemical substances*, Philos. Mag. London, vol. 39, no. 223, p. 32–35.

Heim, A. (1882) *Über Bergstürze*, Naturf. Gesell. Zürich, Neujahrsblatt 84.

Terzaghi, K. (1925) *Erdbaumechanik*, Franz Deuticke, Wien, 399 pages.

———— (1936) *Stability of slopes on natural clay*, 1st Inter. Conf. Soil Mech. Found. Eng., Pr., Cambridge, Mass., vol. I, p. 161–165.

———— (1942) *Soil moisture and capillary phenomena in soils*, Physics of the Earth, vol. IX (Hydrology), McGraw-Hill Co., New York, p. 331–363.

———— (1943a) *Measurements of pore-water pressure in silt and clay*, Civil Eng., vol. 13, p. 33–36.

———— (1943b) *Theoretical soil mechanics*, John Wiley and Sons, New York, 510 pages.

———— and Peck, R. B. (1948) *Soil mechanics in engineering practice*, John Wiley and Sons, New York, 566 pages.

27

Reprinted from pp. 193, 197-200 of *Amer. Soc. Civil Eng. Trans.*, 114(2368), 193-214; discussion, 215-289 (1949)

THE SIGNIFICANCE OF PORE PRESSURE IN HYDRAULIC STRUCTURES

By L. F. Harza, M. ASCE

* * * * * *

This principle (that gravel, sand, and earth, submerged in still water, are fully buoyant—in other words, are reduced in weight to the full extent of the displacement of individual particles thereof) is quite generally accepted. The popular acceptance of this principle rests on the belief that the point to point contacts of each particle with the adjacent ones do not reduce the

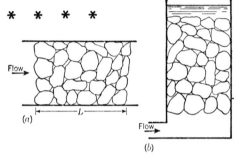

Fig. 3.—Liquid Flow Through Loose Granular Material

effective surface area of each particle subjected to hydrostatic pressure. In other words, each particle is fully surrounded by water like the submerged object in Fig. 1, regardless of these point to point contacts.

If it be assumed that the particles in Figs. 3(a) and 3(b) are all cemented together at their points of contact without increasing the areas of contact, then obviously no change in pressure will result and the solid body thus created would tend to move as a whole, not because of pressure against the upstream face, but because of progressively reducing internal pore pressure causing buoyant and seepage forces to act against each particle.

That all the principles previously enumerated would be just as true if the pores are filled with cement paste to form concrete is, however, not generally accepted by the profession and needs demonstration. Popular opinion, with few exceptions, regards as a fact the belief that the cementing together of particles to form concrete reduces the top and bottom areas of particles subjected to water pressure and thus excludes water pressure from so much of the surface area of each original particle as to preclude the action of buyoancy on the individual particles of the mass. It will be shown, on the contrary, that internal buoyancy continues to be fully as effective in concrete as in loose granular material.

In Fig. 4(a) it is first considered that there is no flow, so that only static pressure is effective. It has been proved herein that concrete is quite porous and it is well known that seepage under pressure passes through any concrete. It has likewise been proved experimentally[7] that a submerged specimen of concrete cannot be measurably compressed by subjecting the liquid in which it is submerged to a high pressure. This proves that all pores communicate and that the static pressure in Fig. 4(a) must exist in all pores. If the cross

---

[6] "Uplift and Seepage Under Dams on Sand," by L. F. Harza, *Transactions*, ASCE, Vol. 100, 1935, p. 1352.

[7] "Simple Tests Determine Hydrostatic Uplift," by Karl Terzaghi, *Engineering News-Record*, June 18, 1936, p. 872.

[*Editor's Note:* A row of asterisks indicates that material has been deleted.]

section of any one pore, such as a, is projected upward, it will eventually end in another pore, such as b, or in greater distance, as c to d; or a large pore, as at e, may require continuation until two or more pores, such as f and g, are encountered before the entire cross section of the cylinder from e is terminated in pores. In any event,

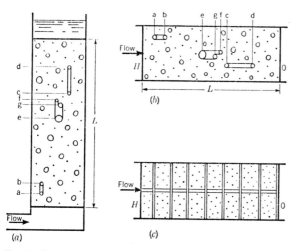

the probability is so great, in any length of concrete greater than a fraction of an inch, as to become a certainty that each such cylinder will terminate in pores at both ends. Moreover, the entire volume of concrete will thus subdivide into such cylinders. Since the differences in hydrostatic head to which these pores will be subjected are equal to their differences in elevation, each column will be

FIG. 4.—INTERRELATION OF PORES AND FLOW THROUGH CONCRETE

subjected to buoyancy just as is the elementary column in Fig. 1. Likewise, since each column is buoyant, and the whole object is composed of such columns, then the entire internal structure is buoyant, and the effect of cementation is nullified.

Even though the popular belief may be shared that internal buoyancy cannot exist in concrete because water pressure cannot be effective over the entire top and bottom areas of the original sand and gravel particles of aggregate because of the cement, yet it is only necessary thus to dismiss consideration of the original aggregate particles and visualize a new conception of a different elementary unit of material to obtain cylindrical particles that are exposed to pressure at top and bottom, and thus to restore the conception of full buoyancy, as in an uncemented mass. The entire volume of submerged concrete may be thought of, therefore, as built up of vertical cylinders averaging very much less than 1 in. in height, each terminating at a pore at top and bottom and subjected to buoyancy—the entire mass therefore weighing only from about 90 to 95 lb per cu ft.

Fig. 4(b) indicates the same pores as illustrated in Fig. 4(a). It has been previously shown that they are sufficient, even in the best of concrete, to form layers of pores filling the entire cross section at intervals of a fraction of an inch, as shown in Fig. 4(c). If water pressure is applied to the concrete filled pipe in Fig. 4(b), the pressure will reduce from H at entry to 0 at the point of escape. Intermediate pores will contain pressures reduced proportionally to distance, the pressure in these pores decreasing from left to right. Each solid cylinder between pores is therefore subject to a force equal to the differential pressure between the pores at the ends of such cylinders. Thus the differential pressure on cylinders ab and cd, if H is the total head lost in the distance L, will be, respectively, $\frac{\overline{ab}}{L} H$ and $\frac{\overline{cd}}{L} H$. The total volume under pressure is the sum of the elementary cylinder volumes; the total cross section is equal to the total cross section of the specimen; and the total summation of pressure is H. The total force therefore is $H A$, in which A is the cross section of the entire specimen. The volume can also be divided into vertical cylinders at right angles to the seepage forces, indicating internal buoyancy. If seepage forces are added to Fig. 4(a) by an applied water pressure, then they become additive to the buoyancy forces instead of normal, as in Fig. 4(b). Ideally the situation is the same as if the pores were to be concentrated into layers as in Fig. 4(c), and each layer of pores were connected by a fine seepage channel as shown, so that the water pressure would reduce in steps from H to 0 and be exerted over the entire area of each concrete disk in each layer of pores. Each disk would thus be subject to a thrust equal to the difference in pressure on its two faces.

The principle is identical with that of flow through cohesionless sand or gravel, and filling the voids with cement paste thus has no influence except to decrease the percolation factor—just as cementing the contact points had no effect on the loose granular material (Fig. 3(b)).

The only difference between buoyant and seepage forces is in the rate of dissipating pressure through the specimen. In buoyancy the rate of increase of pressure with increasing depth is fixed by the weight of water, whereas in Figs. 3(a) and 4(b) it is at the rate of $\frac{H}{L}$, whatever that may be. Also, the pressure is applied not at the entrance face, but along the route of seepage and in the direction of seepage.

It remains only to summarize the principles of static pressure and flow in concrete in comparison with these phenomena in loose granular material, based upon arguments herein and other known and recognized facts.

These principles are:

(a) Concrete is a porous material with all the pores communicating as in loose material.

(b) Hydrostatic pressure will develop in the pores of submerged concrete, equal in magnitude to the depth of a given pore below static water surface, except that a greater time element may elapse in reaching a static condition, analogous to the case of clay rather than to that of coarse loose material.

(c) Flow of water through concrete obeys Darcy's law as it does through loose material.

(d) Internal buoyancy exists in submerged concrete as truly as in loose material; to visualize this it is only necessary to conceive a different unit of material than the original aggregate pieces composing the concrete.

(e) There is no difference hydrostatically or hydraulically between the original loose aggregate and the finished concrete except for the reduction in the percentage and size of voids. These differences have no effect on the laws of flow, but merely on the permeability constants.

(f) The only difference in principles is that the particles adhere to each other in concrete and do not in the original aggregate, but the laws of hydraulics and hydrostatics do not concern themselves with this difference.

Editor's Comments
on Paper 28

28　　HUBBERT and RUBEY

Role of Fluid Pressure in Mechanics of Overthrust Faulting:
I. Mechanics of Fluid-Filled Porous Solids and Its Application to
Overthrust Faulting

We have thus traced, to the mid-1950s, the historical development of concepts known from the perspective of hindsight to be of significance to the supposed "mechanical paradox of overthrust faulting." We therefore now turn to the focal manuscripts of this volume, the important Hubbert–Rubey and Rubey–Hubbert collaborations. Space limitations have regretfully forced us to entirely delete Part II of the Hubbert–Rubey collaboration from this volume; the Geological Society of America has, however, recently made this publication available in complete reprint form (see also p. 245–283 in Hubbert, 1972.)

William Rubey's geological interest in fluids developed rather early in his career and led to fundamental discoveries (e.g., in sediment transport and in the origin of the oceans). Extensive field campaigns covering over three decades in Idaho and Wyoming are of more direct relevance to the thrust-fault question, inasmuch as they led him to search for the riddle posed by the "coefficient-of-friction" approach (cf. Longwell, Paper 20). An artesian fluid-pressure mechanism (cf. Paper 26, p. 92) seemed reasonable to Rubey and was explored in 1948, but it seemed insufficient as a generally applicable explanation. However, Rubey's recognition of the plausibility of other fluid-pressurization mechanisms in the mid-1950s rekindled his consideration of the tectonic role of fluid pressure and led to collaboration when Hubbert's mutual interest became known.

Meanwhile, Hubbert had been pursuing two apparently independent lines of inquiry, one of which, concerning the mechanics of rock behavior, has previously been sufficiently discussed. The other line of investigation concerned the mechanics of fluid flow in porous media, which directly led to what may have been Hubbert's finest contribution, his "theory of ground water motion." Only subsequently, in investigations associated with the problems of circulation loss in oil well drilling and hydraulic fracturing, did Hubbert discover that the mechanics of solids and mechanics of fluid flow were, in fact, necessarily interrelated topics in mechanical considerations of porous geo-logical materials. Hubbert's important paper "Mechanics of Hydraulic Fracturing," coauthored with David Willis, was one result of these investigations, and first demonstrated that hydraulic fracturing could be employed as a stress-measurement device (cf. Paper 43). The possibility of applying the effective-stress concept in overthrust mechanics came into Hubbert's focus somewhat later, appropriately enough, in Canton Glarus of Escher von der Linth.

Reprinted from *Geol. Soc. America Bull.*, **70**(2), 115-166 (1959)

ROLE OF FLUID PRESSURE IN MECHANICS OF OVERTHRUST FAULTING

I. Mechanics of Fluid-Filled Porous Solids and Its Application to Over-thrust Faulting

By M. King Hubbert and William W. Rubey

Abstract

Promise of resolving the paradox of overthrust faulting arises from a consideration of the influence of the pressure of interstitial fluids upon the effective stresses in rocks. If, in a porous rock filled with a fluid at pressure p, the normal and shear components of total stress across any given plane are S and T, then

$$\sigma = S - p, \tag{1}$$

$$\tau = T, \tag{2}$$

are the corresponding components of the *effective* stress in the solid alone.

According to the Mohr-Coulomb law, slippage along any internal plane in the rock should occur when the shear stress along that plane reaches the critical value

$$\tau_{\text{crit}} = \tau_0 + \sigma \tan \phi; \tag{3}$$

where σ is the normal stress across the plane of slippage, τ_0 the shear strength of the material when σ is zero, and ϕ the angle of internal friction. However, once a fracture is started τ_0 is eliminated, and further slippage results when

$$\tau_{\text{crit}} = \sigma \tan \phi = (S - p) \tan \phi. \tag{4}$$

This can be further simplified by expressing p in terms of S by means of the equation

$$p = \lambda S, \tag{5}$$

which, when introduced into equation (4), gives

$$\tau_{\text{crit}} = \sigma \tan \phi = (1 - \lambda)S \tan \phi. \tag{6}$$

From equations (4) and (6) it follows that, without changing the coefficient of friction $\tan \phi$, the critical value of the shearing stress can be made arbitrarily small simply by increasing the fluid pressure p. In a horizontal block the total weight per unit area S_{zz} is jointly supported by the fluid pressure p and the residual solid stress σ_{zz}; as p is increased, σ_{zz} is correspondingly diminished until, as p approaches the limit S_{zz}, or λ approaches 1, σ_{zz} approaches 0.

For the case of gravitational sliding, on a subaerial slope of angle θ

$$T = S \tan \theta, \tag{7}$$

where T is the total shear stress, and S the total normal stress on the inclined plane. However, from equations (2) and (6)

$$T = \tau_{\text{crit}} = (1 - \lambda)S \tan \phi. \tag{8}$$

Then, equating the right-hand terms of equations (7) and (8), we obtain

$$\tan \theta = (1 - \lambda) \tan \phi, \tag{9}$$

which indicates that the angle of slope θ down which the block will slide can be made to approach 0 as λ approaches 1, corresponding to the approach of the fluid pressure p to the total normal stress S.

Hence, given sufficiently high fluid pressures, very much longer fault blocks could be pushed over a nearly horizontal surface, or blocks under their own weight could slide down very much gentler slopes than otherwise would be possible. That the requisite pressures actually do exist is attested by the increasing frequency with which pressures as great as $0.9S_{zz}$ are being observed in deep oil wells in various parts of the world.

CONTENTS

Introduction and Acknowledgments

The present collaboration on this paper is an integration of the results of two lines of inquiry which, during the preceding 30 years, the authors have pursued separately. During this period they have each benefitted from the written and oral discussions of many people which have influenced their subsequent thinking with respect to various aspects of the problem now to be considered. It appears, therefore, that an effective way of acknowledging these obligations would be to give brief résumés of the evolution of each of the authors' thinking with regard to this subject.

Since the mid-1920's Hubbert has had a continuing interest in many of the problems of the mechanics of rock deformation. His first acquaintance with these problems arose during field trips into the regions of highly folded Precambrian rocks in southwestern Wisconsin and the United States–Canadian boundary north of Lake Superior. These field observations were supplemented by the lectures in structural geology of Professor Rollin T. Chamberlin at the University of Chicago, and by extensive reading of the writings of Charles R. Van Hise, C. K. Leith, and W. J. Mead. These cumulative experiences served as an introduction to many of the phenomena of the mechanics of rock deformation, but at the same time led to the conclusion that an understanding of these phenomena would probably not be forthcoming until they had been examined more fully from the viewpoint of the mechanics of deformable bodies. This subject existed already in the theory of elasticity and the hydrodynamics of viscous fluids, but other aspects appropriate to the problems of geology remained to be developed.

Soon after this (1931) Hubbert had the good fortune to meet A. Nadai who had recently arrived in the United States from the University of Göttingen in Germany. Nadai was the author of a book on *Der bildsame Zustand der Werkstoffe* (1927) which had just been translated into English and reissued as an Engineering Societies Monograph under the title *Plasticity* (1931). The study of this book and associated journal articles, plus many personal conferences with Doctor Nadai, including several visits with him in the Westinghouse Research Laboratories in East Pittsburgh, provided a great deal of the theoretical foundation upon which Hubbert's subsequent inquiries into the mechanics of rock deformation have been based (Hubbert, 1951).

In the mid-1930's a long-standing interest in the mechanical properties of structures as a function of their size was precipitated by the prompting of the Interdivisional Committee on the Borderland Fields between Geology, Physics, and Chemistry of the National Research Council into a study of the theory of scale models (Hubbert, 1937) and its implications to the large-scale deformation of the earth. This led to the conclusion that the strong rocks of the earth, when regarded on a reduced-size scale, should behave rather like such weak materials as soft muds. This, in turn, focused attention upon the newly evolving science of soil mechanics with particular attention being paid to the treatises on this subject by Karl Terzaghi and associates, since it appeared that the phenomena of soil mechanics represented in many respects very good scale models of the larger diastrophic phenomena of geology.

An independent line of inquiry concerned the mechanics of ground water (Hubbert, 1940; 1953; 1956), but it was not until about 1954 that the implications of the interlinkage between the mechanical properties of porous rocks and those of their contained water began to be appreciated. This arose from a study which was conducted jointly by Hubbert and a colleague, David G. Willis (1957), in which it was sought to analyze what the mechanical behavior of the rocks in deep oil wells should be in response to hydraulic pressures applied inside the wellbore by means of fluids which either were free to flow into the rock or could be excluded by an impermeable barrier. During the study of this problem, it was confirmed that Terzaghi's (1936) resolution of the total stress into a *neutral* component equal to the fluid pore pressure, and an *effective* component, was a physically significant resolution; but this appeared to be true for reasons differing from those put forward by Terzaghi.

Prior to that study, Hubbert had also been concerned in one way or another with the accumulating evidences of very high fluid pressures observed during the drilling of deep wells in the Texas–Louisiana Gulf Coastal region and in other parts of the world, where pressures were being found capable of supporting columns of water which would extend as high above the ground as the well was deep. These pressures were plainly dynamical phenomena, and they were considered most likely to be the result of a dynamical compression of the water-filled rocks; this, in turn, could be caused either by gravitational loading, as in the Gulf Coast, or by an orogenic compression in tectonically active areas.

In studying these abnormally high pressures, the resolution of the total stresses into a neutral part p and a residual effective stress with which the rock strength is associated led immediately to a recognition that, when p was very large, many rocks at great depths must be very much weaker and more deformable than previously had been supposed. This realization came into focus in another way when, during the summer of 1955, Hubbert was viewing the large Glarus overthrust in Switzerland under the guidance of a young Swiss geologist, Thomas Ch. Locher. The question suddenly occurred: How thick must this plate have been at the time of the overthrusting?, to which the probable answer seemed to be of the order of several kilometers. Then came the second question: In view of the fact that these were porous, water-filled sedimentary rocks which were being subjected to the compressive stresses of the intense Alpine orogeny, what must have been the pressure of the interstitial water? It seemed hardly conceivable that this could have been much less than the maximum amount possible—that equal to the total weight per unit area of the overburden. If this were true, then here, it

seemed, was the answer to the enigma of the large overthrusts: the block could be moved by a very small force since it must have been in a state of incipient flotation.

Rubey's interests in the phenomena bearing upon this problem also date back to the early 1920's when, as a young geologist on the United States Geological Survey, he was assigned the task of preparing for the Director, George Otis Smith, a reply to an inquiry from F. H. Lahee as to whether any cases were known where the ground-water pressures were equal to or in excess of that of the total weight of the overburden. During this early period Rubey also worked in the physics laboratory of P. G. Nutting where he developed a continuing interest in fluid mechanics of surface streams and of ground water and associated properties of the rocks such as porosity and permeability.

It was also about this time that the well-known paper by W. J. Mead (1925) on *The geologic rôle of dilatancy* was published in which evidence was presented in support of the conclusion that the deformation of a porous rock would increase rather than decrease its porosity. Soon, thereafter, the paper of Hedberg (1926) on *The effect of gravitational compaction on the structure of sedimentary rocks* appeared concerning which Rubey (1927) published a discussion. Mead's paper, although plausible for sandstones and many other rocks, gave exactly the wrong answer for shale, which led Rubey (1930) to make a series of porosity measurements of the shales in the Black Hills area, and into a long discussion of the effect of horizontal compression and deformation upon the pore space in fine-grained sediments.

In 1931, under the initial direction of G. R. Mansfield, Rubey began a campaign of field work in the overthrust belt of southern Idaho and southwestern Wyoming which has continued to the present time. Although he has subsequently parted from some of the concepts and conclusions held by Mansfield about the structure of the region, he still nevertheless owes to Mansfield a debt of sincere gratitude for a firm grounding in the problems of the region.

By 1945, both from his own field work and from reading, Rubey had become almost certain that some kind of easy-gliding mechanism must be operative in overthrust faulting. But the paper of Chester Longwell (1945) on *The mechanics of orogeny* which appeared at this time showed that any supposed gravitational gliding down geologically acceptable slopes would be incompatible with known values of the coefficient of friction of rocks. Longwell's paper convinced Rubey that there must be something fundamentally wrong with the "coefficient-of-friction" approach, and that some other way out must be sought.

The general idea that there might be some relation between excess fluid pressures and overthrust faulting suggested itself and in fact was discussed with O. E. Meinzer several times before the latter's death in 1948. However, the only way at that time he was able to account for such pressures was by high intake areas, and these raised more problems than they solved, so the idea came to nought.

Finally, Hatten S. Yoder, Jr. (1955), in his paper on *Role of water in metamorphism*, suggested that in rocks of very low permeability with essentially isolated pores the water pressure should approximate that of the total weight of the overburden. This suggestion revived Rubey's interest in the possibility of the occurrence of high fluid pressures in overthrust fault zones. He discussed with Hollis D. Hedberg the occurrence of such high pressures in deep oil wells and received citations to the several published papers on the occurrence of such pressures in the Texas–Louisiana Gulf Coastal region. The most useful of these was the paper on *Geological aspects of abnormal reservoir pressures in Gulf Coast Louisiana* by George Dickinson (1953).

At this stage of development of the authors' separate lines of inquiry, they accidentally discovered their mutual interest in this subject and agreed upon the present collaboration.

To those who have been mentioned in the foregoing paragraphs, the authors are indebted for important contributions to their present views. A more specific indebtedness, however, is owed to those who have assisted directly in the assembly of the present paper.

Those to whom Hubbert is especially obligated include: David G. Willis, who was a collaborator and coauthor of the study leading to the present paper; John Handin, who performed the experiments on the Berea sandstone, and also read critically a preliminary draft of the manuscript; J. K. O'Brien, who assembled the apparatus and performed the concrete-block experiment; M. A. Biot, who suggested the beer-can experiment, and whose written and oral discussions on consolidation theory and related phenomena have clarified the understanding of questions raised in the present study; Thomas B. Nolan and James Gilluly,

for 10 days of instruction on the geology of central Nevada as the guest of Nolan in the summer of 1957; M. M. Pennell, for the pressure data on Iranian oil fields and wells given in Tables 4 and 5; John J. Prucha, Donald V. Higgs, Charles H. Fay, Willy Hafner, George A. Thompson, Gordon J. F. MacDonald, Clarence R. Allen, David T. Griggs, and Lymon C. Reese, for critical reading of the manuscript; and, finally, Miss Marjorie Marek and Mrs. Martha Shirley, who have contributed invaluable editorial and technical assistance in the assembly of the papers.

Rubey wishes to acknowledge his personal indebtedness to O. E. Meinzer, P. G. Nutting, and G. R. Mansfield for their contributions to his background interest in the problems to be discussed; to Hollis D. Hedberg, L. F. Athy, Karl Terzaghi, George Dickinson, W. J. Mead, Chester R. Longwell, and Hatten S. Yoder, Jr., whose writings have influenced his own thinking on the current subject; and, finally, to David T. Griggs, John Handin, Steven S. Oriel, Walter S. White, and others too numerous to mention for help in the present paper.

Both authors wish to express their sincere appreciation for the sympathetic interest and administrative assistance on the part of the officials of both Shell Development Company and the United States Geological Survey that have made this study possible.

Phenomenon of Overthrust Faulting

Definition of "Overthrust Faulting"

The term "overthrust faulting," as it will be understood in the present paper, is in substantial agreement with the following definition taken from Marland P. Billings' (1954, p. 184) textbook on *Structural Geology*:

> "*Overthrusts* are spectacular geological features along which large masses of rock are displaced great distances. An *overthrust* may be defined as a thrust fault with an initial dip of 10 degrees or less and a net slip that is measured in miles."

Early Recognition of Overthrusting

The phenomenon of overthrusting, according to Bailey Willis (1923, p. 84) quoting Rothpletz, appears first to have been recognized by Weiss in 1826 near Dresden, Germany, where an ancient granite was found to be lying flat upon Cretaceous strata.

E. B. Bailey (1935, p. 15–16, 36–37, 45–50) credits Arnold Escher von der Linth with the first working out of an example of large over-thrusting in Canton Glarus, Switzerland. R. I. Murchison (1849, p. 246–253), later Director of the Geological Survey of Great Britain, visited the area, was convinced that Escher's interpretation was correct, and published a brief account of this new structural feature and a tribute to its discoverer.

In 1843 the Rogers brothers, W. B. and H. D., described the structure of the Appalachian Mountains in Virginia and Pennsylvania and clearly recognized the northwestward overturning and thrusting of the strata. J. M. Safford (1856), in his first report as State Geologist of Tennessee, and subsequently (1869), recognized "eight great faults" in eastern Tennessee, the "long ribbon-like masses or blocks . . . crowded, one upon another, like thick slates or tiles on a roof, the edge of one overlapping the opposing edge of the other." In 1860, W. E. Logan (1860; Barrande, Logan, and Hall, 1861), first Director of the Geological Survey of Canada, discovered the "great break" near Quebec City, along which metamorphic rocks were thrust northwestward over younger sedimentary rocks. He indicated the course of this fault through southern Canada and its probable connection with the series of great dislocations traversing eastern North America that had "been described by Messrs. Rogers and by Mr. Safford." "Logan's line" was subsequently traced southward into Vermont, New Hampshire, and New York; and the Taconic allochthon or thrust sheet, as it is now called, has recently been interpreted by Cady (1945) to have a displacement of 50 miles or more.

The thrust faults in the Northwest Highlands of Scotland were first reported by Nicol in 1861, who compared the structural relations there with those previously observed in the Alps. It is ironical that Murchison, the man who years before had been among the first to accept Escher's evidence for overthrusting in the Swiss Alps, rejected Nicol's interpretation and thereby precipitated a long controversy (Bailey, 1952, p. 65–67, 90–94, 108–115, 134–135). Eventually Callaway (1883a; 1883b) and Lapworth (1883) worked out the relationship—low-angle overthrusts between metamorphic rocks above and unmetamorphosed rocks below—in two separate areas. Geikie, then Director of the Geological Survey of Great Britain, immediately sent his best field men—Peach, Horne, and several associates—into the Highlands to do detailed mapping in key areas. The

following year these men, with an Introduction by Geikie, announced their findings (Peach and Horne, 1884); and in a series of papers over the next 25 years made classic the great flat Moine and associated overthrusts (Peach et al., 1907). The horizontal displacement of the Moine and associated thrusts was found to be in excess of 10 miles.

The foregoing synopsis has been limited to the chronology of the first recognition of overthrusting in various localities. Space here does not permit of a review of complementary literature wherein each of these early interpretations was severely challenged by the original investigator's contemporaries. Perhaps the greatest significance of the work by Peach, Horne, and associates, on the thrusts of the Scottish Highlands, was that it established beyond any further doubt that large overthrusts were in fact geological phenomena and not the controversial subject they had been theretofore.

Occurrence of Overthrusting

While the overthrusting in the Scottish Highlands was being studied in Great Britain, Törnebohm (1883; 1888; 1896) was working out the great overthrust in the Scandinavian Peninsula, which was eventually shown to have a horizontal displacement of more than 130 km or 80 miles (1896, p. 194). It is also interesting to note that, although the overthrusts of the Scottish Highlands and the Scandinavian thrust were both formed during the Caledonian orogeny, the displacements of the Scottish faults were toward the northwest, whereas that of the Scandinavian fault was toward the southeast.

In North America, McConnell (1887) discovered the overthrusts along the east margin of the Canadian Rockies (horizontal displacement of about 7 miles). In the same geological province just south of the Canadian border, Willis (1902) discovered the Lewis overthrust, which, from the amplitude of the variations of its eastern margin, was shown to have a displacement of at least 7 miles. Daly (1912), in studying the Lewis overthrust along the International Boundary, found a minimum bodily movement along the Lewis overthrust of 8 miles and thought that it might be as much as 40 miles. Campbell (1914) found the displacement on the Lewis overthrust to be at least 15 miles; and recently Hume (1957) reported that a well drilled in the Flathead valley some 20 miles west of the eastern front of the fault had penetrated about 4500 feet of Precambrian

Beltian strata of the upper block and had then entered the underlying Mississippian strata.

Following Willis' (1902) recognition of the Lewis overthrust, the next important event in this general region was Veatch's (1907) discovery and naming of the Absaroka, Crawford, and other overthrusts in southwestern Wyoming and adjacent parts of Utah. Then came Blackwelder's (1910) recognition of the eastward-dipping Willard overthrust north of Ogden, Utah, followed a few years later by the Bannock (Richards and Mansfield, 1912) and Darby (Schultz, 1914) overthrusts of southeastern Idaho and western Wyoming, respectively. The minimum displacement along the Bannock overthrust is reported to be about 12 miles (Mansfield, 1927).

In the Bighorn basin of Wyoming, Dake (1918) mapped a number of outliers, including Heart Mountain, of sediments ranging in age from Ordovician to Mississippian, which rest upon an overthrust fault surface truncating beds ranging from Paleozoic to Tertiary. These constitute the remnants of the Heart Mountain and associated overthrusts, which have subsequently been studied by Hewett (1920), Bucher (1933), and Pierce (1941; 1957).

Pierce, in his 1941 paper (p. 2028), reported a displacement from west to east of at least 34 miles. In his 1957 paper he concluded that both the Heart Mountain and the near-by South Fork overthrusts were detachment thrusts or décollements—blocks that had broken loose and moved long distances, probably by gravitational sliding.

Longwell (1922) discovered the Muddy Mountain overthrust in southeastern Nevada, in which a block of Paleozoic strata with a stratigraphic thickness of about 25,000 feet had overridden the same section for about 15 miles.

In central Nevada the Roberts Mountains overthrust, first recognized by Merriam and Anderson (1942), has been extensively mapped by Gilluly (1957) in the Battle Mountain area, and by T. B. Nolan (Personal communication, 1957) in the Eureka area some 60 miles to the southeast. In this fault of post Early (?) Mississippian age, clastic and volcanic rocks of Ordovician, Silurian, and Devonian age have been thrust from west to east over carbonates of Cambrian to early Mississippian age. According to Gilluly, as determined by exposures in erosional windows, this fault has a minimum displacement of 50 miles. Also, according to Gilluly (Personal communication during field trip in area in 1957), the thickness of rocks in the upper plate at the time of the orogeny could possibly have been as great as 5 miles.

In the Appalachian Mountains, the "great break" near Quebec City discovered by Logan and "Logan's line" have already been mentioned. The Taconic thrust near Albany, New York, according to Cady (1945), has a horizontal displacement of 50 miles or more. Hayes (1891) extended the mapping of the overthrusts, which had earlier been recognized in Tennessee, southward into Georgia where he found that the Cartersville thrust had a displacement from the southeast of not less than 11 miles.

Another major Appalachian overthrust is the Pine Mountain overthrust and the Cumberland thrust block first described by Wentworth (1921a; 1921b) and subsequently studied in more detail by Butts (1927), Rich (1934), and more recently with the benefit of well data by Miller and Fuller (1947), Stearns (1954), and Young (1957). The Cumberland thrust block is a single block, lying along the Kentucky, Tennessee, and Virginia boundaries, about 110 miles along the strike by 25 miles in the direction of the movement, with an average horizontal displacement of about 6 miles. Wentworth estimated the thickness of the block to be about 0.5 mile, and Butts estimated it to be about 2 miles. According to Young (1957), the fault surface has been intersected by numerous gas wells which show that it occurs near the base of a Devonian shale at an average depth of about 5500 feet, or a little more than a mile. It is also of interest that high pressures encountered in the fault zone cause troublesome blowouts while drilling. No pressure measurements were given, but since the drilling was with cable tools this does not necessarily imply that the pressure was abnormally high.

No detailed review of the evidences for large overthrust faulting in South America has been made, but the faulting in one locality is of outstanding interest. This is an area of Tertiary sediments on the westernmost point of Peru and at Ancon in the near-by Santa Elena peninsula of Ecuador (Baldry, 1938; Brown, 1938). According to Baldry and to Brown, these nearly horizontal sediments, which form a plateau between the mountains to the southeast and the Pacific Ocean to the west, are intricately dissected by breccia zones cutting across the strata at low angles and dipping seaward; from the geological correlations in the numerous oil wells drilled in the areas, no interpretation other than gravitational sliding satisfactorily accounts for the observed phenomena.

Overthrusting in the Himalaya Mountains in India was discovered by Oldham (1893). More recently Auden (1937, p. 420–421, 432, sec. 3, Pl. 37) has found horizontal translations of 20 to 50 miles; West (1939, Fig. 11, p. 162) has reported a displacement of at least 30 miles; and Heim and Gansser (1939, p. 13–14, 16, sec. on geol. map and sec. 2, Pl. 1) have reported a displacement of more than 50 miles in Central Himalaya.

Returning to the Alps in Switzerland, one of the most striking nappe complexes, the Mürtschen-Glarus, is bounded by a great fault which for convenience is here referred to as the "Glarus overthrust." This overthrust is easily seen in the field in the area of the Helvetian Alps between Glarus and Elm. In this fault, as described by Oberholzer (1933, profiles 8 and 9, Tafel 3, and a tectonic map, Tafel 8; see also Oberholzer, 1942), Permian sediments from the south have overridden Tertiary Flysch sediments for an aggregate distance of at least 40 km. The fault is plainly visible in the area around the village of Elm, where it intersects successive mountain peaks. Near the town of Schwanden, a few kilometers to the northwest, it has descended to near road level where it is readily accessible for close inspection. The fault zone comprises a thin limestone a few tens of centimeters thick, which is said to be Jurassic or Triassic in age. This occurs beneath and parallel to the bedding of the Permian conglomerate of the upper plate of the fault and rests upon Flysch sediments which are truncated at an angle of about 30 degrees. The fault surface proper occurs as a discrete polished surface in a clay layer about 1 cm thick, within the fault-zone limestone. In view of the distance of the displacement, the trivial amount of deformation evident a meter or so away from the fault zone is most impressive.

MECHANICAL PARADOX OF LARGE OVERTHRUSTS

Earlier Calculations of the Maximum Length of Overthrust Block

Since their earliest recognition, the existence of large overthrusts has presented a mechanical paradox that has never been satisfactorily resolved. Two distinct methods of propulsion of the upper plate over the lower are possible, corresponding to the application, respectively,

of surface forces, or of body forces, as these terms are understood in the theory of elasticity.

The simplest case of propulsion by means of a surface force would involve the pushing of a block along a horizontal surface by means of a surface force applied along the rearward edge of the block. The maximum length of a block of uniform thickness that can be moved in this manner depends jointly upon the coefficient of friction of rock on rock and upon the strength of the rock. Smoluchowski (1909) discussed the problem of sliding a rectangular block along a plane horizontal surface. Taking b as the length of the block parallel to the motion, a its breadth, c its thickness, w its weight per unit volume, and e the coefficient of friction, he showed that the force required to slide the block would be

$$F = (abc)we.$$

Hence the pressure distributed over the end ac would be

$$\frac{F}{ac} = wbe,$$

which would be equal to the weight of a column of the rock having a height equal to be. Assuming $e = 0.15$ (the coefficient of friction for iron on iron), and $b = 100$ miles, he found the required strength of the rock would have to be that capable of supporting a column 15 miles high; the crushing strength of granite will support a column only about 2 miles high.

From this he concluded that, under the mechanical conditions assumed, it would be impossible to move such a block. However, rather than to condemn the theory of Alpine overthrusts, he suggested two alternatives: (1) that the thrusts may have occurred down an inclined plane, the slope of which, for the coefficient of friction assumed, would need to be but 1:6.5; or (2) that the rocks involved were plastic with a still lower coefficient of friction than that assumed.

Considering the same problem, R. D. Oldham (1921, p. lxxxix) assumed for the coefficient of friction the value of 0.6 and a rock strength sufficient to support a column 4.8 km (3 miles) high. With these data he obtained 8 km (5 miles) as about the maximum length of a fault block which could be pushed.

A. C. Lawson (1922, p. 341–342) attempted to analyze a composite of the two cases. He considered the fault block to be a wedge with its upper surface horizontal and its lower surface formed by the inclined fault plane up which the block would be pushed. An attempt was made to determine the maximum length of the wedge. The result obtained was between 32 and 52 km when the thrust plane approached horizontality as a limit, and less than this if the thrust plane were inclined.

Regrettably, the equations on which this conclusion was based do not appear to be entirely valid, so this result must be discounted.

New Estimate of Maximum Length of Thrust Block

Review of recent data on mechanics of rocks.— Within recent years many new experimental data on the mechanical properties of rocks have become available as the result of tests made under conditions of triaxial stress (Handin and Hager, 1957, p. 10–12). These tests are made by placing jacketed cylindrical specimens of rock in a testing machine in which a radial stress can be applied by means of a fluid pressure while an axial stress is simultaneously applied by a piston.

If we let the radial stress be the least stress, σ_3 (compressive stresses are here regarded as positive), then for any given value of σ_3, the axial stress σ_1 can be increased until the specimen fails. These tests can be repeated for successive specimens of the same rock with increasing values of σ_3, and thus a whole family of pairs of values of σ_1 and σ_3 for which the rock fails can be obtained.

One of the more informative methods of representing the results of such tests is by means of the Mohr diagram (Mohr, 1882; Terzaghi, 1943, p. 15–44; Nadai, 1950, p. 94–108; Hubbert, 1951) in which the normal stress σ and the shear stress τ are chosen as the co-ordinate axes of abscissas and ordinates, respectively. On an internal plane within the stressed specimen, perpendicular to the $\sigma_1 \sigma_3$-plane and making an angle α with the σ_3-axis, the shear and normal stresses are given by the following respective equations:

$$\tau = \frac{\sigma_1 - \sigma_3}{2} \sin 2\alpha \qquad (1)$$

$$\sigma = \frac{\sigma_1 + \sigma_3}{2} + \frac{\sigma_1 - \sigma_3}{2} \cos 2\alpha . \qquad (2)$$

The representation of these quantities graphically on the Mohr $\sigma\tau$-diagram is shown in Figure 1. Here the principal stresses σ_3 and σ_1 are represented by two points at successive distances on the σ-axis. The quantity $(\sigma_1 + \sigma_3)/2$ is the abscissa of a point half way between σ_3 and σ_1; and $(\sigma_1 - \sigma_3)/2$ is half the distance between the points σ_1 and σ_3.

If, from the mid-point between σ_3 and σ_1, a radius vector of length $(\sigma_1 - \sigma_3)/2$ is erected at a counterclockwise angle 2α with the positive direction of the σ-axis, the co-ordinates of its extremity will be σ and τ as given in equations (1) and (2). Then, for fixed values of σ_1 and σ_3, the values of σ and τ satisfying equations (1) and (2), when the angle α is made to vary, will all lie on the circle of radius $(\sigma_1 - \sigma_3)/2$ which passes through the points σ_3 and σ_1. This is known as the Mohr stress circle, and its significance is that it gives at a glance the values of the normal and shear stresses across the plane of reference for any angle α and any combination of values of the greatest and the least stress.

From the results of triaxial testing, for each pair of values σ_3 and σ_1 for which failure occurs, a particular Mohr circle may be plotted, and the successive tests for increasing values of σ_3 will be a family of such circles whose centers lie at successively greater distances out on the σ-axis. At the start of a test, for a given value of σ_3, the axial stress σ_1 may be taken initially as equal to σ_3 and then gradually increased until failure occurs. The Mohr circle corresponding to this operation would thus start as a point at σ_3 and would gradually increase in diameter, passing always through σ_1 and σ_3, as σ_1 is increased. This implies that for a given value of σ_3, and a corresponding value of σ_1 short of failure, there is no combination of shear and normal stress, τ and σ, within the specimen which has reached the limit of the strength of the material. Finally, when failure does occur, such a critical combination must have been reached; but for a single test there is an infinity of different pairs of values of σ and τ (represented by points on the Mohr circle) for which this could have occurred. Hence, the critical pair is not known. However, when a series of such tests is made, the envelopes of the successive overlapping Mohr circles must be the loci of these critical values since failure occurs only when the Mohr circle becomes tangent to this envelope.

In the testing of rocks, within the stress range of present interest, the Mohr envelopes of stress can usually be approximated by the equation

$$\tau = \tau_0 + \sigma \tan \phi, \qquad (3)$$

which, because it was anticipated by Coulomb (1776), is commonly referred to as Coulomb's law of failure (Fig. 2). Here, τ_0 is the ordinate of the Mohr envelope when the normal stress σ is equal to zero, and it represents physically the initial shear strength of the rock when the

For the case where the Mohr envelopes are a pair of straight lines, in accordance with the Coulomb law of failure of equation (3), a simple relation exists between the values of σ_3 and σ_1

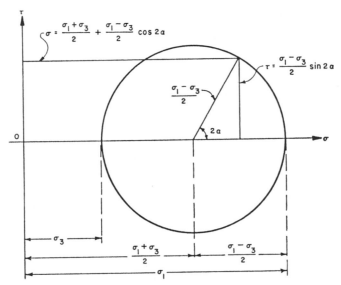

FIGURE 1.—MOHR DIAGRAM FOR NORMAL AND SHEAR STRESSES PRODUCED BY GREATEST AND LEAST PRINCIPAL STRESSES

normal stress is zero. The constant angle ϕ is a property of the rock, and, since

$$\tan \phi = \frac{\tau - \tau_0}{\sigma},$$

it is of the nature of a coefficient of friction with respect to the incremental shear stress $\tau - \tau_0$ and the normal stress σ. It is known accordingly as the *coefficient of internal friction* and ϕ as the *angle of internal friction*.

For the more plastic rocks such as marble or rock salt, the Mohr envelopes at higher stresses tend to become parallel to the σ-axis, but, for most rocks within the range of stresses occurring in the outer 10 km of the earth, the Coulomb law of equation (3) can be taken as a good approximation of the results of a large amount of test data.

The magnitude of the angle ϕ, which is the slope of the envelope, varies somewhat for different rocks, but a representative mean value would be about 30 degrees, or $\tan \phi$ about 0.6. Tests made by Handin and Hager (1957, p. 35–37, and personal communication) on a variety of sedimentary rocks have shown the mean value of τ_0 for such rocks to be about 200 bars (2×10^8 dynes/cm²).

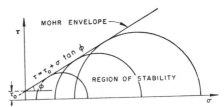

FIGURE 2.—MOHR ENVELOPE OF STRESS CIRCLES FOR A SERIES OF TESTS SHOWING FAILURE ACCORDING TO COULOMB'S LAW

corresponding to failure. From Figure 3

$$\frac{\sigma_1 - \sigma_3}{2} = \left[\frac{\sigma_1 + \sigma_3}{2} + \frac{\tau_0}{\tan \phi} \right] \sin \phi.$$

Upon multiplying both sides by 2 and transposing terms, this becomes

$$\sigma_1 - \sigma_3 - \sigma_1 \sin \phi - \sigma_3 \sin \phi = \frac{2\tau_0}{\tan \phi} \sin \phi,$$

which simplifies to

$$\sigma_1 (1 - \sin \phi) = 2\tau_0 \cos \phi + \sigma_3 (1 + \sin \phi),$$

or to

$$\sigma_1 = \frac{2\tau_0 \cos \phi}{1 - \sin \phi} + \sigma_3 \left[\frac{1 + \sin \phi}{1 - \sin \phi} \right]. \quad (4)$$

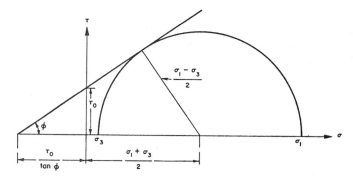

FIGURE 3.—DIAGRAM FOR THE TRANSFORMATION OF $\sigma\tau$- TO $\sigma_1\sigma_3$-AXES

However,

$$\frac{\cos\phi}{1-\sin\phi} = \sqrt{\frac{1-\sin^2\phi}{(1-\sin\phi)^2}} \left.\begin{array}{l} \\ = \sqrt{\frac{(1-\sin\phi)(1+\sin\phi)}{(1-\sin\phi)(1-\sin\phi)}} \\ = \sqrt{\frac{1+\sin\phi}{1-\sin\phi}}. \end{array}\right\} \quad (5)$$

Substituting this into equation (4) then gives

$$\sigma_1 = 2\tau_0 \sqrt{\frac{1+\sin\phi}{1-\sin\phi}} + \left[\frac{1+\sin\phi}{1-\sin\phi}\right]\sigma_3. \quad (6)$$

Then, since τ_0 and ϕ are constants of the material, it follows that equation (6) is of the linear form (Fig. 4)

$$\sigma_1 = a + b\sigma_3 \quad (7)$$

in which

$$a = 2\tau_0\sqrt{b}, \quad (8)$$

$$b = \frac{1+\sin\phi}{1-\sin\phi}. \quad (9)$$

Thus, the results obtained from triaxial tests may be plotted either in the form of the Mohr diagram, in which τ_0 and ϕ are the constants, or in the form of a linear graph of σ_1 versus σ_3, for which a and b are the constants. Conversion from either set of the constants τ_0 and ϕ, or a and b to the other, can then be effected by means of equations (8) and (9).

Maximum length of thrust block.—With this background, consider now the block of rock shown in Figure 5. This rests upon a horizontal surface and has a thickness z_1, and length x_1, which we shall take to be the maximum length the block can have and still be capable of being pushed. Let σ_{xx} be the normal stress applied

to the block on its rearward edge, and σ_{zz_1} and τ_{zx} the normal and shear stresses along its base. Since the horizontal acceleration will be taken to be negligible, the sum of the forces in the x-direction applied to the block must be zero. Hence the equation of equilibrium of the forces in the x-direction is

$$\int_0^{z_1} \sigma_{xx}dz - \int_0^{x_1} \tau_{zx}dx = 0. \quad (10)$$

Referring to equation (3), it will be seen that the total shear stress required to initiate a fracture on a small specimen is the sum of the initial shear strength τ_0 and the shear resistance to sliding. Once the fracture is produced, only the latter resistance remains. On an area as large as the surface of a regional fault, it appears to be most likely that the fracture will be propagated as a dislocation. If so, the area over which the stress τ_0 would be involved at any one time should comprise but a minute fraction of the total area of the fault surface. We shall accordingly assume that the effect of τ_0 is negligible. Then, the value of the shear stress τ_{zx} at which sliding will occur is given from the law of frictional sliding by

$$\tau_{zx} = \sigma_{zz_1} \tan\phi; \quad (11)$$

and the vertical stress σ_{zz_1} at the base of the block is

$$\sigma_{zz_1} = \rho g z_1,$$

ρ being the density of the rock and g the acceleration of gravity.

With these substitutions

$$\int_0^{x_1} \tau_{zx}dx = \int_0^{x_1} \rho g z_1 \tan\phi\, dx \left.\begin{array}{l}\\ = \rho g z_1 x_1 \tan\phi. \end{array}\right\} \quad (12)$$

Before we can evaluate the left-hand integral in equation (10), we must know the value of σ_{xx} as a function of the depth z, where σ_{xx} represents the maximum stress which the rock

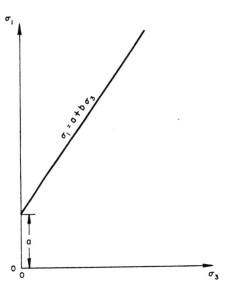

FIGURE 4.—FAILURE ACCORDING TO THE COU-
LOMB LAW IN TERMS OF $\sigma_1\sigma_3$-AXES

can sustain short of failure. Making use of equation (7), and noting that in this case σ_{xx} is the greatest and σ_{zz} the least stress, we obtain

$$\sigma_{xx} = a + b\sigma_{zz} = a + b\rho gz. \qquad (13)$$

Then

$$\left.\begin{array}{l} \int_0^{z_1} \sigma_{xx}dz = \int_0^{z_1} (a + b\rho gz)dz \\[2mm] \qquad = az_1 + \dfrac{b\rho gz_1^2}{2}. \end{array}\right\} \qquad (14)$$

Substituting equations (12) and (14) into equation (10) then gives

$$az_1 + \frac{b\rho gz_1^2}{2} - \rho gz_1 x_1 \tan\phi = 0,$$

which, when solved for x_1, gives

$$x_1 = \frac{a}{\rho g \tan\phi} + \frac{b}{2\tan\phi}\cdot z_1. \qquad (15)$$

Hence the maximum length of block that can be pushed by the stress σ_{xx} is made up of two terms, the first a constant and the second proportional to the thickness of the block. The first term represents the maximum length

of a block that could be pushed by the constant stress

$$\sigma_{xx} = a,$$

which is the surface value of σ_{xx}, and also the ordinary crushing strength of the rock. The second term represents the additional length of block which could be pushed by the incremental stress resulting from the increase of the strength of the rock with depth. If the block were longer than the value of x_1 given in equation (15), and the stress σ_{xx} were increased until failure occurred, the result would be to shear off the rearward end of the block along an inclined plane as a conventional reverse fault. (*See* Hubbert, 1951, Figs. 1, 2.)

Supplying data for the constants of equation (15), the approximate magnitude of x_1 can be determined. Taking 30 degrees for ϕ, then

$$\tan\phi = 0.577,$$

and

$$b = \frac{1 + \sin\phi}{1 - \sin\phi} = 3.$$

Then from equation (8)

$$a = 2\sqrt{3}\cdot\tau_0;$$

and, with 2×10^8 dynes/cm^2 as a representative value of τ_0, we obtain

$$a = 7 \times 10^8 \text{ dynes/cm}^2.$$

Then, taking 2.31 gm/cm^3 as a representative value of the density, and 980 dynes/gm for g, equation (15) becomes

$$x_1 = \frac{7 \times 10^8 \text{ dynes/cm}^2}{2.31 \text{ gm/cm}^3 \times 9.8 \times 10^2 \text{ dynes/gm} \times 0.58}$$

$$+ \frac{3}{1.15} z_1 \text{ cm}$$

$$= 5.36 \times 10^5 + 2.60z_1 \text{ cm}$$

or, approximately (Fig. 6),

$$x_1 = 5.4 + 2.60z_1 \text{ km}.$$

Thus, for a rock having the properties assumed, the maximum length of a thrust block 1 km thick would be 8.0 km (4.9 miles); and that for a block 5 km (3.1 miles) thick would be 18.4 km (11.4 miles). These figures could be changed slightly by a variation in the assumed rock properties, but the permissible variation of such properties falls within rather narrow limits. Consequently, for the conditions assumed, the pushing of a thrust block, whose length is of the order of 30 km or more, along a horizontal surface appears to be a mechanical impossibility.

The first "conjecture" that gravitational sliding might explain overthrusting in the Rocky Mountains appears to have been made by Ransome (1915, p. 367–369). Reeves (1925;

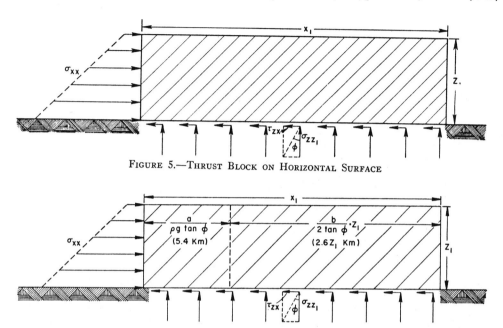

FIGURE 5.—THRUST BLOCK ON HORIZONTAL SURFACE

FIGURE 6.—ALGEBRAIC AND APPROXIMATE NUMERICAL VALUES OF THE TERMS OF EQUATION (15) FOR MAXIMUM LENGTH OF BLOCK

Gravitational Sliding

The difficulty of sliding a block in the foregoing manner has been appreciated since the earliest recognition of large overthrusts, and various hypotheses have been employed to circumvent it. Perhaps the most favored of these is gravitational sliding. The idea of gravitational sliding is an old one and goes back at least to Schardt (1898), the man who, with his 1893 interpretation of the far-traveled nature of the Prealps, really started the nappe theory which for years has dominated the concepts of Alpine tectonics. Schardt pictured the Prealps as being pulled free from their southern roots and, in Bailey's paraphrase, "gliding down the forward slope of an advancing wave [like] a Hawaiian surf-rider mounted on his board."

The idea of gravitational sliding to account for the Hohe Tauern in the Eastern Alps and the draping of folds over continental margins toward the oceanic basins is expressed cautiously at a few places in Suess' (1909, p. 171, 177, 582) monumental *Das Antlitz der Erde*.

1946) attributed the belt of faulted folds around the Bearpaw Mountains of Montana to downhill sliding off a structural dome. Daly (1925, p. 295–297) remarked that "the strong lateral compression of the Alpine and similar geosynclines during mountain building [can be understood] if the crust has had energy of position, large blocks of the crust *sliding* toward the geosynclines" and in the following year (1926, p. 271–290) expanded the concept to apply to continental tectonics.

Haarmann (1930) developed at some length the hypothesis that folds, overthrusts, the arclike pattern of many ranges, and other evidence of horizontal movement in the sedimentary cover are secondary down-slope gliding effects of large-scale vertical movements in the crust. A year later, Jeffreys (1931) requested geologists to re-evaluate, in the light of some such gravitational hypothesis, the amount of horizontal shortening to which major mountain ranges have been subjected. Since the early 1930's, a "revisionist" school of European geologists have written extensively (Migliorini, 1933; Ampferer, 1934; Van Bemmelen, 1936, p. 975–976; Lugeon, 1940; Merla, 1952) on the part played by gravitational sliding in the tectonics of the Alps and elsewhere. A symposium of six articles on *la tectonique*

The implications of this are clear when we compute the angle of tilt which the fault surface must assume before gravitational sliding can occur. Consider the block shown in

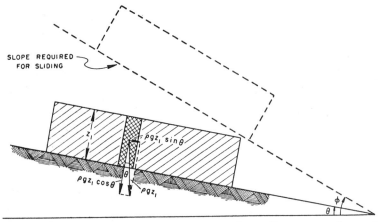

SLOPE REQUIRED FOR SLIDING

FIGURE 7.—NORMAL AND SHEAR STRESSES ON BASE OF BLOCK INCLINED AT ANGLE θ, AND ANGLE ϕ REQUIRED FOR SLIDING

d'écoulement par gravité (De Sitter, 1950a; 1950b; Gignoux, 1950; Goguel, 1950; Tercier, 1950; Van Bemmelen, 1950) presents a wide range of opinion on the subject. Aspects of this extensive literature have been reviewed recently by De Sitter (1954). From a somewhat different point of view, that of the bearing of laboratory experiments on the interpretation of folds and nappes, Bucher (1956) has also recently analyzed the place of gravitational spreading in tectonics and critically appraised current thinking on this topic.

By thus substituting a body force for the originally supposed surface force, the limitation imposed by the insufficient strength of the rock is eliminated, but what appears to be an equally insuperable difficulty still remains in the form of the measured values of the coefficient of friction of rock on rock. From a wide variety of tests, both of rocks and of the loose materials of soil mechanics, the angle of internal friction ϕ consistently yields values close to 30 degrees, which is also about the mean value of the angle of ordinary sliding friction of rock on rock. This correspondingly gives for the ratio of the shear stress to the normal stress along the surface of sliding the approximate value

$$\frac{\tau}{\sigma} = \tan \phi \cong 0.6.$$

Figure 7 which is resting upon a plane of inclination θ. If we consider a column of unit cross-sectional area perpendicular to the base of the block, the weight of this column will be $\rho g z_1$, and the normal and shear stresses along the base will be

$$\sigma = \rho g z_1 \cos \theta, \qquad (16)$$

$$\tau = \rho g z_1 \sin \theta, \qquad (17)$$

from which

$$\frac{\tau}{\sigma} = \tan \theta. \qquad (18)$$

However, for the block to slide, it is necessary for

$$\frac{\tau}{\sigma} = \tan \phi.$$

Consequently, the angle of tilt of the surface must be increased until

$$\theta = \phi, \qquad (19)$$

where ϕ, on the basis of measurements, must have a value not far from 30 degrees.

Thus we are left in almost as much difficulty as we were in before, since the field evidence precludes a mean angle of inclination of the overthrust fault surfaces anywhere near this magnitude.

Necessity for Reduction of Frictional Resistance

Hence, regardless of how the block is propelled, we are confronted with the logical necessity of somehow reducing the frictional resistance to sliding of the upper fault block over the lower to a small fraction of that which we have assumed. This too has long been appreciated, and various suggestions have been made as to how it may have been accomplished. Since the rocks are water wet, an obvious suggestion has been that water acts as a lubricant and so reduces the coefficient of sliding friction. Other related suggestions have been that clays and fault gouge may have acted as lubricants. Smoluchowski (1909) postulated plasticity, and M. P. Rudzki, in his book *Physik der Erde* (1911, p. 242–246), supposed that this may have resulted from the occurrence of the faulting before the rocks were uplifted, while they were at a higher temperature. R. D. Oldham (1921, p. xc–xci) suggested that the frictional drag could be greatly reduced if the overthrust block did not move simultaneously throughout its whole extent, but piecemeal in the manner of the locomotion of a caterpillar.

With the exception of the lubricating effect of water, and the plasticity of such materials as rock salt, each of these postulated effects has some intuitive plausibility and some qualitative support experimentally; yet each is sufficiently vague or indefinite as to leave considerable doubt as to whether it may have been the actual mechanism that was effective in any given instance. Concerning the lubricating effect of water, Terzaghi (1950, p. 91) has shown that water definitely is not a lubricant on rock materials, and its presence, if anything, tends to increase the coefficient of friction.

MECHANICS OF FLUID-FILLED POROUS SOLIDS

Buoyancy Effect

Pressure–depth relationships.—What promises to be a satisfactory solution to the foregoing difficulties has emerged during recent years as the result of various inquiries into the effect of the fluid pore pressures upon the mechanical properties of porous rocks. In particular, the rocks comprising the outer few kilometers of the lithosphere commonly have either intergranular or fracture porosity; and below depths of a few tens of meters these pore spaces are filled with water, or, exceptionally, with oil or gas.

Usually, within depths of 1 or 2 km, the pressure of the water as a function of the depth z is closely approximated by the equation

$$p = \rho_w g z, \tag{20}$$

where ρ_w is the density of the water and g the acceleration of gravity, which is the hydrostatic

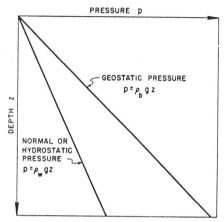

FIGURE 8.—HYDROSTATIC AND GEOSTATIC PRESSURE VARIATIONS WITH DEPTH

pressure of a column of water extending from the surface of the ground to that depth. Sometimes, however, pressures are encountered which differ markedly from this value, being occasionally less than, but more often greater than, this amount; and in a number of instances they have been known to approach the magnitude

$$p = \rho_b g z,^* \tag{21}$$

where ρ_b is the bulk density of the water-saturated rock.

In oil-well drilling practice, a pressure–depth variation in accordance with equation (20) is referred to as a *normal* or "hydrostatic" pressure, whereas the pressure given by equation (21), which is equal to that of the entire weight of the overburden, is known as the *geostatic* (or sometimes as the "lithostatic") pressure (Fig. 8). To refer to the normal pressure as being also hydrostatic is not strictly correct since the equipressure surfaces in the hydrostatic state must be horizontal, while, by definition, those in the normal-pressure case must follow approximately the relief of the topography. The existence of such pressure states results from the relief of the ground-water table and is also

---

*The quantity ρ_b in equation (21) must not be confused with the product ρb. The same is true for ρ_l in equation (22).

indicative of the general ground-water circulation. However, as contrasted with high abnormal pressures approaching the values given by equation (21), no significant error results

Consequently, the surfaces of equal pressure will be horizontal, and the pressure will increase downward at a uniform rate.

By definition, the gradient of p is the rate of

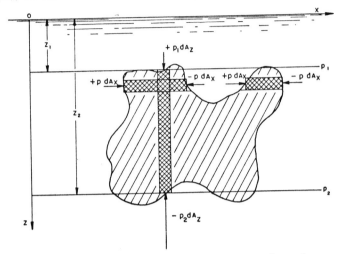

FIGURE 9.—BUOYANCY OR UPLIFT FORCE ON COMPLETELY CLOSED SOLID IMMERSED IN LIQUID

from regarding the normal-pressure state as being essentially also one of hydrostatics.

Archimedes principle.—The problem of concern is this: What will be the effect of this interstitial water pressure on the properties of the rocks themselves? In what manner may this be expected to modify the calculations of the preceding sections?

One of the oldest known effects of such pressures is, of course, the buoyant effect exerted by a fluid upon a completely immersed solid, as expressed by the principle of Archimedes. According to this principle, the force by which an immersed body is buoyed up is equal to the weight of the displaced fluid; and the reason why this should be so can be shown analytically in the following manner:

Consider a static liquid of density ρ_l, and choose co-ordinate axes with the origin at the liquid surface and the z-axis directed vertically downward. Within this liquid the pressure p at any point (x, y, z) will be given by

$$p = \rho_l g z \qquad (22)$$

from which

$$\left.\begin{array}{l} \dfrac{\partial p}{\partial x} = 0, \\[2mm] \dfrac{\partial p}{\partial y} = 0, \\[2mm] \dfrac{\partial p}{\partial z} = \rho_l g. \end{array}\right\} \qquad (23)$$

increase of the pressure with distance along the normal to the equipressure surface. It is a vector which is normal to the equipressure surface and directed toward the side of the higher pressure. From equation (23) it is accordingly seen that

$$\text{grad } p \equiv \frac{\partial p}{\partial n} = \frac{\partial p}{\partial z} = \rho_l g, \qquad (24)$$

where n is the normal in the direction of increase to the equipressure surface.

Now let a solid of arbitrary shape, exterior surface A, and volume V be immersed in this liquid. The force exerted by the liquid upon the solid will be

$$\mathbf{F} = -\iint_A p d\mathbf{A}, \qquad (25)$$

where $d\mathbf{A}$ is the vectorial representation of the surface element dA, having a magnitude equal to the area dA and a direction parallel to its outward normal. The integration of equation (25) can easily be accomplished by resolving the force $\mathbf{F}$ into components F_x, F_y, and F_z; and the element of area $d\mathbf{A}$ into components dA_x, dA_y, and dA_z, where

$$\left.\begin{array}{l} dA_x = dA \cos \alpha, \\[2mm] dA_y = dA \cos \beta, \\[2mm] dA_z = dA \cos \gamma; \end{array}\right\} \qquad (26)$$

and α, β, and γ are the angles between the outward normal to the surface element dA and the positive directions, respectively, of the x-, y-, and z-axes.

Then,

$$\left.\begin{array}{l} F_x = -\iint_{A_x} p\,dA_x, \\[2mm] F_y = -\iint_{A_y} p\,dA_y, \\[2mm] F_z = -\iint_{A_z} p\,dA_z. \end{array}\right\} \quad (27)$$

For any prism parallel to either the x- or the y-axis (Fig. 9), the pressure p will be constant, and the surface element dA_x or dA_y will occur an even number of times with alternately positive and negative signs. It will be positive, for example, when its normal makes an acute angle with the respective axis and negative when the angle is obtuse. Because of this the first two integrals of equation (27) are zero, and

$$F_x = F_y = 0. \quad (28)$$

The third integral, however, is not zero because the pressures p, acting upon the upper and lower ends of any vertical prism, will be different. Consequently, for each vertical column of cross-sectional area dA_z

$$\left.\begin{array}{l} F_z = -\iint_{A_z} (p_2 - p_1)\,dA_z \\[2mm] = -\iint_{A_z} \dfrac{\partial p}{\partial z} (z_2 - z_1)\,dA_z \\[2mm] = -\iiint_V \dfrac{\partial p}{\partial z}\,dV. \end{array}\right\} \quad (29)$$

Then, since $\partial p/\partial z = \rho_l g$, which is constant, the integration of equation (29) gives

$$F_z = -\rho_l g V; \quad (30)$$

and, since this is also the total force exerted on the solid by the fluid, it may be expressed vectorially by

$$\mathbf{F} = -\rho_l g V = -V\,\mathrm{grad}\,p, \quad (31)$$

according to which the force of buoyancy exerted by the fluid on the immersed solid is equal in magnitude, but opposite in direction, to the weight of the displaced fluid.

Equation (31) is but a special case of a more general proposition. Here we deal with a fluid in which

$$\mathrm{grad}\,p = \rho_l \mathbf{g} = \mathrm{const.}$$

However, there are cases, for instance when the fluid is subjected to centrifugal forces or has a variable density, where grad p is not a constant but varies in both magnitude and direction with position in space. If we let a solid, as in Figure 9, be suspended in a static liquid having a variable grad p, the components of the total force exerted by the fluid pressure will still be those given by equation (27). Beyond this, however, the evaluation of the components must proceed in a slightly different manner. Integrating equation (27) gives

$$\left.\begin{array}{l} F_x = -\iint_{A_x} \left[\int_{z_1}^{z_2} \dfrac{\partial p}{\partial x}\,dx \right] dA_x \\[2mm] = -\iiint_V \dfrac{\partial p}{\partial x}\,dV, \\[4mm] F_y = -\iint_{A_y} \left[\int_{y_1}^{y_2} \dfrac{\partial p}{\partial y}\,dy \right] dA_y \\[2mm] = -\iiint_V \dfrac{\partial p}{\partial y}\,dV, \\[4mm] F_z = -\iint_{A_z} \left[\int_{z_1}^{z_2} \dfrac{\partial p}{\partial z}\,dz \right] dA_z \\[2mm] = -\iiint_V \dfrac{\partial p}{\partial z}\,dV. \end{array}\right\} \quad (32)$$

Then, noting that

$$\mathbf{F} = \mathbf{i}F_x + \mathbf{j}F_y + \mathbf{k}F_z; \quad (33)$$

and that

$$\mathbf{i}\frac{\partial p}{\partial x} + \mathbf{j}\frac{\partial p}{\partial y} + \mathbf{k}\frac{\partial p}{\partial z} = \mathrm{grad}\,p, \quad (34)$$

where $\mathbf{i}$, $\mathbf{j}$, and $\mathbf{k}$ are unit vectors parallel to the x-, y-, and z-axes, we obtain from equations (32) and (33)

$$\mathbf{F} = -\iiint_V \left(\mathbf{i}\frac{\partial p}{\partial x} + \mathbf{j}\frac{\partial p}{\partial y} + \mathbf{k}\frac{\partial p}{\partial z} \right) dV; \quad (35)$$

which, in view of equations (25) and (34), becomes

$$\mathbf{F} = -\iiint_V \mathrm{grad}\,p\,dV = -\iint_A p\,dA. \quad (36)$$

This we may regard as being a generalized statement of the principle of Archimedes.

Equation (31) is the classical explanation of buoyancy, and equation (36) is an extension of the same principle to a more general pressure field; but, when attempts have been made to apply an analogous procedure to the porous rocks underground, in order to compute the so-called "uplift" force exerted by the water filling the pore spaces, a certain amount of confusion has resulted. The difficulty, as was first pointed out by Terzaghi (1932), arises from the circumstance that the force of buoyancy, as derived above, results from the action upon the solid of the liquid pressure applied over an external surface that is *completely closed*; whereas, in bodies of porous rocks underground, no closed external surface can be drawn around any given macroscopic volume can be drawn. In fact, if any surface of large radii of curvature is passed through such rocks, this surface will intersect solid- and fluid-occupied spaces alternately.

Extension of Archimedes principle to water-filled porous rocks.—Let us now consider whether the principle of Archimedes can be extended to include the case of the porous water-filled rocks of present interest. In this case let us require only that the pore spaces of the rocks all be interconnected, that they be fluid-filled, and that the pressure p of this fluid and its first directional derivatives vary continuously in the fluid-filled space. Within this fluid-filled, porous–solid space, let us now enclose a volume V, which is large as compared with the pore or grain structure of the rock, by means of a surface A. We then wish to determine the force of buoyancy exerted by the fluid of the system upon the rock and the water contained within this volume.

Let

$$f_A = \frac{\Delta A_p}{\Delta A} \tag{37}$$

be the surface porosity on a plane passed through the system, where ΔA is an element of area, which is large as compared with the grain or pore size of the rock, and ΔA_p the fraction of ΔA occupied by pore area. If the rock is homogeneous and the surface porosity is constant, then, by integrating ΔA_p and ΔA along the normal to ΔA, we obtain the volumetric porosity

$$f = \frac{\Delta V_p}{\Delta V} = \frac{\int \Delta A_p dn}{\int \Delta A dn} = \frac{\Delta A_p n}{\Delta A n} = f_A, \tag{38}$$

which is also equal to the surface porosity. In an analogous manner it can be shown that the porosity along a line is also equal to the surface and to the volumetric porosity.

Now let us consider the forces which are exerted separately upon the solid and the liquid contents of the volume V. In this case we have to modify equations (32) to allow for the fact that the integrals of the form

$$\int_{x_1}^{x_2} \frac{\partial p}{\partial x} dx$$

shall now apply only to the solid or the liquid segments of the respective prisms. In view of the fact that the liquid fraction of the length Δx of the prism is given by $f\Delta x$, and the solid fraction by $(1 - f)\Delta x$, then we may write

$$\int_{x_1}^{x_2} (1 - f) \frac{\partial p}{\partial x} dx$$

for the solid integration, and

$$\int_{x_1}^{x_2} f \frac{\partial p}{\partial x} dx$$

for the liquid.

With these modifications, equations (32), in conjunction with (33), (34), and (35), reduce to

$$\left. \begin{aligned} \mathbf{F}_s &= -\iiint_V (1 - f)(\text{grad } p)dV, \\[2mm] \mathbf{F}_l &= -\iiint_V f(\text{grad } p)dV, \end{aligned} \right\} \tag{39}$$

where the subscripts s and l signify the solid and the liquid, respectively.

Now, according to the theorem expressed by equations (25) and (31),

$$\left. \begin{aligned} \mathbf{F}_s &= -\iiint_V (1 - f)(\text{grad } p)dV \\[2mm] &= -\iint_{A_s} (1 - f)p_s\, d\mathbf{A} - \iint_{A_{is}} p d\mathbf{A}_{is}, \\[2mm] \mathbf{F}_l &= -\iiint_V f(\text{grad } p)dV \\[2mm] &= -\iint_{A_l} f p_l\, d\mathbf{A} - \iint_{A_{il}} p d\mathbf{A}_{il}, \end{aligned} \right\} \tag{40}$$

where A_s and A_l are the parts of the exterior surface A passing through solid and liquid spaces, respectively; p_s and p_l the normal stresses on these surface elements; and A_{is} and A_{il} the total solid–fluid interface inside the macroscopic volume V, regarded as enclosing the solid volume or the fluid volume, respectively. The surface A_{is} is therefore identical with the surface A_{il} except that the outward-drawn normals are on opposite sides of the surface. Because of this the surface elements, when represented vectorially, have opposite signs. Thus

$$d\mathbf{A}_{is} = -d\mathbf{A}_{il}.$$

Then, when equations (40) are added, the last two integrals to the right cancel each other, and we obtain

$$\mathbf{F} = -\iiint_V (\operatorname{grad} p)dV$$
$$= -\iint_A [(1-f)p_s + fp_l]d\mathbf{A}. \tag{41}$$

But, by equation (36),

$$\iiint_V (\operatorname{grad} p)dV = \iint_A pd\mathbf{A}. \tag{36}$$

Therefore, equating the integrands of the surface integrals of equations (36) and (41), and noting that $p_l = p$, we obtain

$$(1-f)p_s + fp = p,$$

from which

$$p_s = p. \tag{42}$$

Then, substituting this value of p into equation (40), and adding, we obtain

$$\mathbf{F} = \mathbf{F}_s + \mathbf{F}_l = -\iiint_V (\operatorname{grad} p)dV$$
$$= -\iint_A pd\mathbf{A}. \tag{43}$$

We are thus led to the conclusion that the force exerted by the pressure of the contained fluid upon the combined fluid and solid content of the volume V is given by either of the integrals of equation (43). Hence, the force may be computed either by the integration of the fluid pressure p over the *entire* external area of the space without regard to the value of the surface porosity or by integrating the gradient of the pressure with respect to the volume over the total volume of the space considered.

These results have been obtained with the implicit assumption that the fluid is at rest with reference to the solid. However, if the fluid is flowing through the solid, it will exert a frictional drag on the interior solid–fluid boundaries, and *vice versa*. By Newton's third law of motion these two forces are equal and opposite to each other, so that, when both the fluid and the solid content of the volume are considered simultaneously, these forces cancel and so produce no external effect. Consequently, for this more general case also, equation (43) still remains valid. Since equation (43) is a generalized statement of the principle of Archimedes, we may say that this principle is applicable to the porous rocks underground, and that the pressure p is effective over the entire area of any surface which may be passed through the system irrespective of the fraction of the area lying within the solid-filled part of the space.

Illustrations.—For two illustrative examples consider the vertical cylinders immersed in a liquid shown in Figure 10, the first being ideally nonporous and the second having a surface porosity intermediate between 0 and 1. Let the density of the nonporous cylinder be ρ_s, that of the water-filled porous cylinder ρ_b, and the density of the liquid ρ_l.

Now let us apply equation (43) to a volume of each of these cylinders extending from the top of the cylinder at depth z_1 to depth z_2. According to equation (43), the force of buoyancy on the part of the nonporous cylinder will be

$$\mathbf{F} = -\iiint_V (\operatorname{grad} p)dV, \tag{44}$$

and since in this case

$$\operatorname{grad} p = \rho_l \mathbf{g} = \text{const}$$

then

$$\mathbf{F} = \rho_l \mathbf{g}V. \tag{45}$$

The submerged weight of that part of the cylinder enclosed in the volume V will accordingly be

$$\mathbf{w}' = (\rho_s - \rho_l)\mathbf{g}V, \tag{46}$$

which is the weight of the solid less that of the displaced liquid. The additional stress in excess of the pressure of the water which this weight will impose on the part of the cylinder below will therefore be

$$\sigma = w'/A = (\rho_s - \rho_l)g(V/A)$$
$$= (\rho_s - \rho_l)g(z_2 - z_1). \tag{47}$$

The total stress S at the depth z_2 will therefore be the sum

$$S = p + \sigma \qquad (48)$$

or

$$S = \rho_l g z_2 + (\rho_s - \rho_l) g (z_2 - z_1). \qquad (49)$$

That this is a correct result can easily be seen by computing S directly in terms of the mass of

Making a similar analysis with respect to the porous cylinder, by Archimedes' principle the total stress at depth z_2 will be

$$S = p + \sigma = \rho_l g z_2 + (\rho_b - \rho_l) g (z_2 - z_1). \qquad (54)$$

However, by direct analysis based on the weight of the composite column,

$$S = \rho_l g z_1 + \rho_b g (z_2 - z_1), \qquad (55)$$

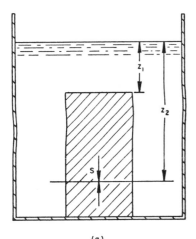

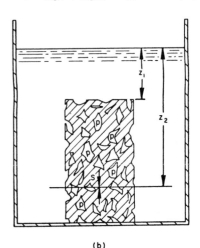

(a)   (b)

FIGURE 10.—EFFECT OF BUOYANCY ON SUBMERGED, NONPOROUS COLUMN AND ON POROUS COLUMN
(a) Effect on submerged, nonporous column. (b) Effect of buoyancy on porous column.

the composite solid and liquid cylinder extending from the depth z_2 to the surface of the liquid. Obtained in this manner

$$S = w/A = \rho_l g z_1 + \rho_s g (z_2 - z_1), \qquad (50)$$

where w is the weight of the total column. However, we may regard the density ρ_s of the solid to be the sum of two partial densities: one part that of the liquid ρ_l, and the other the excess $(\rho_s - \rho_l)$ of the solid density above that of the liquid. We may accordingly write the identity

$$\rho_s = \rho_l + (\rho_s - \rho_l). \qquad (51)$$

Then, when this is substituted into equation (50), we obtain

$$\left.\begin{aligned}
S &= \rho_l g z_1 + \rho_l g (z_2 - z_1) + (\rho_s - \rho_l) g (z_2 - z_1) \\
&= \rho_l g z_2 + (\rho_s - \rho_l) g (z_2 - z_1);
\end{aligned}\right\} \qquad (52)$$

or, as heretofore,

$$S = p + \sigma, \qquad (53)$$

which is the same result as that obtained originally in equations (48) and (49) from the direct application of Archimedes' principle.

which, by the same operation as that applied to equation (50), reduces to the same result as equation (54) above.

These two illustrations have been chosen to show specifically that the principle of Archimedes does not require for its validity that the body acted upon by the fluid pressure be completely enclosed by an exterior surface. In fact, in neither of the cases chosen can such a surface be drawn, yet the buoyancy in each instance is exactly the same as if the volume considered were entirely surrounded by the liquid.

Following the usage introduced by Terzaghi (1936, p. 874), the components p and σ, into which the total normal stress S has been resolved in equations (53) and (54), may be referred to, respectively, as the *neutral* and *effective* components of the total stress. The significance of this resolution will be discussed in more detail later, but for the present it may be noted that the component p is simply the pressure of the ambient fluid regarded as extending through both the fluid and the solid components of the system. The stress σ is therefore the excess of the stress in the solid (regarded as force per unit of macroscopic area) above the neutral stress p. Since the stress p has no shear components, any deformation of the solid (change of shape only) must be in response to σ.

The significance of σ will perhaps be clearer by reference to Figure 10 and to equations (47) and (53). In these two equations, for the cases of the nonporous and the porous columns of Figure 10, σ has the value, respectively, of

$$\sigma = (\rho_s - \rho_l)g(z_2 - z_1),$$

$$\sigma = (\rho_b - \rho_l)g(z_2 - z_1).$$

It will be noted that when ρ_s or ρ_b is greater than ρ_l, σ is positive (that is, compressive); when $\rho_s = \rho_b = \rho_l$, σ is zero; and when ρ_s and ρ_b are less than ρ_l, σ becomes negative, or tensile. That these are correct results can easily be seen by imagining the solids to liquefy in each case. In the first case the two columns would collapse; in the second they would remain unchanged; and in the third they would rise, stretching vertically and contracting horizontally.

Application to geology.—Applying the results expressed by equation (43) to a geological situation, let us compute the vertical component of the force of buoyancy exerted upon a block of porous rock and its water content, comprising a vertical cylinder of cross-sectional area A_z, extending to the depth z_1. Taking the z-component of the force of buoyancy given by equation (43),

$$\left. \begin{aligned} F_z &= -\iint_{A_z} \left[\int_0^{z_1} \frac{\partial p}{\partial z}\, dz \right] dA_z \\ &= -\iint_{A_z} p_1 dA_z, \end{aligned} \right\} \tag{56}$$

where p_1 is the fluid pressure at depth z_1; then, if the equipressure surfaces at depth z_1 are horizontal, p_1 will be a constant, and

$$F_z = -p_1 A_z. \tag{57}$$

Now let

$$S_1 = \rho_b g z_1, \tag{58}$$

which is the total weight per unit area of the overburden, be the vertical component at depth z_1 of the total stress. Then it follows that the fraction of the total weight of the overburden that can be regarded as being supported by the rock at depth z_1 is

$$\sigma_1 A_z = (\rho_b g z_1 - p_1)A_z;$$

or, per unit area,

$$\sigma_1 = S_1 - p_1. \tag{59}$$

Then, since the value of S_1 at depth z_1 is fixed, it follows that, as $p_1 \to S_1$, $\sigma_1 \to 0$.

Comparison with the results of others.—The results which we have just derived, although obtained by a different method of reasoning, confirm the conclusions of Hubbert and Willis (1957, p. 161) and are in substantial agreement with those reached by L. F. Harza (1949) in a study of the significance of pore pressures in hydraulic structures, published in the TRANS-ACTIONS of the American Society of Civil Engineers. The fact that Harza's paper of 22 pages provoked 55 pages of dominantly adverse discussion, contributed by 21 authors, is eloquent evidence that these conclusions are not in accord with the majority opinion of the civil-engineering profession—at least that in the United States in 1949.

The opposing view, which appears to be accepted almost universally in the soil-mechanics branch of civil engineering, and in other branches as well, is that which was originally put forward by Terzaghi (1932; 1936; 1945) and has subsequently been elaborated upon by many others, notably McHenry (1948) and Bishop and Eldin (1950–1951). According to this view, the fluid pressure acting upon the horizontal base of an engineering structure, such as a dam, is effective only over the pore area; the so-called "uplift pressure" p_u is given by

$$p_u = f_b p, \tag{60}$$

where f_b is the "boundary porosity" of the surface across which the fluid pressure is applied.

However, extensive experimental tests (Terzaghi, 1932; 1936, p. 785; 1945, p. 786, 788, 790–791; McHenry, 1948, p. 12–13, 16, 21; Taylor, 1948, p. 126; Bishop and Eldin, 1950–1951) made on such materials as loose soils and concrete have led invariably to the result that

$$f_b \cong 1. \tag{61}$$

If one assumes the validity of equation (60), then, in view of the fact shown in equation (38) that the surface porosity on a plane surface is the same as the volumetric porosity, the result of equation (61) is difficult to explain. The porosity of a loose sand is about 0.35; whereas, according to Rall, Hamontre, and Taliaferro (1954, p. 19–21), the mean porosity of more than 4800 core samples of oil-reservoir sandstones taken from wells in nine States is about 0.15. The mean porosity of concrete, according to Harza (1949, p. 194), is about 0.12. Therefore, from equation (60), one would expect the uplift pressure in loose sands to be about one-third, that in natural sandstones about one-seventh, and that in concrete about one-eighth of the fluid pressure itself.

To circumvent this difficulty, it has been found necessary to adopt the expedient of considering, instead of a plane surface, a sinuous surface which is made to pass through the presumed point contacts between individual particles. On this surface the area of these contacts is assumed to be very small as compared with the total area of the surface, thus leading to the surface porosity of approximately unit value. This is clearly stated by Terzaghi (1945, p. 786) as follows:

"To account for the test results, it is necessary to assume that the value of n_b in Eq. 9 is almost equal to unity. In other words, the area of actual contact or merging between the individual grains could not have exceeded a small fraction of the total intergranular boundaries. Yet, since the volume porosity of both materials is very low, the voids must consist of very narrow, but continuous slits."

While this procedure has a certain plausibility when applied to a loose sand, its validity is seriously jeopardized when applied to solid rocks and concrete whose tensile strengths negate the existence of point contacts, or of a surface porosity of unity, even on a sinuous surface.

Fallacy of hypothesis.—Since the hypothesis stated by equation (60) does not agree with the experimental results expressed by equation (61), when credible values of surface porosity are used, we are obliged to seek for the source of the discrepancy in the hypothesis itself.

This can conveniently be done by examining Terzaghi's (1932) experimental test of this hypothesis made on a cake of clay immersed in water to which pressure was applied. Terzaghi reasoned that, because of the finite areas of contact between the particles, the area of the clay particles exposed to water pressure on the outside of the cake must be greater than that of the same particles inside the cake. Then, since the fluid pressure was presumed to be operative only over the areas exposed to the water, he concluded that if the cake were immersed in water and the pressure were raised the outside force on the cake produced by the water pressure would be greater than the opposing inside force, and the cake would become compacted.

To test this hypothesis the cake was placed in water in a pressure vessel and subjected to a hydrostatic pressure of 6 kg/cm², yet no measurable compaction was produced; whereas, on a companion speci.. n subjected to a direct pressure of 0.5 kg/cm², applied by a piston, a compression of 100 scale divisions on the same strain gage resulted. From this it was concluded that the area of contact between the clay particles must have been substantially zero.

That no compaction should have occurred in the case of the immersed specimen, regardless of the area of contact between the grains, becomes evident when we examine the stress state inside the solid part of a porous solid when immersed in a fluid at pressure p. Neglecting for the present the body force due to gravity, let a porous solid of arbitrary shape, except for the exclusion of closed cavities, be completely immersed in a liquid at the constant pressure p (Fig. 11). What will be the magnitudes of the shear stress τ' and the normal stress σ' inside the solid across any given plane?

By a method analogous to that employed in the derivation of equations (25) to (31), let us consider the equilibrium of all the forces acting upon the part of the porous solid lying wholly on one side of a plane passed through the solid. Let us choose a co-ordinate system with the xy-plane coinciding with the intersecting plane, and the positive z-direction on the side of the xy-plane containing the part of the body under consideration. The equations of equilibrium of this mass will then be

$$\left.\begin{aligned} F_{zx} - p \iint_{A_x} dA_x &= 0, \\[6pt] F_{zy} - p \iint_{A_y} dA_y &= 0, \\[6pt] F_{zz} - p \iint_{A_z} dA_z &= 0, \end{aligned}\right\} \quad (62)$$

where F_{zx} and F_{zy} are the tangential components, and F_{zz} the normal component of force acting on the body across the xy-plane; and dA_x, dA_y, and dA_z the x-, y-, and z-components of the surface element dA as defined in equations (26).

From the geometry of the solid, for every positive element dA_x and dA_y there occurs a matching negative element. Consequently

$$\left.\begin{aligned} F_{zx} = p \iint_{A_x} dA_x &= 0, \\[6pt] F_{zy} = p \iint_{A_y} dA_y &= 0. \end{aligned}\right\} \quad (63)$$

For the dA_z's, however, only that fraction which coincides with the pore areas of the xy-plane has equal positive and negative values. The complementary fraction which terminates

which would exist in the liquid at the same place were the solid removed, it follows that this stress will have no tendency to produce any deformation in the porous solid.

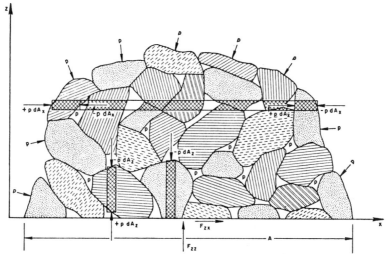

FIGURE 11.—STRESS INSIDE POROUS SOLID IMMERSED IN LIQUID AT CONSTANT PRESSURE

on the solid areas of the xy-plane has only positive values. Hence

$$F_{zz} = p \iint_{A_z} dA_s = pA_s = p(1 - f)A, \quad (64)$$

where A is the macroscopic area of the solid intersected by the xy-plane, A_s the net solid area, and f the surface porosity or volumetric porosity.

The microscopic stress components inside the solid, τ' and σ', on the xy-plane are then given by

$$\left.\begin{array}{l} \tau'_{zx} = \dfrac{F_{zx}}{A_s} = 0, \\[2mm] \tau'_{zy} = \dfrac{F_{zy}}{A_s} = 0, \\[2mm] \sigma'_{zz} = \dfrac{F_{zz}}{A_s} = \dfrac{p(1-f)A}{(1-f)A} = p. \end{array}\right\} \quad (65)$$

Thus, in the interior of the solid space, we see that the stress induced by the exterior fluid pressure has no shear stress components, and its normal component is equal to the fluid pressure p. Consequently, the pressure p does not terminate at the fluid–solid boundary but is continuous throughout the total space occupied by the fluid and solid combined. Since the stress state inside the solid has no shear components, and is in fact exactly the same as that

Equations (65) are strictly valid only for the case specified—namely, when body forces, such as gravity, are nonexistent. However, even in the actual cases where gravity cannot be neglected, they are still valid provided the density of the solid is the same as that of the liquid.

Returning now to Terzaghi's experiment, it is clear that, except for a very small second-order effect resulting from the slight difference of density between the clay and water, the stress state inside the clay consisted of the same hydrostatic pressure as existed in the water outside; and that this pressure, because of the absence of shear stresses, was incapable of producing deformation (change of shape). Moreover, provided only that the clay be saturated with water, this would continue to be true whatever might be the value of the surface porosity.

Since the pressure p is thus common to both the water and the clay, when we re-examine equation (60), the source of the error becomes immediately apparent: The pressure p acts over the whole of any surface passed through the porous solid, and the correct statement of this equation should have been

$$p_u = p, \quad (66)$$

with the surface porosity being in no way involved.

Resolution of the Total Stress Field into Partial Components

Definition of total stress.—The foregoing considerations enable us now to deal with the problem of the effect of the pressure of the interstitial fluids upon the deformational properties of porous solids. The total stress field in such a solid can be specified in terms of its normal and tangential components across given plane surfaces.

On a plane surface area ΔA, which is large as compared with the pore diameters of the solid but small as compared with its macroscopic dimensions, let ΔF_n be the normal component of the total force exerted by the solid and its interstitial fluid combined, and let ΔF_t be the tangential component. The corresponding normal and tangential components of total stress across the area ΔA will then be defined by

$$S = \frac{\Delta F_n}{\Delta A}, \left.\begin{array}{c} \\ \\ \end{array}\right\} \quad (67)$$
$$T = \frac{\Delta F_t}{\Delta A}.$$

The complete field of total stress will be a tensor having different normal and shear components on surfaces of different orientations passing through any given point, and it can be completely specified at any point in terms of its three components upon each of three mutually perpendicular surfaces passing through that point. Let Σ signify the tensor of total stress; its expansion in terms of components parallel to the x-, y-, and z-axes is given by

$$\Sigma \equiv \begin{vmatrix} S_{xx} & T_{xy} & T_{xz}, \\ T_{yx} & S_{yy} & T_{yz}, \\ T_{zx} & T_{zy} & S_{zz}. \end{vmatrix} \quad (68)$$

Here the first subscript signifies the axis normal to the plane upon which the stress component acts, and the second the axis to which the component is parallel. Thus S_{xx} signifies the stress component normal to the yz-plane and parallel to the x-axis, and T_{xy} the shear component on the same plane parallel to the y-axis. Contrary to the usual conventions, compressive stresses are here regarded as being positive so that all stress components are positive in the negative directions of their respective axes (Fig. 12).

Since turning moments on a volume element inside a body must be zero, then it follows that

$$\begin{aligned} T_{xy} &= T_{yx}, \\ T_{xz} &= T_{zx}, \\ T_{yz} &= T_{zy}. \end{aligned} \right\} \quad (69)$$

Neutral stress Π.—In view of our prior discussion, the component of this total stress attributable to the interstitial fluid is the pressure p, which can be regarded as being continuous throughout the total space. The pressure p can

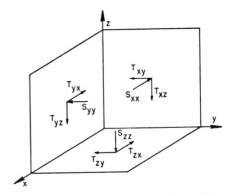

FIGURE 12.—COMPONENTS OF THE TOTAL-STRESS TENSOR Σ

be regarded as a special kind of tensor, which for comparison with that of the total stress Σ we shall designate as Π. In general, the fluid may be in very slow motion through the interconnecting pore spaces of the solid. When this occurs we must deal with the continuous deformation of a viscous liquid within the pore spaces of the solid, and the stress tensor Π, seen on this microscopic scale, will have the normal and shear stress components given by

$$\Pi \equiv \begin{vmatrix} p_{xx} & p_{xy} & p_{xz} \\ p_{yx} & p_{yy} & p_{yz} \\ p_{zx} & p_{zy} & p_{zz} \end{vmatrix}, \quad (70)$$

where, as before, $p_{ij} = p_{ji}$.

However, even in this case, when viewed on a macroscopic scale, the pressure which can be measured by a manometer whose terminal opening is large compared to the pore dimensions is a scalar that approximates the local mean value of the quantity

$$p = (p_{xx} + p_{yy} + p_{zz})/3. \quad (71)$$

The microscopic shear stresses are expended locally against fluid–solid boundaries, so that their only macroscopic effect is to transmit by viscous coupling to the solid skeleton whatever net impelling force may be applied to the fluid. Hence macroscopic shear stresses resulting from the tensor Π do not exist.

Consequently, on a macroscopic scale, the fluid pressure in three-dimensional space, whether the fluid is stationary or flowing, can be represented by the tensor

$$\mathbf{\Pi} \equiv \begin{vmatrix} p & 0 & 0 \\ 0 & p & 0 \\ 0 & 0 & p \end{vmatrix} \tag{72}$$

whose shear stresses are zero.

Effective stress Ω.—We now have a field of total stress Σ, which includes the stresses of the solid and fluid combined, and a partial field Π, which also permeates the entire space and whose value at any point is equal to that of the macroscopic fluid pressure. The latter has no shear stresses and hence tends to produce no deformation in the solid. The difference of these two fields defines a second partial field Ω given by

$$\Omega = \Sigma - \Pi. \tag{73}$$

Expanding Σ and Π into their respective components, we then obtain

$$\begin{aligned}\Omega &\equiv \begin{vmatrix} \sigma_{xx} & \tau_{xy} & \tau_{xz} \\ \tau_{yx} & \sigma_{yy} & \tau_{yz} \\ \tau_{zx} & \tau_{zy} & \sigma_{zz} \end{vmatrix} \\ &= \begin{vmatrix} (S_{xx}-p) & T_{xy} & T_{xz} \\ T_{yx} & (S_{yy}-p) & T_{yz} \\ T_{zx} & T_{zy} & (S_{zz}-p) \end{vmatrix} \end{aligned} \tag{74}$$

from which it is clear that across any arbitrary surface

$$\left. \begin{aligned} \sigma &= S - p, \\ \tau &= T. \end{aligned} \right\} \tag{75}$$

As explained earlier, following Terzaghi (1936, p. 874), the stress Π, because it produces no deformation, will be referred to as the *neutral stress*; the residual stress Ω will be known as the *effective stress.*

Experimental Tests

Theory of triaxial tests.—It has been inferred that although the total stress field Σ exists at any point interior to a porous rock, the deformation of this rock should be almost wholly related to the partial stress Ω and should be relatively insensitive to the complementary component Π. It remains for us to verify this inference experimentally. The most direct and effective way to do this is by means of triaxial tests made on fluid-filled specimens enclosed in flexible, impermeable jackets. Exterior to the jacket the total stresses, S_1 and S_3, can be applied while interiorly a pressure p of arbitrary magnitude can be added independently.

Consider first the total stress on the speci-

men; the normal and shear components of this stress on an interior plane making an angle α with the S_3-axis, in accordance with equations (1) and (2), will be

$$\left. \begin{aligned} S &= \frac{S_1 + S_3}{2} + \frac{S_1 - S_3}{2} \cos 2\alpha, \\ T &= \frac{S_1 - S_3}{2} \sin 2\alpha. \end{aligned} \right\} \tag{76}$$

However, since the pressure p inside the jacket will completely permeate both the solid- and fluid-filled spaces, it will oppose the exterior stresses over the whole of the outside area of the specimen. Hence, of the applied total stresses S_1 and S_3, only the fractions $(S_1 - p)$ and $(S_3 - p)$ will be effective in producing deformation in the specimen. In response to these fractional stresses, new normal and shear stresses S' and T', given by

$$\left. \begin{aligned} S' &= \frac{(S_1 - p) + (S_3 - p)}{2} \\ &\quad + \frac{(S_1 - p) - (S_3 - p)}{2} \cos 2\alpha, \\ T' &= \frac{(S_1 - p) - (S_3 - p)}{2} \sin 2\alpha, \end{aligned} \right\} \tag{77}$$

will be produced inside the specimen. Upon simplification these equations reduce to

$$\left. \begin{aligned} S' &= \left[\frac{S_1 + S_3}{2} + \frac{S_1 - S_3}{2} \cos 2\alpha - p \right] \\ &= S - p, \\ T' &= \frac{S_1 - S_3}{2} \sin 2\alpha = T. \end{aligned} \right\} \tag{78}$$

Noting now that, by definition,

$$\begin{aligned} S_1 - p &= \sigma_1, \\ S_3 - p &= \sigma_3, \\ T &= \tau, \end{aligned}$$

it follows that, by virtue of the diminution of the exterior stresses by the opposing pressure p, the normal and shear stresses inside the specimen become

$$\left. \begin{aligned} S' &= S - p = \sigma, \\ T' &= T = \tau, \end{aligned} \right\} \tag{79}$$

indicating again that the deformation of the specimen should be related only to the stress Ω of components σ and τ (Fig. 13).

Excellent experimental confirmation of this has been afforded by an elaborate series of tests on concrete made by Douglas McHenry (1948) of the U. S. Bureau of Reclamation in Denver and more recently, at a higher range of stresses, on rocks, by John Handin (1958) of the Shell Development Company in Houston.

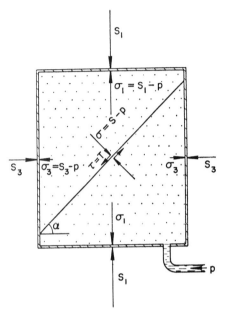

FIGURE 13.—TOTAL AND PARTIAL STRESSES ON JACKETED SPECIMEN WITH INTERNAL FLUID PRESSURE

Experiments of McHenry.—McHenry's experiments were made on a large testing machine capable of handling specimens up to 15 cm in diameter. Axial stresses were applied by means of a piston, and radial stresses by the pressure of kerosene for jacketed specimens and nitrogen gas for unjacketed specimens. In most of the tests the fluid pressure inside the jacketed specimens was zero (atmospheric), though in a few tests higher internal pressures were used. In the unjacketed tests, the nitrogen gas permeated the specimens giving an interior pressure $p = S_3$.

In all cases the effective stresses in these tests were

$$\left.\begin{aligned} \sigma_1 &= S_1 - p, \\ \sigma_3 &= S_3 - p; \end{aligned}\right\} \quad (80)$$

but in the special cases of jacketed specimens with zero internal pressure, and of unjacketed specimens, the principal effective stresses were, respectively,

Jacketed: $\left.\begin{aligned} \sigma_1 &= S_1, \\ \sigma_3 &= S_3, \end{aligned}\right\} \quad (81)$

Unjacketed: $\left.\begin{aligned} \sigma_1 &= S_1 - S_3, \\ \sigma_3 &= 0. \end{aligned}\right\} \quad (82)$

Test results on the jacketed specimens with zero internal pressure, when plotted as a Mohr diagram, were in accordance with the Coulomb law of equation (3); or, when σ_1 was plotted against σ_3, the linear form of equation (7)

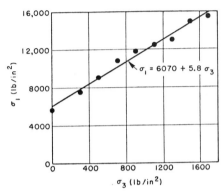

FIGURE 14.—TESTS BY MCHENRY ON JACKETED SPECIMENS OF CONCRETE WITH ZERO INTERNAL PRESSURE

shown in Figure 14 was obtained with the constants

$$a = 6070 \text{ lb/in}^2,$$

$$b = 5.8.$$

Then, from equations (8) and (9), the values of τ_0 and ϕ are

$$\tau_0 = 1265 \text{ lb/in}^2,$$

$$\phi = 45 \text{ degrees.}$$

In the tests on the unjacketed specimens, through a range of nitrogen pressure S_3 from 0 to 1581 lb/in², the values of $\sigma_1 = S_1 - S_3$ at which the failure occurred, when plotted as a function of S_3, clustered about a horizontal straight line (Fig. 15) corresponding to a mean value of σ_1 of 4926 lb/in². This indicated that the failure of the specimens under these conditions occurred at a nearly constant value of σ_1 (with $\sigma_3 = 0$), which corresponds to the ordinary crushing strength of the material at atmospheric pressure.

In those specimens tested with a pore pressure p intermediate between 0 and S_3, the values of σ_1 plotted against σ_3 fell approximately on the straight-line graph shown in Figure 16, independent of the magnitude of the pore pressure.

Thus the results of McHenry's tests on concrete show that, when this material has an internal fluid pressure p, its behavior with respect to the effective stresses σ and τ is in substantial agreement with its behavior in response to total stresses when $p = 0$. It is worthy of note, however, that the analysis that has here been made of McHenry's results has a quite different theoretical basis from that

which differs insignificantly from 1. With the introduction of this value, equation (83) then reduces to

$$S' = S - p,$$

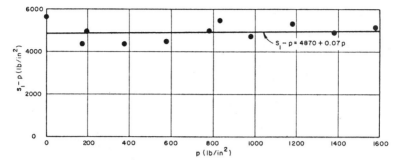

FIGURE 15.—TESTS BY McHENRY ON UNJACKETED SPECIMENS OF CONCRETE

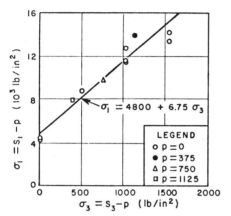

FIGURE 16.—TESTS BY McHENRY ON JACKETED SPECIMENS OF CONCRETE WITH VARIOUS INTERNAL PRESSURES

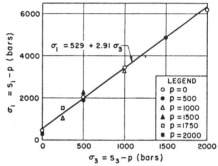

FIGURE 17.—TESTS BY HANDIN ON BEREA SANDSTONE WITH VARIOUS INTERNAL PRESSURES

which he, himself, used. His object in making the experiments was, in fact, to determine experimentally the value of the parameter f_A in the equation

$$S' = S - f_A p, \qquad (83)$$

where S' was the effective normal stress across the surface of slip at failure, and f_A was interpreted literally as the porosity of this surface, as postulated by Terzaghi. Despite the fact that the volumetric porosity of the concrete can hardly have been greater than about 0.15, and that the surface porosity on any plane surface must be equal to the volumetric porosity, the values obtained by McHenry for f_A from tests of 337 specimens all fell within the range 0.78 and 1.18; and a least-square solution for the best value yielded the result

$$f_A = 1.02 \pm 0.019,$$

which is the σ of current usage. For reasons pointed out previously, the value of unity obtained for the factor f_A has nothing to do with the porosity of the material, but does confirm eloquently that the fluid pressure permeates the entire space and reduces the effectiveness of the total stress accordingly.

Experiments of Handin.—Tests similar to those of McHenry, but with much higher ranges of stress and internal pressure, have recently been made by Handin (1958) on natural rocks. His test data for Berea sandstone are shown graphically in Figure 17. There the values of $\sigma_1 = S_1 - p$ up to 6300 bars, corresponding to failure, have been plotted against $\sigma_3 = S_3 - p$ for a range of pore pressures from 0 to 2000 bars (1 bar = 10^6 dynes cm² ≅ 15 lb/in²). These tests include both axial extension and shortening. Throughout this range the data plot as an excellent straight line with the constants

$$a = 529 \text{ bars},$$

$$b = 2.91,$$

from which

$$\tau_0 = 154 \text{ bars},$$

$$\phi = 29.25 \text{ degrees}.$$

A Mohr diagram of the same data is shown in Figure 18.

These experiments of McHenry on concrete, and of Handin on rock, supplement the earlier tests made on the loose materials of soil mechanics, in establishing the result that for porous fluid-filled solids the Mohr-Coulomb criterion for failure must now be modified to the form

$$\tau_{\text{crit}} = \tau_0 + (S - p) \tan \phi. \qquad (84)$$

Also, since Handin's tests have shown that this relationship continues to be valid up to pore pressures of 2000 bars, corresponding to stress conditions in the earth to a depth of

where $\bar{\rho}_b$ is the average value of the bulk density of the overburden. Then, if p is the fluid pressure at depth z, we may consider the overburden to be jointly supported by the fluid pressure p and the residual solid stress σ_{zz} in accordance with the equation

$$p + \sigma_{zz} = S_{zz}. \qquad (86)$$

Thus, as p increases from 0 to S_{zz}, σ_{zz} decreases from S_{zz} to 0; and, as p approaches S_{zz}, the overburden is in a state of incipient flotation. In order for p to become greater than S_{zz}, it would be necessary for the rocks to be able to sustain a state of tensile stress. In view of the large number of joints by which any large body of rock is intersected, the tensile strength of such a body should be zero, so that values of p greater than S_{zz} appear to be precluded. This also agrees with observations made in oil-well drilling where fluid pressures

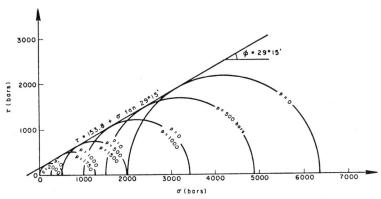

FIGURE 18.—MOHR DIAGRAM OF HANDIN TEST DATA ON BEREA SANDSTONE
The data include the results of both axial extension and axial compression.

20 km, it may be used with considerable confidence as a basis for the mechanical analysis of tectonic deformation of porous rocks.

APPLICATION TO OVERTHRUST FAULTING

Total and Partial Stresses Underground

Let us return now to the problem of overthrust faulting. On a horizontal surface at a depth z, which is large in comparison with the local topographic relief, the vertical normal component S_{zz} of total stress will be approximately equal to the weight of the total overburden—solids and fluids combined—above that depth. Hence

$$S_{zz} \cong \bar{\rho}_b g z, \qquad (85)$$

definitely greater than S_{zz} have never been reported.

Our further analysis can be somewhat simplified if the pressure p be expressed in terms of S_{zz} by means of a parameter λ defined by

$$p = \lambda S_{zz}, \qquad (87)$$

whose value ranges between the limits 0 and 1 as p ranges between 0 and S_{zz}.

Maximum Length of Horizontal Overthrust Block

We now reconsider the problem of the maximum length an overthrust block can have and still be capable of propulsion over a horizontal surface by a stress applied to its rearward edge; only, in this case, the rock is to have an interstitial fluid pressure p which varies as a function of the depth z. Consequently, the strength and frictional properties of the rock will be related to the partial stresses σ and τ, whereas the propulsion will be in response to the total stress. The stress tending to move the block

(89), we obtain

$$S_{xx} = a + b(1 - \lambda)\rho_b gz + \lambda\rho_b gz \\ = a + [b + (1 - b)\lambda]\rho_b gz. \quad (93)$$

Similarly, when the block is at the state of slippage, the shear stress τ_{zx} along its base is

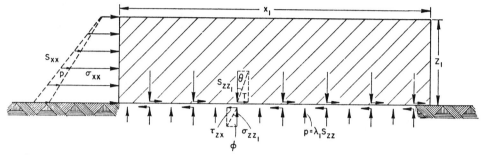

FIGURE 19.—BOUNDARY STRESSES ON HORIZONTAL BLOCK WITH INTERNAL FLUID PRESSURE AT STATE OF INCIPIENT MOTION

(Fig. 19) will accordingly be the total stress component S_{xx} applied to its rearward edge, and that opposing its motion will be the shear stress τ_{zx} along its base.

The equation of equilibrium of the x-components of force when the block is at a state of incipient motion will be

$$\int_0^{z_1} S_{xx}dz - \int_0^{x_1} \tau_{zx}dx = 0; \quad (88)$$

but, before these integrals can be evaluated, it will be necessary to express S_{xx} and τ_{zz} in terms of quantities whose values may be specified. Thus, from equation (75)

$$S_{xx} = \sigma_{xx} + p. \quad (89)$$

Furthermore, the horizontal component σ_{xx} is approximately the greatest principal stress, and the vertical component σ_{zz} the least. Then, by equation (7), when the rock is at incipient failure,

$$\sigma_{xx} = a + b\sigma_{zz}; \quad (90)$$

and by equations (85), (86), and (87)

$$\sigma_{zz} = S_{zz} - p = (1 - \lambda)S_{zz} \\ = (1 - \lambda)\rho_b gz, \quad (91)$$

where ρ_b (assumed constant) is the water-saturated bulk density of the rock.

Substituting this into equation (90) then gives

$$\sigma_{xx} = a + b(1 - \lambda)\rho_b gz; \quad (92)$$

when this in turn is substituted into equation

given by

$$\tau_{zx} = \sigma_{zz_1} \tan\phi \\ = (1 - \lambda_1)S_{zz_1} \tan\phi = (1 - \lambda_1)\rho_b gz_1 \tan\phi, \quad (94)$$

where λ_1 is the value of λ at the depth z_1 of the base of the block.

Substituting the expressions for S_{xx} and τ_{zx} from equations (93) and (94) into equation (88), and performing the integrations, we obtain

$$az_1 + \frac{b\rho_b gz_1{}^2}{2} + (1 - b)\rho_b g\int_0^{z_1} \lambda z\, dz$$

$$- (1 - \lambda_1)\rho_b gz_1 x_1 \tan\phi = 0,$$

from which x_1, the maximum length of the block, is given by

$$x_1 = \frac{1}{(1 - \lambda_1)}\left[\frac{a}{\rho_b g\tan\phi} + \frac{b}{2\tan\phi}\cdot z_1 \\ + \frac{(1 - b)\int_0^{z_1} \lambda z\, dz}{z_1\tan\phi}\right], \quad (95)$$

in which x_1 can be evaluated when the variation of λ as a function of the depth z is specified.

In principle, λ may have any value between 0 and 1, and it may also vary as a function of depth. However, since slippage is favored along surfaces on which λ is a maximum, there is some justification for assuming that the value of λ at depths less than z_1 shall not exceed λ_1, the value at the base of the block. Then, in view of the fact that b is taken to have the value 3, the quantity $(1 - b)$ in the last term of equation (95) has the numerical value -2, which renders that term negative. Conse-

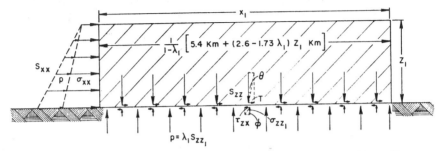

FIGURE 20.—MAXIMUM LENGTH OF HORIZONTAL BLOCK WITH INTERNAL FLUID PRESSURE WHICH CAN BE MOVED BY PUSH FROM REAR

quently, the minimum length of block given by equation (95) will result when the last term has a maximum value, and this will occur when λ for all depths is assigned its maximum value λ_1. Making this assignment, the last term can then be integrated, and we obtain

$$
x_1 = \frac{1}{1 - \lambda_1} \left[\frac{a}{\rho_b g \tan \phi} + \frac{b + (1 - b)\lambda_1}{2 \tan \phi} \cdot z_1 \right] \quad (96)
$$

From this it can be seen by inspection that, when $\lambda_1 = 0$, equation (96) reduces, as it should, to

$$
x_1 = \frac{a}{\rho_b g \tan \phi} + \frac{b}{2 \tan \phi} \cdot z_1, \quad (96a)
$$

which (for the same bulk density) is the same as equation (15) for the dry-rock case obtained earlier.

Supplying numerical data, let:

$a \quad = 7 \times 10^8$ dynes/cm²

$\tan \phi = \tan 30° = 0.577$

$b \quad = 3$

$g \quad = 980$ dynes/gm

$\rho_b \quad = 2.31$ gm/cm³.

Then, for the general case given by equation (96)

$$
x_1(\text{cm}) = \frac{1}{1 - \lambda_1} [5.4 \times 10^5 + (2.6 - 1.73\lambda_1)z_1]
$$

when lengths are measured in cm; or

$$
x_1(\text{km}) = \frac{1}{1 - \lambda_1} [5.4 + (2.6 - 1.73\lambda_1)z_1]
$$

when lengths are measured in km (Fig. 20).

It is seen from equation (96) that the maximum length of block x_1 is a function both of the thickness z_1 and of the pressure–overburden ratio λ_1. Values of x_1, in km, computed for

TABLE 1.—MAXIMUM LENGTH (IN KM) OF HORIZONTAL OVERTHRUST FOR VARIOUS THICKNESSES AND VALUES OF PRESSURE–OVERBURDEN RATIO

z_1 (km)	λ_1						
	0	0.465	0.5	0.6	0.7	0.8	0.9
1	8.0	13.4	14.2	17.3	22.5	32.9	64.0
2	10.6	16.7	17.6	21.2	27.1	39.0	74.4
3	13.2	20.1	21.1	25.1	31.8	45.1	84.8
4	15.8	23.5	24.6	29.0	36.4	51.2	95.2
5	18.4	26.8	28.0	32.9	41.0	57.3	106
6	21.0	30.2	31.5	36.8	45.6	63.4	116
7	23.6	33.6	34.9	40.7	50.3	69.5	126
8	26.2	36.9	38.4	44.6	54.9	75.6	137

thicknesses from 1 to 8 km and for λ_1 from 0 to 0.90 are given in Table 1. For an increase of λ_1 from 0 to 0.465, the latter corresponding to normal hydrostatic pressure, the length of the block is increased by about 50 per cent. A dry block 6 km thick, for example, would have a maximum length of 21 km; whereas, if the water pressure were hydrostatic, the length would be increased to 30 km. For abnormal pressures with λ_1 in the range from 0.7 to 0.9, and for blocks of thicknesses from 5 to 8 km, the maximum length would range between 41 and 137 km.

Gravitational Sliding

General discussion.—Let us now re-examine the problem of gravitational sliding. Let a block of porous rock of constant thickness and indefinite length rest on a potential plane of slip inclined at a small angle θ. Let S be the normal component of total stress across this plane, T

the tangential component, and p the fluid pressure. Then, at a given point on the plane, the corresponding components of effective solid stress will be

where z is the depth below sea level. From the base of this block let a vertical column of unit cross section be extended to sea level (Fig. 21).

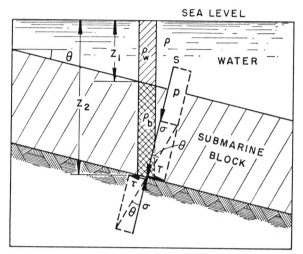

FIGURE 21.—STRESSES AT BASE OF SUBMARINE BLOCK

stress will be

$$\left.\begin{array}{l} \sigma = S - p, \\ \tau = T; \end{array}\right\} \quad (97)$$

and the condition necessary for sliding is that

$$\tau = \sigma \tan \phi. \quad (98)$$

Thus σ and τ are determined by the values of S, T, and p at the bottom of the block; but S and T also depend upon the distribution of p in space. Let us consider, therefore, two separate situations: the first where the block will be submerged in the sea, and the second where the block is on land.

However, before considering these two cases in detail, let us point out two important differences between them. In the submarine block, because of the horizontal sea-level reference surface, it will be convenient to use a vertical and horizontal co-ordinate system; whereas, in the land block, a tilted system with the z-axes perpendicular to the tilted block will be used. Another very important difference between the two cases will be the inclination of the equipressure surfaces corresponding to "normal" pressure. For the submarine block these surfaces will be horizontal, while for the terrestrial block they will have the same inclination as the block.

Submarine block.—For the submerged block consider first the case for which the pressure p has everywhere the hydrostatic value

$$p = \rho_w g z,$$

The normal component S of the total stress at the base of the block will be equal to the normal component of the total weight of this column. Hence, if z_1 be the depth to the top of the block, and z_2 that to its base along the column, then

$$S = [(\rho_b - \rho_w)g(z_2 - z_1) \cos \theta] \cos \theta + p. \quad (99)$$

The shear stress T at the base of the column, however, will be only the tangential component of the submerged weight of the rock column (with its fluid content), and this will be given by

$$T = [(\rho_b - \rho_w)g(z_2 - z_1) \cos \theta] \sin \theta. \quad (100)$$

Then, at the base of the block, the components of effective stress will be

$$\left.\begin{array}{l} \sigma = S - p \\ = [(\rho_b - \rho_w)g(z_2 - z_1) \cos \theta] \cos \theta; \end{array}\right\} \quad (101)$$

and

$$\tau = T = [(\rho_b - \rho_w)g(z_2 - z_1) \cos \theta] \sin \theta, \quad (102)$$

which are simply the corresponding components of the weight per unit area of the submerged block. Also, from equations (101) and (102),

$$\tau/\sigma = \sin \theta/\cos \theta = \tan \theta. \quad (103)$$

However, from equation (98) the necessary condition for slippage is that

$$\tau/\sigma = \tan \phi.$$

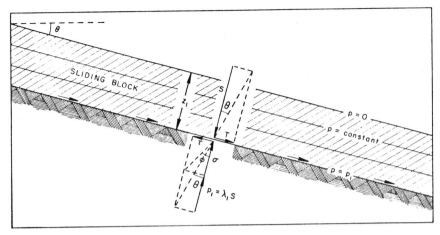

FIGURE 22.—GRAVITATIONAL SLIDING OF SUBAERIAL BLOCK

Then, if we compare equations (98) and (103), it will be seen that for submarine sliding to occur under conditions of normal hydrostatic pressure it will be necessary for

$$\tan \theta = \tan \phi, \qquad (104)$$

or

$$\theta = \phi, \qquad (105)$$

which would require a slope of about 30 degrees.

If, however, in addition to the hydrostatic pressure of the sea water, there were a superposed anomalous pressure p_a within the block of rock, and if the surfaces p_a = constant were parallel to the base of the block, this pressure would not affect the shear stress component at the base of the block but would further diminish the effective normal stress. In this case let us take σ and τ of equations (101) and (102) as reference stresses and let σ' and τ' be the new values of the effective stress components at the base of the block as a result of the anomalous pressure p_a. Then

$$\left. \begin{array}{l} \sigma' = \sigma - p_a = (1 - \lambda_a)\sigma, \\ \tau' = \tau, \end{array} \right\} \qquad (106)$$

in which $\lambda_a = p_a/\sigma$, and

$$\tau'/\sigma' = \tan \phi. \qquad (107)$$

Then, if we equate τ and τ' in equations (103) and (107), we obtain

$$\sigma' \tan \phi = \sigma \tan \theta, \qquad (108)$$

and, when the value of σ' from equation (106) is substituted into equation (108), the latter

becomes

$$(1 - \lambda_a)\sigma \tan \phi = \sigma \tan \theta \qquad (109)$$

from which the slope θ necessary for sliding to occur is given by

$$\tan \theta = (1 - \lambda_a) \tan \phi. \qquad (110)$$

Hence, as λ_a approaches 1, corresponding to p_a approaching σ, the angle of slope θ required for submarine sliding approaches zero.

Subaerial block.—Consider now a subaerial block of slope θ and constant thickness, only in this case let us use inclined axes with the thickness of the block represented by z_1. Let the fluid pressure p be constant (atmospheric) along the upper surface of the block, and let p within the block be a function of the depth only so that the surfaces of constant pressure will be inclined at the same angle θ as the block (Fig. 22).

The components of total stress at the bottom of the block will be equal to the normal and tangential components of the weight of a perpendicular column of unit cross section, and length z_1. Consequently,

$$\left. \begin{array}{l} S = \rho_b g z_1 \cos \theta, \\ T = \rho_b g z_1 \sin \theta, \end{array} \right\} \qquad (111)$$

and

$$T/S = \tan \theta. \qquad (112)$$

The corresponding components of effective stress are

$$\left. \begin{array}{l} \sigma = S - p_1 = (1 - \lambda_1)S, \\ \tau = T, \end{array} \right\} \qquad (113)$$

where p_1 and λ_1 are the values of p and λ at the base of the block. Again, the condition for sliding is that

$$\tau/\sigma = \tan \phi. \qquad (114)$$

Equating τ and T from equations (112), (113), and (114) gives

$$\sigma \tan \phi = S \tan \theta; \qquad (115)$$

when the value of σ from equation (113) is introduced, equation (115) becomes

$$(1 - \lambda_1)S \tan \phi = S \tan \theta, \qquad (116)$$

or

$$\tan \theta = (1 - \lambda_1) \tan \phi. \qquad (117)$$

In this case, if the pressure at the base of the block were everywhere at the value of the normal pressure,

$$p_1 = \rho_w g z_1, \qquad (118)$$

then from equations (118) and (111)

$$\lambda_1 = \frac{p_1}{S} = \frac{\rho_w g z_1}{\rho_b g z_1} = \frac{\rho_w}{\rho_b}. \qquad (119)$$

TABLE 2.—ANGLE θ OF SLOPE DOWN WHICH BLOCK WILL SLIDE AS FUNCTION OF λ_1

λ_1	$1 - \lambda_1$	$\tan \theta =$ $(1 - \lambda_1) \tan \phi$	θ (degrees)
0.0	1.0	0.577	30.0
0.2	0.8	0.462	24.7
0.4	0.6	0.346	19.1
0.6	0.4	0.231	13.0
0.8	0.2	0.115	6.6
0.9	0.1	0.058	3.3
0.91	0.09	0.0519	3.0
0.92	0.08	0.0462	2.6
0.93	0.07	0.0404	2.3
0.94	0.06	0.0346	2.0
0.95	0.05	0.0288	1.6

With this substitution equation (117) becomes, for the case of normal fluid pressures,

$$\tan \theta = (\rho_b - \rho_w)/\rho_b \cdot \tan \phi \cong 0.6 \tan \phi.$$

Then, for $\tan \phi = 0.58$ corresponding to $\phi = 30$ degrees, $\tan \theta = 0.35$, which corresponds to a reduction of the critical angle of sliding from 30 degrees, when the pressure is zero, to about 20 degrees when the vertical pressure gradient is normal.

However, according to equation (115) and as shown in Table 2, if the fluid pressure is greater than normal, as λ_1 approaches unity, corresponding to the approach of the pressure p_1 to S, the critical angle for sliding approaches the limit zero. In fact, Terzaghi (1950, p. 92–94), in his study of low-angle landslides, has made explicit use of this principle and has shown that the measured water pressures also agree with what is expected from the theory.

Combination of Gravitational Sliding and Push from Rear

As a last example, let us now consider the case of a subaerial block pushed from the rear down an inclined plane whose angle of slope is θ (Fig. 23). For this purpose it will be convenient to use a set of inclined co-ordinate axes parallel to the sides of the block with the thickness and length of the block represented by z_1 and x_1, respectively.

In this case the surfaces $\lambda = $ const will be parallel to the top and base of the block, and the block will be propelled jointly by the total stress S_{xx} applied to its rearward edge and the component of its weight parallel to the slope. Its motion will be opposed by the shear stress τ_{zx} along its base, and the equation of equilibrium of the forces acting upon a section of unit thickness parallel to the y-axis when the

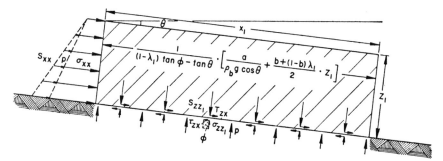

FIGURE 23.—LENGTH OF BLOCK WHICH CAN BE MOVED DOWN SLOPE θ BY COMBINATION OF PUSH FROM REAR AND GRAVITATIONAL PROPULSION

TABLE 3.—MAXIMUM LENGTH (IN KM) OF A BLOCK 6 KM THICK, WITH A PRESSURE–OVERBURDEN RATIO λ_1, WHICH CAN BE PUSHED DOWN A θ-DEGREE SLOPE*

θ (degrees)	λ_1							
	0	0.465	0.5	0.6	0.7	0.8	0.9	0.95
0	21.0	30.2	31.5	36.8	45.6	63.4	116	222
0.5	21.3	31.0	32.5	38.2	48.1	68.8	137	320
1.0	21.6	31.8	33.4	39.7	50.6	74.4	163	
2.0	22.3	33.9	35.8	43.3	57.2	91.1		
3.0	23.0	36.2	38.4	47.5	65.3	116		
4.0	23.9	39.0	41.6	52.8	76.7			
5.0	24.7	41.9	45.0	59.0				
6.0	25.7	45.7	49.5	67.5	Values have been omitted when $x_1 \sin \theta > 8$ km.			
7.0	26.7	50.2	54.9					
8.0	27.8	55.5						
9.0	28.9							
10.0	30.3							

* The data in this table have been plotted as Figure 11 in Part II.

block is at the point of incipient slippage will be

$$\int_0^{z_1} S_{xx}\, dz + \rho_b g z_1 x_1 \sin \theta - \int_0^{x_1} \tau_{zx}\, dx = 0. \quad (120)$$

Then

$$S_{xx} = \sigma_{xx} + p,$$

and

$$\left. \begin{aligned} \sigma_{xx} &= a + b\sigma_{zz} \\ &= a + b(1 - \lambda)\rho_b g z \cos \theta. \end{aligned} \right\} \quad (121)$$

Introducing equation (121) into equation (120), we obtain

$$\left. \begin{aligned} S_{xx} &= a + b(1 - \lambda)\rho_b g z \cos \theta + \lambda \rho_b g z \cos \theta \\ &= a + \rho_b g \cos \theta [b + (1 - b)\lambda]z. \end{aligned} \right\} \quad (122)$$

Likewise

$$\left. \begin{aligned} \tau_{zx} &= \sigma_{zz1} \tan \phi \\ &= (1 - \lambda_1)\rho_b g z_1 \cos \theta \tan \phi. \end{aligned} \right\} \quad (123)$$

Then, introducing equations (122) and (123)

into equation (120) and integrating, we obtain

$$\left. \begin{aligned} a z_1 &+ \rho_b g \cos \theta \left[\frac{b}{2} \cdot z_1^2 + (1 - b) \int_0^{z_1} \lambda z\, dz \right] \\ &- [(1 - \lambda_1)\rho_b g \cos \theta \tan \phi + \rho_b g \sin \theta] z_1 x_1 \\ &= 0. \end{aligned} \right\} \quad (124)$$

As was pointed out in the discussion of the horizontal block, provided the values of λ at depths less than z_1 are not permitted to be greater than λ_1, the minimum length of block will correspond to $\lambda = \lambda_1$. Using this value of λ, and performing integration in the second term of equation (124) and then solving the result for x_1, gives

$$\left. \begin{aligned} x_1 = \frac{1}{(1 - \lambda_1)\tan \phi - \tan \theta} \\ \cdot \left[\frac{a}{\rho_b g \cos \theta} + \frac{b + (1 - b)\lambda_1}{2} \cdot z_1 \right]. \end{aligned} \right\} \quad (125)$$

The maximum length of block which can be moved in this manner is thus a function of the angle of slope θ, the thickness z_1, and the fluid pressure–overburden ratio λ_1. Values of x_1 are given in Table 3 for a block of the fixed thickness of 6 km, values of θ between 0 and 10 degrees, and values of λ_1 from 0 to 0.95. The difference of elevation between the lower and upper ends of the block is $x_1 \sin \theta$. Computations have arbitrarily been discontinued for the cases where this quantity exceeds 8 km. In these computations the values $a = 7 \times 10^8$ dynes/cm^2 and $b = 3$ are still the ones employed.

The first row in Table 3 is the same as that of Table 1 for the horizontal block 6 km in thickness. Under these conditions, if λ_1 were 0.8, a block 63 km long could be moved; and if λ_1 were 0.9, the length could be increased to 116 km. For slopes between 0 and 3 degrees, and values of λ_1 from 0.465, corresponding to hydrostatic conditions, to 0.9, the maximum length of block within the limits of Table 3 would vary from 30 to 163 km.

Summary

From this review of the effect upon the mechanical properties of rocks produced by the pressure of the interstitial ground water, it appears that a means has indeed been found of reducing the frictional resistance of the sliding block by the amount required. It is to be emphasized, however, that this is not accomplished, as has been postulated heretofore, by reducing the coefficient of friction of the rock; rather, the reduction of friction is accomplished by reducing the normal component of effective stress which correspondingly diminishes the critical value of the shear stress required to produce sliding.

ABNORMAL PRESSURES

Abnormal Pressures in Oil and Gas Reservoirs

Before reviewing the evidence for the existence of abnormal pressures, let us consider briefly some of the possible causes of such pressures. Here a distinction must be made with regard to whether the fluid exhibiting such pressure is a hydrocarbon—petroleum or natural gas—or ground water, because the mechanism of maintaining abnormal pressures is different in the two cases.

A body of petroleum or natural gas in its undisturbed state is usually in essentially hydrostatic equilibrium in a space underground completely surrounded by ground water. As has been shown by Hubbert (1953, p. 1956–1957),

the space is characterized by being a position of minimum potential energy for the given hydrocarbon (liquid or gas) with respect to three environmental fields: the earth's gravity field, the pressure field of the surrounding water, and a capillary energy field dependent upon the difference of capillary pressure between hydrocarbon and water at any given point.

The minerals of sedimentary rocks are hydrophilic, so that, if oil or gas is injected into water-saturated rocks, the interfacial menisci will be concave toward the hydrocarbon, and across this interface the pressure in the hydrocarbon will exceed that in the water by an amount

$$p_c = C\gamma/d, \qquad (126)$$

where γ is the interfacial tension between water and hydrocarbon, d the mean grain diameter of the rock, and C a dimensionless factor of proportionality having a value of about 16 (Hubbert, 1953, p. 1975–1979). From this it follows that the capillary pressure of water against hydrocarbon is small in coarse-textured rocks but becomes very large as the grain diameter of the sediment approaches clay sizes. Across an interface between coarse- and fine-textured sediments, the difference in the capillary pressures of water against hydrocarbon will accordingly be

$$\Delta p_c = C\gamma \left(\frac{1}{d_2} - \frac{1}{d_1} \right), \qquad (127)$$

where d_1 is the grain diameter of the coarse sediment, and d_2 that of the fine.

The value of γ for crude oil and water is about 25 dynes/cm. Then, letting d_2 for clay be 10^{-5} cm and d_1 for a sand be 10^{-2} cm, equation (127) gives 4×10^7 dynes/cm^2, or 40 atmospheres, as a representative value of Δp_c between sand and clay. It is realized that clay particles may be somewhat larger than this, but their platy shape renders them more nearly equivalent to spheres of this smaller size. This figure agrees as to order of magnitude with a measurement recently made at the research laboratory in Houston of Shell Development Company. Oil pressure was applied against a shale core taken from the Ventura Avenue field in California. Up to a maximum pressure of 2000 lb/in^2, or about 135 atmospheres, no injection could be detected by observing the displacement of the oil in a capillary tube having a volumetric capacity of 1 cm^3 per 180 cm of length. The permeability of this shale to brine was 10^{-5} millidarcies, or 10^{-11} cm^2 (Personal communication from Harold J. Hill).

It is for this reason that hydrocarbons tend to accumulate in coarse-textured rocks, and the interface between a coarse- and a fine-textured rock acts as a unilateral barrier to petroleum migration; across such a barrier water alone is free to move in either direction.

such as some of those in Iran, quite high pressures could be produced. For example, the abnormal pressure at the top of a column of oil having a density of 0.8 gm/cm³ and a vertical extent of 1000 m would be about 20 bars, or 300 lb/in²; the pressure of a similar column of gas

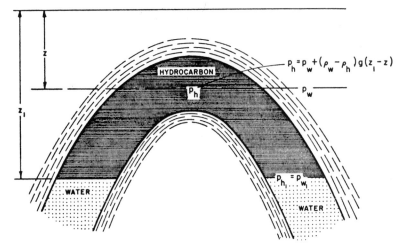

FIGURE 24.—ABNORMAL PRESSURES IN HYDROCARBON ACCUMULATION IN HYDROSTATIC
WATER ENVIRONMENT

Consider then a body of hydrocarbon surrounded by water with both fluids in hydrostatic equilibrium (Fig. 24). At the base of the hydrocarbon in the coarse-textured rock, the capillary pressure difference between the hydrocarbon and the subjacent water will be negligible. Let this depth be z_1, and let the two pressures be that of the water

$$p_{h_1} = p_{w_1} = \rho_w g z_1 .$$

At a shallower depth z, the pressures in the two fluids will be

$$\left. \begin{array}{l} p_w = p_{w_1} - \rho_w g(z_1 - z), \\ p_h = p_{h_1} - \rho_h g(z_1 - z). \end{array} \right\} \quad (128)$$

Consequently, at this depth the hydrocarbon pressure will exceed the water pressure by the amount

$$\Delta p = p_h - p_w = (\rho_w - \rho_h)g(z_1 - z), \quad (129)$$

and this difference at the depth z can be maintained stably by the capillary pressure between the hydrocarbon and the adjacent water.

In oil or gas reservoirs of small vertical amplitude, the abnormal pressures produced in this manner would not be very significant; but, in anticlinal structures of very large amplitude,

with a mean density of 0.2 gm/cm³ would be about 80 bars, or 1200 lb/in².

In view of this fact it is unavoidable that the pressures measured within hydrocarbon accumulations must always be abnormally high with respect to pressures measured at the same depths in the local water; and the greater the density contrast $(\rho_w - \rho_h)$ and the vertical extent of the hydrocarbon $(z_1 - z)$, the greater will this abnormality become. However, if the surrounding waters are at normal pressure, the anomalous hydrocarbon pressures are of strictly local significance.

Anomalous Water Pressures

Whereas anomalous hydrocarbon pressures are compatible with a hydrostatic state in the surrounding ground water, anomalous water pressures invariably imply a hydrodynamic state. The reason for this is that the pressures for the hydrostatic state (which is approximated by our "normal" pressures) are, by definition, those of equilibrium. Hence any departure of the pressures from such a state must represent a disturbance of the equilibrium.

This can be shown more formally by noting that the equation of flow of underground water is given by Darcy's law of which one form of statement is the following:

$$\mathbf{q} = -K \text{ grad } h, \qquad (130)$$

where $\mathbf{q}$ is a vector in the direction of the macroscopic flow of the fluid whose magnitude

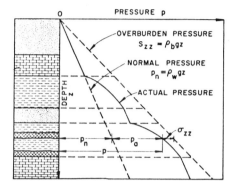

FIGURE 25.—TYPICAL VARIATION OF PRESSURE WITH DEPTH, SHOWING RELATIONS AMONG p_n, p_a, σ_{zz}, AND S_{zz}

is given by the volume of the fluid crossing unit area normal to the flow direction in unit time, h is the height above the reference level $z = 0$ to which the water will rise in a manometer terminated at any given point, and K is a dimensional factor of proportionality depending upon the properties of the fluid and the permeability of the rock. If the fluid properties may be assumed to be constant, then K would vary as a function of the permeability only.

From hydrostatics, the value of h corresponding to a point at depth z with pressure p would be

$$h = p/\rho_w g - z, \qquad (131)$$

where h is the height *above* the reference level and z the depth *below*.

Accordingly,

$$\text{grad } h = (1/\rho_w g) \text{ grad } p - \text{grad } z. \qquad (132)$$

Now, if we let the total pressure p be resolved into the components (Fig. 25)

$$p = p_n + p_a, \qquad (133)$$

where p_n is the normal pressure and p_a a superposed anomalous pressure, then

$$\text{grad } p = \text{grad } p_n + \text{grad } p_a. \qquad (134)$$

When this is introduced into equation (132), that becomes

$$\text{grad } h = [(1/\rho_w g) \text{ grad } p_n - \text{grad } z] \\ + (1/\rho_w g) \text{ grad } p_a. \qquad (135)$$

Noting further that

$$\text{grad } p_n = \rho_w g,$$

and that grad z is a unit vector directed downward, the bracketed term in equation (135) becomes zero, and that equation simplifies to

$$\text{grad } h = (1/\rho_w g) \text{ grad } p_a. \qquad (136)$$

Introducing this into equation (130), we obtain

$$\mathbf{q} = -(K/\rho_w g) \text{ grad } p_a \qquad (137)$$

as an alternative expression of Darcy's law.

Thus, the rate of flow of the water does not depend upon the magnitude of the anomalous pressure p_a, but upon its gradient or rate of change with distance. We could accordingly have a high-permeability sand completely imbedded in a low-permeability shale with a high value of p_a inside the sand. Yet, if the water inside the sand were in extremely slow motion, p_a in the sand would have an essentially constant value, and grad p_a would differ negligibly from zero.

On the other hand, if exterior to the same shale there occurred a sand at normal pressure with $p_a = 0$, then across the shale between the two sands there would exist a gradient in the anomalous pressure whose mean value would be

$$\text{grad } p_a = \Delta p_a/\Delta z;$$

by equation (134), unless the intervening shale were ideally impermeable, the water would flow through the shale from the sand having the higher value of p_a to that having the lower.

Again, if it were possible for p_a to have a constant value throughout space, no flow would occur. This, however, is precluded by the circumstance that at the surface of the ground or, more strictly, at the water table, the pressures p, p_n, and p_a are all zero. Consequently, an anomalous pressure greater than zero anywhere in space implies the existence of values of grad p_a other than zero. Then, unless the rocks are ideally impermeable, it follows that away from any region of greater than normal pressures the water must be flowing and must continue to do so until the excess pressure is dissipated.

Causes of Abnormal Pressures

Because of this continuous leakage, abnormal water pressures are thus transient phenomena and require some dynamical activity to bring them into existence and to maintain them. Let us now consider what some of these may be.

The first and most obvious is the classical picture of the cause of artesian flow from a sand

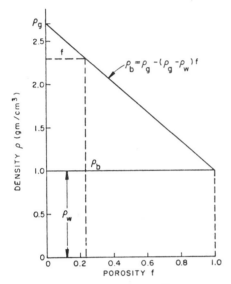

FIGURE 26.—RELATION BETWEEN POROSITY AND BULK DENSITY OF WATER-SATURATED ROCK

at some depth. The fact that the well flows is itself evidence of the existence of an abnormal pressure, and this may often be shown to result from the fact that the sand receives its water from a region of higher elevation where it crops out. In this case the dynamical mechanism for maintaining the pressure is the continuous renewal of the water at the outcrop by rainfall.

A second mechanism for producing abnormal pressures is that of mechanical compression of water-filled porous rocks. We have already seen that, when a porous rock filled with water is subjected to a mechanical compressive stress S, the reaction to this stress is jointly divided between the solid support σ and the fluid support p. Then, if the fluid is allowed slowly to escape by leakage, the pressure p will gradually diminish, and the solid stress σ will increase, until eventually σ will be carrying the entire load. Accompanying this process, the porosity of the rock will diminish, and the bearing surfaces between the grains and the strength of the rock will increase.

Thus, if a readily compressible rock of large porosity and low permeability, initially filled with water at normal hydrostatic pressure, were

to have a mechanical load suddenly imposed upon it, the water, because of its slowness of leakage, would initially assume a corresponding increase of pressure almost equal to the entire increment of load. Then, gradually, the water would be expelled, and the rock compacted until a new equilibrium was reached, and the water pressure was again at its normal value.

Of all the common rocks, those composed of clays lend themselves most pronouncedly to this process. Clays are weak, plastic minerals with moisture-free density of about 2.7 gm/cm³, which are capable of forming porous solid aggregates with porosities ranging from 0 to more than 0.8. The degree of compaction of such a clay whose pore space is filled with water may be inferred from its bulk density, which is related to its porosity by the equation

$$\rho_b = f\rho_w + (1 - f)\,\rho_g,$$

where f is the porosity and ρ_b, ρ_w, and ρ_g are the bulk density (water-saturated) and the water and grain densities, respectively. With a slight rearrangement, this becomes

$$\rho_b = \rho_g - (\rho_g - \rho_w)f, \qquad (138)$$

which shows that the bulk density ρ_b varies linearly from the mineral density ρ_g when the porosity is 0 to that of water when the porosity is 1 (Fig. 26).

The significance of this is that, although the porosities of claystones are measured only occasionally, bulk densities have been measured extensively. In the Tertiary rocks of the Texas and Louisiana Gulf Coastal region, according to Nettleton (1934), the mean density of the sediments, which are dominantly clays, increases from 1.9 gm/cm³ at the surface to about 2.3 at a depth of 5000 feet (1520 m). According to equation (138), using 2.7 gm/cm³ as the mineral density of moisture-free clay, this would correspond to a porosity of 0.47 at the surface and 0.235 at 5000 feet. Hence, roughly half the water contained in the clays when near the surface must have been expelled by the time they were buried to this depth. This indicates two things: (1) that the clays are not ideally impermeable, and (2) that during the expulsion process the anomalous pressure inside the clay must have been higher than that outside, for otherwise no flow of water could have occurred.

From this it follows that the progressive loading of a sedimentary section by continually adding new sediments must create a continuing hydrodynamical condition characterized by anomalous pressures inside the compacting clays, and also inside less compactible rocks such as sands that may be completely embedded in such clays or otherwise isolated by faulting.

In contrast to that produced by simple gravitational loading, more drastic compression can be produced by horizontal compressive stresses of tectonic origin. By "stresses of tectonic origin" is simply meant those stresses which have been produced by causes other than those directly attributable to gravitational loading. In this case the greatest principal stress S_1 is horizontal, while the least principal stress S_3 is vertical and equal to S_{zz}, the weight per unit area of the overburden. The effective greatest and least principal stresses are, therefore,

$$\left.\begin{aligned} \sigma_1 &= \sigma_{xx} = S_{xx} - p, \\ \sigma_3 &= \sigma_{zz} = S_{zz} - p. \end{aligned}\right\} \qquad (139)$$

Since, at a given depth, S_{zz} is fixed, it follows that, if σ_1 is increased more rapidly than the pressure can be dissipated by the leakage of water, p will increase until it reaches its maximum possible value of $p = S_{zz}$. When this occurs λ will have the value of unity, whereby the superincumbent material could be moved tangentially with negligible frictional resistance. At the same time σ_3 would be zero, and σ_1 corresponding to failure of the rock, as given by equation (7), would reduce to

$$\sigma_1 = a,$$

the ordinary crushing strength of the material. Hence an incompletely compacted water-filled clay or shale, or other weak, low-permeability rock, responds to deformational stresses by an increase of the pressure of the water, and this, in turn, weakens the rock by reducing the values of the effective stresses necessary to cause the rock to fail.

The magnitude which the pressure p may reach has the upper limit $p = S_{zz}$ corresponding to $\lambda = 1$; but whether this limit may be reached or approached depends upon the relative rates of two opposing processes: the rate of application of the deformational stress, and the rate of pressure dissipation by leakage of the contained water. Therefore, of the two processes for increasing the deformational stresses—that of sedimentary loading in a tectonically quiescent geosyncline, and that of the application of orogenic stresses—the latter appears to be the more effective. In the case of sedimentary loading, the water pressure should be able to approach the value $\lambda = 1$ only in the limiting case where the clay is so completely uncompacted that the effective stress σ_{zz}, which it is able to provide in support of the overburden, is small as compared with the total load S_{zz}.

In the orogenic compression, on the other hand, since the pressure of the overburden is now the least stress, no such limitation exists, so that, if the rate of increase of the applied stresses is sufficiently rapid as compared with the rate of pressure dissipation, there is nothing (except the possible presence of stronger rocks such as sandstone or quartzite, which may act as reinforcing plates) to prevent p from becoming equal to S_{zz}, and λ from becoming unity.

These inferences have in fact been extensively investigated in soil mechanics (Taylor, 1948, p. 234–249; Terzaghi and Peck, 1948, p. 233–242) in studies of consolidation of clays under near-surface conditions. It has been found that if a water-saturated clay is placed in a vertical cylinder, with means for the escape of water provided at the top and bottom, and a vertical load is applied to the clay by means of a piston, the water pressure inside the clay, as measured by manometers, will rise abruptly upon application of the load and will then slowly decline—more rapidly near the ends than in the middle—as the water escapes from the interior of the clay body. A more detailed theoretical treatment of the consolidation process is found in the following series of papers by M. A. Biot (1935a; 1935b; 1941; 1955; 1956).

Measurement of Pressures of Interstitial Fluids

The pressures of the interstitial fluids in rocks penetrated by drilling may be measured in various ways, but these fall into two principal classes: (1) direct measurement of the pressure by an instrument lowered into the well with the depth interval in which the pressure is desired isolated from the intervals above and below; and (2) indirect measurements by means of the surface pressure, density, and length of a column of fluid in the well which is in pressure equilibrium with the interstitial fluids in the isolated interval below.

Of these two methods, that of lowering the instrument in the well is the more precise with a potential accuracy of better than 0.1 per cent. This also is time consuming and expensive, so only occasionally are such measurements made. Measurements of less accuracy, the order of 10 per cent, are those that result from determining what density the drilling mud must be given in order to prevent the formation fluids from entering the well. For various reasons, including expense, the density of drilling mud is kept only slightly greater than that necessary to retain the formation fluids. Consequently, as higher anomalous pressures are encountered for which the mud density is too low, this fact is indicated by evidences of contamination as the circulating mud returns to the surface, and also by occasional pressure surges. It is thus possible to determine empirically what mud density is required to balance the interstitial fluid pressure and from this to obtain an approximate measure of these pressures.

Since this balancing of the mud weight against the pressure of the formation fluids during drilling is a continuous process, it provides a fairly accurate log of the pressures of the formation fluids throughout the depth of the well. When in balance with the mud pressure, that of the interstitial fluids is given by

$$p = p_0 + \rho_m g z, \qquad (140)$$

where ρ_m is the density of the mud, and p_0 is the pressure at the wellhead.

A few decades ago it was difficult to obtain stable drilling muds with densities much greater than 1.5 gm/cm^3. At that time if pressures greater than those of a column of mud of this density were encountered, it was necessary to increase the pressure by adding an additional pressure p_0 at the wellhead. Muds are now available, however, which can be given any density from just above that of water to the mean density of the total overburden, that is from about 1.1 to 2.3 gm/cm^3. With such muds the density can be so adjusted as to make an additional well-head pressure unnecessary.

Occurrence of Abnormal Pressures

Since the geological occurrences of abnormal pressures will be reviewed in some detail in Part II, they will here be mentioned only summarily.

Abnormal pressures in geosynclinal basins with values of λ at least as high as 0.9, which presumably are the result of sedimentary loading, occur in the Gulf Coastal region of Texas and Louisiana (Cannon and Craze, 1938; Cannon and Sullins, 1946; McCaslin, 1949; Dickinson, 1951; 1953; Anonymous, 1957), north Germany (Thomeer, 1955, p. 7), East Pakistan (Sekules, 1958), and have been reported without documentation in a number of other basins of thick Tertiary sediments.

Abnormal pressures in tectonically active areas occur in the deeper zones of the Ventura anticline in California (Watts, 1948) where λ

reaches 0.9, in Trinidad (Reed, 1946; Suter, 1954, p. 98, 118–119), in Burma (Abraham, 1937), in the Tupungato oil field in the Andean foothills of Argentina (Baldwin, 1944), in Iran (Mostofi and Gansser, 1957), and in the Khaur field (Keep and Ward, 1934) and the Jhatla well (Anderson, 1927, p. 708) on the Potwar plateau of the Punjab in Pakistan, immediately south of the overturned folds in the foothills of the Himalaya Mountains.

Of these, two of the best documented are the Khaur field in Pakistan and the Qum field in Iran. A detailed account of the drilling of the Khaur field has been given by Keep and Ward. In this field abnormal pressures were encountered at shallow depths, and the abnormality increased with depth (Fig. 27). Water and oil sands at crestal depths of 5100 to 5500 feet were penetrated in which the pressure was so great that with muds weighing 88 to 90 lb/ft^3, corresponding to densities of 1.41 to 1.45 gm/cm^3, it was necessary to apply well-head pressures of 1800 to 2000 lb/in^2. Three specific sets of data from which bottom-hole pressures, and hence λ, could be computed are the following:

Well-head pressure (lb/in^2)	Specific weight of mud (lb/ft^3)	Depth (ft)	Bottom-hole fluid pressure (lb/in^2)	Over-burden pressure (lb/in^2)	λ
2000	88	5215	5187	5528	0.94
1800	90	5215	5059	5528	0.92
2000	90	5478	5424	5807	0.93

The computation of the pressure of the overburden is based on the authors' statement that the mean of the measured densities of the rocks was about 2.45 gm/cm^3, which corresponds to an overburden pressure gradient of 1.06 (lb/in^2)/ft.

According to the account by Mostofi and Gansser (1957), the discovery well, Alborz No. 5, of the Qum oil field in central Iran required mud weighing 135 lb/ft^3 ($\rho_m = 2.16$) in drilling an evaporite section from 7000 feet to below 8500. Before drilling into the reservoir rock, the mud weight was reduced to 129 lb/ft^3 ($\rho_m = 2.06$ gm/cm^3), which would balance a fluid pressure at $\lambda = 0.89$ based on an overburden density $\rho_b = 2.31$. After drilling 2 inches into limestone, one of the most spectacular blowouts in the history of the petroleum industry occurred, and the well ran wild, producing oil with little gas, for 3 months before being brought under control.

All that can be said from the data is that the

pressure in the reservoir corresponded to a value of λ greater than 0.89. This can in part be accounted for by the hydrostatic abnormality at the crest of an oil structure over that of the water at the same depth. However, the fact that

certain to an extent of 10 per cent. The density of the overburden is uncertain by an amount of possibly 5 per cent, and a comparable uncertainty exists with respect to the pressure measurements.

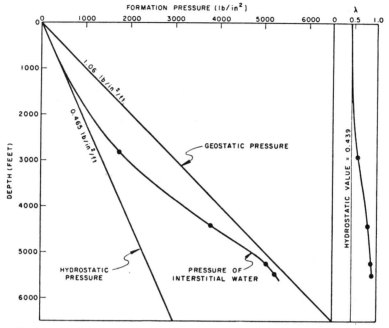

FIGURE 27.—VARIATION OF PRESSURE, AND OF CORRESPONDING VALUES OF λ, WITH DEPTH IN KHAUR FIELD, PAKISTAN
Keep and Ward, 1934

135 lb/ft³ mud, corresponding to a fluid pressure with λ = 0.94, was required in the water-bearing section above leaves little doubt that the abnormality of the water pressure in this locality must be of the order of λ = 0.90 or higher.

Abnormal pressures arising at least in part from the difference in pressure gradients between hydrocarbons and water occur conspicuously at the crests of the structures of many of the Iranian oil fields. Data for eight of these fields, obtained from British Petroleum Company Limited sources, which have kindly been made available by M. M. Pennell of W. C. Connel, New York, are shown in Table 4. Since a pressure gradient of 1 lb/in²/ft corresponds to an overburden density of 2.31 gm/cm³, which is very close to the mean bulk density of water-filled sediments, then the value of λ is given directly by the ratio of the pressure in (lb/in²)/ft to the depth in feet. The values of λ in Tables 4 and 5 are probably un-

However, as in the case of the Qum field, evidence that the abnormal pressures in this region are not solely due to the difference in pressure gradients in oil and water is provided in Table 5. Here, for 8 wells in the Agha Jari field, and for 7 in the Naft Safid, the pressures of water shows encountered while drilling gave values of λ ranging from 0.84 to 1.0, and for 10 of the 15 wells the values fell within the range from 0.90 to 1.0. All these occurrences were in the so-called Lower Fars formation, which is composed of marls, anhydrite, some limestones, and a great deal of salt.

Another instance of abnormal water pressures in a drilling well, whose location was not specified by Cooke, but according to M. M. Pennell (Personal communication) was in Chia-Surkh, Iraq, has been described by Cooke (1955). In this well, high abnormal pressures were encountered throughout a depth interval of 8000 feet, in which drilling muds with densities seldom less than 2.0 and frequently greater than

2.25 gm/cm³ were required. From the data given the approximate variation of pressure with depth is shown in Figure 28. It is interesting to note that in a high-permeability limestone

TABLE 4.—DEPTH OF COVER AND ORIGINAL PRESSURE AT CREST OF IRANIAN OIL RESERVOIRS

Field*	Cover (ft)	Ratio of pressure to depth (lb/in²/ft)	λ
Masjid-i-Sulaiman	640	0.88	0.88
Lali	3900	0.76	0.76
Haft Kel	1900	0.74	0.74
Naft Safid	3000	0.98	0.98
Agha Jari	4800	0.71	0.71
Pazanun	5700	0.68	0.68
Gach Saran	2600	0.83	0.83
Naft-i-Shah	2300	0.95	0.95

* All these fields except Pazanun had active seepages. Data from British Petroleum Company Limited, courtesy M. M. Pennell, W. C. Connel, New York.

section extending from 5500 to 9000 feet, the pressure, although abnormal, has an essentially hydrostatic gradient.

Still another instance of an abnormal pressure at the crest, in this case, of a gas field is provided by the Lacq field in the Aquitaine basin near the Pyrenees Mountains in France (Berger, 1955). The crest of the gas accumulation occurs at a depth of about 3450 m, and the deepest well drilled reached 4350 m without reaching the gas–water contact. The pressure at 3450 m was 645 kg/cm²; that at a depth of 4350 m was approximately 677 kg/cm² (p. 1454, Fig. 1). Assuming a bulk density of 2.31 gm/cm³ for the overburden, these data give 0.81 and 0.68 as the values of λ at the top and bottom of the interval. The pressure gradient in the gas was 0.355 kg/cm²/10 m. If the pressure in the water at the gas–water contact were normal, and the pressure gradient of the water is 1.08 kg/cm²/10 m, then this contact would have to occur at a depth of 7150 m, giving a gas column 3700 m in vertical extent. Since a column of this length appears unlikely, it is probable that the subjacent water has some degree of abnormal pressure also.

Most of the foregoing evidences of abnormal pressures have become available during the last 25 years as the result of drilling progressively deeper for oil in widely separated geographical areas and the development of more reliable methods of measuring the fluid pressures en-

countered. They confirm beyond a doubt that pressures almost great enough to float the total overburden do in fact exist, and that these appear to originate both from gravitational

TABLE 5.—WATER SHOWS ENCOUNTERED IN WELLS DRILLED AT AGHA JARI AND NAFT SAFID FIELDS, IRAN

Well	Depth of show* (ft)	Ratio of pressure to depth (lb/in²/ft)	λ
AJ 5	8354	0.85	0.85
AJ 10	4703	0.84	0.84
AJ 18	3010	0.90	0.90
AJ 22	6314	0.97	0.97
AJ 24	4681	0.83	0.83
AJ 25	7012	1.00	1.00
AJ 27	2445	0.86	0.86
AJ 36	4846	0.91	0.91
W 7	4870	1.00	1.00
W 11	4999	0.94	0.94
W 13	3147	0.97	0.97
W 14	4885	0.91	0.91
W 16	4538	0.87	0.87
W 21	5226	0.98	0.98
W 24	3712	0.91	0.91

* All these shows occurred in the so-called Lower Fars composed of marls, anhydrite, some limestones, and a great deal of salt. Data from British Petroleum Company Limited, courtesy M. M. Pennell, W. C. Connel, New York.

loading, in the case of geologically young sedimentary basins, and from active tectonic compression. Then, if we consider the epochs of intense orogeny which have occurred during the geologic past during which geosynclinal sediments having aggregate thicknesses of the order of 10 km have been involved, and consider further that the pore spaces of these rocks have been filled with water, it is difficult to escape the conclusion that in response to the tectonic compression the pressure of this water must often have been raised to the limit. If so, not only would the conditions favorable for faulting have existed, but the same conditions would greatly facilitate other types of deformation as well.

EXPERIMENTAL CONFIRMATION

Recent Tests on Clay

In addition to the triaxial tests on concrete by McHenry, and on rock by Handin, described earlier, several other types of experimental evidence strongly support our earlier inferences. It is a common observation in field geology that bedding-plane faults occur more often within shale sections than in beds of harder rocks. In

internal pore pressure is completely dissipated. These are the so-called "undrained" and "drained" tests.

For the undrained tests, as the total stress is increased, the pore pressure increases by the

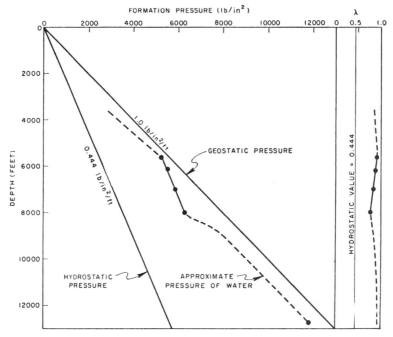

FIGURE 28.—APPROXIMATE PRESSURE–DEPTH RELATIONS OF WATER IN WELL IN CHIA-SURKH, IRAQ
Cooke, 1955

view of the greater mechanical weakness of shales than of sandstones and limestones, this has led to the reasonable inference that wet clays facilitate fault motion by acting as lubricants.

Earlier triaxial tests on clays made in various soil-mechanics laboratories tended to support this viewpoint since the angles ϕ of the Mohr envelopes were commonly very much less than 30 degrees, with correspondingly small values for the coefficient of friction. Moreover, different tests gave different results so that no consistent relationship in the mechanical behavior of wet clays was apparent.

Work done during the last decade (Bishop and Eldin, 1950–1951; Bishop and Henkel, 1957), however, has largely dispelled this anomaly and has shown that the angle of internal friction is not significantly different from that of other rocks. This has resulted from tests on water-saturated specimens in which the contained water may be completely retained by an impermeable jacket, or else may be expelled as the specimen is slowly compressed so that the

same amount so that the effective normal stress σ remains constant. Thus, if, for a given value of the least effective stress σ_3, failure occurs at a given value of σ_1, then the stress difference $\sigma_1 - \sigma_3$ corresponding to failure will be

$$\sigma_1 - \sigma_3 = (S_1 - p) - (S_3 - p) = S_1 - S_3 = \text{const}$$

for all values of S_1 and S_3.

If a Mohr diagram be plotted for a succession of pairs of values of S_1 and S_3 corresponding to failure, the result will be a series of circles of constant radius whose centers lie at increasing distances out on the S-axis. The angle ϕ of the envelope of these circles will be zero, and the Coulomb equation (3) for the shear stress along the plane of slippage

$$T = T_0 + S \tan \phi$$

will reduce to the constant shear stress

$$T = T_0 = (S_1 - S_3)/2.$$

Thus the clay will have an initial shear strength T_0 but a zero coefficient of friction with respect to increments of the total normal stress (Fig. 29a).

On the other hand, the drained tests (Fig. 29b) on the same materials, in which the stresses

fall away from the geostatic curve with increasing depth.

Thus, as shown in Figure 25, in a stratigraphic section there are certain favored localities for

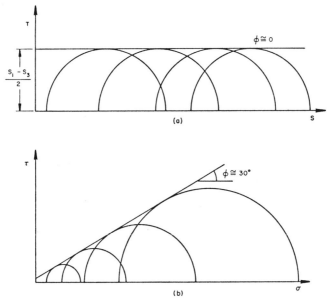

FIGURE 29.—CHARACTERISTIC BEHAVIOR OF CLAYS
(a) In undrained tests, and (b) in drained tests (Bishop and Henkel, 1957, p. 10–21).

are increased very slowly and the water allowed to escape and completely dissipate the internal pore pressure, give results which are in all essential respects like those for other rocks, including values for the angle ϕ of about 30 degrees.

The earlier inconsistencies arose from the failure to distinguish between total stresses and effective stresses when testing water-saturated clay specimens. When this distinction is made and only the effective stresses are plotted on the Mohr diagrams, the anomaly vanishes.

The significance of this to the overthrust problem is that clays, because of their low permeabilities, are among the most favorable habitats for abnormal pressures since, when these rocks are compressed either by gravitational or by tectonic compaction, the slowness with which the water is able to escape prevents the pressures inside the clays from being rapidly dissipated. In a high-permeability sand or fractured limestone, on the other hand, even if the pressure at the upper surface were at the maximum value corresponding to $\lambda = 1$, the pressure gradient within the rock would have the normal value $\rho_w g$, and the pressure would

the fluid pressures to approach most closely to the geostatic pressures, and these appear to be inside thick shale or evaporite sections, whereas in a highly permeable rock the largest value of λ should be at its upper surface, and the lowest value at its base. These circumstances not only favor shales and evaporite sections as surfaces of least frictional resistance to slippage, but they also show that the supposed lubricating effects of such rocks are in reality the consequence of the high pore pressures of the contained water rather than of an intrinsically low coefficient of internal friction.

Terzaghi (1950, p. 88–105, 119–120) has made explicit use of this property in his explanation of low-angle landslides in loose soils. In several instances he has shown that sliding occurred on gentle slopes when the water pressures in the clays became great enough to reduce the frictional resistance to sliding until it became equal to the gravitational force acting on the block down the given slope.

Concrete-Block Experiment

For a direct confirmation of the theory, the laboratory experiment shown in Figure 30 has been devised. This consists of a base of porous concrete, in a water-tight box, with a plane horizontal upper surface, upon which rest one or two blocks of cast concrete. Surrounding the

the O-ring and the water. For pressures below p_1 the blocks are supported jointly by the O-ring, the concrete, and the water. Let us consider the range from p_2 to p_1 as constituting

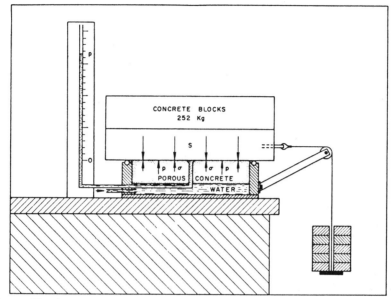

FIGURE 30.—APPARATUS FOR MEASURING REDUCTION OF SLIDING FRICTION AS UPLIFT PRESSURE IS INCREASED

top of the base block is an O-ring, made of laboratory rubber tubing, which is inflated to retain water admitted under pressure through the base of the bottom block. The object of the experiment is to determine the effect upon the shearing force required to slide the upper blocks over the lower as the fluid pressure p at the plane of contact is gradually increased.

The problem is somewhat complicated by the necessity of having to use the rubber O-ring to prevent leakage of the water. This results in a division of the total shear force F_t required to slide the block into two components, F_{tr} and F_{tc}, required, respectively, to overcome the frictional resistance of the rubber and of the concrete. The component F_{tr} was made as small as possible by lubrication, but in no instance could it be made negligible.

The experiment accordingly divides itself into two phases. The water pressure may have any value from 0 to p_2, the pressure required to support the total weight of the upper concrete blocks. Within this range there is an intermediate pressure p_1 above which contact is lost between the upper and lower concrete blocks, and the upper blocks are supported entirely by

Phase I of the experiment, and the range from p_1 to 0 as Phase II.

For Phase I we deal only with friction between the upper block and the O-ring. If we let F_t be the horizontal shearing force applied to the upper block, F_{nr} be the normal force between this block and the O-ring, then from the law of sliding friction

$$F_{tr}/F_{nr} = \tan \phi_r \qquad (141)$$

will be the coefficient of friction between the rubber and the concrete. If W is the weight of the block, and A is the area over which the pressure p is applied, then

$$F_{nr} = W - pA, \qquad (142)$$

which, when introduced into equation (107), gives

$$F_t = \tan \phi_r(W - pA). \qquad (143)$$

This indicates that, within the range of pressures represented by Phase I, the values of the total shearing force F_t, required to slide the blocks, when plotted graphically against the quantity $(W - pA)$, should fall along a straight

245

line passing through the origin with slope $\tan \phi_r$.

When the pressure is reduced to p_1, contact is made between the upper and lower concrete blocks, preventing any further compression of

then obtain, for the pressure range from p_1 to 0,

$$F_t = F_{t_1} + (W - pA - F_{nr_1}) \tan \phi_c,$$

which, when rearranged, becomes

$$F_t = (F_{t_1} - F_{nr_1} \tan \phi_c) + \tan \phi_c (W - pA), \quad (147)$$

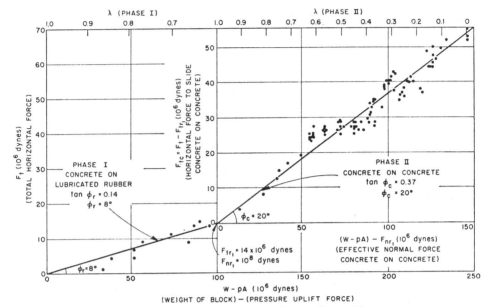

FIGURE 31.—OBSERVATIONAL DATA FROM SLIDING-BLOCK EXPERIMENT

the rubber O-ring. Consequently, during Phase II both the uplifting force and the frictional force exerted by the rubber O-ring retain their constant values F_{t_1} and F_{nr_1} reached when the pressure was reduced to p_1. In this phase then the frictional resistance of concrete on concrete can be evaluated by our being able to eliminate the previously determined effect of the O-ring. In this case

$$\tan \phi_c = \tau/\sigma = F_{tc}/F_{nc},$$

or

$$F_{tc} = F_{nc} \tan \phi_c, \quad (144)$$

where $\tan \phi_c$ is the coefficient of friction between the concrete blocks, and F_{tc} and F_{nc} the shearing force and normal force, respectively, between the blocks. The total shearing force required to move the upper block will accordingly be

$$F_t = F_{t_1} + F_{tc}, \quad (145)$$

and the normal force exerted by the lower concrete block on the upper will be

$$F_{nc} = (W - pA - F_{nr_1}). \quad (146)$$

Combining equations (145) and (146), we

which again is a linear equation between F_t and $(W - pA)$, with a slope, in this instance, of $\tan \phi_c$.

Hence, if the experimental data for F_t are plotted against $(W - pA)$, the curve should follow two successive straight-line segments, the first according to equation (143) passing through the origin and continuing with an angle of slope ϕ_r to the abscissa $(W - p_1 A)$. From this point the curve should continue with an abrupt change of slope from $\tan \phi_r$ to $\tan \phi_c$.

A graph of the experimental data based on 131 measurements is given in Figure 31. Notwithstanding the fact that the scattering of the points caused by experimental error is fairly wide, the data agree very satisfactorily with the theoretical predictions. As the fluid pressure was increased, the force required to slide the upper blocks diminished, so that at the higher pressures the 252-kg (550 lb) blocks could easily be slid with one hand. It should be kept in mind that at those higher pressures the experiment was in Phase I, and the coefficient of friction in that phase was that of concrete on rubber O-ring. Nevertheless, at those pressures, where the λ of Phase I was approaching unity, the force required to move the block was that of only a few kilograms of weight, whereas, with the full weight of the blocks resting on the O-ring, the force required would have been that of about 36 kg of weight.

between the metal and the glass without affecting the tangential component. The can stops at the edge of the glass because the pressure is released.

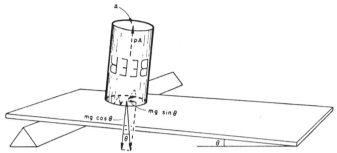

FIGURE 32.—BEER-CAN EXPERIMENT

Beer-Can Experiment

A more elegant demonstration, originally suggested as a result of an accidental observation by M. A. Biot (Personal communication), can be made with a piece of plate glass and an empty beer can open at one end.

A piece of plate glass a meter or so in length is cleaned with a liquid or nonabrasive detergent so that it will retain a continuous film of water. On this glass, which is first wet with water, is placed in upright position an empty beer can. The glass is then tilted until the critical angle ϕ is reached at which the can will slide down the surface. This angle of approximately 17 degrees gives for the coefficient of friction of metal on wet glass a value of about 0.3.

Next, the beer can is chilled, either by being placed in the freezing compartment of a refrigerator or in a container of solid carbon dioxide, and the experiment is repeated. The can is first placed on the glass with its open end *upward* and the angle of sliding redetermined. It is found to be the same as that previously, indicating that the coefficient of friction of the metal on wet glass is not temperature sensitive within this range of temperature.

Finally, with the angle of slope of the glass fixed at about 1 degree, the can (rechilled if necessary) is placed on the glass with its open end *downward* (Fig. 32). In this case it will slide down the slope the full length of the plate, but will stop abruptly at the edge.

The physical reason for this behavior is exactly the same as that which we have deduced for the case of overthrust faulting. As the cold can becomes warm the air inside expands and causes the pressure to increase. This, in turn, partially supports the weight of the can and so reduces the normal component of force

The formal theory of the experiment is the following: Let F_n be the normal component of force exerted by the beer can on the glass, and F_t the tangential component. Sliding will occur when

$$F_t/F_n = \tan \phi. \tag{148}$$

If θ be the angle of tilt, p the excess pressure of the air inside the can over that outside, A the base of the can, and m its mass, then

$$F_t = mg \sin \theta, \tag{149}$$

$$F_n = mg \cos \theta - pA. \tag{150}$$

Then, if we let

$$pA = \lambda(mg \cos \theta) \tag{151}$$

and introduce this into equation (150), we obtain

$$\frac{F_t}{F_n} = \tan \phi = \frac{mg \sin \theta}{(1 - \lambda)mg \cos \theta} = \frac{1}{1 - \lambda} \cdot \tan \theta,$$

or

$$\tan \theta = (1 - \lambda) \tan \phi. \tag{152}$$

Since this is the same equation and has the same physical basis as that derived earlier for the case of gravitational sliding of fault blocks [equation (117)], the beer-can experiment constitutes a legitimate confirmation of the validity of the earlier equation.

Bearings of the Mount Palomar Telescope

A striking example of the use of substantially the same principle occurred in the design of the bearings for the 200-inch telescope at the Mount Palomar Observatory (Karelitz, 1938). The telescope and its frame weighs approximately a million pounds (or 450 metric tons), yet the frictional forces opposing rotation that could be tolerated in the design were very much less than those which would be produced by any known bearing material. The problem was solved by supporting the telescope and its mount on five large bearings, each consisting of an open wide-rimmed box upon which the load rested. When in operation oil is pumped into each box at a pressure great enough to support the load without metal-to-metal contact, with the oil continually escaping through the narrow space between the rim of the bearing box and the load.

According to Karelitz, this gave an equivalent coefficient of friction of 2.6×10^{-6} at the driving speed of the telescope, and a $\frac{1}{12}$-horsepower motor is more than adequate as a source of power.

The support in this case differs from the hydraulic support for fault blocks with which we are concerned only in that the pressure p is maintained at such a value that λ is always unity. At smaller values of p and λ, the frictional resistance of the bearings would be given by an equation analogous to (117) and (152).

GENERAL SUMMARY AND CONCLUSION

Despite the widespread empirical evidence for the existence of large-scale overthrust faults which has been accumulating for more than a century, a satisfactory mechanical explanation of this phenomenon has proven to be extraordinarily difficult to develop. However, as a result of an understanding which currently is evolving concerning the mechanics of fluid-filled porous solids, it appears that we now have a simple and adequate means of reducing by the required amount the frictional resistance to the sliding of large overthrust blocks.

This arises from the circumstance that the weight of such a block per unit area S_{zz} is jointly supported by a solid stress σ_{zz} and by the pressure p of the interstitial fluids, whereby

$$p + \sigma_{zz} = S_{zz};$$

whereas the critical shear stress required to slide the block depends only upon σ_{zz} in accordance with the equation

$$\tau_{\text{crit}} = \sigma_{zz} \tan \phi.$$

Then, when the value of σ_{zz} from the preceding equation is substituted into the one immediately above, we obtain

$$\tau_{\text{crit}} = (S_{zz} - p) \tan \phi,$$

from which it is clear that as the fluid pressure p approaches S_{zz}, corresponding to a flotation of the overburden, the shear stress required to move the block approaches zero.

Most rocks are to some degree porous, and beneath shallow depths their pores are usually filled with water. In a nearly static state the pressure of this water at any given depth would be approximately equal to that of a column of water extending from this depth to the surface of the ground. The pressure corresponding to such a state is taken as a reference or "normal" pressure field. However, if bodies of porous and mechanically weak, rocks of low permeability, such as clays and interbedded clays and evaporites, which were initially filled with water at normal pressure, should be subjected to an increased compressive stress, the pressure of the water would be raised, and the water would tend to be expelled, and the rock to become compacted.

Two ways geologically in which such compressive stresses can be applied are by means of gravitational loading by continued sedimentation in a subsiding geosynclinal basin, or by the application of tectonic compressive stresses during periods of orogeny. The maximum value that the pressure p can attain is $p = S_{zz}$. In the case of gravitational loading, the maximum principal stress is vertical so that the pressure p can only approach S_{zz} as a limit. Under tectonic compression, however, the maximum stress is horizontal and hence greater than S_{zz}. In this case the pressure p can readily be raised until it is equal to S_{zz}.

Observations made during recent decades in oil wells drilled in various parts of the world have shown that fluid pressures of the order of $0.9S_{zz}$, and greater, occur both in geosynclinal basins and in areas which are probably being compressed tectonically at present.

It therefore appears that, during periods of orogeny in the geologic past, which often have affected sedimentary sections many kilometers thick, the pressure in the water contained in large parts of these sediments must have been raised to, or approaching, the limit of flotation of the overburden. This would greatly facilitate the deformation of the rocks involved, and the associated great overthrusts, whether motivated by a push from the rear or by a gravitational pull down an inclined surface, would no longer pose the enigma they have presented heretofore.

REFERENCES CITED

Abraham, W. E. V., 1937, Geological aspects of deep drilling problems: Inst. Petroleum Technologists Jour. (London), v. 23, p. 378–387; Discussion, p. 387–390.

Ampferer, Otto, 1934, Über die Gleitformung der Glarner Alpen: Akad. Wiss. Wien, mat.-nat. Kl., Sber. Abt. I, Bd. 143, H. 3–4, p. 109–121.

Anderson, Robert Van Vleck, 1927, Tertiary stratigraphy and orogeny of the Northern Punjab: Geol. Soc. America Bull., v. 38, p. 665–720.

Anonymous, 1957, Deepest well plugs back, but still champ: Oil and Gas Jour., v. 55, Feb. 4, p. 82.

Auden, J. B., 1937, The structure of the Himalayas in Garhwal: Geol. Survey India Records, v. 71, pt. 4, p. 407–433.

Bailey, Edward B., 1935, Tectonic essays, mainly Alpine: Oxford, Clarendon Press, 200 p.

—— 1952, Geological Survey of Great Britain: London, Murby, 278 p.

Baldry, R. A., 1938, Slip-planes and breccia zones in the Tertiary rocks of Peru: Geol. Soc. London Quart. Jour., v. 94, p. 347–358.

Baldwin, Harry L., 1944, Tupungato oil field, Mendoza, Argentina: Am. Assoc. Petroleum Geologists Bull., v. 28, p. 1455–1484.

Barrande, Joachim, Logan, William, and Hall, James, 1861, Correspondence on the Taconic system and the age of the fossils found in the rock of northern New England, and the Quebec group of rocks: Am. Jour. Sci., 2d ser., v. 31, p. 210–226.

Berger, Y., 1955, Lacq profond: Gisement de gaz a pression anormalement élevée. Problèmes de forage: Revue de l'Institut Français de Pétrole et annales des combustibles liquides, v. X, p. 1453–1466.

Billings, Marland P., 1954, Structural geology: 2d ed., New York, Prentice-Hall, 514 p.

Biot, M. A., 1935a, Un cas d'intégrabilité de l'équation non linéaire de la chaleur et de la consolidation des sédiments argileux: Ann. Soc. Sci. Bruxelles, Sér. B, t. LV, p. 106–109.

—— 1935b, Le problème de la consolidation des matières argileuses sous une charge: Ann. Soc. Sci. Bruxelles, Sér. B, t. LV, p. 110–113.

—— 1941, General theory of three-dimensional consolidation: Jour. Applied Physics, v. 12, p. 155–164.

—— 1955, Theory of elasticity and consolidation for a porous anisotropic solid: Jour. Applied Physics, v. 26, p. 182–185.

—— 1956, General solutions of the equations of elasticity and consolidation for a porous material: Jour. Applied Mechanics, v. 23, p. 91–96.

Bishop, Alan W., and Eldin, Gamal, 1950–1951, Undrained triaxial tests on saturated sands and their significance in the general theory of shear strength: Géotechnique, v. II, p. 13–32.

Bishop, Alan W., and Henkel, D. J., 1957, The measurement of soil properties in the triaxial test: London, Edward Arnold, 190 p.

Blackwelder, Eliot, 1910, New light on the geology of the Wasatch Mountains, Utah: Geol. Soc. America Bull., v. 21, p. 517–542; Discussion, p. 767.

Brown, C. Barrington, 1938, On a theory of gravitational sliding applied to the Tertiary of Ancon, Ecuador: Geol. Soc. London Quart. Jour., v. 94, p. 359–368; Discussion, p. 368–370.

Bucher, W. H., 1933, Volcanic explosions and overthrusts: Am. Geophys. Union Trans., 14th Ann. Meeting, p. 238–242.

—— 1956, Role of gravity in orogenesis: Geol. Soc. America Bull., v. 67, p. 1295–1318.

Butts, Charles, 1927, Fensters in the Cumberland overthrust block in southwestern Virginia: Va. Geol. Survey Bull. 28, 12 p.

Cady, W. M., 1945, Stratigraphy and structure of west-central Vermont: Geol. Soc. America Bull., v. 56, p. 515–587.

Callaway, Charles, 1883a, The age of the newer gneissic rocks of the northern Highlands: Geol. Soc. London Jour., v. 39, p. 355–414; Discussion, p. 414–422.

—— 1883b, The Highland problem: Geol. Mag., new. ser., dec. 2, v. 10, p. 139–140.

Campbell, Marius R., 1914, The Glacier National Park: a popular guide to its geology and scenery: U. S. Geol. Survey Bull. 600, 54 p.

Cannon, G. E., and Craze, R. C., 1938, Excessive pressures and pressure variations with depth of petroleum reservoirs in the Gulf Coast region of Texas and Louisiana: Am. Inst. Min. Metall. Engineers Trans., v. 127, p. 31–37; Discussion, p. 37–38.

Cannon, G. E., and Sullins, R. S., 1946, Problems encountered in drilling abnormal pressure formations: Am. Petroleum Inst. Drill. Prod. Practice, p. 29–33; Discussion, p. 33.

Cooke, P. W., 1955, Some aspects of high weight muds used in drilling abnormally high pressure formation: 4th World Petroleum Cong. Proc., Rome, sec. II, p. 43–57; Discussion, p. 57–58.

Coulomb, C. A., 1776, Essai sur une application des règles de maximis et minimis à quelques problèmes de statique, relatifs à l'architecture: Acad. Royale Sci., Paris, Mèm. Math. Phys., v. 7 (1773), p. 343–382.

Dake, C. L., 1918, The Hart Mountain overthrust and associated structures in Park County, Wyoming: Jour. Geology, v. 26, p. 45–55.

Daly, R. A., 1912, Geology of the North American Cordillera at the 49th Parallel: Geol. Survey Canada Mem. 38, pt. 1, 546 p.

—— 1925, Relation of mountain-building to igneous action: Am. Philos. Soc. Proc., v. 64, p. 283–307.

—— 1926, Our mobile earth: New York, Chas. Scribner's Sons, 342 p.

de Sitter, L. U., 1950a, Introduction au symposium sur la tectonique d'écoulement par gravité, et quelques conclusions: Geol. Mijnbouw, v. 12, new ser., p. 329–330.

—— 1950b, La tectonique d'écoulement dans les Alpes bergamasques: Geol. Mijnbouw, v. 12, new ser., p. 361–365.

—— 1954, Gravitational gliding tectonics, an essay on comparative structural geology: Am. Jour. Sci., v. 252, p. 321–344.

Dickinson, George, 1951, Geological aspects of abnormal reservoir pressures in the Gulf Coast region of Louisiana, U.S.A.: 3d World Petroleum Cong. Proc., The Hague, sec. I, p. 1–16; Discussion, p. 16–17.

—— 1953, Geological aspects of abnormal reservoir pressures in Gulf Coast Louisiana: Am. Assoc. Petroleum Geologists Bull., v. 37, p. 410–432.

Gignoux, Maurice, 1948, Méditations sur la tectonique d'écoulement par gravité: Travaux du Laboratorie de Géologie de l'Université de Grenoble, Tome XXVII, p. 1–34.

—— 1950, Comment les géologues des Alpes françaises conçoivent la tectonique d'écoulement: Geol. Mijnbouw, v. 12, new ser., p. 342–346.

Gilluly, James, 1957, Transcurrent fault and over-turned thrust, Shoshone range, Nevada (Abstract): Geol. Soc. America Bull., v. 68, p. 1735.

Goguel, Jean, 1950, L'influence de l'échelle dans les phénomènes d'écoulement: Geol. Mijnbouw, v. 12, new ser., p. 346–351.

Haarmann, E., 1930, Die Oszillationstheorie: Stuttgart, Ferdinand Enke, 260 p.

Handin, John, 1958, Effects of pore pressure on the experimental deformation of some sedimentary rocks (Abstract): Geol. Soc. America Bull., v. 69, p. 1576.

Handin, John, and Hager, Rex V., Jr., 1957, Experimental deformation of sedimentary rocks under confining pressure: Tests at room temperature on dry samples: Am. Assoc. Petroleum Geologists Bull., v. 41, p. 1–50.

Harza, L. F., 1949, The significance of pore pressure in hydraulic structures: Am. Soc. Civil Engineers Trans., v. 114, p. 193–214; Discussion, p. 215–289.

Hayes, C. Willard, 1891, The overthrust faults of the Southern Appalachians: Geol. Soc. America Bull., v. 2, p. 141–152; Discussion, p. 153–154.

Hedberg, H. D., 1926, The effect of gravitational compaction on the structure of sedimentary rocks: Am. Assoc. Petroleum Geologists Bull., v. 10, p. 1035–1072.

Heim, Arnold, and Gansser, August, 1939, Central Himalaya. Geological observations of the Swiss Expedition 1936: Schweiz. Naturf. Gesell., Denkschrift, v. 73, Mém. 1, 245 p.

Hewett, D. F., 1920, The Heart Mountain overthrust, Wyoming: Jour. Geology, v. 28, p. 536–557.

Hubbert, M. King, 1937, Theory of scale models as applied to the study of geologic structures: Geol. Soc. America Bull., v. 48, p. 1459–1520.

—— 1940, The theory of ground-water motion: Jour. Geology, v. XLVIII, p. 785–944.

—— 1951, Mechanical basis for certain familiar geologic structures: Geol. Soc. America Bull., v. 62, p. 355–372.

—— 1953, Entrapment of petroleum under hydrodynamic conditions: Am. Assoc. Petroleum Geologists Bull., v. 37, p. 1954–2026.

—— 1956, Darcy's law and the field equations of the flow of underground fluids: Am. Inst. Min. Metall. Engineers Trans., v. 207, p. 222–239.

Hubbert, M. King, and Willis, David G., 1957, Mechanics of hydraulic fracturing: Am. Inst. Min. Metall. Engineers Trans., v. 210, p. 153–166; Discussion, p. 167–168.

Hume, G. S., 1957, Fault structures in the foothills and eastern Rocky Mountains of southern Alberta: Geol. Soc. America Bull., v. 68, p. 395–412.

Jeffreys, Harold, 1931, On the mechanics of mountains: Geol. Mag., v. 68, p. 435–442.

Karelitz, M. B., 1938, Oil-pad bearings and driving gears of 200-inch telescope: Mech. Engineering, v. 60, p. 541–544.

Keep, C. E., and Ward, H. L., 1934, Drilling against high rock pressures with particular reference to operations conducted in the Khaur field, Punjab: Inst. Petroleum Technologists Jour. (London), v. 20, p. 990–1013; Discussion, p. 1013–1026.

Lapworth, Charles, 1883, The secret of the Highlands: Geol. Mag., new ser., dec. 2, v. 10, p. 120–128, 193–199, 337–344.

Lawson, Andrew C., 1922, Isostatic compensation considered as a cause of thrusting: Geol. Soc. America Bull., v. 33, p. 337–351.

Logan, W. E., 1860, Remarks on the fauna of the Quebec group of rocks and the Primordial zone of Canada: Canadian Naturalist, v. 5, p. 472–477.

Longwell, C. R., 1922, The Muddy Mountain overthrust in southeastern Nevada: Jour. Geology, v. 30, p. 63–72.

—— 1945, The mechanics of orogeny: Am. Jour. Sci., v. 243-A, Daly Volume, p. 417–447.

Lugeon, Maurice, 1940, Sur la formation des Alpes franco-suisse: Soc. Géol. France, C. R., p. 7–11.

Mansfield, G. R., 1927, Geography, geology, and mineral resources of part of southeastern Idaho: U. S. Geol. Survey Prof. Paper 152, 453 p.

McCaslin, Leigh S., Jr., 1949, Tide Water's record-breaking well—Bottom-hole pressure 12,635 psi: Oil and Gas Jour., v. 48, Sept. 8, p. 58–59, 94.

McConnell, R. G., 1887, Report on the geological structure of a portion of the Rocky Mountains, accompanied by a section measured near the 51st Parallel: Geol. Survey Canada Ann. Rept. 1886, v. 2, Rept. D, p. 5–41.

McHenry, Douglas, 1948, The effect of uplift pressure on the shearing strength of concrete: 3d Congres des Grands Barrages, Stockholm, C.R., v. 1, Question no. 8, R. 48, p. 1–24.

Mead, W. J., 1925, The geologic rôle of dilatancy: Jour. Geology, v. 33, p. 685–698.

Merla, Giovanni, 1952, Geologia dell'Appennino settentrionale: Pubblicazione n. 14 del Centro di Studi per la Geologia dell'Appennino del Consiglio Nazionale delle Ricerche; reprinted from Bollettino della Società Geologica Italiana, v. LXX, 1951, p. 95–382.

Merriam, C. W., and Anderson, C. A., 1942, Reconnaissance survey of the Roberts Mountains, Nevada: Geol. Soc. America Bull., v. 53, p. 1675–1728.

Migliorini, C. I., 1933, Considerazioni su di un particolare effetto dell'orogenesi: Soc. Geol. Italiana Boll., v. 52, p. 293–304.

Miller, Ralph L., and Fuller, J. Osborn, 1947, Geologic and structure contour maps of the Rose Hill oil field, Lee County, Virginia: U. S. Geol. Survey Oil and Gas Invest. Prelim. Map 76.

Mohr, O. C., 1882, Ueber die Darstellung des Spannungszustandes und des Deformationszustandes eines Körperelementes und über die Anwendung derselben in der Festigkeitslehre: Der Civilingenieur, v. XXVIII, p. 113–156.

Mostofi, B., and Gansser, August, 1957, The story behind the 5 Alborz . . . most spectacular discovery of 1956: Oil and Gas Jour., v. 55, Jan. 21, p. 78–84.

Murchison, R. I., 1849, On the geological structure of the Alps, Apennines and Carpathians, more especially to prove a transition from Secondary to Tertiary rocks, and the development of Eocene deposits in southern Europe: Geol. Soc. London Quart. Jour., v. 5, p. 157–312.

Nadai, A., 1927, Der bildsame Zustand der Werkstoffe: Berlin, Julius Springer, 171 p.

—— 1931, Plasticity: New York, McGraw-Hill Book Co., 349 p.

Nadai, A., 1950, Theory of flow and fracture of solids: v. 1, 2d ed., New York, McGraw-Hill Book Co., 572 p.

Nettleton, L. L., 1934, Fluid mechanics of salt domes: Am. Assoc. Petroleum Geologists Bull., v. 18, p. 1175–1204.

Nicol, James, 1861, On the structure of the North-Western Highlands and the relations of the gneiss, red sandstone, and quartzite of Suterland and Ross-shire: Geol. Soc. London Quart. Jour., v. 17, p. 85–113.

Oberholzer, J., 1933, Geologie der Glarneralpen: Beiträge Geologischen Karte der Schweiz, Neue Folge, Lieferung 28, 626 p.

—— 1942, Geologische Karte des Kantons Glarus, 1:50,000: Geol. Kommiss. Schweiz., Spezialkarte no. 117.

Oldham, R. D., 1893, A manual of the geology of India. Stratigraphic and structural geology: Calcutta, 2d ed. rev., 543 p.

—— 1921, ΓΝΩΘΙ ΣΕΑΥΤΟΝ, "Know your faults" (Presidential address): Proc. Geol. Soc. London, in Geol. Soc. London Quart. Jour. v. LXXVII, pt. I, p. lxxvii–xcii.

Peach, B. N., and Horne, John, 1884, Report on the geology of North-West Sutherland: Nature, v. XXXI, p. 31–35.

Peach, B. N., Horne, John, Gunn, W., Clough, C.,T., and Hinxman, L. W., 1907, The geological structure of the Northwest Highlands of Scotland; with petrological chapters and notes by J. J. H. Teall, edited by Sir Archibald Geikie: Geol. Survey Great Britain Mem., 668 p.

Pierce, W. G., 1941, Heart Mountain and South Fork thrusts, Park County, Wyoming: Am. Assoc. Petroleum Geologists Bull., v. 25, p. 2021–2045.

——, 1957, Heart Mountain and South Fork detachment thrusts of Wyoming: Am. Assoc. Petroleum Geologists Bull., v. 41, p. 591–626.

Rall, Cleo G., Hamontre, H. C., and Taliaferro, D. B., 1954, Determination of porosity by a Bureau of Mines method: A list of porosities of oil sands: U. S. Bur. Mines Rept. of Investigations 5025, 24 p.

Ransome, F. L., 1915, The Tertiary orogeny of the North American Cordillera and its problems, p. 287–376 in Problems of American geology: New Haven, Yale University Press, 505 p.

Reed, Paul, 1946, Trinidad Leaseholds applies advanced methods in drilling and production: Oil and Gas Jour., v. 45, Oct. 5, p. 44–46, 139.

Reeves, Frank, 1925, Shallow folding and faulting around the Bearpaw Mountains: Am. Jour. Sci., 5th ser., v. 10, p. 187–200.

—— 1946, Origin and mechanics of thrust faults adjacent to the Bearpaw Mountains, Montana: Geol. Soc. America Bull., v. 57, p. 1033–1047.

Rich, John L., 1934, Mechanics of low-angle overthrust faulting as illustrated by Cumberland thrust block, Virginia, Kentucky, and Tennessee: Am. Assoc. Petroleum Geologists Bull., v. 18, p. 1584–1596.

Richards, R. W., and Mansfield, G. R., 1912, The Bannock overthrust, a major fault in southeastern Idaho and northeastern Utah: Jour. Geology, v. 20, p. 681–709.

Rogers, W. B., and H. D., 1843, On the physical structure of the Appalachian chain, as exemplifying the laws which have regulated the elevation of great mountain chains generally: Assoc. Am. Geologists Rept., p. 474–531.

Rubey, William W., 1927, The effect of gravitational compaction on the structure of sedimentary rocks: a discussion: Am. Assoc. Petroleum Geologists Bull., v. 11, p. 621–632.

—— 1930, Lithologic studies of fine-grained Upper Cretaceous sedimentary rocks of the Black Hills region: U. S. Geol. Survey Prof. Paper 165-A, 54 p.

Rudzki, M. P., 1911, Physik der Erde: Leipzig, Chr. Herm. Tauchnitz, 584 p.

Safford, J. M., 1856, A geological reconnaissance of the state of Tennessee: Nashville, Mercer, 164 p.

—— 1869, Geology of Tennessee: Nashville, Mercer, 550 p.

Schardt, Hans, 1898, Les régions exotiques du versant nord des Alpes suisses: Bull. Soc. vaudoise des Sci. Nat., v. 34, p. 114–219.

Schultz, A. R., 1914, Geology and geography of a portion of Lincoln County, Wyoming: U. S. Geol. Survey Bull. 543, 141 p.

Sekules, Walter, 1958, Outlook good for East Pakistan gas field: World Oil, v. 147, no. 1, p. 154–155.

Smoluchowski, M. S., 1909, Some remarks on the mechanics of overthrusts: Geol. Mag., new ser., dec. V, v. VI, p. 204–205.

Stearns, R. G., 1954, The Cumberland Plateau overthrust and geology of the Crab Orchard Mountains area, Tennessee: Tenn. Div. Geology, Bull. 60, 47 p.

Suess, Eduard, 1909, Das Antlitz der Erde (English translation: The face of the earth, Sollas edition): v. 4, Oxford, Clarendon Press, 673 p.

Suter, Hans H., 1954, The general and economic geology of Trinidad, B.W.I.: (Reprinted with amendments from Colonial Geology and Mineral Resources, v. 2, nos. 3, 4; v. 3, no. 1, 1952) London, H.M.S.O., 134 p.

Taylor, Donald W., 1948, Fundamentals of soil mechanics: New York, John Wiley and Sons, 700 p.

Tercier, J., 1950, La tectonique d'écoulement dans les Alpes suisses: Geol. Mijnbouw, v. 12, new ser., p. 330–342.

Terzaghi, K., 1932, Tragfähigkeit der Flachgründungen (Bearing Capacity of Shallow Foundations): Prelim. Pub. First Cong., Internat. Assoc. Bridge and Structural Eng., p. 659–683.

—— 1936, Simple tests determine hydrostatic uplift: Engineering News-Record, v. 116, June 18, p. 872–875.

—— 1943, Theoretical soil mechanics: New York, John Wiley and Sons, 510 p.

—— 1945, Stress conditions for the failure of saturated concrete and rock: Am. Soc. Testing Materials Proc., v. 45, p. 777–792; Discussion, p. 793–801.

—— 1950, Mechanism of landslides, p. 83–123 in Paige, Sidney, Editor, Application of geology to engineering practice (Berkey Volume): Geol. Soc. America, 327 p.

Terzaghi, K., and Peck, Ralph B., 1948, Soil mechanics in engineering practice: New York, John Wiley and Sons, 566 p.

Thomeer, J. H. M. A., 1955, The unstable behaviour of shale formations in boreholes and its control by properly adjusted mudflush quality: 4th World Petroleum Cong. Proc., Rome, sec. II, p. 1–8; Discussion, p. 8–10.

Törnebohm, A. E., 1883, Om Dalformationens geologiska ålder: Geol. Fören. Stockholm Förhandl., v. 6, p. 622–661.
—— 1888, Om fjällproblemet: Geol. Fören. Stockholm Förhandl., v. 10, p. 328–336.
—— 1896, Grunddragen af det centrala Skandinaviens bergbyggnad: K. Sv. Vet. Akad. Handl., v. 28 (5), 201 p.
van Bemmelen, R. W., 1936, The undation theory of the development of the earth's crust: 16th Internat. Geol. Cong. Rept., Washington, v. 2, p. 965–982.
—— 1950, Gravitational tectogenesis in Indonesia: Geol. Mijnbouw, v. 12, new ser., p. 351–361.
Veatch, A. C., 1907, Geography and geology of a portion of southwestern Wyoming: U. S. Geol. Survey Prof. Paper 56, 178 p.
Watts, E. V., 1948, Some aspects of high pressures in the D-7 zone of the Ventura Avenue field: Am. Inst. Min. Metall. Engineers Trans., v. 174, p. 191–200; Discussion, p. 200–205.
Wentworth, Chester K., 1921a, Russell Fork fault in Giles, A. W., The geology and coal resources of Dickenson County, Virginia: Va. Geol. Survey Bull. 21, p. 53–67.
—— 1921b, Russell Fork fault of southwest Virginia: Jour. Geology, v. 29, p. 351–369.

West, W. D., 1939, The structure of the Shali "window" near Simla: Geol. Survey India Records, v. 74, pt. 1, p. 133–163.
Willis, Bailey, 1902, Stratigraphy and structure, Lewis and Livingston ranges, Montana: Geol. Soc. America Bull., v. 13, p. 305–352.
—— 1923, Geologic structures: 1st ed., New York, McGraw-Hill Book Co., 295 p.
Yoder, Hatten S., Jr., 1955, Role of water in metamorphism, p. 505–523 in Poldervaart, Arie, Editor, Crust of the earth: Geol. Soc. America Special Paper 62, 762 p.
Young, David M., 1957, Deep drilling through Cumberland overthrust block in southwestern Virginia: Am. Assoc. Petroleum Geologists Bull., v. 41, p. 2567–2573.

SHELL DEVELOPMENT COMPANY, EXPLORATION AND PRODUCTION RESEARCH DIVISION, HOUSTON, TEXAS; UNITED STATES GEOLOGICAL SURVEY, WASHINGTON, D. C.
MANUSCRIPT RECEIVED BY THE SECRETARY OF THE SOCIETY, NOVEMBER 17, 1958
PUBLICATION AUTHORIZED BY SHELL DEVELOPMENT COMPANY AND THE DIRECTOR, U. S. GEOLOGICAL SURVEY

ROLE OF FLUID PRESSURE IN MECHANICS OF OVERTHRUST FAULTING

Errata

1. The paragraph on page 121 which begins:

   "Longwell (1922) discovered the Muddy Mountain overthrust..."

   should be changed to read:

   "Longwell (1922, 1945, 1949) discovered and has extensively studied the Muddy Mountain overthrust in southeastern Nevada. Here a block of Paleozoic and (now eroded) Mesozoic strata of an estimated thickness of about 13,500 feet over-rode the same section with the fault surface on Jurassic sandstone, for a distance of at least 15 miles."

Added Reference

Longwell, Chester R., 1949, Structure of the northern Muddy Mountain Area, Nevada: Geol. Soc. America Bull. v. 60, p. 923-968.

2. The last sentence beginning on page 132 and continuing on page 133 should be changed to read:

   "Now, according to the theorem expressed by equations (25) and (31),

$$\begin{aligned}
\vec{F}_s &= -\iiint_V (1 - f)(\text{grad } p)dV \\
&= -\iint_A (1 - f)p_s \, d\vec{A} - \iint_{A_{is}} p\,d\vec{A}_{is}, \\
\vec{F}_l &= -\iiint_V f(\text{grad } p)dV \\
&= -\iint_A f p_l \, d\vec{A} - \iint_{A_{il}} p\,d\vec{A}_{il},
\end{aligned} \right\} \quad (40)$$

where $\underline{A}$ is the total exterior surface of the volume, $\underline{p}_s$ and $\underline{p}_l$ are the normal macroscopic stresses across that surface inside the solid and the fluid elements, respectively; and $\underline{A}_{is}$ and $\underline{A}_{il}$ represent the total solid-fluid interface inside the macroscopic volume $\underline{V}$, regarded as inclosing the solid-filled space or the fluid-filled space, respectively."

Part V

REFINEMENT AND REFLECTION

Editor's Comments
on Papers 29 Through 37

Upon publication in 1959 the Hubbert and Rubey collaboration generated considerable recognition and rather widespread acceptance of the tectonic role of fluid-pressure enhancement, despite some adverse criticism. Five technical discussions and four replies were published in the period 1960–1966; these have not been included herein, primarily because of space limitations. This criticism was chiefly attached to individual topics raised in the Hubbert and Rubey paper, but generally not to their fundamental point—that application of the effective-stress principle could provide a key to mechanical problems suggested by overthrusts. Hans Laubscher, for example, accepted as valid the suggestion that abnormal fluid pressure could reduce the friction at the base of overthrust sheets and indeed subsequently applied and extended the concept in his study of Jura deformation (Paper 29); but he argued in 1960, as did Wally Moore (1961), that the force of uplift is proportional to Terzaghi's surface porosity. From the point of view of application to fault mechanics, the surface-porosity question seems merely of academic interest; as a problem in mechanical theory, however, the question achieves greater importance, and the replies by Hubbert and Rubey (1960, 1961a, 1961b) did much to clarify their position in this matter. Nonetheless, the question has remained controversial (Geertsma, 1957; Skempton, 1960; Nur and Byerlee, 1971; Garg and Nur, 1973).

For the field geologist, issues of significance were raised by Francis Birch (1961). Among other comments, Birch pointed out (1) some consequences of the deletion of the cohesion intercept (i.e., τ_0) in the Coulomb equation, $\tau = \tau_0 + (1 - \lambda)S \tan \phi$, used to describe basal friction of the thrust block, and (2) internal inconsistencies in the thrust-block model (e.g., the neglect of shear stresses at block edges). In their reply, Hubbert and Rubey agreed to several "second-order" points raised by Birch, but (1) declared him in error in his insistence upon retention of τ_0; and (2) suggested that his "self-consistency" challenge to the validity of their approximate calculations of length of overthrust blocks was based on unessential details.

In the same year, Laubscher devised an expression for maximum permissible shear stress at the sole of a horizontal thrust sheet (Paper 29), based on an equilibrium equation (Hubbert and Rubey, Paper 28, p. 143, Eq. 88) and the Coulomb-Terzaghi equation. In applying it to the case of Jura deformation, the values of τ_{max} obtained fall within the range 30–90 bars; but what, Laubscher remarks, is "die Natur der basalen Schmierung: Kristallplastizität oder Porendruck"? Or some combination thereof? Must it be assumed that the cohesion intercept will totally vanish as a consequence of "dislocation propagation" along the sole of an overthrust block?

Certainly, any geologist familiar with Alpine nappes, or, for that matter, nappes of the northern Appalachians, will only with great reluctance accept a mechanical idealization which requires a smooth, cohesionless fracture as the overthrust surface. Indeed, even if one were to accept a "block model" as a mathematically convenient (if geometrically unrealistic) first approximation, the concept of a cohesionless basal boundary would intuitively seem to be a special rare case rather than a general one.

Mathematically, there is no problem in considering the effect of the cohesion intercept; the Hubbert and Rubey calculations can be easily modified so as to include a τ_0 term. More relevant is the question whether the inclusion of a finite cohesion intercept would lead to much different conclusions than those suggested by Hubbert and Rubey. This is basically the theme developed by Ken Hsü in 1961. Hsü, a former member of the excellent Shell Development Company research group in Houston (a program, incidentally, initiated by Hubbert in the 1940s), had recently joined the staff of the Geological Institute at ETH in Zürich; as befits a man attempting to follow in the footsteps of such as Escher and Albert Heim, Hsü shifted his focus from the curious Franciscan rocks along the vertebrae of the California Coast Ranges to questions associated with the Glarus nappes. Fortunately, the geometry and kinematics of the Glarus structures had been the subject of recent analysis by Hsü's new Swiss colleague, Rudolf Trümpy, and Hsü was thus able to concentrate his attention upon mechanical aspects of the problem. Confronted by the ductilely smeared *Lochseitenkalk* at the Glarus sole thrust, Hsü (Paper 36; cf. 1969b) returned to the question of the cohesion intercept first raised by Birch, and systematically considered the mechanical implications of omitting (and of not omitting) that term (cf. discussion by Hubbert and Rubey, 1969; Hsü, 1969a).

Behavior of the thrust sole was, however, not the only point to receive attention. Soon after the appearance of Hubbert and Rubey's 1959 papers, it was recognized that, at least in the early stages of movement, the thrust block must develop a *toe* at the leading edge, thus producing an opposing force to thrust movement (Hubbert and Rubey, 1961a, p. 1449; cf. Paper 19). Consideration of the trailing edge came a decade later, leading to a concept of *plastic (or viscous) wedges* in which gravitationally induced lateral pressures acting at the rear of a glide block provide a significant propulsive force (Voight, 1973, Paper 31). Both considerations can be regarded for the most part, as refinements, or "second approximations," to Hubbert and Rubey's collaborative effort. It should, however, be pointed out that all these concepts had been considered previously in the literature of soil mechanics (e.g., Odenstad, 1951; Zaruba and Mencl, 1969), and indeed, as I discovered in the preparation of this volume, the "wedging" role ascribed to fluid injections in the rear detachment area of glide blocks (Voight, 1973; p. 118-120; Paper 31) seems to be basically identical to notions expressed over a century and a half ago by Sir James Hall (Paper 3).

Progress in soil mechanics has continued, naturally enough, to the present time, and it hardly needs saying that many of these concepts seem more or less directly applicable to problems of thrust faulting. For example, Skempton's work (Paper 32) seems of direct concern to questions of cohesion and friction mobilized along overthrusts.

Skempton will be recalled from our historical discussion of the development of the effective-stress principle; one of the pioneers in the vital formative period of British soil mechanics in the 1930s, he had been selected to present the fourth Rankine Lecture, perhaps the outstanding international award in geotechnical engineering. His work revitalizes Haefeli's concept of *residual strength*. [The term *ultimate strength* has also been used synonymously by soil mechanicists; this usage differs, however, with its widely employed meaning in rock mechanics (i.e., as the maximum differential stress sustained by rock in a given strength experiment).] Skempton demonstrated that with continued displacements, the coefficients of friction of geological materials could markedly decrease, and the cohesion intercept could disappear completely! Were Hubbert and Rubey correct in their crucial assumption, after all?

The determination of residual strength parameters requires rather extensive and time-consuming experiments. Fortunately, there appears to be a relation between residual friction and the Atterberg plasticity (consistency) limits; the latter are determinable by simple tests and are conducted on a regular basis in all soil mechanics laboratories. Thus, the figure in Paper 33 may prove useful for field geologists who wish to pursue the Hubbert–Rubey approach to specific field areas. Of significance is the wide range of appropriate frictional coefficients, e.g., from the 0.6 (0.577) assumed appropriate by Hubbert and Rubey, to 0.1 for montmorillonitic clay shales of the Great Plains (cf. Paper 19)!

Another significant development of the mid-1960s was the recognition given to the roughness of shear surfaces, illustrated here by Frank Patton's presentation to the First International Congress of the newly formed International Society of Rock Mechanics (Paper 34). The effect of roughness is to produce, at significant confining pressures, an *apparent cohesion intercept* if the (linear) Coulomb criterion is adopted. This concept, too, seems to fall into the category of a "rediscovery," inasmuch as knowledge of the phenomenon dates back to Leonhard Euler in two papers presented in 1750 to the Royal Academy of Sciences in Berlin (cf. Bowden and Tabor, 1964).

Meanwhile, additional geologic pore pressurization mechanisms were being discovered (or rediscovered). Lucian Platt (1962), for example, considered the effects of the rise of a pluton into a sedimentary pile; as one example, he suggested that the movement of Prealpine nappes was synchronous with (and caused by) high fluid pressure induced by recrystallization of Pennine nappe cores, which in turn had been produced by granitic intrusion. Rather than referring to the pluton as "syntectonic," Platt suggests it is more appropriate to say that the thrusts are "syn-granitic."

Hanshaw and Zen (1965) next proposed osmotic equilibrium across shale beds as an important cause of enhanced fluid pressures, particularly in association with evaporites; E-an Zen was, at that time, deeply involved in his extensive field investigations of Taconian nappes in eastern North America. The important effect of phase changes on fluid pressurization was treated in detail by Hugh Heard (then at UCLA), and Rubey, with emphasis on the effects of gypsum dehydration. A correlation between evaporites and décollement had long been recognized in the Alps, Jura, Mediterranean Atlas, Pyrenees, and Iranian ranges (cf. Schardt, 1893; Termier, 1906; Buxtorf, 1908; Suess, 1909, p. 178, 221; de Böckh et al., 1929, and subsequent workers). In North America, although not widely celebrated, the association of evaporite and thrusts also seems significant, both in Appalachian and Cordilleran fold thrust belts. For an explanation of this association most workers have appealed to the "weakness" of evaporite rocks; Laubscher, however, suggested further that pore pressures could be directly enhanced by the conversion of gypsum to anhydrite plus water, and that any abnormal pressure thus derived could be sustained by the low permeability of clay and salt layers typically interlaminated with gypsum-anhydrite (Paper 29, p. 248–251). At about the same time Heard had barely completed an extremely important study (his Ph.D. dissertation) on the effect of strain rate on the deformation of marble; this work (Heard, 1963) for the first time provided a reasonable basis for extrapolation of short-time laboratory test data on rock to geologic time periods. The special apparatus thus developed was next employed to deform alabaster (polycrystalline gypsum), leading to a collaboration with Rubey, who meanwhile had noted that, for about half of the total length (255 miles) of their outcrop, the Idaho–Wyoming thrusts follow evaporite-bearing horizons (Paper 35).

Analogous phase change–fluid pressure mechanisms can occur in peletic sequences; in the Gulf Coast, for example, alteration of montmorillonite (smectite) to illite occurs at 180–250°F (Powers, 1967; Burst, 1969; Perry and Hower, 1972), leading to the development of décollement and "plastic wedges" (cf. Bruce, 1973, Figs. 4, 7; Voight, 1973, Paper 31).

Hsü's article (Paper 36), which reactivated Birch's (1961) criticism concerning the neglect of τ_0 in the Hubbert–Rubey hypothesis, followed soon after; his paper was followed in turn by that of Forristal (Paper 37), who also reactivated a point previously made by Birch (1961, p. 1442; cf. Hubbert and Rubey, 1961a, p. 1449): i.e., that neglect of shear stresses at block boundaries led to an overestimation of possible lengths of overthrust blocks. By taking into account the complete elastic state of stress in the block, Forristal demonstrated the important consequence of a seemingly insignificant omission in previous analyses; for that reason, the "maximum length" data previously given by Hubbert and Rubey (Paper 28) and Hsü (Paper 36) are about 50 percent too large!

Finally, a question is asked that may surprise the reader: Why are the Hubbert–Rubey papers commonly regarded as of classical importance? An adversely critical overview might be expressed as follows: What new information, for example, does Part I of that work really present? The "mechanical paradox" as presented by Hubbert and Rubey follows directly, with minor revision, from Smoluchowski (Paper 2); Smoluchowski's proposed resolution of that paradox, however, seems to have been incompletely appreciated. Coulomb's law, Mohr's stress circle, and associated derivations previously had been presented in detail and made use of by others, e.g., Anderson (Paper 12), Nádai (Paper 13), and even by Hubbert himself (Paper 22). The effective-stress principle and the theory of consolidation dated from Terzaghi's work of the 1920s and 1930s (Paper 14), and indeed these concepts had already been applied to rock, as well as to soil, both experimentally and in the field (Terzaghi, Papers 15 and 26). The early recognition of enhanced pore pressure as a mechanism for low-angle landslides seems, perhaps literally, hardly a stone's throw from its adoption as an explanation of overthrust faulting. Although the Hubbert–Rubey theoretical assessment of the buoyancy question differs somewhat from Terzaghi's boundary porosity "idealization," this difference is not of practical significance from the viewpoint of overthrust faults; moreover, the Hubbert–Rubey view was preceded by Harza's elaboration of, substantially, the same conclusions (Paper 27). Furthermore, the views of Harza and Hubbert–Rubey seem insufficient as a general law (Geertsma, 1957; Skempton, 1960; Nur and Byerlee, 1971; Garg and Nur, 1973). The Hubbert–Rubey calculation of maximum theoretical length of an overthrust block appears to be significantly in error (Paper 37), and their assumption that τ_0 could be neglected also appears to be, discounting exceptional circumstances, erroneous. The geological occurrence of abnormal fluid pressures are reviewed in detail in Part II, as well as summarily in Part I, but these statements represent merely an inexhaustive effort to compile information already available in the literature. Finally, if one accepts in general the essence of the theoretical development and takes note of the prior experimental and field oriented (e.g., landslide) work reviewed herein, the concrete block and beer can experiments become, however interesting, merely demonstrations of the obvious.

One could suppose that the entire hypothesis could have been presented as a short note, rather than, practically, as an entire issue of the *Bulletin of the Geological Society of America*. Is not the essence of the argument merely this: *that Terzaghi's effective-stress principle, heretofore essential in "explaining" the occurrence of low-angle land-slides, also seems useful as an "explanation" of the occurrence of many overthrusts?* Would not a short paper with appropriate literature cita-tions and a discussion of geological mechanisms of fluid-pressure en-hancement have sufficed to get the point across?

Perhaps, but possibly that is not the point. Any new interpretation of nature emerges first in the mind of one or a few individuals; it is they who somehow first learn to see the world of nature differently, and their ability in this regard is affected by factors not common to most members of their profession. Their attention has been intensely concentrated upon "crisis-provoking" problems (such problems are illustrated in our case by, for example, the writings of Mellard Reade, Oldham, and Longwell). In addition, they are usually so new to the "crisis-ridden" field that practice has committed them less deeply to old explanations (Kuhn, 1962), or they are armed with information (in this case, notions of effective stress) not widely circulated among their contemporaries. The time must be ripe, if we are to agree with Kuhn that "crises" are a necessary precondition for the emergence of novel theories (i.e., paradigms). *But are we, in fact, dealing with a paradigm?* Or (merely!) traditional, incremental, "normal science"?

It should be specifically appreciated that Hubbert and Rubey were not writing for the rock mechanics specialist; their intended audience was comprised of rank-and-file field geologists as well as specialists in geological mechanics for, ultimately, they clearly recognized that it must be by examination of critical evidence from the field that geologi-cal hypotheses stand or fall. Their writing is lucid, their composition accomplished; their discussion of the overthrust belt of western Wyoming remains intact as a superb example of the integration of field and theoretical research. Smoluchowski's admirable but brief report suffered from its timing and perhaps by limited circulation. It did little to arrest the "crises" that arose in association with the em-ployment by subsequent workers of a "frictional" mechanical idealiza-tion. In contrast, Hubbert and Rubey took pains not to be denied; their collaboration led to one of the most influential investigations in struc-tural geology of this century.

29

Reprinted from pp. 221, 223-224, 243-246, 278-279 of *Eclogae Geol. Helv.*, 54(1), 221-282 (1961)

Die Fernschubhypothese der Jurafaltung[1])

Von **Hans P. Laubscher** (Basel)

Mit 17 Textfiguren

———

ABSTRACT

Basement structure. The relative positions of major synclines and borehole data in the Jura mountains region permit mapping of the basement within a limit of error not exceeding a few hundred meters. It turns out that the Jura basement has the shape of an elevated platform which is dissected into a number of antithetic fault blocks rotated to the southeast. They form part of a system of extensional tectonics which had developed largely in the lower Tertiary and survived upper Tertiary folding of the sediments without being visibly affected. On the other hand, geometrical and mechanical considerations show that, had basement been compressed to an extent equalling that of the sedimentary cover, a block of material averaging several kilometers in height would have been squeezed out, or else crustal downdrag of geosynclinal character would have to be postulated. Neither one of these happened in the Jura mountains. Both, new and old facts more emphatically than ever demand acceptance of the view that basement is not involved in the upper Tertiary folding.

Mechanics. The mechanical possibilities of thrusting of an extensive thin sheet of sediments are controlled by the amount of friction at its base. From the geometrical evidence of borehole data in the Lons-le-Saunier area a specific friction of only 30 kg/cm² may be computed for the base of the plateaus of Lons-le-Saunier, Champagnole, and Nozeroy. Further numerical estimates indicate that maximum permissible friction at the Triassic base of the Molasse basin would have been three times this value. Transmission of the push from the Alps to Besançon thus appears to have been possible. These extremely low values of specific friction may have been due to plastic yielding of salt or to the existence of abnormally high pore pressures in the Triassic evaporite series or both. At any rate, they must have helped in developing an exceedingly unstable position of the entire sedimentary cover north of the Alpine Central Massifs. The situation was aggravated by the rise of these massifs and by overloading of their steepened northern flank by Alpine thrust masses. In the sedimentary cover a tendency developed to rotate northward and thereby to push the northern parts uphill onto the Jura platform. Rotation may have started spontaneously under the sole influence of gravity for an average specific basal friction below 30 kg/cm². For higher values, stability was reduced to an extent where even a slight push from the massifs would have initiated rotation. In the autochthonous sedimentary cover gaps have been known for a long time and may be interpreted as scars left by the sliding away of the decollement nappe.

Paleogeography. It is hardly feasible to connect slight Mesozoic epicontinental warps with the folding of the Jura mountains which is a comparatively local affair that occurred more than 100 million years later. Lower Tertiary structural elements including a few fold-like features may in most instances be shown to form part of an extensional fault system which developed at that time. They can hardly be considered as embryonic forerunners of the upper Tertiary folding, as this produced numerous folds of large amplitude without discernible block movements. As to the folding

[1]) Gedruckt mit Unterstützung der Freiwilligen Akademischen Gesellschaft der Stadt Basel.

[*Editor's Note:* A row of asterisks indicates that material has been deleted.]

itself, there is weighty regional evidence for its having developed in one phase instead of in two phases separated by a period of peneplanation, as is usually assumed.

Kinematics. The thrust sheet is composed of a number of comparatively rigid blocks, deformation being concentrated along their borders. Consequently, the direction of folds often is not perpendicular to mass transport, and there are systems of wrench faults forming acute angles with fold axes. Furthermore, by mapping estimated direction and amount of displacement with respect to basement, it is found that the thrust sheet apparently rotated around its northeastern tip by an estimated 7°. Location of folds was determined largely by pre-existing irregularities, particularly flexures and faults of the lower Tertiary system (frame tectonics). However, faults in the sedimentary cover had to be detached from their roots when decollement took place, and subsequently developed independently. To overcome basement obstacles, lower strata in the thrust sheet in many instances may have been forced into chaotic structures unforeseeable from present exposures. Wrench faulting at the southwestern end of the Jura mountains is believed to have enabled the Jura folds to form many kilometers to the northwest of their presumable continuation in the Savoyan subalpine chains. Furthermore, this displacement of folding to the northwest may be regarded as due to the comparatively competent Molasse basin being embodied into the thrust sheet – in conjunction with the development or increase in thickness of Triassic evaporite beds north of the Ile Crémieu.

2. Abschätzung der basalen Reibung

HUBBERT und RUBEY (1959) benützen die obigen Gleichgewichtsbeziehungen, um die maximale abscherbare Blocklänge für durchschnittliche Werte der Materialkonstanten und für verschiedene z_1 und λ zu berechnen, wobei $\lambda = $ const. für den ganzen Block angenommen wird. Im Jura ist anderseits mit einer basalen Gleitschicht zu rechnen, die aus Salz, Salzton oder Gips besteht, und von der man jedenfalls gegenüber den andern Gesteinen des Blockes eine bedeutend verminderte Reibung erwarten darf. Man wird deshalb mit Vorteil umgekehrt rechnen und aus den lokal messbaren geometrischen Daten die maximal zulässige basale Reibung bestimmen. Dem Material der Gleitschicht entsprechend (Salz, Gips), das Kristallplastizität besitzt, sei versuchsweise für die Basis des Blocks statt der COULOMB-schen Reibung eine plastische Fliessgrenzspannung oder Plastizitätsschwelle eingesetzt, deren Wert konstant, unabhängig vom Überlagerungsdruck ist; inwiefern dieser Ansatz berechtigt ist, wird weiter unten diskutiert. Für das Blockinnere hingegen gelte nach wie vor die COULOMBsche innere Reibung. Man erhält nach Integration

$$\tau_{max} = (az_1 + Bz_1{}^2)/x_1 , \quad \text{wo } B = [b + (1 - b)\lambda]\varrho g/2 = 2{,}3 \cdot 10^3 \text{ dyn/cm}^3 \text{ für } \lambda = 0{,}5$$

Als Materialkonstanten seien die von HUBBERT und RUBEY verwendeten Durchschnittswerte eingesetzt. Weiterhin sind Breite und Dicke des Blockes festzulegen. Um einen möglichst hypothesenfreien Wert für die Reibung zu erhalten, sei zunächst von der Fernschubhypothese Abstand genommen, und es seien nur solche Blöcke berücksichtigt, für deren Tangentialschub wenig Zweifel bestehen, also für die Plateaux zwischen der Bresse und dem Faisceau Helvétique. Für diese ergaben die Bohrungen von Lons-le-Saunier, dass sie brettartig, fast en bloc, subhorizontal geschoben wurden. Sie entsprechen also besonders gut dem Modell, das den benützten Formeln zugrunde liegt, und sollten folglich besonders zuverlässige Werte für die basale Reibung liefern.

Eine gewisse Unsicherheit besteht noch hinsichtlich der Dicke der Blöcke zur Zeit des Schubes; denn an sich konnte sie ja seither durch Erosion beträchtlich vermindert werden. Nun gibt es viele Anhaltspunkte, die vermuten lassen, dass dies nicht der Fall war. Vor allem ist zu bedenken, dass ja sozusagen die gesamte Niveaudifferenz zwischen Grabengebiet und Hochblock während des Oligozäns geschaffen und durch die jungtertiäre, fast rein horizontale Verschiebung nicht verändert wurde. Die Erosion musste also schon im Oligozän wirken, sie dürfte damals sogar am kräftigsten gewesen sein. Ausserdem war vermutlich das Zeitintervall vom Beginn der alttertiären Bewegungen bis zur Faltung um ein Mehrfaches länger als das seit der Faltung verstrichene, vielleicht 30 Millionen Jahre gegenüber 10 Millionen. Stratigraphisch zeigt sich die kräftige oligo-miozäne Erosion in den Konglomeratschüttungen des Bressegrabens, vor allem aber auch in den Bohrbefunden: vermutlich miozäne «Gompholithe pralinée» transgrediert über den Grabenrand hinaus auf das Paläozoikum des Hochblockes (LEFAVRAIS et al., 1957).

Es sei nur im Vorbeigehen bemerkt, dass diese Tatsachen von ungeheurer Bedeutung für die Morphogenese des ganzen Juragebirges sind.

Der zwischen der Bresse und der Euthekette gelegene Block wäre dann 17 km lang und 0,6 km hoch, und daraus errechnet sich eine maximale spezifische basale Reibung $\tau_{max} = 29{,}6 \sim 30$ kg/cm² (32 kg/cm² für $\lambda = 0$). Für die ganze Breite der Plateaux zwischen der Bresse und dem Faisceau Helvétique ergibt sich bei einer Blockdimension von 30 km auf 1 km (geschätzte durchschnittliche Dicke) $\tau_{max} = 31$ kg/cm² (35 kg/cm² für $\lambda = 0$), bei einer Dimension von 35 km auf 1 km $\tau_{max} = 26{,}6 \sim 27$ kg/cm² (30 kg/cm² für $\lambda = 0$).

Es ist bemerkenswert, dass diese Werte fast genau der von GOGUEL (1948, p. 406 ff.) auf ganz anderem Wege erhaltenen Plastizitätsschwelle für die Basis der abgescherten Falten im Querschnitt von Gex entsprechen (40 kg/cm²).

Geht man einen Schritt weiter und nimmt an, der gesamte Faltenjura stelle eine Abscherungsdecke dar, so sind bei der Rechnung verschiedene Komplikationen zu berücksichtigen. Die heutige maximale Breite des Juras zwischen den Avant Monts am Ognon und dem Innenrand beträgt etwa 75 km, sie ist aber gegenüber der ursprünglichen Breite vor der Faltung um 10 bis 20 km verkürzt. Zu Beginn der Faltung lag also ein Block von etwa 90 km Breite vor, aber er wurde wohl nicht gleichzeitig bewegt. Kinematische Betrachtungen (s.u.) zeigen, dass mit einem Fortschreiten der Faltung von Süden nach Norden zu rechnen ist, so dass der Block schon um ein paar Kilometer verkürzt sein mochte, bevor die Abscherung die nördlichsten Partien erfasste. Die Faltung im Süden wiederum bedeutete eine tektonische Verdickung der Sedimenthaut, die bei der Abschätzung der Blockdicke zu berücksichtigen ist. Die Schubbahn ist im ganzen wenig geneigt und kann als eben vorausgesetzt werden, aber es ist zu bedenken, dass die inneren Ketten aus dem Bereich der Nordwest-«Flexur» des Molassebeckens stammen und über ihrem ansehnlichen Mesozoikum auch noch eine bedeutende Molasseschicht trugen. Schätzungsweise dürfte eine Dimensionierung von 80 km auf 1,6 km (Durchschnitt) den Verhältnissen gerecht werden. Daraus würde sich ein $\tau_{max} = 21,4 \sim 21$ kg/cm² ergeben. Wie jedoch aus den nachfolgenden Betrachtungen hervorgehen wird, sollte man statt der durchschnittlichen Mächtigkeit eher die des südlichen Abschnittes in Rechnung setzen. Sie beträgt mindestens 2 km, und daraus ergibt sich τ_{max} zu mindestens 29 kg/cm². Es ist demnach im ganzen Jura mit einer maximalen spezifischen basalen Reibung oder einer Plastizitätsschwelle von rund 30 kg/cm² zu rechnen. Liess sich also die Sedimenthaut im Gebiet von Lons-le-Saunier abscheren – und dies ist nach den Bohrergebnissen anzunehmen – so konnte sie dies ebensogut über die ganze Breite des Juras geschehen lassen. Aus dieser Perspektive gesehen, treten jedenfalls im Jura selbst für die Fernschubhypothese keine Schwierigkeiten auf.

Wie aber steht es mit dem Schub durch das Molassebecken? Dieses hat ungefähr die Form eines stumpfen Keiles und weicht damit vom bisher betrachteten Modell ab. Eigentlich müsste also die Spannungsverteilung von neuem nach der Elastizitätstheorie berechnet werden, doch ist die Abweichung für den hier speziell untersuchten Fall gering (2° 24' Divergenz), und es darf angenommen werden, die Abweichung der Spannungsverteilung von der eines rechteckigen Blockes sei ebenfalls gering. Sie sei deshalb im folgenden vernachlässigt. Dann lässt sich ein idealisiertes einfaches Verfahren zur numerischen Abschätzung der basalen Reibung anwenden.

Die Blockdicke nehme also linear mit der Länge zu (Fig. 10). Man kann diese Form angenähert wiedergeben durch eine Reihe von rechteckigen Blöcken. Für jeden davon lässt sich dann der HUBBERT-RUBEYsche Ansatz wie für den Jura ver-

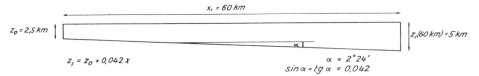

Fig. 10. Geometrisches Modell des Molassebeckens
wie es den Berechnungen zugrunde liegt.

wenden, nur muss durch den Schub zusätzlich zur basalen Reibung des jeweils betrachteten Teilblockes auch noch jene aller davorliegenden Teilblöcke überwunden werden. Die hinteren Blöcke müssen also grössere Widerstände überwinden, dafür sind sie aber auch dicker, sie können grösseren Kompressionskräften standhalten. Es ist deshalb die schwächste Stelle x aus $\partial\,\tau_{max}/\partial\,x = 0$ zu ermitteln, und für sie ist – sofern sie ins Innere des Modelles fällt – τ_{max} zu berechnen. Eine weitere Komplikation ergibt sich für geneigte Schubbahnen wie die des Molassebeckens, da in diesem Fall nicht nur der jeweils betrachtete Teilblock, sondern auch alle davor-liegenden hangauf gestossen werden müssen. Die in die Schubbahn fallende Ge-wichtskomponente beträgt

$$G' = x/2\ (2\,z_0 + x\,\mathrm{tg}\alpha)\,\varrho\,g\,\sin\alpha = x\,z_0\,\varrho\,g\,\sin\alpha + x^2/2\ \varrho\,g\,\sin\alpha\,\mathrm{tg}\alpha.$$

Zudem ändern sich die Ausdrücke für Druckfestigkeit und basale Reibung (sofern sie COULOMBscher Art ist), da immer nur Komponenten der Schwerkraft senkrecht zur Schubrichtung wirken. Doch können die hier auftretenden Faktoren $\cos\alpha$ für den betrachteten Fall von vornherein gleich 1 gesetzt werden (cos $2°\ 24' = 0{,}9991 \sim 1$), desgleichen ist $\sin\alpha = \mathrm{tg}\alpha$. Endlich ist zu berücksichtigen, dass das Molassebecken auch den gesamten Jura vor sich herstossen, also eine zu-sätzliche basale Reibung $R_{Jura} = 30.10^6$ dyn cm^{-2}. 80.10^5 cm$^2 = 24 \cdot 10^{13}$ dyn überwinden musste. Auch diese ist für das Gleichgewicht der Kräfte in Rechnung zu setzen, und man erhält insgesamt

$$\tau_{max} \cdot x + G' + R_{Jura} = az_1 + Bz_1^2,\quad \text{wo}\quad z_1 = z_0 + x\,\mathrm{tg}\alpha$$
$$\tau_{max} = \mathrm{tg}\alpha\,[a + z_0\,(2B - \varrho g)] + x\,\mathrm{tg}^2\alpha\,(2B - \varrho g)/2 + 1/x\,(az_0 + Bz_0^2 - R_{Jura})$$

Die schwächste Stelle tritt da auf, wo

$$\partial\,\tau_{max}/\partial\,x = 0 = \mathrm{tg}^2\alpha\,(2B - \varrho g)/2 - 1/x^2\,(az_0 + Bz_0^2 - R_{Jura})$$

also bei

$$x = [2\,(az_0 + Bz_0^2 - R_{Jura})/\mathrm{tg}^2\alpha\,(2B - \varrho g)]^{1/2} = 62{,}5\ \text{km}$$

d.h. knapp ausserhalb der Rückwand unseres Modells. Für diese Stelle beträgt $\tau_{max} = 91{,}4$ kg/cm^2, so dass man für das Molassebecken mit einem Wert von rund 90 kg/cm^2 rechnen darf. Es ergibt sich also das Resultat, dass die Schubübertragung durch das Mittelland um ein Mehrfaches leichter ist als jene durch den Jura selbst, sofern die Abscherung in einem kristallplastischen Milieu stattfindet, dessen Plastizitätsgrenze konstant, unabhängig vom Überlastungsdruck ist.

LITERATURVERZEICHNIS

GOGUEL, J. (1948): *Introduction à l'étude mécanique des déformations de l'écorce terrestre.* Mém. pour servir à l'expl. de la carte géol. dét. de la France (Paris, Imprimerie Nationale, 2e éd.).

HUBBERT, M. K., & RUBEY, W. W. (1959): *Role of fluid pressure in mechanics of overthrust faulting, 1. Mechanics of fluid-filled porous solids and its application to overthrust faulting.* Geol. Soc. Amer. Bull. *70*, 115–166.

LEFAVR IS, A., LIENHARDT, G., MONOMAKHOFF, C., & RICOUR, J. (1957): *Données nouvelles sur le c evauchement de la bordure du Jura sur la Bresse dans la région de Lons-le-Saunier (Jura).* Bull. Soc. géol. France [sér. 6] *7*, 1157–1166.

Reprinted from *Geol. Soc. America Bull.*, 74(7), 819–830 (1963)

Effect of the Toe in the
Mechanics of Overthrust Faulting

C. B. RALEIGH *Dept. Geophysics, Australian National University, Canberra, Australia*
D. T. GRIGGS *Institute of Geophysics, University of California, Los Angeles, Calif.*

Abstract: Hubbert and Rubey's theory (1959) that overthrust faults are facilitated by abnormal fluid pressures is extended to include the effect of the "toe" of the thrust plate. Three types of toes are considered: an eroding toe, a noneroding toe or "riser," and a composite thrust sheet with a riser and toe as exemplified by Rich's model of the Pine Mountain thrust. Equations are derived for each type of toe in three cases: gravity sliding, horizontal overthrusting, and tectonic thrusting down a slope.

In the absence of a toe, a thrust sheet of any length will slide down the slope of critical angle according to Hubbert and Rubey. The presence of an eroding toe sets a minimum length-to-thickness ratio for a thrust sheet which will slide down a given slope with a given fluid-to-overburden pressure ratio. Typical geological examples with an eroding toe require roughly twice the slope derived without consideration of the toe. A riser demands twice the

slope again. A Pine Mountain type thrust is intermediate between thrusts having an eroding toe and thrusts with a riser.

The maximum length of a horizontal thrust with an eroding toe is a little more than half that derived without the toe for typical geological cases. The riser again doubles the effect of the toe, and the Pine Mountain thrust is intermediate. Similar effects are found for thrusts tectonically pushed down a slope.

While consideration of the toe requires either greater slopes or higher fluid pressures than those derived by Hubbert and Rubey, these increases are not such as to invalidate their hypothesis except in certain instances. For example, a very thick thrust plate with a noneroding toe seems impossible. For the most part, however, these considerations serve as a refinement, or second approximation to their very attractive theory.

CONTENTS

INTRODUCTION

Hubbert and Rubey (1959) and Rubey and Hubbert (1959) advanced the hypothesis that overthrusts are made possible by the presence of abnormally high fluid pressures at the base of the thrust sheet, which have the effect of reducing the sliding friction by a very large factor. This paper amplifies their treatment of the problem regarding the effect of the "toe" on the thrust plate, which was not considered in their papers.

Birch (1961) advanced trenchant criticisms of the Hubbert-Rubey theory which in our opinion were not wholly dismissed by the reply (Hubbert and Rubey, 1961). Birch raised three important points: (1) The Hubbert-Rubey analysis is inconsistent with a fracture trajectory parallel to the thrust surface; (2) the assumption that the cohesive strength is negligible because of dislocation propagation ($\tau_0 = 0$) is unwarranted and leads to ambiguities in the analysis if applied equally to the overthrust block and the underlying layer; and (3) the Hubbert-Rubey stress analysis lacks rigor in the neglect of some boundary shear stresses which can materially affect the result.

With respect to the third point, Birch remarks that "This error diminishes as the ratio of the length to height of block increases. . . ." He and Hubbert and Rubey (1961) agree that the correction involved is very small for the geometry of typical overthrusts. Hence, for simplicity the authors have adopted the approximate analytic treatment of Hubbert and Rubey (1959).

With respect to the first point, the present authors were unable to satisfy themselves by analysis. In addition to the effect of stratification advanced by Hubbert and Rubey, there might be an effect due to a discontinuity or steep gradient in fluid pressure at the base of the overthrust block, but we were unable to predict the nature of this effect. In an attempt to clarify this problem, a sand-box model of the Hubbert (1951) type was set up which incorporated an impermeable sand layer at the base of the (piston-pushed) overthrust block. It was found that when λ in the underlying sand layer was low, fracture followed the Coulomb-Mohr trajectory as observed by Hubbert in dry sand. When, however, the overthrust layer was weakened at a large distance from the piston and λ in the underlying layer exceeded a critical value depending on the degree of this weakening, the fracture developed along the top of the underlying layer out to the zone of weakness, whence it followed a Coulomb-Mohr trajectory to the surface. This behavior is consistent with Birch's (1961, p. 1442) observation,

"The mechanism of pore pressure can be helpful, however, if the high pore pressure is confined in the *underlying* layer, rather than in the thrust block. In this case, as pore pressure increases, the condition of failure of the underlying medium approaches the Tresca condition of maximum shear stress, equal to the τ_0 of the underlying medium. This may be no more than 100 bars or so, even for relatively strong rocks such as the Berea sandstone."

In our case, $\tau_0 = 0$ in the underlying sand layer. This experiment and Birch's reasoning constitute support for Hubbert and Rubey's contention that the fracture will follow the stratigraphic boundary of high fluid pressure, even if there is little difference in the dry strength of the rocks across this pressure boundary, provided other necessary conditions are also attained (*e.g.*, a zone of weakness at the forward terminus of the thrust).

With respect to Birch's second point, the authors had difficulty following Hubbert and Rubey's dislocation hypothesis of reduction of τ_0, because it seemed that this hypothesis should apply equally to all megascopic cases, whereas, in fact, the τ_0 is derived from laboratory measurement on a megascopic scale. The authors were able, however, to construct a mathematical model which illustrates that, if the fracture trajectory is horizontal, then on a large scale, dislocation type propagation is possible and τ_0 approaches zero as the size increases.

These considerations, which are intended for separate publication, have persuaded us that Birch's criticisms are not fatal to the Hubbert-Rubey hypothesis and that it is profitable to probe more deeply into the geological details using the Hubbert-Rubey simplified analytical methods.

ACKNOWLEDGMENTS

We are indebted to Drs. W. W. Rubey and M. King Hubbert for valuable discussions and to Mrs. M. L. Shirley for her helpful comments and criticism.

EFFECT OF THE TOE OF A THRUST PLATE

Gravity Sliding

In their papers on the role of abnormal fluid pressures in overthrusting, Hubbert and Rubey (1959) and Rubey and Hubbert (1959) reconsidered the possibility of gravity sliding as a mechanism by which large overthrusts might be produced. They showed that abnormal fluid pressure reduces the effective sliding friction, and they derived the following relation for gravity sliding of a thrust plate (Hubbert and Rubey, 1959, p. 146):

$$\tan \theta = (1 - \lambda_1) \tan \phi, \qquad (1)$$

where θ is the dip of the thrust plane, $\tan \phi$ is the coefficient of dry friction (typically $\phi = 30°$, $\tan \phi = 0.577$ for rocks) and λ_1 is the ratio of the fluid pressure beneath the thrust plane to the normal stress on that plane owing to the weight of the overburden.

bending of the rocks where the thrust plane changes dip would require a finite but unknown amount of energy and is neglected in the following calculations. Both of these assumptions favor the case for gravity sliding; i.e., they act in the direction of underestimating the necessary slope for sliding to take place.

Hubbert and Rubey's notation for the subaerial case is used throughout. This is appropriate when sea level is at or below the surface of the toe. When sea level is higher than this, a redefinition of λ is necessary, as prescribed by Hubbert and Rubey (1959, p. 145).

In Figure 1, the main portion of the block, with a volume V_1, rests on a slope with an inclination of angle θ. The toe, with a volume V_2

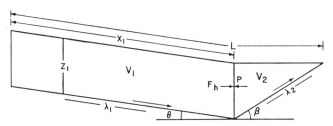

Figure 1.   Gravitational sliding of subaerial block of volume, V_1, with toe of volume, V_2

The equation given above was derived for the theoretical case in which the sliding block was unobstructed at the downslope end. Such is not the case, however, in geological situations, since a large moving thrust plate must somewhere along its length shear off and ride up over a stationary block of rock (e.g., Rubey and Hubbert, 1959, Figs. 6, 7). The resistance to movement provided by this toe of the thrust plate will require either a larger angle of the slope (θ) or a greater fluid-overburden pressure ratio (λ) for sliding to occur.

Let us now examine the magnitude of this effect of the toe, using the methods of Hubbert and Rubey. The model of the thrust plate illustrated in Figure 1 is essentially the same as the idealized one given by Rubey and Hubbert (1959, Fig. 9) with the addition of a toe. Following their interpretation of the geologic evidence, the toe of the sliding plate is assumed to be slowly eroding away as it emerges from near the axis of the geosyncline above base level. Thus, the mass of the toe can be assumed to remain approximately constant during sliding, and no thrusting of the sheet out over the opposite flank of the geosyncline takes place. The

lies on a slope dipping in the opposite direction at an angle β. If we assume a frictionless contact between the blocks at the plane P (again minimizing the required slope), the downhill motion of block V_1 will be opposed by friction at its base plus a horizontal force, F_h, required to slide the toe uphill. The equation of forces for sliding equilibrium of V_1 is then given by

$$\rho_b g V_1 \sin \theta = (1 - \lambda_1) \tan \phi (\rho_b g V_1 \cos \theta + F_h \sin \theta) + F_h \cos \theta \qquad (2)$$

where ρ_b is the density of the water-saturated sediments, and g the acceleration due to gravity. Similarly, for uphill sliding equilibrium of the toe, V_2,

$$F_h \cos \beta = (1 - \lambda_2) \tan \phi (\rho_b g V_2 \cos \beta + F_h \sin \beta) + \rho_b g V_2 \sin \beta, \qquad (3)$$

where λ_2 is the ratio of the average fluid pressure beneath the toe to the average normal stress on the toe. The normal force on the base of the toe is the sum of the normal component of the weight of the toe and the normal component of the force, F_h, due to push of the main block. The ratio of these two normal forces, $F_h \sin \beta / V_2 \rho_b g \cos \beta$, can be evaluated with respect to given values of λ_2, if F_h is considered to be a function of λ_2. As λ_2 varies from 1.00 to 0.465, and $\beta = 30°$, $F_h \sin \beta / V_2 \rho_b g \cos \beta$ varies from 1/3 to 2/3. Although λ_1 is defined in the same way, at small angles of θ, $F_h \sin \theta$ will be negligible compared to the normal force owing to the normal component of the weight of the main thrust block.

Solving equations (2) and (3) for $\tan \theta$ we obtain:

$$\tan \theta = \frac{(1 - \lambda_1) \tan \phi V_1 + V_2 \left[\dfrac{(1 - \lambda_2) \tan \phi + \tan \beta}{1 - (1 - \lambda_2) \tan \phi \tan \beta} \right]}{V_1 - (1 - \lambda_1) \tan \phi V_2 \left[\dfrac{(1 - \lambda_2) \tan \phi + \tan \beta}{1 - (1 - \lambda_2) \tan \phi \tan \beta} \right]}. \quad (4)$$

Since $(1 - \lambda_1)$ must be small and V_1 several times greater than V_2 for sliding to occur, then in the denominator of the right hand term in equation (4)

$$V_1 \gg (1 - \lambda_1) \tan \phi V_2$$
$$\times \left[\frac{(1 - \lambda_2) \tan \phi + \tan \beta}{1 - (1 - \lambda_2) \tan \phi \tan \beta} \right] \quad (5)$$

so that with less than 1 per cent error in $\tan \theta$, equation (4) reduces to

$$\tan \theta = (1 - \lambda_1) \tan \phi$$
$$+ \frac{V_2}{V_1} \left[\frac{(1 - \lambda_2) \tan \phi + \tan \beta}{1 - (1 - \lambda_2) \tan \phi \tan \beta} \right]. \quad (6)$$

For low angles of slope (θ), z_1 will be approximately equal to the thickness of the block so that

$$\frac{V_2}{V_1} = \frac{z_1}{2 x_1 \tan \beta} \quad (7)$$

and

$$\tan \theta = (1 - \lambda_1) \tan \phi$$
$$+ \frac{z_1}{2 x_1 \tan \beta} \left[\frac{(1 - \lambda_2) \tan \phi + \tan \beta}{1 - (1 - \lambda_2) \tan \phi \tan \beta} \right]. \quad (8)$$

When $\phi = \beta = 30°$ (typical value for both ϕ and β):

$$\tan \theta = 0.5774(1 - \lambda_1) + \frac{3 z_1}{2 x_1} \left(\frac{2 - \lambda_2}{2 + \lambda_2} \right). \quad (9)$$

Notice that equation (8) differs from Rubey and Hubbert's relation (Equation 1) by the addition of the second term on the right which is zero if there is no toe.

Now in order to examine the variation of θ with respect to λ_1 and λ_2, we must choose values for the dimensions of the thrust plate and for β, the angle of slope of the toe, and hold these constant[1]. The dips of the toes of the major thrusts as shown in Rubey and Hubbert's structure sections (1959, Fig. 6) range from 20° to 30°. Since it can be shown that the effect produced upon θ by varying the angle β between 20° and 30° is negligible, $\beta = 30°$ will be used throughout for the sake of convenience; for example, when $\lambda_1 = 0.92$, $\lambda_2 = 0.465$ and $x_1 = 50$ km, varying β between 20° and 50° changes θ by only 0.5°. The thickness (z_1) of the main blocks of the Wyoming thrusts are nearly constant everywhere at 6–7 km. The lengths (L) of the thrusts in the direction of sliding are not known, but various possibilities ranging from 50 to 280 km may be considered (Rubey and Hubbert, 1959, p. 197). The former figure is a minimum based on the distance from the thrust belt to the nearest intermontane valley possibly representing a pull-apart gap; the latter is the distance from the thrust belt to the Idaho batholith. The figure of 100 km is chosen here on the basis of a personal communication by Rubey as a likely guess for the length of the thrust plates in this region.

For purposes of calculation, x_1 may be obtained from the length, L, by subtracting the length of the toe, $z_1 / \tan \beta$. For the case under consideration where $\phi = \beta = 30°$, $z_1 = 6$ km, and $L = 100$ km; $x_1 = 90$ km. Table 1 is derived by inserting these figures into equation (9) and assigning different values for λ_1 and λ_2 to get θ, the slope necessary for sliding to occur. For all cases in which λ_1 is 0.90 or less, the elevation at the rear edge of the plate ($x_1 \sin \theta$) is greater than 8 km which is taken as the highest probable elevation by Hubbert and Rubey.

The examples cited by Hubbert and Rubey of high pore pressures in oil wells justify the high values chosen for λ_1 since the main portion of a thrust plate sliding under gravity could be assumed to be riding on a single zone of high fluid pressure corresponding to a stratigraphic

---

[1] The thrusts of western Wyoming, discussed by Rubey and Hubbert (1959), will be considered in the following discussions of the effect of the toe. The geologic evidence assembled by Rubey indicates that, at the time of thrusting, the sediments dipped toward the center of the geosyncline at an angle of 1°–5°; a dip of 2½°–3° was taken as a reasonable guess for the average slope (Rubey and Hubbert, 1959, p. 146). The fluid-over-burden pressure ratio required for sliding by Rubey and Hubbert's equation (1) is $\lambda = 0.92$ for that slope.

TABLE 1. ANGLE θ FOR GRAVITY SLIDING WITH AND
WITHOUT TOE, FOR VARIOUS VALUES OF
FLUID-OVERBURDEN PRESSURE RATIOS—
λ_1 BENEATH MAIN BLOCK, AND λ_2 UNDER TOE
$L = 100$ km., $x_1 = 90$ km., $z_1 = 6$ km., $\phi = \beta = 30°$
(For $\lambda_1 = 0.9$, the height at the rear end of the block
exceeds 8 km for all values of λ_2.) ($\phi \geq 5.25°$)

	$\lambda_1 = 0.95$	$\lambda_1 = 1.00$
λ_2	$\theta°$	$\theta°$
0.465	$x_1 \sin \theta > 8$ km	3.6
0.60	4.7	3.1
0.70	4.4	2.8
0.80	4.1	2.5
0.90	3.8	2.2
0.95	3.7	2.1
1.00	3.6	1.9
No Toe	1.6	0

horizon. The thrust plane beneath the toe, however, transects a stratigraphic column of varied lithology in which the pore pressures might show considerable variation. Table 1 shows the effect of various values of the fluid-overburden pressure ratio, λ_2, beneath the toe.

The other important variable which affects θ is the length-to-thickness ratio of the main block. The curves in Figure 2 were derived by

plotting this ratio, L/z_1, against θ for different values of $\lambda_1 = \lambda_2$. The arrows along the abscissa indicate three possibilities for the Wyoming thrust belt of lengths of 50, 100, and 280 km at a constant thickness of 6 km. The column on the right gives Hubbert and Rubey's values of θ for the case in which the toe is absent.

Comparing these results with those given by Hubbert and Rubey, a 50-km long block with a toe requires an increase in slope of more than 4° at a given fluid-overburden pressure ratio over that for the block alone. If λ_2 beneath the toe is less than λ_1 an additional increase in slope is required. If the Wyoming thrusts were 100 km in length, however, the necessary increase in slope has a minimum value of slightly less than 2°. If we assume that $\lambda_1 = \lambda_2 = 0.95$ beneath the Wyoming thrust blocks, then the minimum slope required for gravity sliding is 6.3° for a 50-km block and 3.7° for a 100-km block. In the former case, the slope is considerably greater than the 2½–3° postulated by Rubey and Hubbert; in the latter case the elevation at the rear of the thrust plate would be almost 20,000 feet above sea level. If the thrust plate extended all the way back to the Idaho batholith, at fluid-overburden pressure ratios of less than $\lambda_1 = \lambda_2 = 0.97$ the elevation

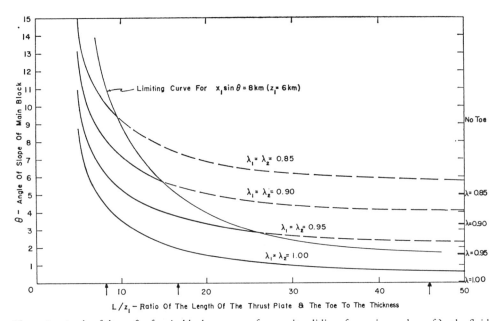

Figure 2. Angle of slope, θ, of main block necessary for gravity sliding, for various values of λ, the fluid-overburden pressure ratio, and L/z_1, the ratio of the length of the thrust plate and toe to the thickness

at the rear of the block would be greater than 25,000 feet.

For thrust blocks which thicken or thin toward the rear, the expression for $\tan \theta$ will be different because of change in volume of the main block, V_1. If, as in Figure 3, z_1 is the thickness of the main block at the toe and z_2 the thickness at the rear edge, then

$$V_1 = \frac{x_1 y_1 (z_1 + z_2)}{2} \qquad (10)$$

and

$$\frac{V_2}{V_1} = \frac{z_1^2}{x_1 (z_1 + z_2) \tan \beta}. \qquad (11)$$

well, and it will be about the same shape as in the case of pure gravity sliding. Before examining this combination of forces, let us look at the case of a horizontal push from the rear.

Maximum Length of Horizontal Overthrust Block with Toe

For the case of a plate of uniform thickness, without a toe, pushed on a horizontal plane surface, Hubbert and Rubey (1959, p. 144) derive the maximum length, x_1, of the block:

$$x_1 = \frac{1}{(1 - \lambda_1)} \left[\frac{a}{\rho_b g \tan \phi} + \frac{b + (1 - b)\lambda_1}{2 \tan \phi} \cdot z_1 \right]. \qquad (13)$$

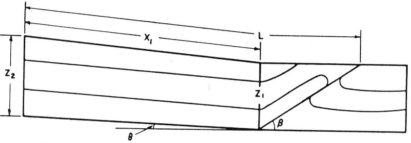

Figure 3.　Gravitational sliding of thrust sheet, thickening toward rear

The length of the main block, x_1, is taken to be equal to the perpendicular distance between z_1 and z_2 with negligible error for small values of θ. The slope, θ, required for gravity sliding is given by

$$\tan \theta = 0.5774(1 - \lambda_1) + \frac{3z_1^2}{x_1(z_1 + z_2)} \left(\frac{2 - \lambda_2}{2 + \lambda_2} \right) \qquad (12)$$

where $\phi = \beta = 30°$.

Compared with blocks of constant thickness, thrusts which thicken toward the rear require somewhat lower slopes for gravity sliding but have greater elevations at the rear edge; thrusts which thin to the rear require higher slopes for sliding but have lower elevations at the rear.

Thickening toward the rear would permit longer thrust blocks to be pushed from the rear because of the increase in strength owing to thickening at the pushing edge. The opposite would be true for thrusts which thin toward the rear.

Rubey and Hubbert (1959, p. 200) suggest that a combination of gravity forces and a tectonic push from the rear may be a likely mechanism. A toe is required in that case as

Consider now the case of such a horizontal thrust plate with a toe (Fig. 4). Again we assume that the toe is constantly eroded to maintain zero relief on the upper surface; that the toe moves as a rigid block with zero friction on the vertical contact between the blocks V_1 and V_2, neglecting the bending which would be present in the real case. These assumptions tend to minimize the effect of the toe, and to overestimate the length of the thrust plate which could actually be pushed.

For uphill sliding equilibrium of block V_2:

$$F_h \cos \beta = (1 - \lambda_2) \tan \phi (\rho_b g V_2 \cos \beta + F_h \sin \beta) + \rho_b g V_2 \sin \beta \qquad (14)$$

and

$$V_2 = \frac{z_1^2 y_1}{2 \tan \beta}$$

$$F_h = \frac{\rho_b g z_1^2 y_1}{2 \tan \beta} \left[\frac{(1 - \lambda_2) \tan \phi + \tan \beta}{1 - (1 - \lambda_2) \tan \phi \tan \beta} \right]. \qquad (15)$$

Following Hubbert and Rubey (1959, p. 143, Equation 88), the equation of equilibrium of the two blocks is:

$$\int_0^{z_1} S_{xx} dz - \int_0^{x_1} \tau_{zx} dx - \frac{F_h}{y_1} = 0. \qquad (16)$$

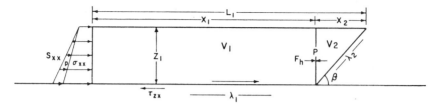

Figure 4. Thrust block, V_1, on horizontal surface with toe, V_2

Using their nomenclature and substituting from (15), the maximum value of the length of the thrust plate is:

$$L_1 = x_1 + x_2 = \frac{1}{(1 - \lambda_1)} \left[\frac{a}{\rho_b g \tan \phi} + \frac{b + (1 - b)\lambda_1}{2 \tan \phi} \cdot z_1 \right]$$

$$- \frac{z_1}{\tan \beta} \left\{ \frac{\tan \beta + (1 - \lambda_2) \tan \phi}{2 \tan \phi (1 - \lambda_1)[1 - (1 - \lambda_2) \tan \phi \tan \beta]} - 1 \right\}. \tag{17}$$

The first term on the right is the length without the toe, as in equation (13). The second term is the reduction in length caused by the toe. For typical values ($a = 7 \times 10^8$ dynes/cm^2, $b = 3$, $\rho_b = 2.31$gm/cm^3, $\lambda_1 = \lambda_2 = 0.90$, $\phi = \beta = 30°$) this has the magnitude:

$$L_1 = 53.6 + 2.3 z_1, \tag{18}$$

where L_1 and z_1 are measured in km. Without the toe,

$$L_1 = 53.6 + 10.4 z_1. \tag{19}$$

It is thus seen that the effect of the toe is to diminish greatly the term which depends on z_1,

the thickness of the thrust plate. The amount of this reduction is sensitive to the fluid pressure at the base of the toe. Figure 5 shows, for the above case, the reduction in length per unit thickness of the thrust plate ($\Delta x/z_1$) as a function of λ_2, the average fluid pressure to overburden pressure ratio at the base of the toe. If $\lambda_2 = 0.73$, then $\Delta x/z_1 = 10.4$ and $L_1 = 54$ km, independent of thickness.

When $\phi = \beta = 30°$, equation (17) reduces to:

$$L_1 = \frac{1}{(1 - \lambda_1)} \left\{ 5.36 + z_1 \left[4.33 - 3.46\lambda_1 \right. \right.$$

$$\left. \left. - 2.60 \left(\frac{2 - \lambda_2}{2 + \lambda_2} \right) \right] \right\}. \tag{20}$$

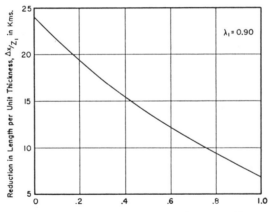

Figure 5. Reduction in length per unit thickness of thrust, $\Delta x/z_1$, for different values of fluid-overburden pressure ratio, λ_2, beneath toe

TABLE 2. MAXIMUM LENGTH OF THRUST PLATE (L_1, km) AS A FUNCTION OF FLUID-OVERBURDEN PRESSURE RATIO ON THE TOE (λ_2)

$\phi = \beta = 30°, \lambda_1 = 0.90$

λ_2	L_1 $z_1 = 1$ km	L_1 $z_1 = 6$ km
0	40	neg.
0.465	50	30
0.6	52	43
0.7	53	52
0.8	55	60
0.9	56	67
0.95	56	71
1.00	57	75
Length without toe	64	116

Gravity Sliding Plus Push From Rear

Hubbert and Rubey (1959, p. 148, equation 125) derive the following equation for the maximum length of a subaerial block pushed from the rear down an inclined slope of angle θ:

$$x_1 = \frac{\dfrac{a}{\rho_b g \cos \theta} + \dfrac{b + (1 - b)\lambda_1}{2} \cdot z_1}{(1 - \lambda_1) \tan \phi - \tan \theta}. \quad (21)$$

Using the same methods and assumptions as in the preceding sections, the maximum length of an overthrust block with a toe pushed down a slope is:

$$L_1 = \frac{\dfrac{a}{\rho_b g \cos \theta} + \dfrac{b + (1 - b)\lambda_1}{2} \cdot z_1 - \dfrac{z_1}{2 \tan \beta}\left[\dfrac{(1 - \lambda_2) \tan \phi + \tan \beta}{1 - (1 - \lambda_2) \tan \theta \tan \beta}\right]}{(1 - \lambda_1) \tan \phi - \tan \theta} + \frac{z_1}{\tan \beta}. \quad (22)$$

Table 2 shows the effect of varying λ_2 for the typical value of $\lambda_1 = 0.90$. Thicker thrust plates are much more sensitive to λ_2 than are thin plates as illustrated in this example.

For the special case in which $\lambda_1 = \lambda_2$ and $\phi = \beta = 30°$, Table 3 shows the maximum length of thrusts for various values of λ and thickness of the thrust plate. For comparison, values (x_1) from Hubbert and Rubey (1959, p. 144, Table 1) are included to show the length without the toe. The toe of the thrust plate reduces the maximum length of a thrust 6 km thick by roughly a factor of 2 for high values of λ, as compared to the values given by Hubbert and Rubey.

Table 4 shows the maximum length with and without toe, for the case in which $\lambda_1 = \lambda_2$, $\phi = \beta = 30°$, $z_1 = 6$ km.

Rubey and Hubbert (1959, p. 200) considered the Idaho batholith, 170 miles away from the easternmost thrust of the Western Wyoming belt, as a possible source of compression. Choosing a 1-degree slope as a maximum for that distance they calculated a required fluid overburden-pressure ratio of 0.93 using the same values of crushing strength, etc., as before. With a toe, $\lambda_1 = \lambda_2 = 0.95$ for the same slope. The presence of a toe, then, does not greatly increase the difficulties in this case, although as Rubey and Hubbert (1959, p. 200)

TABLE 3. MAXIMUM LENGTH OF HORIZONTAL OVERTHRUST WITHOUT TOE (x_1, KM) AND WITH TOE (L_1, KM) FOR VARIOUS THICKNESSES AND VALUES OF FLUID-OVERBURDEN PRESSURE RATIO, FOR THE CASE: $\lambda_1 = \lambda_2$, $\phi = \beta = 30°$

z_1 (km)	$\lambda = 0.465$ x_1	L_1	$\lambda = 0.7$ x_1	L_1	$\lambda = 0.8$ x_1	L_1	$\lambda = 0.9$ x_1	L_1	$\lambda = 0.95$ x_1	L_1
1	13.4	12.1	22.5	20.1	32.9	29.1	64.0	55.9	126	109
2	16.7	14.1	27.1	22.2	39.0	31.3	74.4	58.2	145	112
3	20.1	16.2	31.8	24.5	45.1	33.6	84.8	60.4	164	114
4	23.5	18.3	36.4	26.6	51.2	35.9	95.2	62.7	183	116
5	26.8	20.3	41.0	28.8	57.3	38.2	106	65	203	119
6	30.2	22.4	45.6	31.0	63.4	40.4	116	67	222	121
7	33.6	24.6	50.3	33.2	69.5	42.7	126	70	241	123
8	36.9	26.6	54.9	35.4	75.6	45.0	137	72	260	126

have pointed out, such fluid-overburden pressure ratios must exist over the entire area of the thrust block, at least 17,000 square miles.

EFFECT OF AN OVERRIDING TOE

Thrusts have been recognized in which the toe, instead of eroding as it emerges, is presumed to have overridden part or all of the section and advanced out on a new thrust plane or over the surface. The Pine Mountain overthrust (J. L. Rich, 1934) in the Appalachians and the Livingstone thrust (R. J. W. Douglas, 1950; J. C. Scott, 1953) are interpreted as

ness. The term, f, equals 1 when the thrust breaks across the complete section and out over the surface. In this case the riser will have twice the volume of an eroded toe and the effect will be correspondingly twice as great. Where $f = \frac{1}{2}$, the riser will have the same effect as an eroding toe, except that the riser is half the length of the toe.

Let us consider a composite thrust block having a riser and an eroding toe, taking Figure 6 as a model. This figure is a diagrammatic cross section through Rich's hypothetical Pine Mountain thrust plate after a small amount of

TABLE 4. MAXIMUM LENGTH (KM) OF A THRUST PLATE 6 KM THICK, WITH FLUID-OVERBURDEN PRESSURE RATIO, λ, WHICH CAN BE PUSHED DOWN A θ-DEGREE SLOPE, WITHOUT TOE (x_1) AND WITH TOE (L_1)

$\phi = \beta = 30°$ (Values omitted when $x_1 \sin \theta > 8$ km)

| θ | $\lambda_1 = \lambda_2 = 0.7$ | | $\lambda_1 = \lambda_2 = 0.8$ | | $\lambda_1 = \lambda_2 = 0.9$ | | $\lambda_1 = \lambda_2 = 0.95$ | | $\lambda_1 = 0.95, \lambda_2 = 0.8$ | |
	x_1	L_1	x_1	L_1	x_1	L_1	x_1	L_1	x_1	L_1
0	46	31	63	40	116	67	221	121	221	98
0.5	48	32	68	43	137	77	317	169	317	136
1.0	51	33	74	45	166	92	—	290	—	232
2.0	57	36	91	53	—	154				
3.0	65	40	116	65						
4.0	76	45	—	86						
5.0	92	52								
6.0	—	63								

bedding plane thrusts which have broken across to higher stratigraphic positions in a stepwise fashion (Fig. 6). The Ochre Mountain and North Pass thrusts of the Gold Hill District, Utah (Nolan, 1935), have apparently carried their thrust plates out over the surface for some distance. The overriding toe or "riser" as it is called here will affect the slope angle required for gravity sliding and the maximum lengths of blocks that can be pushed. The effect produced will, in general, be different in magnitude from the effect caused by an eroding toe.

This difference is a result of the different volumes of rock in an eroding toe, V_2, and a riser, V_2' for the same vertical thickness of the thrust block. For an eroding toe (Fig. 1)

$$V_2 = \frac{z_1^2}{2 \tan \beta}$$

and for a riser (Fig. 6)

$$V_2' = \frac{fz_1^2}{\tan \beta}$$

where f is the ratio of the thickness of the section crossed by the thrust to the total thick-

movement has taken place. By considering the thrust at this stage and neglecting the slight thickening of the block ahead of the riser, the values of λ necessary are maximized. The equations for sliding equilibrium are derived by summing up the forces acting over the entire thrust plate and setting them equal to zero. The same assumptions and approximations employed previously are used here, and it will be assumed that $\phi = \beta = 30°$. Making use of the fact that

$$z_3 = (1 - f)z_1 ,$$

the minimum slope (θ) down which the thrust plate may be pushed is given by

$$\tan \theta = \frac{z_1}{x_1 + (1-f)x_3} \left\{ \frac{x_1}{z_1}(1 - \lambda_1) \tan \phi \right.$$

$$+ (1 - f) \frac{x_3}{z_3} (1 - \lambda_3) \tan \phi$$

$$+ 3f \left(\frac{2 - \lambda_2}{2 + \lambda_2} \right) + \frac{3(1-f)^2}{2} \left(\frac{2 - \lambda_4}{2 + \lambda_4} \right)$$

$$\left. - \left[\frac{a}{\rho_b g z_1} + \frac{b + (1 - b)\lambda_1}{2} \right] \right\} . \qquad (23)$$

The slope angles required for gravity sliding may be derived from the above equation by letting the last two terms bracketed together on the right go to zero. These terms represent the contribution of the pushing force, which is assumed equal to the crushing strength of the rock, using Hubbert and Rubey's formulation. For the case in which the block is pushed along a horizontal surface the desired equation may be obtained from equation (23) by setting tan θ equal to zero.

85 km at λ = 0.90; at a thickness of 3 km the maximum length is 283 km.

There is a maximum thickness of a riser set by the strength of the rocks. We shall not take the space here to derive this maximum thickness. When f = 1, the maximum thickness is 6.2 km when λ = 1.0, and 3.7 km when λ = 0.465, for the values of strength used by Hubbert and Rubey. When f = ½, or in the case of an eroding toe, there is no maximum thickness.

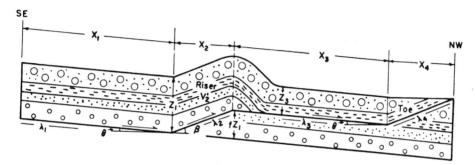

Figure 6.   Idealized cross section of Pine Mountain overthrust after beginning of thrusting (Modified from J. L. Rich, 1934, Figs. 3, 4)

Equation (23) may be used for the consideration of the special case in which the thrust cuts across the entire section at the stage before the thrust plate has moved out over the surface (a noneroding toe). The coefficient, $(1 - f)$, is 0 in this case, and the terms in equation (23) which are concerned with the block to the right of the riser drop out. The slope, θ, required for gravity sliding of a block with a riser which transects the complete section is approximately twice that required for a block with an eroding toe when λ_1 is close to 1.00. Referring to Table 1, when λ_1 = 1.00 and λ_2 = 0.95, for example, θ for a thrust block with an eroding toe is 2.1°; for a block with a riser $(f = 1)$, θ = 4.2°. The maximum length of a horizontal thrust block with a riser $(f = 1)$ can be derived from Table 4 by doubling L_1 and subtracting from that the length without a toe, x_1 and the length of the toe, z_1/tan β. The maximum length with a riser $(f = 1)$ is very sensitive to the thickness, z_1. For example, a horizontal block 5 km thick at λ_1 = λ_2 = 0.90 has a maximum length of 15.3 km; whereas with z_1 = 1 km, the maximum length is 46.1 km. Similarly, for a 5-km thick thrust block and riser $(f = 1)$ pushed down a 3-degree slope the maximum length is

PINE MOUNTAIN OVERTHRUST

Let us examine the Pine Mountain thrust as a specific example of a composite thrust sheet. Rich (1954) and Miller and Fuller (1954) consider the thrust plate to have moved across the section from a stratigraphic position above Cambrian shale to another shale horizon halfway up the section. The block thus has a riser $(f = ½)$ and is considered here, for the purpose of discussion, to terminate in an eroding toe where it emerges to the northwest. The thickness, z_1 = 3.8 km, and z_3 = 1.9 km are taken from Miller and Fuller (1954). The length, x_3, is approximately 30 km. The length, x_1, is not known but here will be taken to be 30 km also. The average dip, β, of the thrust under both the toe and riser is approximately 30°, and therefore x_2 = x_4 = 3.3 km. For simplicity, we also assume that λ_1 = λ_2 and λ_3 = λ_4.

In order to assign values to λ_1 and λ_3 it will be necessary to know how λ might be expected to vary with depth. Rubey and Hubbert (1959, Fig. 4, equation 18) give a relationship between depth, porosity, and λ for an average shale or mudstone and elsewhere plot (Rubey and Hubbert, 1959, Fig. 2) a curve of the depth-porosity relations in the Gulf Coast sediments. Combining these data gives the desired relationship between λ and depth in an area where high abnormal fluid pressures are common. This relationship can be expressed empirically by

$$\left(\frac{1}{1-\lambda}\right)^2 = 25z \qquad (24)$$

where z is the depth in kilometers and varies between 1.6 and 4.8. By assuming a state of compaction in the shales of Pine Mountain block comparable to that in the present-day Gulf Coast region we may relate λ_1 to λ_3 as follows:

$$1.4(1 - \lambda_1) = (1 - \lambda_3) \qquad (25)$$

where λ_1 and λ_3 are fluid-overburden pressure ratios at the depths z_1 and z_3, and $z_1 = 2z_3$.

Table 5 gives the values of λ_1 and λ_3 necessary for gravity sliding, horizontal thrusting, and gravity sliding plus tectonic push from the rear. A single thrust block with an eroding toe, having the same length and average thickness, would require about half the slope for gravity sliding. It would, however, be very nearly the same as the Pine Mountain thrust in the case of horizontal thrusting or thrusting down a slope.

From the above, it is clear that a thrust riding out on the surface presents difficulties. A thrust riding up over part of the section requires a substantially higher slope for gravity sliding than one with an eroding toe but may be very similar in the case of horizontal thrusting or downhill thrusting.

CONCLUSIONS

A thrust of large displacement must ultimately come to the surface of the rocks, thus a toe is required. The effect of the toe was not considered quantitatively by Hubbert and Rubey, although an eroding toe was favored and its importance discussed (Rubey and Hubbert, 1959, p. 194). Using the same methods of analysis as Hubbert and Rubey, this paper expands their treatment to include quantitatively the cases of a thrust with an eroding toe, a thrust which rides out over the surface, and a composite thrust of the Pine Mountain type. The analysis presented here indicates that large thrusts can only form when the toe is continually eroded, as suggested by Rubey and

Hubbert. Although the slope or fluid pressure required in the case of the eroding toe is increased as compared to Hubbert and Rubey's values derived without a toe, these slopes and pressures seem well within the geological possibilities. Thrusts of the Pine Mountain type require still higher slopes or fluid pressure if they originate by gravity sliding but are similar to simple thrusts with an eroding toe if they originate by tectonic forces. Thick thrust plates emerging on the surface without erosion seem precluded.

TABLE 5. NECESSARY CONDITIONS FOR PINE MOUNTAIN OVERTHRUST

λ_1	λ_3	θ	
(a) The slope, θ, required for gravity sliding at various values of $\lambda_1 = \lambda_2$ and $\lambda_3 = \lambda_4$			
1.00	1.00	3°	
.95	.93	5°	
.91	.87	6.8°	($L_1 \sin \theta = 8$ km)
(b) The minimum slope, θ, down which the thrust plate may be pushed at various values of $\lambda_1 = \lambda_2$, $\lambda_3 = \lambda_4$			
.91	.87	0°	(Horizontal thrust)
.88	.83	1°	
.86	.80	2°	
.83	.76	3°	
.78	.69	5°	

Rubey and Hubbert (1959, p. 200, 202) considered a combination of gravity sliding and a tectonic push from the rear to be the most likely mechanism for producing the western Wyoming thrusts. They also discussed the problem of the origin of the tectonic push, but it bears restatement here (Rubey and Hubbert, 1959, p. 200).

"Like all other hypotheses suggested to explain overthrusting, except pure gravitation sliding, the concept of compressive and gravitational forces acting in combination requires the full amount of superficial horizontal shortening that is indicated by the observed overthrusting and folding."

Thus, resort to tectonic forces simply transfers the difficulties from the exposed thrust plates to the unexposed depths, without making the problem any easier. On the other hand, the hypothesis of gravity sliding, if true, would constitute a nearly complete explanation, since only regional doming is required, which presents no fundamental problems.

REFERENCES CITED

Birch, Francis, 1961, Role of fluid pressure in mechanics of overthrust faulting: Discussion: Geol. Soc. America Bull., v. 72, p. 1441–1444

Douglas, R. J. W., 1950, Callum Creek, Langford Creek, and Gap map-areas, Alberta: Geol. Survey Canada Mem. 255, 124 p.

Hubbert, M. K., 1951, Mechanical basis for certain familiar geologic structures: Geol. Soc. America Bull., v. 62, p. 355–372

Hubbert, M. K., and Rubey, W. W., 1959, Role of fluid pressure in mechanics of overthrust faulting: I, Mechanics of fluid-filled porous solids and its application to overthrust faulting: Geol. Soc. America Bull., v. 70, p. 115–166

—— 1961, Role of fluid pressure in mechanics of overthrust faulting: I, Mechanics of fluid-filled porous solids and its application to overthrust faulting: Reply to discussion by Francis Birch: Geol. Soc. America Bull., v. 72, p. 1445–1451

Miller, R. L., and Fuller, J. O., 1954, Geology and oil resources of the Rose Hill district: Virginia Geol. Survey Bull. 71, 383 p.

Nolan, T. B., 1935, The Gold Hill mining district: U. S. Geol. Survey Prof. Paper 177, 172 p.

Rich, John L., 1934, Mechanics of low-angle overthrust faulting as illustrated by Cumberland thrust block, Virginia, Kentucky, and Tennessee: Am. Assoc. Petroleum Geologists Bull., v. 18, p. 1584–1596

Rubey, W. W., and Hubbert, M. K., 1959, Role of fluid pressure in mechanics of overthrust faulting: II, Overthrust belt in geosynclinal area of western Wyoming in light of fluid-pressure hypothesis: Geol. Soc. America Bull., v. 70, p. 167–205

Scott, J. C., 1953, Savanna Creek structure: Alberta Soc. Petroleum Geologists Guidebook, p. 134–138

31

Reprinted from pp. 395, 399, 400, 406, 407, 408, 417, 418, 419 of
Approaches to Taphrogenesis (Inter-Union Commission on Geodynamics
Sci. Rept. 8), 1974

Thin-skinned Graben, Plastic Wedges, and Deformable-plate Tectonics

by

Barry Voight

* * * * * *

Fig. 5, modified from Hansen (1965, Fig. 24) is a schematic block diagram in which the essential structural elements, characteristic of all major Anchorage slides, are shown; these include the graben zone at the head of the slide, the horizontal slip "zone", and the pressure ridge at the toe.

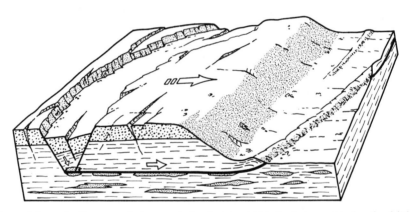

Fig. 5. Schematic diagram of characteristics for the 1964 Anchorage, Alaska, landslides.

* * * * * *

By process of elimination, therefore, some lateral pressure mechanisms seem required. At least two can be envisaged, and they are related; (1) open fractures, filled wholly or in part with water or water-sediment mixtures, exert significant lateral hydrostatic or hydrodynamic pressures against the fracture walls; in the hydrostatic case, the force arising from this "fluid", or "viscous-wedge" mechanism is proportional to the square of the height of the fluid prism; (2) the weight of material within the graben exerts lateral pressure on the boundary walls. Thus as a graben subsides, a continuous (but decreasing) lateral pressure is applied to the moving glide block; the subsiding ground under the graben behaves as a "plastic wedge" which actively contributes to block propulsion.

Horizontal force components resulting from both mechanisms (1) and (2) are

[*Editor's Note:* A row of asterisks indicates that material has been deleted.]

significant. The principal distinction between them appears to be in their respective
abilities to su s t a i n lateral pressures once slide motion commences; in general,
fluid-filled fractures can be extremely transient sources of propulsion inasmuch as
the fluid typically escapes with relatively small lateral block motion. Plastic wedges
exhibit a more delayed form of transient behavior, principally because the
"volume" is larger. A case has been presented concerning the significance of the
plastic wedge mechanism in the development of the Anchorage landslides (VOIGHT
1973).

Attention will now be given to the possibility of application of these "wedge"
mechanisms to geometrically analogous structures on continental and global scales.
Are the classic continental-scale graben and rift valleys necessarily passive features
produced by crustal extension which permit the blocks to fall into place? Or could
they be ductile wedges of stone, hammered into the earth's crust by the hand of
gravity, helping to pry apart the crustal blocks with an unrelenting horizontal
pressure?

Further consideration of this question may lead to what may be termed "thick-
skinned" and "thin-skinned" taphrogenetic hypotheses, wholly analogous to "décollement"
versus "basement involvement" hypotheses for Appalachian and Jura thrust and fold belts.
The graben which developed in association with the Anchorage slides, for example, are
clearly of the "thin-skinned" variety, requiring a basal structural discontinuity (e. g.,
décollement) for their formation. Similar features have been reported by RYBÁŘ (1971)
from the Tertiary brown coal basins of Bohemia, Bulgaria and Germany. Analogous
extension zones, on a larger scale, occur in association with the Heart Mountain "fault"
of western Wyoming, U.S.A. (PIERCE 1973; VOIGHT 1973). That structure, which seems to
be a great rockslide (10^3 km³), is bounded in the area of slab release by multiple-level
décollement horizons which follow stratigraphic boundaries. Graben zones extend to
décollement level and no further (Fig. 9), and were instrumental in the disruption of the
glide mass. To extend the structural association to yet another order of magnitude, reference is made to Fig. 10, which illustrates the RUBEY & HUBBERT (1959) concept of a
major bedding-plane glide surface on the flank of a geosyncline. Section A shows the central and deepest part of the geosyncline in which thick, rapid sedimentation has caused
abnormal fluid pressures to develop at depth; at some depth at which the critical relationship between fluid pressure and the lateral component of stresses acting on the glide surface is exceeded, a thick section of sedimentary rock would part from its foundation and
begin to move (RUBEY & HUBBERT 1959, p. 194). Rocks would buckle at the forward
edge of the moving plate (Section B), causing increased resistance which could be overcome by subsequent erosion, hence raising a possibility of renewed movement. The status
of the hypothesis can be tested for a given area by consideration of body and surface
forces acting on a free-body diagram, in which a force balance is taken which includes
resistive components of toe and glide horizon, and propulsive components, e. g., the net
downslope weight component of the glide mass. The same question arises that has been
previously posed: is the graben zone an area of tensile resistance, or one of compressive
propulsion in association with the plastic wedge mechanism? For the geosynclinal area of
western Wyoming, Z may be taken as approximately 6.0 km, $\varrho = 2.3$ g/cm² (RUBEY &
HUBBERT 1959, p. 196—200). The crushing strength of the rock mass, particularly as influenced by shale sections, cannot have been large, e. g., $C_0 \cong 10^2$ bars. Under the circumstances, even with hydrostatic pore pressure, tension could not be sustained by the rock
mass; a pull-apart zone such as illustrated in Section B can be regarded as a physical impossibility, and would have to be replaced by a zone of plastic deformation; this zone
could be a regionally-extensive domain of plastic deformation, or a localized zone in
association with well-defined grabens. In either case, the plastic wedge force would develop
and would be additive to other "propulsive" forces considered in the force balance. As
movement progressed, graben areas would tend to fill with younger sediments, with the
consequence that the active pressure value of plastic wedges could be maintained. In any
event, consideration of the plastic wedge mechanism decreases the fluid pressure-overburden
ratio theoretically required for a gravitational sliding hypothesis.

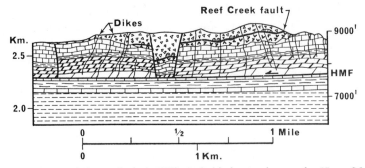

Fig. 9. Cross-section of the Cathedral Cliffs fault block complex on the Heart Mountain décollement, western Wyoming, U.S.A. (modified after PIERCE 1963). The graben zone terminates at the décollement boundary. HMF refers to "Heart Mountain Fault".

* * * * * *

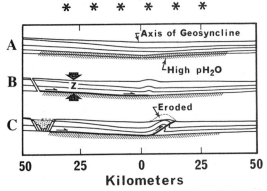

Fig. 10. Idealized diagram to show zone of abnormally-high fluid pressure, bedding plane glide surface on flank of a geosyncline, buckling at forward edge and graben development at the rear edge of a glide block, and erosion and "break-through" of the frontal thrust (after RUBEY & HUBBERT 1959).

* * * * * *

References cited

HANSEN, W. R., 1965: Effects of the earthquake of March 27, 1964, at Anchorage, Alaska. — U.S. Geol. Surv. Prof. Paper **542 A**, 68 p.

PIERCE, W. G., 1963: Reef Creek detachment fault, northwestern Wyoming. — Geol. Soc. Amer. Bull., **74**, 1225—1236.

— 1973: Principal features of the Heart Mountain Fault and the Mechanism problem. — In: DE JONG, K. & SCHOLTEN, R. (eds.), Gravity and Tectonics, J. Wiley and Sons, New York.

RUBEY, W. W. & HUBBERT, M. K., 1959: Role of fluid pressure in mechanics of overthrust faulting, part II. — Geol. Soc. Amer. Bull., **70**, 167—206.

RYBÁŘ, J., 1971: Tektonisch beeinflußte Hangdeformationen in Braunkohlenbecken. — Rock Mechanics, **3**, 139—158.

VOIGHT, B., 1973: The mechanics of retrogressive block-gliding, with emphasis on the evolution of the Turnagain Heights landslide, Anchorage, Alaska. — In: DE JONG, K. & SCHOLTEN, R. (eds.), Gravity and Tectonics, J. Wiley and Sons, New York.

32

Reprinted from pp. 77, 78-81, 100-101 of *Geotechnique*, 4, 77-102 (1964)

Fourth Rankine Lecture

LONG-TERM STABILITY OF CLAY SLOPES

by

A. W. SKEMPTON, D.Sc., M.I.C.E., F.R.S.

*   *   *   *   *   *

PEAK AND RESIDUAL STRENGTHS

Frequent reference will be made to "over-consolidated" clay and it is well to be clear as to the meaning of this term.   In Fig. 2, point (a) represents a clay immediately after deposition, for example on the bed of an estuary.   The deposition of more clay will cause an increase in effective pressure and a decrease in water content.   At a stage represented by point (b) the clay is "normally-consolidated", in the sense that it has not been subjected to a pressure greater than the present overburden.   The shear strength of normally-consolidated clay is proportional to the effective pressure, and the graph expressing the relation between strength and pressure is therefore a straight line passing through the origin of axes.

Many post-glacial clays are normally-consolidated; having been deposited during the eustatic rise of sea level consequent upon the melting of the ice sheets in Late Pleistocene times.   But the great majority of clays are much older, and during their geological history have been subjected to very considerable pressure, corresponding to depths of overburden of

[*Editor's Note:* A row of asterisks indicates that material has been deleted.]

several hundred, or even several thousand feet of sediments, which have subsequently been removed by erosion. The clay is then left in an "over-consolidated" state represented in Fig. 2 by point (d).

The removal of pressure is accompanied by an increase in water content, but this increase is far less than the decrease in water content during consolidation. Thus although the clay at point (d) is under the same effective pressure as the clay at point (b), the water content of the over-consolidated clay is considerably smaller. The particles are therefore in a denser state of packing and, not surprisingly, the shear strength is greater than that of the normally-consolidated clay.

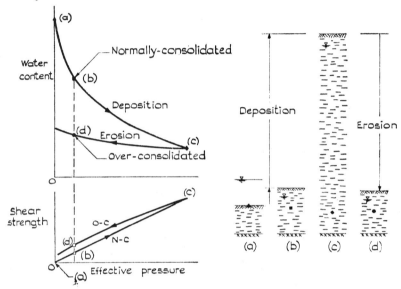

NORMALLY – AND OVER –CONSOLIDATED CLAY

Fig. 2

Examples of over-consolidated clays include the Lias, the Gault and the London Clay. These, and many other clays in this category, are fissured and jointed. They occasionally contain slickensides as well; probably as a result of tectonic movements or unequal expansion during the erosion cycle. Boulder clays are also over-consolidated, by the weight of ice rather than the weight of sediments, but they often show very little in the way of structural discontinuities; being intact, non-fissured materials without joints or slickensides. Most normally-consolidated clays are also free from such imperfections, though fissures can some times be seen in clays of this type.

In this Lecture I shall be concerned mainly with over-consolidated clays, since they give rise to much greater problems than the normally-consolidated clays, and they are more widespread.*

Let us, therefore, examine the shear strength characteristics of an over-consolidated clay, as shown in Fig. 3. This figure typifies the results obtained by carrying out slow drained tests in a shear box apparatus, in which the clay is subjected to displacements amounting to several inches.

As the clay is strained, so it builds up an increasing resistance. But, under a given effective pressure, there is a definite limit to the resistance the clay can offer, and this is the

---

* Over-consolidated clays of high sensitivity, such as the Leda clay in eastern Canada, represent a special class the properties of which may differ appreciably from those of the typical over-consolidated clays considered here.

"peak strength" s_f.   In ordinary practice the test is stopped shortly after the peak strength has been clearly defined, and s_f is referred to simply as the "shear strength" of the clay (under the given effective pressure) without further qualification except, of course, the statement that the test has been carried out under drained conditions, i.e. without the development of pore pressures.

If, however, the test is continued, then we find that as the displacement increases so the resistance or strength of the clay decreases.   But this process, which may be called "strain-softening", is not without limit, for ultimately a certain "residual strength" s_r is reached which the clay maintains even when subjected to large displacements.*   In the comparatively small number of clays so far investigated from this point of view, the strength falls to the residual value after displacements past the peak of the order of 1 in. or 2 in.   But there is field evidence that the strength is not less than this laboratory value (or only slightly less) even at displacements of many feet.

A second shear box test could now be carried out on another specimen of the same clay, but under a different effective pressure.   The results would be of the same type as those

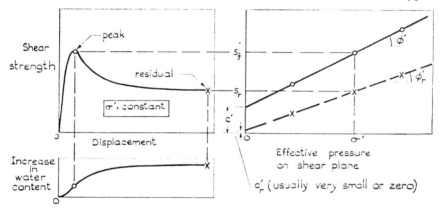

SHEAR   CHARACTERISTICS   OF   OVER-CONSOLIDATED   CLAY

Fig. 3

already described.   And if we suppose that three such tests be made, the peak and residual strengths, when plotted against effective pressure, would show a relationship approximately in accordance with the Coulomb-Terzaghi law.   The peak strengths can therefore be expressed by the equation:

$$s_f = c' + \sigma'. \tan \phi'$$

and the residual strengths by the equation:

$$s_r = c_r' + \sigma'. \tan \phi_r'$$

The test results at present available show almost invariably that c_r' is very small, and probably not significantly different from zero.   Thus I shall assume throughout the rest of the Lecture that for the residual strength we may write:

$$s_r = \sigma'. \tan \phi_r'$$

* The first determination of the residual strength of an undisturbed clay (from the Weser-Elbe canal) was published by Dr B. Tiedemann in 1937.   He carried out drained tests in a ring shear apparatus.   Tests of the same type were made by Dr M. J. Hvorslev (1937) on two clays (Wiener Tegel and Little Belt Clay) consolidated from the slurry condition.   Professor R. Haefeli introduced the term "residual" in 1938 (in German) and in 1950 (in English).   Credit is due to him for insisting on the practical importance of residual strength; at a period when scarcely anyone was interested in this aspect of the shear properties of clays.   Tests leading to an approximate determination of the residual strengths of some compacted soils have been reported by Mr J. MacNeil Turnbull (1952).

In other words, in moving from the peak to the residual, the cohesion intercept c' disappears completely. During the same process the angle of shearing resistance also decreases; in some clays by only 1° or 2°, but in others by as much as 10°.

It is notable that during the shearing process over-consolidated clays tend to expand, especially after passing the peak. Part of the drop in strength from the peak value is therefore due to an increasing water content. Of comparable importance, however, is the development of thin bands or domains in which the flaky clay particles are orientated in the direction of shear. The shear strength of a mass of such particles in random orientation must be greater than when the particles are lying parallel to each other. And while it is probable that the formation of orientation domains begins at relatively small strains (Goldstein et al., 1961), there is decisive evidence for the presence of continuous bands of almost perfectly orientated particles in clays subjected to large strains; both in the laboratory (Astbury, 1960) and in the field (as will be described later in this Lecture).

Irrespective of the physical explanation of the drop in strength after passing the peak, the existence of this decrease in strength (especially in over-consolidated clays) must be accepted as a fact which has been fully established. Thus, if for any reason a clay is forced to pass the peak at some particular point within its mass, the strength at that point will decrease. This action will throw additional stress on to the clay at some other point, causing the peak to be passed at that point also. In this way a progressive failure can be initiated and, in the limit, the strength along the entire length of a slip surface will fall to the residual value. Obviously, in any given case, a slip may occur before the residual strength is attained throughout the clay, but once a progressive failure has started the average strength of the clay will decrease inexorably towards the limiting residual value.

Now it is well-known that the strength of solid materials is greatly diminished by the presence of microscopic cracks, holes and other imperfections. In the simplest terms these act as stress concentrators, and they are responsible for fracture taking place at an average stress which is far less than the ideal strength of the material. And there would seem to be no reason to suppose that the macroscopic fissures, joints and slickensides, present in so many clays, do not also act as stress concentrators in a roughly analogous manner; quite apart from the fact that they act as discontinuous planes of weakness—for it is unlikely that the strength on a fissure or joint can be appreciably higher than the residual value.

Hence we may logically expect that a fissured or jointed clay would not be able to develop its peak strength along the full length of a slip surface. Not only will the fissures and joints reduce the average strength of the clay mass, but they can cause the peak to be crossed, as a result of local over-stressing, and a progressive decrease in strength will follow.

Fissures and other physical discontinuities may not be the only explanation of landslips in clays taking place at strengths well below the conventional peak value. The peak strength as measured in laboratory tests lasting, at the most, a few weeks might be appreciably greater than the strength which could be developed in some clays when the stresses are applied over periods of years, decades or centuries. In other words the effects of shear creep have to be considered; though at present very little quantitative information is available to assess these effects so far as the peak strength is concerned.

✱ ✱ ✱ ✱ ✱ ✱

REFERENCES

ASTBURY, N. F., 1960. "Science in the ceramic industry". *Proc. Roy. Soc. A.*, 258:27–46.

GOLDSTEIN, M. N., V. A. MISUMSKY *and* L. S. LAPIDUS, 1961. "The theory of probability and statistics in relation to the rheology of soils". *Proc. 5th Int. Conf. Soil Mech., Paris*, 1:123–126.

HAEFELI, R., 1938. "Mechanische Eigenschaften von Lockergesteinen". *Schweiz. Bauzeitung*, 111:321–325.

HAEFELI, R., 1950. "Investigation and measurements of the shear strengths of saturated cohesive soils". *Géotechnique*, 2:3:186–208.

HVORSLEV, M. J., 1937. "Uber die Festigkeitseigenschaften gestörter bindiger Böden". *Ingenior. Skrifter A., Copenhagen*, No. 45.

TIEDEMANN, B, 1937. "Uber die Schubfestigkeit bindiger Böden". *Bautechnik*, 15:433–435.

TURNBULL, J. McN., 1952. "Shearing resistance of soils". *Proc. 1st Aust.-N.Z. Conf. Soil Mech., Melbourne*, pp. 48–81.

33

Reprinted from *Geotechnique*, **23**, 265-267 (June 1973)

Correlation between Atterberg plasticity limits and residual shear strength of natural soils

B. VOIGHT*

The purpose of this Note is to call attention to a practical correlation which seems to exist between residual shear strength and plasticity index (Fig. 1). Other workers (e.g. Kenney, 1967) have claimed that no relation existed between residual strength and soil plasticity (or water content at the residual state). This conclusion was perhaps based on liquid or plastic limit distributions, rather than plasticity index.

Drained direct shear tests by numerous workers on natural soils, pure minerals and mineral mixtures have shown that residual shear strength is primarily dependent on mineral composition and system chemistry (Kenney, 1967). For massive minerals such as quartz, feldspar, calcite and dolomite, $\mu'_r \geq 0.58$; for micaceous minerals (e.g. hydrous mica, illite), $\mu'_r \geq 0.30$; for montmorillonitic minerals, $\mu'_r < 0.20$. These mineralogical factors affect Atterberg index parameters, so it does not seem surprising that strength and plasticity can be correlated.

* Formerly at the Technological University, Delft, Netherlands, now at the Pennsylvania State University, U.S.A.

Sample localities are:
1. Selnes
2. Manglerud
3. Asrum
4. Labrador
5. Ottawa
6, 7. Sandnes
8. Little Belt
9. Bear paw
10. Pierre
11. Pepper
12. Cucharacha
13–18. Vaiont
19. Walton Wood
20, 21. Guildford
22–24. Atherfield
25, 26. Weald
27–28. Manglea
29. Wraysbury
30. London
31, 32. Gault
33. Chalk
34–36. Keuper marl
37. Lias
38–40. Appalachian colluvium
39. Upper Coal Measures

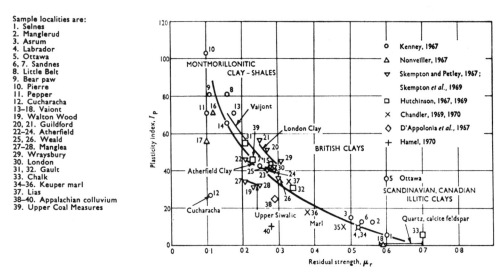

Fig. 1. Plasticity index, I_p, plotted against residual strength coefficient, μ'_r

Small values of μ'_r are thus characteristic of soils having large amounts of montmorillonite or mixed-layer minerals containing montmorillonite; hence $I_p > 50$. Soils having large values of μ'_r typically contain large quantities of massive, chloritic or micaceous minerals, and relatively small quantities of the montmorillonite-type minerals; generally $I_p < 20$.

A certain amount of scatter in empirical diagrams such as Fig. 1 is to be expected; values which seem especially anomalous include unexpectedly low I_p values for the Cucharacha (Panamá) sample (12)[1] and for Appalachian clay–shale of Pennsylvanian (Carboniferous) age (40). These cases appear to be explicable; the latter sample, for instance, contained appreciable quantities of fine sand-size and silt-size particles, many of which were platy claystone or shale fragments. This could account for their low measured plasticity. When these particles become orientated parallel to the failure surface in the shearing process, the measured residual friction angle can be close to that of their clay mineral constituents (Hamel, 1970, p. 284). A similar argument can be given for the Cucharacha sample (12), for which an anomalous clay-fraction determination was reported by Kenney (1967, p. 125). The grain-size error was attributed to either flocculation or to insufficient breakdown of particle aggregates during pre-test preparation (Kenney, 1967, p. 125). The anomalously high I_p value for the Ottawa clay must owe its explanation to some other, unknown, mechanism.

Individual mineral fractions tested in various grain-size ranges and in various homoionic states with different pore fluid salt concentrations (Kenney, 1967, p. 126 and Table 1) also appear to exhibit a correlation of strength with plasticity index. However, the values of such mineral fractions are not directly comparable to those of natural soils, and the overall natural soil trend of increasing I_p for decreasing μ'_r is not observed in all cases.

A similar, although less distinctive, correlation was obtained with a plot of water content at residual strength (W_r) against residual strength coefficient. Overlap existed for W_r in the range of 20–30%, with equivalent μ'_r values varying over the (large) range of 0·25–0·60.

---

[1] Numbers refer to sample locations cited in Fig. 1.

Soils with $\mu'_r < 0\cdot20$ consistently had $W_r > 40\%$. Anomalous values were present for the Ottawa clay.

The principal conclusion of this survey is that plasticity index appears to be a useful field guide to the important engineering property of residual strength of natural soils, and that further examination of this correlation is warranted.

REFERENCES

Chandler, R. J. (1969). The effect of weathering on the shear strength properties of Keuper Marl. *Géotechnique* **19**, No. 3, 321–334.

Chandler, R. J. (1970). A shallow slab slide in the Lias clay near Uppingham, Rutland. *Géotechnique* **20**, No. 3, 253–260.

D'Appolonia, E., Alperstein, R., & D'Appolonia, D. J. (1967). Behavior of a colluvial slope. *Jnl Soil Mech. Fdns Div. Am. Soc. Civ. Engrs* **93**, SM4, 447–473.

Hamel, J. V. (1970). *Stability of slopes in soft, altered rocks.* Ph.D. thesis, University of Pittsburgh.

Hutchinson, J. N. (1967). Written discussion. *Proc. Geotech. Conf., Oslo* **2**, 183–184.

Hutchinson, J. N. (1969). A reconsideration of the coastal landslides at Folkestone Warren, Kent. *Géotechnique* **19**, No. 1, 6–38.

Kenney, T. C. (1967). Influence of mineral composition on the residual strength of natural soils. *Proc. Geotech. Conf., Oslo* **1**, 123–129.

Nonveiler, E. (1967). Shear strength of bedded and jointed rock as determined from the Zalesina and Vajont slides. *Proc. Geotech. Conf., Oslo* **1**, 289–294.

Skempton, A. W. & Petley, D. J. (1967). The strength along structural discontinuities in stiff clays. *Proc. Geotech. Conf., Oslo* **2**, 29–46.

Skempton, A. W., Schuster, R. L. & Petley, D. J. (1969). Joints and fissures in the London Clay at Wraysbury and Edgware. *Géotechnique* **19**, No. 2, 205–217.

34

Reprinted from pp. 509, 511-513 of *Proc. 1st Congr. Internat. Soc. Rock Mechanics*, 1, 509-513 (1966)

Multiple modes of shear failure in rock

by F. D. Patton

[*Editor's Note:* In the original, material precedes this excerpt.]

IV. Definition of terms

ϕ is the angle of sliding or shearing resistance. It is used where a more specific term does not seem warranted.

ϕ_μ is the angle of frictional sliding resistance. Its value changes with the surface characteristics of the rock. For most practical problems involving rocks, the appropriate value of ϕ_μ can apparently be obtained after large displacements have occurred along macroscopically smooth and flat but microscopically irregular (i. e., unpolished) wet surfaces.

ϕ_r is the angle of residual shearing resistance of materials which initially were partly or completely intact. It is obtained from the asymptotic minimum values of shear strength following large displacements.

i is the angle of inclination of the failure surfaces with respect to the direction of application of the shearing force. It is also used in a graphical sense as a particular angle on a shear strength diagram.

V. Results

The results presented here are from the tests on specimens of kaolinite-plaster. Similar results were obtained from tests on the sand-plaster specimens.

1) Specimens with flat surfaces

Figure 2 shows a typical failure envelope from a series of direct shear tests on relatively flat, unpolished, surfaces.

Failure envelopes from these specimens were straight lines passing through the origin and inclined at an angle ϕ_μ from the horizontal. The angle ϕ_μ for the specimens of the stronger mix (kaolinite-plaster 1:2) was 31°. For the weaker mix ϕ_μ was 27½°.

2) Specimens with inclined teeth at low normal loads

Figure 3 shows two failure envelopes typical of those obtained from tests *at low normal loads* on specimens with inclined teeth. The maximum strengths recorded for a number of specimens were used to form the maximum strength envelope (line *A*). The residual strengths remaining in these same specimens after large displacements had occurred were the basis for the residual strength envelope (line *B*).

The equation describing the maximum strength envelope is $S = N \tan (\phi_\mu + i)$ where S is the total shearing strength and N is the total normal load. The inclination of the residual envelope is ϕ_r and the envelope can be described by the equation $S = N \tan \phi_r$. For the various plaster specimens, the angle ϕ_r was always within 1½° of ϕ_μ and the two were often identical.

Line *A* of Figure 3 represents two different types of strengths. It represents the value of the external frictional resistance along the inclined planes, and it represents the internal strength of the teeth at the point of failure. When failure occurs these two strengths are equal.

It may be noted from line *A* that although intact material was sheared there was no cohesion intercept indicated when the results were plotted. Yet the internal «cohesive» strength of the teeth still contributed to the total strength by making possible the development of increased frictional resistance along the surface of the teeth. The precise contribution of the internal «cohesive» strength of the teeth at any given normal load is the difference in strengths between the maximum and residual strength envelopes. A cohesion intercept would occur if the sum of $\phi_\mu + i$ became equal to or greater than 90°.

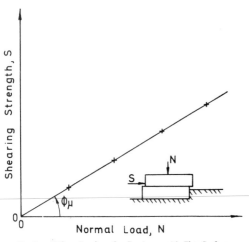

Fig. 2 — *Failure Envelope for Specimens with Flat Surfaces*

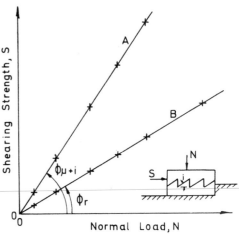

Fig. 3 — *Failure Envelopes for Specimens with Irregular Surfaces*

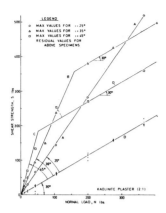

Fig. 4 — *Failure Envelopes for Specimens with Different Inclinations of Teeth*

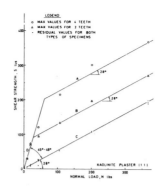

Fig. 5 — *Failure Envelopes for Specimens with Different Numbers of Teeth*

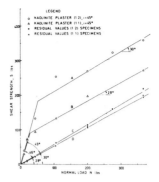

Fig. 6 — *Failure Envelopes for Specimens with Different Internal Strengths*

3) *Different inclinations of teeth*

Results from three series of tests, each made on specimens with different inclinations of teeth, are shown in Figure 4. The failure envelope for specimens with $i = 25°$ is a straight line — line A. For specimens with $i = 35°$ and $i = 45°$ the failure envelopes are curved but each envelope can be approximated by two straight lines as are envelopes B and C, respectively. Line D is drawn through the residual shear strengths of all three series of specimens.

The inclinations of the lower or primary portions of lines A, B, and C are equal to, or within one degree of, $\phi_\mu + i$. The inclinations of the upper or secondary portions of lines B and C are very close to the value ϕ_r. The abrupt changes in the slopes of lines B and C are related to changes in the mode of failure. Below the changes in slope the maximum shearing strength is related to the frictional resistance along the inclined surfaces. Above the transition in slope the maximum strength is unrelated to the increased surface friction due to the inclination of the teeth.

The cross-sectional area of the intact material at the base of the 35° teeth is greater than for the 45° teeth. This explains why the transition in the mode of failure for the two inclinations of teeth occurred at different normal loads. Line A is straight because the range of normal loads used was not high enough to reach the transition for the specimens with 25° teeth.

4) *Varying the number of teeth*

Figure 5 shows the effect of doubling the number of teeth from two to four and keeping the specimens identical in other respects. Each maximum strength failure envelope, although curved, is approximately described by two straight lines. The secondary portion of the failure envelope for specimens with four teeth (line A) is about twice as far above the residual envelope (line C) as the envelope for specimens with two teeth (line B).

The steeply sloping primary portions of the failure envelopes are approximately equal to $\phi_\mu + i$. The inclinations of the secondary portions of the failure envelopes are approximately ϕ_r. The change in slope again is related to a change in the mode of failure associated with the initial displacements.

The effect of having additional teeth is to move the abrupt change in slope of the failure envelope to a higher normal load and to move the secondary portion of the failure envelope about twice as far above the residual envelope as the failure envelope for two teeth.

This diagram illustrates the difficulties encountered in attaching any real meaning to the average shearing stresses computed for tests on real rocks. In rocks the number, size, and shape of the irregularities are unknown; hence the real shearing and normal stresses are also unknown.

From tests made on higher strength specimens it was found that specimens with four teeth often gave failure envelopes that were only slightly greater than the envelopes for specimens with two teeth. This was interpreted as evidence of progressive failure.

5) *Varying the strength of the teeth*

Figure 6 shows the results of tests on two series of specimens with identical surface configurations but different internal strengths. Line A is the failure envelope for the stronger specimens and line B for the weaker specimens. Lines C and D are their respective residual strength envelopes.

Slopes of the primary and secondary portions of the failure envelopes are slightly different for each series of tests. These differences reflect a change in ϕ_μ and ϕ_r for the two strengths of specimens. The change in mode of failure occurs at a higher normal stress for the stronger specimens than for the weaker ones. Thus, increasing the strength of the specimen teeth has an effect similar to that of increasing the number of teeth.

VI. Conclusions

Three general conclusions can be drawn from the results of the tests on plaster specimens: 1) failure envelopes for specimens with irregular failure surfaces are curved, 2) changes in the slope of the failure envelope reflect changes in the mode of failure, and 3) changes in the mode of failure are related to the physical properties of the irregularities along the failure surface.

These conclusions, together with the fact that ϕ does not vary throughout a wide range of normal loads (although $\phi + i$ does vary), have many practical applications. In particular, they facilitate the interpretation of curved failure envelopes.

VII. Interpretation of tests on real rocks

From the results of shear tests on real rocks one would expect to obtain a superposition of the effects of the separate variables investigated for the plaster specimens. For example, in the same sample of rock the irregularities along the failure surface would have different sizes, inclinations, internal

291

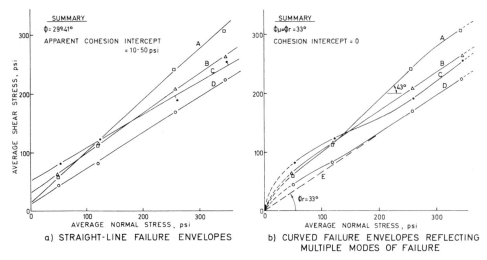

Fig. 7 — *Two Interpretations of Direct Shear Tests on Rock Samples with Irregular Surfaces*

strengths, and coefficients of friction. Thus, failure envelopes for rocks would not reflect a simple change in the mode of failure but changes in the «intensities» of different modes of failure occurring simultaneously.

Figure 7 illustrates two interpretations that can be given to four series of tests (*A* , *B* , *C* , and *D*) on different surfaces of the same rock. Figure 7a shows the shear test results interpreted as forming straight-line failure envelopes. This is equivalent to saying that only one mode of failure ocurred during the tests at different stress levels. From Figure 7a it would also appear that the value of ϕ was different for each series of tests and was not a relatively constant property of the material. In addition, the straight-line envelopes could lead some designers to conclude that an appreciable amount of cohesive strength exists at zero normal load. These errors are avoided in Figure 7b in which the same data is used to form curved failure envelopes.

The curved failure envelopes in Figure 7b also provide more information on the geometry and effectiveness of the surface irregularities than is offered in Figure 7a. For example, at a given normal stress the vertical distance between a point on any maximum strength failure envelope and the residual envelope (line *E*) gives the internal strength contributed by the irregularities. This strength is the strength that is lost when significant displacements occur along the failure plane.

From Figure 7b the rocks of the test series outlined by line *A* can be interpreted as having small relatively steep irregularities which were effective between a normal stress of 0 to 40 psi. Above 40 psi these small irregularities failed before displacements could occur along them. Between a

normal load of 120 and 270 psi some larger irregularities which had inclinations of 10° (43° minus 33°) became effective. Above 270 psi these larger irregularities began to fail before displacements could occur.

For some engineering design purposes straight-line failure envelopes are adequate. But to facilitate an understanding of the failure mechanisms curved failure envelopes reflecting the multiple modes of shear failure appear to be a necessity.

VIII. Acknowledgments

This paper is based upon a thesis submitted in partial fulfillment of the requirements for a Ph. D. in Geology at the University of Illinois. The thesis was completed under the direction of Dr. D. U. Deere, professor of civil engineering and geology, who made many valuable contributions to the study.

References

NEWLAND. P. L., and B. H. ALLELY — 1957, *Volume changes in drained triaxial tests on granular materials*, Geotechnique, Vol. VII, pp. 17-34.
PATTON, F. D. — 1966, *Multiple Modes of Shear Failure in Rock and Related Materials*, Ph. D. Thesis, Univ. of Illinois, 282 pp.
RIPLEY, C. F., and K. L. LEE — 1961, *Sliding friction tests on sedimentary rock specimens*, Communication 8, 7th Congress of Large Dams, Vol. IV, pp. 657-671.
WITHERS, J. H. — 1964, *Sliding Resistance Along Discontinuities in Rock Masses*, Ph. D. Thesis, Univ. of Illinois, 124 pp.

35

Reprinted from *Geol. Soc. America Bull.*, **77**(7), 741–760 (1966)

Tectonic Implications of Gypsum Dehydration

HUGH C. HEARD   *Exploration and Production Research Division, Shell Development Company, Houston, Texas*

WILLIAM W. RUBEY   *Dept. Geology and Institute of Geophysics and Planetary Physics, University of California, Los Angeles, California*

Abstract: The Hubbert-Rubey overthrust hypothesis suggests that overthrusting can occur by gravity sliding and/or a push from the rear if the pore pressure approaches that of the overburden, thus lowering the effective normal stress across the potential failure plane. One suggested means of increasing the fluid pressure above ambient is through thermal decomposition of hydrous minerals in a relatively impermeable rock sequence with progressive burial. Gypsum and/or anhydrite have been commonly observed at the sole of many thrusts; abnormally high fluid pressures have been encountered in drilling such evaporites interbedded with salt and clay. Extreme geothermal gradients from the Gulf Coast geosynclinal area place the calculated gypsum $\rightleftharpoons$ anhydrite plus H_2O reaction at depths of 2500–6000 feet, corresponding with estimates of some thrust sheets (*e.g.*, the Jura thrust: 2000–7000 feet).

In a triaxial compression test at 5 kb pressure, we observe a tenfold strength decrease (from 2.7 to 0.25 kb) in sealed polycrystalline gypsum cylinders if the temperature is increased from 100° to 150° C. This marked decrease is interpreted as being due to the dehydration of gypsum with consequent rise in fluid (pore) pressure to a value approaching the confining (overburden) pressure. Microscopic and X-ray observations reveal that gypsum is converted to hemihydrate and/or anhydrite plus water at these temperatures, occasionally with rehydration to gypsum. Similar stress-strain data at 2-kb confining pressure

show identical results but at slightly lower temperatures. Longer preheating periods and a thousand-fold decrease in strain rate (to $3 \cdot 10^{-7}$/sec.) depress this strength-sensitive region further: 80°–130° C. It is expected that over longer equilibration periods (geologic) these gross strength decreases would coincide with the equilibrium gypsum-anhydrite transition.

The effect of dehydration on the strength of gypsum was chosen for investigation partly because of the known occurrence of evaporites along many thrust faults but chiefly because this dehydration was considered to be typical of, but more tractable to laboratory investigation than, the hydrous-anhydrous reactions of many metamorphic rocks. The recent experimental results of Raleigh and Paterson show a large strength decrease of jacketed specimens of serpentinite when they are heated above the dehydration temperature. Essentially the same dehydration mechanism, with consequent increase of fluid pressure and decrease of rock strength, that is postulated for gypsum may help also to explain the occurrence of ultramafic rocks along tectonic boundaries in many mountain belts and the type of deformation that is widespread in pelitic and carbonate-bearing schists. The fact that at least one wave of metamorphism must be syntectonic in order for the mechanism to operate affords a basis for testing the hypothesis.

CONTENTS

INTRODUCTION

The experimental study reported in the first portion of this paper illustrates the mechanical behavior of a rock type common in evaporite sequences of sedimentary rocks, namely poly-crystalline gypsum, in response to temperature changes sufficient for conversion to anhydrite and water. These properties are correlated with the equilibration transition boundary which occurs at crustal depths consistent with observed shallow thrust faulting. The following section documents the common association of evaporites with thrust-faulting. Observed associations of mafic and ultramafic rocks with tectonic boundaries coupled with Raleigh and Paterson's recent results on large strength decreases in serpentine at dehydration further reinforce the implications suggested by the present gypsum study. The argument is then logically extended to expected syntectonic metamorphic effects in the more generally massive pelitic and carbonate sequences.

Previous experiments show that the strength (differential stress, $\sigma_1 - \sigma_3$, as measured at any given strain) and ductility of several different rock types are strongly influenced by the interstitial fluid pressure within the pores (Robinson, 1959, p. 184–193; Heard, 1960, p. 216–223; Handin and others, 1963, p. 724–743). In jacketed triaxial tests, the confining (or lithostatic) pressure and pore (or formation) pressure are externally and independently controlled. Thus, when the ratio of pore pressure to confining pressure, (λ), ranges from about 0.45 to 1.0 (at the appropriate temperature), the test simulates conditions which are known to occur in thick sedimentary sections. All these earlier tests of which the writers are aware have been performed on materials that are physically-chemically stable in the test environment. If the sample is permeable to the pore fluid but the fluid cannot escape and the pore pressure remains constant during the test, then the strength decreases progressively to the standard crushing strength (strength at atmospheric conditions) as λ is changed from 0 to 1.0, consistent with the effective stress concept: the deformational behavior of a given material depends only on the difference between the total stress and the interstitial fluid pressure (Terzaghi, 1936; Hubbert and Rubey, 1959, p. 129–139). For a relatively impermeable sample, the strength may decrease in some other unpredictable fashion as λ increases (Heard, 1960, p. 216–221; Handin and others, 1963, p. 737–739).

In previous tests, the component minerals did not undergo any phase changes, and hence individual grain strength and intergranular cohesion were unchanged. When a deforming material does undergo a phase transformation and at the same time λ is 1.0, it is not *a priori* clear just what the aggregate strength would be. Griggs and Handin (1960, p. 360–361) suggested three changes which might grossly affect mechanical properties: solid-solid, solid-melt, and solid (hydrous)-solid (anhydrous) plus fluid. For a solid-solid reconstructive phase change, the relative strengths between phases have been shown in at least one case to be relatively insensitive to the transformation: calcite-aragonite (Griggs and others, 1960, p. 97–99). Solid-melt transformations are obviously associated with large strength changes for a wide variety of materials. In considering the third type of phase change, one might intuitively expect that the strength differences between solid (hydrous) and solid (anhydrous) plus fluid would be great, the strength of the latter being of the same order as the standard crushing strength of the anhydrous phase if the released water were retained in the immediate sample region ($\lambda = 1.0$). If the fluid escapes ($\lambda < 1.0$), the effects would be less clear. Here, the strength would probably depend on such parameters as confining pressure, λ, grain size, grain cohesion, *etc.*

ACKNOWLEDGMENTS

The authors wish to express their sincere appreciation to David T. Griggs, W. Gary Ernst, and Robert C. Newton for many helpful criticisms and suggestions in the preparation of this paper. Most of the experimental work reported here was carried out at the Institute of Geophysics and Planetary Physics, University of California at Los Angeles, in the laboratory of Prof. D. T. Griggs, and generously supported through his NSF grant G-16195. Heard is indebted to the Shell Development Company for permission to complete this work and to his colleagues at Shell for discussion and support. Rubey is equally indebted to other friends and colleagues, especially to John Rodgers, for criticism and discussion of the concepts involved in this paper.

APPARATUS AND PROCEDURE

All samples were cored from a block of pure, fine-grained Italian alabaster obtained from Wards Natural Science Establishment, New York. The texture can best be described as porphyroblastic with the interlocking matrix an-

294

hedra ranging from 0.01 to 0.05 mm; the few scattered 0.1–0.8-mm porphyroblasts are commonly euhedral, a few poikilitically enclosing 0.01-mm anhydrite crystals (Pl. 1, fig. 1). About one tenth of the block is composed of stringers of 0.1–1.0-mm gypsum crystals. This coarse-grained material was avoided as much as possible in sample coring. Because cursory microscopic examination of thin sections revealed no obvious preferred dimensional or crystallographic orientations, parallel cores were drilled from several slabs randomly sawed from the block. The ends of the 0.35-inch-diameter cores were lapped with 600-mesh grit to form right cylinders approximately 0.70 inch long. Three mutually perpendicular sets of measurements were recorded of the diameter at each end and in the middle. The length was measured, and the sample was sealed in a 0.01-inch-thick annealed copper tube between solid loading pistons. This assembly was then inserted into the high-pressure test cylinder of an adaptable apparatus, capable of tests at widely different but predetermined rates of strain[1] (10^{-1}–10^{-8}/sec.). This apparatus and test procedures have been described in detail earlier (Heard, 1962, Ph.D. dissert., Univ. California, Los Angeles, p. 12–42; 1963, p. 165–175) and will not be treated further here.

Jacketed alabaster samples were axially compressed at both 2 and 5 kb confining pressure at average axial strain rates of $3.3 \cdot 10^{-4}$ and $3.4 \cdot 10^{-7}$/sec. at different temperatures. Samples were held at test temperatures and pressures for several periods to explore the degree of gypsum dehydration. In all tests, the confining pressure was applied before the sample was heated; preheating to pressure-temperature equilibrium took less than 0.5 hour. After the test, the axial load was removed and the sample quenched at pressure; cooling times (to room temperature) were at most 2.0 hours for the highest temperature tests.

The force borne by the annealed copper jacket can usually be disregarded (*e.g.*, Griggs and Miller, 1951, p. 857; Handin and Hager, 1957, p. 11; Heard, 1960, p. 196–197) because the correction is well within experimental reproducibility. However, for the 0.35-inch-diameter samples and 0.01-inch-thick annealed copper jackets used here, the correction becomes larger than 3 per cent (about equal to the reproducibility: *see* Table 1) for ratios of sample strength to copper strength of less than 3.8. Annealed copper test cylinders were compressed at two confining pressures and strain rates and at several temperatures corresponding to the range of test conditions in Table 1. In all tests, ratios of the alabaster strength to copper strength (measured at the yield stress[2]) were less than 3.8. The necessary correction (at the yield stress) ranged from 4 per cent for the strongest samples (Experiments 333, 334) to 30 per cent for the weakest (Experiments 361, 362). Corrections at strains larger than that corresponding to the yield stress were much greater for the weaker samples because of work hardening in the copper and decrease in sample strength. All stress-strain data reported in Table 1 have been corrected for jacket strength.

Calculations of true stress difference were based on initial sample cross-sectional area and its computed change with progressive strain under the assumptions that deformation was homogeneous and no volume changes occurred. The latter was not strictly correct for those samples that showed evidence of dehydration because the anhydrite plus water reaction involves a volume increase of from 5 to 10 per cent[3]. This implies that the cross-sectional area increased up to 6 per cent (depending on the pressure, temperature, and completeness of transformation) and hence the stress differences (Table 1) would be too high by that amount. However, the mineral phases involved here are not only anhydrite but metastable hemihydrate (*see* footnote 4), and expected volume increases would be somewhat less. No correction was made because even for the very weak samples involving the largest effect, it was only a small fraction of the jacket force and was probably within the experimental uncertainty. The values of corrected stress difference (Table 1) are accurate to ±2 per cent or 35 bars, whichever is greater. Confining pressure and temperature

---

[1] Strain rate, as used here, is the change in length divided by the initial length for unit time ($\Delta L./LT$). Although it is not strictly correct, $\dot{\epsilon}$ (Figs. 1–5) is used synonymously with this definition. Exactly, $\dot{\epsilon}$ is the first derivative of strain with respect to time, $d\epsilon/dt$.

[2] Yield stress, as used in this paper, includes both the stress difference at the knee of the stress-strain curve for materials exhibiting work hardening (and hence is not always definite) and the ultimate stress difference for materials possessing little permanent strain before faulting. For a more complete discussion of terms used in triaxial tests on rocks, *see* Handin and Hager (1957, p. 3–5).

[3] From 5.0 per cent at 77 degrees and 5 kb (Experiment 394) to 9.8 per cent at 138 degrees and 2 kb (Experiment 367) (Sharp, 1962, p. 12, 14, 15)

are known to 0.5 per cent and $\pm 2°$ C, respectively; strain rates were calculated near 10 per cent strain and are accurate to ± 1.5 per cent.

EXPERIMENTAL RESULTS

Figure 1 shows a series of stress-strain curves for the Italian alabaster deformed at different temperatures and 5-kb confining pressure. Each curve represents the average of two tests. Each sample was held at its appropriate test temperature for 0.5 hour and then axially compressed at a strain rate of $3.3 \cdot 10^{-4}$/sec. These curves show large reductions in strength over a relatively narrow temperature range ($100°$–$150°$ C) which are attributed to the transformation of gypsum to metastable hemihydrate[4]-anhydrite plus water with the consequent rise of λ to 1.0. All samples tested up to $101°$ C were not visibly wet when removed from their jackets, nor did thin sections[5] disclose any evidence of dehydration. Figure 2 of Plate 1 (Experiment 338—$101°$ C) shows no hemihydrate nor any textural evidence suggesting a phase transformation. In-

---

[4] X-ray observations definitely establish that hemihydrate is present from one of its principal intensity peaks at $14.8° 2\theta$, which may not be confused with any intense peak of gypsum, anhydrite or other known possible phase. Small amounts of anhydrite cannot be identified with certainty by X ray in the presence of appreciable amounts of hemihydrate because of peak interference at similar 2θ values.

N. L. Carter has determined the optical properties of this subhydrate, giving the following properties: uniaxial positive, maximum birefringence of 0.024, high index, positive elongation, and parallel extinction. The data are consistent only with the reported properties of hemihydrate (bassanite or "soluble anhydrite." Palache and others, 1951, p. 476, 484). He reports that at temperatures in excess of $152°$ C, appreciable amounts of anhydrite are present in samples deformed at $177°$ and $250°$ C (e.g., Experiments 399, 368, 393).

On the time scale we are exploring here, around 100 hours at most, gypsum passes quickly to hemihydrate plus water (metastably) at $90°$ to $177°$ C in our pressure range and inverts very slowly to the stable anhydrite. Over longer durations, and thus at "geologic" times, the stable inversion of gypsum $\rightleftharpoons$ anhydrite plus water would be expected. We believe that although the strength-sensitive region explored in these experiments is associated primarily with gypsum $\rightleftharpoons$ hemihydrate plus water, the demonstrated strength behavior would hold equally well for the mineral pair gypsum-anhydrite in nature at the equilibrium boundary.

[5] All thin sections were prepared using a cold setting cement, Epon 815. Standard thin-section techniques requiring heating of the rock chip were observed to produce various amounts of unidentified subhydrate(s) from the gypsum during thin-section preparation.

tragranular flow features (undulatory extinction, kink bands) mark the only obvious differences between Figure 2 of Plate 1 and the starting material shown on Figure 1 of Plate 1.

Samples tested at $125°$ C were damp when disassembled although no hemihydrate was detected microscopically. At $138°$ C, samples were damp after the test, and small amounts of gypsum pseudomorphous after hemihydrate as well as hemihydrate were identified in the thin section. At $152°$ C and higher, excess water was noted in the jacket after each experiment, and varying proportions of hemihydrate, anhydrite, and gypsum pseudomorphous after hemihydrate were noted, both optically and by X-ray patterns. Figures 3 and 4 of Plate 1 (Experiment 340—$152°$ C) show the total obliteration of the original texture (compare with Figure 1 of Plate 1). The equant anhedral gypsum has been completely replaced by radiating clusters of bladed hemihydrate, some of which have reacted with the water upon cooling to give gypsum in this habit. Where excess water was apparent, the sample appeared as a damp friable paste with slight cohesion. Prominent narrow fault zones were seen in nearly all samples tested at $125°$ C and above although complete loss of cohesion never occurred.

Note that the aggregate strength does not become vanishingly small at the higher temperatures ($> 152°$) but remains at a low value (Fig. 1); evidently internal friction is still appreciable even when as much as 48 per cent of the sample volume is water. Comparison of the strength of samples tested above $150°$ C (where $\lambda = 1.0$) with the crushing strength of Blaine anhydrite, 1270 bars[6] (Handin and Hager, 1957, p. 40), shows the anhydrite plus water paste to be weaker by a factor of 5 to 6. The strength of these samples is also lower by 2 to 3 compared to the crushing strength (540 bars) for Ohio alabaster (Griggs, 1940, Fig. 3, p. 1010). When a single crystal of a mineral undergoes a reconstructive transformation as is the case for gypsum dehydrating to anhydrite plus fluid, most chemical bonds must be broken, and thus its strength must approach zero. In an aggregate of such crystals, the process would not only affect individual crystal strength but would also disrupt intergranular cohesion, so that in polycrystalline gypsum, the strength could range from the crushing strength of a nearly pure, co-

---

[6] Measured at $24°$ C. An increase in temperature to $150°$–$200°$ C would not be expected to alter this value appreciably.

Figure 1. Photomicrograph of undeformed alabaster. Crossed nicols

Figure 2. Photomicrograph of sample after Experiment 338 (101°C at a strain rate of $3.3 \cdot 10^{-4}$/sec., 0.5 hour initial heating time, 10 per cent strain), crossed nicols. Note similarity to Figure 1 of this plate except for small proportion of gypsum grains showing undulatory extinction, kink bands. No dehydration of gypsum is present.

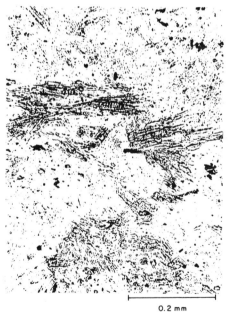

Figure 3. Photomicrograph of sample after Experiment 340 (152°C at a strain rate of $3.3 \cdot 10^{-4}$/sec., 0.5 hour initial heating time, 9.6 per cent strain), crossed nicols. Note textural change from Figure 1 (or Figure 2) of this plate. Thin section consists of about one tenth equant gypsum, two thirds gypsum pseudomorphous after hemihydrate (radiating, bladed aggregates), and one fourth hemihydrate (from original equant gypsum)

Figure 4. Same as Figure 3 of this plate, uncrossed nicols. Scattered high-index bladed material is hemihydrate.

PHOTOMICROGRAPHS OF ITALIAN ALABASTER
BEFORE AND AFTER EXPERIMENTS

TABLE 1. SUMMARY OF EXPERIMENTS

All samples heated at indicated temperature, confining pressure for about 0.5 hour before differential stress was applied, unless otherwise noted.

Experiment number	Temperature (°C)	Confining pressure (kb)	Strain rate (sec.$^{-1}$)	Differential stress (in bars) at				Mineral phases present after test
				2 per cent strain	5 per cent strain	10 per cent strain	Yield point	
333	26	5.00	$3.1 \cdot 10^{-4}$	2660	2940	3150*	2740	..
334	26	5.00	$3.1 \cdot 10^{-4}$	2640	2960	3220*	2840	..
396	26	5.00	$3.4 \cdot 10^{-7}$	2150	2720	2940	2500	..
363	26	2.00	$3.2 \cdot 10^{-4}$	2220	2400	2380	2430	..
335	77	5.00	$3.1 \cdot 10^{-4}$	2600	2930	3160†	2710	G ..
336	77	5.00	$3.1 \cdot 10^{-4}$	2650	2970	3120†	2760	..
394	77	5.00	$3.5 \cdot 10^{-7}$	2140	1860	1470	2300	G ..
370	77	2.00	$3.2 \cdot 10^{-4}$	2040	2270	2320	2180	..
398	90	5.00	$3.6 \cdot 10^{-7}$	1430	920		1450	G ..
338	101	5.00	$3.2 \cdot 10^{-4}$	2380	2780	2800	2640	G ..
357	101	5.00	$3.3 \cdot 10^{-4}$	2340	2710	2580	2610	G ..
373**	101	5.00	$3.2 \cdot 10^{-4}$	2180	2770	2800	2580	G ..
395	101	5.00	$3.4 \cdot 10^{-7}$	1060	580	500	1110	G ..
365	101	2.00	$3.2 \cdot 10^{-4}$	1930	2160	2220	2050	..
392	114	2.00	$3.2 \cdot 10^{-4}$	1860	1960	1840	1900	G ..
339	125	5.00	$3.4 \cdot 10^{-4}$	2100	2080	1680	2260	..
356	125	5.00	$3.5 \cdot 10^{-4}$	1880	2020	1430	2220	G ..
371**	125	5.00	$3.3 \cdot 10^{-4}$	1500	1730	1470	1740	G ..
397	125	5.00	$3.4 \cdot 10^{-7}$	360	220	80	500	G + H
369	125	2.00	$3.3 \cdot 10^{-4}$	1130	1180	890	1230	G + H
342	138	5.00	$3.3 \cdot 10^{-4}$	1850	1290	890	1850	G + H
358	138	5.00	$3.4 \cdot 10^{-4}$	1620	1170	640	1720	..
372**	138	5.00	$3.3 \cdot 10^{-4}$	230	200	130	240	G + H
367	138	2.00	$3.3 \cdot 10^{-4}$	170	140	100	180	..
340	152	5.00	$3.2 \cdot 10^{-4}$	310	330	290	280	G + H
359	152	5.00	$3.2 \cdot 10^{-4}$	300	290	230	270	G + H
341	177	5.00	$3.3 \cdot 10^{-4}$	220	260	240	300	G + H
360	177	5.00	$3.3 \cdot 10^{-4}$	220	170	140	250	G + H
399	177	5.00	$3.3 \cdot 10^{-7}$	60	50		70	G + H + A
368	177	2.00	$3.3 \cdot 10^{-4}$	170	150	90	180	.. H + A
361	250	5.00	$3.3 \cdot 10^{-4}$	160	140	120	170	..
362	250	5.00	$3.3 \cdot 10^{-4}$	140	120	60	170	..
393**	250	5.00	$3.3 \cdot 10^{-4}$	100	60	20	100	.. A

* Extrapolated value from 8.0 per cent strain
† Extrapolated value from 6.9 per cent strain
** Heated for 25 hours at indicated temperature, confining pressure before differential stress was applied.

G = gypsum
H = hemihydrate } Determined by X ray and from thin section
A = anhydrite

herent gypsum aggregate, through zero, to that of a secondary anhydrite aggregate of variable grain size and cohesion[7]. This could result in a

[7] Intergranular cohesion as well as grain size can grossly affect the mechanical properties: compare the well-cemented Oil Creek sandstone (Handin and Hager, 1957, p. 19–23) with its disaggregated equivalent St. Peter sand (Borg and others, 1960, p. 144–147). Also compare the mechanically isotropic Wombeyan marble and Solenhofen limestone, which have a grain size of ~ 1 μ and $5 \cdot 10^{-3}$ mm, respectively (Paterson, 1958, p. 466–468; Heard, 1960, p. 199–203).

lowered initial shear strength τ_0 for the Mohr failure criterion, $\tau = \tau_0 + \sigma \tan \phi$, where τ and σ are the shear and effective normal stresses, respectively, across a potential failure plane and $\tan \phi$ is the coefficient of internal friction. Thus, we can argue that τ_0 may be small; one does not need necessarily to regard a fault propagating as a dislocation and assume therefore, that τ_0 is negligible compared to τ (as did Hubbert and Rubey, 1959, p. 125).

Handin and others (1963, p. 743–745) and Heard (1960, p. 219–221) also noted that their results for several rocks could not be explained by the effective stress concept. Handin and others (1963, p. 753) point out that this concept does adequately explain the mechanical behavior of several different jacketed rocks tested in the laboratory, provided that three conditions are satisfied: (1) the pore fluid is chemically inert with regard to the rock; (2) the rock is an aggregate with interconnected pore spaces such that the pore pressure is transmitted throughout the solid phases; and (3) the permeability is large enough to allow pervasion of the fluid within the jacketed sample and to insure that pore pressure is constant and uniform throughout the rock during the time of the test. Absence of conditions (2) and (3) could account for the results of Handin and others and Heard. All three conditions may not be met for the hemihydrate-anhydrite plus water paste discussed here.

Similar stress-strain data have been gathered at 2-kb confining pressure and otherwise identical test conditions over a comparable temperature range (Fig. 2). Comparison with Figure 1 shows a very similar pattern. Minor differences are a slightly lower strength at temperatures up to 100° C and a large decrease in strength in a slightly lower-temperature region than was noted at 5 kb (this lower temperature would be expected if the large strength decrease were related to the gypsum-anhydrite [hemihydrate] plus water transformation). The samples compressed at 2 kb yielded similar evidence of a phase transformation; the onset of dehydration (as deduced from X ray, thin section, and the presence of water after the test) occurred at about 125 degrees. In order to compare these data, Figure 3 was prepared with the yield stress (as taken from the stress-strain curves) plotted as a function of the test temperature. These curves emphasize the large decrease in strength (by a factor of 10) for both series of tests.

It was recognized early in this work that the 0.5-hour preheating period probably would not be long enough for the attainment of phase equilibrium in the sample. Consequently, several tests were carried out at 5 kb with a 25-hour heating period before the compression was started—all other conditions remained the same. These results, also plotted in Figure 3, suggest that even the metastable phase equilibrium was not attained during the shorter heating period because the temperature region for the strength transition is lowered about 15 degrees. For longer heating periods (as in nature), the strength transition could be expected

to occur at an even lower temperature. From thermochemical data in Kelley and others (1941, p. 43–44), MacDonald (1953, p. 886–887) has calculated the equilibrium boundary between gypsum-anhydrite plus water to 500 bars. Using Posnjak's (1938, p. 268) empirical measurement for the boundary at 1 bar (42° C), MacDonald's calculated slope dP/dT (85.4

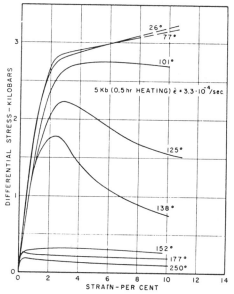

Figure 1.   Stress-strain curves for jacketed Italian alabaster compressed at 5 kb confining pressure and at a strain rate of $3.3 \cdot 10^{-4}$/sec., different temperatures. Each curve is the average of two duplicate tests which have been corrected for jacket strength. All tests maintained at pressure and at indicated temperature for 0.5 hour before sample loaded.

bars/° C), and assuming a linear extrapolation to 5000 bars (which is not strictly valid), one may predict the equilibrium boundary at 65° C and 101° C at 2 and 5 kb, respectively. Zen (1965, p. 144–146, 153–160) re-examined the thermochemical data of Kelley and others with later revised measurements of Kelley (1960, p. 46) and calculated somewhat different values: 46° C for the boundary at 1 bar and a slope dP/dT of 71 bars/° C. The equilibrium boundary based on Zen's intercept and slope would be 74° C at 2 kb and 116° C at 5 kb. Either pair of calculated temperatures can be regarded as upper temperature limits, for dP/dT is expected to increase at the higher pressures. One should be cautious, however, in such extrapolations as no equilibrium data are available on the hemihydrate stability field. Preliminary measurements by G. C. Kennedy (1963, personal communication) suggest that an equilibrium point between gypsum, hemihydrate, anhydrite, and H_2O occurs near 6 kb and 130° C, with the hemihydrate field widening at higher pressures—some-

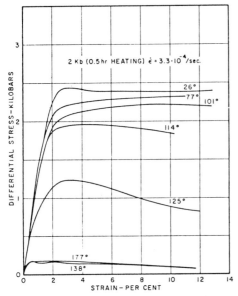

Figure 2.   Stress-strain curves for jacketed Italian alabaster compressed at 2 kb confining pressure and at a strain rate of $3.3 \cdot 10^{-4}$/sec., different temperatures. Each test maintained at pressure and at indicated temperature for 0.5 hour before sample loaded; corrected for jacket strength.

what above the region of interest here. Zen (1965, Fig. 14) calculates this quadruple point to be near 170° C and 9 kb.

Efforts to determine empirically the gypsum-anhydrite equilibrium boundary at 2 kb in conventional hydrothermal apparatus were unsuccessful. X-ray studies of reagent grade $CaSO_4 \cdot 2 H_2O$ and $CaSO_4$ (in the gypsum and anhydrite phases) and intermediate mixtures of the two exposed to both distilled water and a 0.6 weight per cent solution of NaCl (the latter to promote reaction) show no reaction at temperatures to 110° C for durations up to 116 hours.

All tests discussed so far were at a strain rate of $3 \cdot 10^{-4}$/sec., a 10^9 to 10^{10}-fold higher rate

than has been observed (angular distortion 0.1 sec./year; displacements 0.4 to 2 inches/year) from measurements along the San Andreas and other faults in California (Whitten, 1955; 1956; Crowell, 1962, p. 50). Because Heard had recognized in an earlier study (1962, Ph.D. dissert., Univ. California, Los Angeles) that large changes in strain rate can result in enormous changes in the mechanical properties in rocks, the next step was to evaluate the effect of strain rate on strength associated with this phase transformation. Figure 4 illustrates the behavior of the alabaster at a strain rate of $3.4 \cdot 10^{-7}$/sec. (similar to tests of Figure 1 except for strain rate). Comparison of Figures 1 and 4 (and the replotted yield stress data in Figure 3) shows similar large changes in strength over a 50-degree temperature range, but the temperature interval has been lowered about 20 degrees by the thousandfold decrease in strain rate. As already noted for the three series of deformation tests discussed, the onset of gypsum breakdown occurs midway through the strength transition zone: 101 degrees in this case. Extrapolation of the trend from 10^{-4} and 10^{-7}/sec. tests suggests that the strength transition would occur at still lower temperatures at natural strain rates, presumably at the equilibrium gypsum-anhydrite phase boundary.

DISCUSSION

Geologic Significance of Experimental Results

Hubbert and Rubey (1959, p. 142–149) and Rubey and Hubbert (1959, p. 185–200) advance the hypothesis that overthrusting of thin sedimentary sheets over long distances may be accomplished by increase of the value of λ, then sliding of the sheet by a push from the rear or by body forces acting along gentle slopes or by some combination of the two. Rubey and Hubbert (1959, p. 170, 184, 185, 192, 201) suggest that values of λ approaching 1.0 may be due to: (1) sedimentary loading of impermeable strata, (2) compressive stresses of tectonic origin, (3) breakdown of hydrous minerals by heating, or (4) melting of the eutectic fraction of a rock body at essentially lithostatic pressure. The results on gypsum-anhydrite (Heard and Rubey, 1964; this paper) provide the first empirical data to show that the strength of at least one hydrous-anhydrous mineral pair is decreased by a large factor at the transition temperature if λ is 1.0. If the released water is not free to escape, similar behavior would almost certainly be expected when other hydrous minerals—the "clay" family, zeolites, micas, amphiboles, etc. —are heated through their respective dehydration temperatures.[8]

The relationships of pressure and depth to temperature for two extreme geothermal gradients in the present-day Gulf Coast geosyncline are compared with MacDonald's (1953, p. 888) and Zen's (1965, p. 161) boundary between gypsum and anhydrite (Fig. 5). Curve A represents a rather high gradient of 36° C/km, measured in the East Texas onshore area; curve B is a rather low gradient (22° C/km) from the Louisiana offshore area. Curves D and E are the boundaries of the temperature-sensitive strength region for tests at a strain rate of $3 \cdot 10^{-4}$/second at 2 and 5 kb respectively (*from* Figs. 1–3). Curves F and G mark the similar transition region at $3 \cdot 10^{-7}$/second, 5 kb (*from* Fig. 3) under the assumption that the boundaries parallel D and E. As the strain rate decreases, or the preheating time of the gypsum increases, this strength transition region would progressively migrate up the geothermal gradient (somewhere between A and B) toward the surface until it coincided with MacDonald's or Zen's equilibrium boundary, curves C_1 or C_2. While undergoing burial during natural sedimentation, a layer of gypsum deposited within a suitably impermeable sequence travels down some gradient between A and B. When the gypsum-anhydrite transition is reached, λ would approach 1.0, and sliding could occur as shallow as 2000 feet or as deep as 5000 feet (C_1) or 2500 feet to 6000 feet (C_2).

As a comparison, Laubscher (1961, p. 228, 244–245, and Fig. 1) gives the range of thickness of the Jura thrust sheet as 2000–7000 feet. The same argument could equally well apply to sedimentary sequences containing layers of other hydrous minerals where transitions would then be expected at other temperatures and depths.

It is important to recognize that high fluid pressures would not develop from the dehydration of gypsum unless the rate of escape of the released water were significantly lower than the rate at which the gypsum changed to anhydrite. The rate at which this transition runs in nature doubtless depends upon many factors, but it seems unlikely that abnormal fluid pressures can build up where the associated and enclosing rocks are highly permeable, as are many sand-

[8] These expectations have been confirmed for serpentinite by Raleigh and Paterson (1965, p. 3965–3985) and later by Riecker and Rooney (1966, p. 196–198).

stones and highly fractured rocks, and where relatively open channelways extend upward or laterally to the ground surface. On the other hand, where rocks of very low permeability, such as some clay shales, limestones, dolomites, slates, schists, or gneisses, make up a significant part of the rock column, the released water can escape only very slowly and, as a result, high fluid pressures would commonly develop and

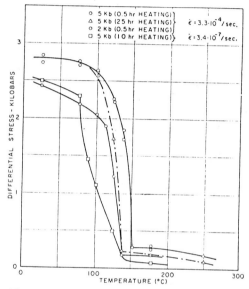

Figure 3.   Differential stress at yield stress *vs* temperature for Italian alabaster compressed at indicated conditions. Replotted from Figures 1, 2, and 4 and Table 1

persist long after the hydrous-anhydrous reaction had gone to completion. It is to be noted that high fluid pressures could develop and persist in a highly permeable sandstone that is enclosed within rocks of low permeability. Such enclosure could be the result either of normal sedimentary deposition of fine-grained clay rocks above the sandstone or of the tectonic superposition above the sandstone of a thrust plate composed of impermeable rocks.

A numerical example will serve to illustrate the role that rocks of low permeability could play in the development of high fluid pressures. Assume that a bed of gypsum 50 feet thick, buried to a depth of 4000 feet and at a temperature of 50° C, undergoes a transition to anhydrite and water. Assume also that the 4000-foot column of overlying rock with its pores filled with water has a mean specific weight of 1.0 lb/in². ft (approximately 2.3 gm/cm³), that it contains beds of clay shale so distributed through the column that interstitial water cannot readily escape upward or laterally, and that the lowest such shale bed has a permeability of 10^{-6} millidarcys (probably about mid-range in the values of permeability of clay shale: Rubey

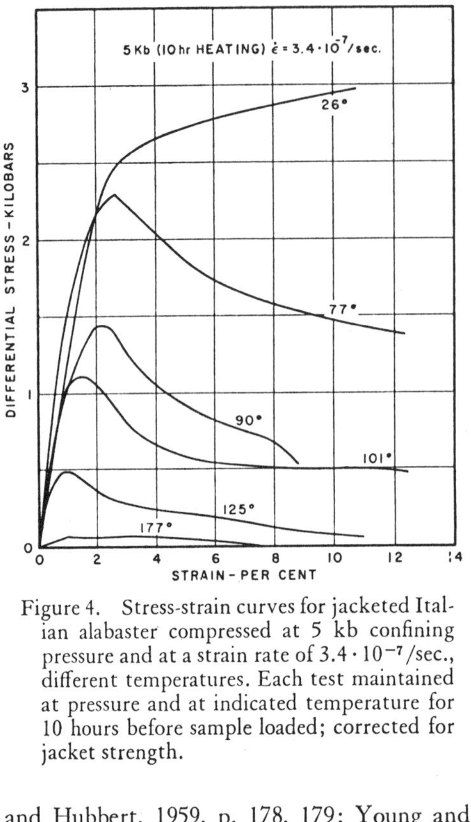

Figure 4. Stress-strain curves for jacketed Italian alabaster compressed at 5 kb confining pressure and at a strain rate of $3.4 \cdot 10^{-7}$/sec., different temperatures. Each test maintained at pressure and at indicated temperature for 10 hours before sample loaded; corrected for jacket strength.

and Hubbert, 1959, p. 178, 179; Young and others, 1964, p. 4239). At this pressure and temperature, gypsum releases 48.5 per cent of its volume as water when it is converted to anhydrite. Inasmuch as this released water cannot escape readily, it comes to support approximately the full weight of the overburden which in our example is 4000 lbs/in² plus atmospheric pressure at the surface.

Under these circumstances, the hydraulic gradient from the evaporite bed undergoing dehydration to the top of the overlying rocks is

$$\frac{p_2 - p_1}{\gamma w(z_2 - z_1)} - 1 = \frac{4014.7 - 14.7}{0.434(4000-0)} - 1,$$

and the volume of flow of interstitial fluid past

a unit cross section in unit time is (Rubey and Hubbert, 1959, p. 172, 178)

$$q = K\left(\frac{p_2 - p_1}{\gamma w(z_2 - z_1)} - 1\right),$$

where

$K = \rho_w/\mu$ coefficient of permeability = 1.26ψ ft/yr (estimated value for average of 35° C and 140 bars)

ψ = number of millidarcys = 10^{-6}

p_2 = fluid pressure at top of evaporite bed = 4014.7 lb/in²

p_1 = atmospheric pressure at ground surface = 14.7 lb/in²

γw = specific weight of water = 0.434 lb/in²·ft (equivalent to 1.0 gm/cm³)

z_2 = depth to top of evaporite bed = 4000 ft

z_1 = depth at surface = 0

$$q = 1.26 \times 10^{-6}\left(\frac{4014.7 - 14.7}{0.434(4000-0)} - 1\right)$$

$$= 1.26 \times 10^{-6}(2.304 - 1) = 1.64 \times 10^{-6} \text{ ft/yr}.$$

The volume of water released over t years is $0.485 \times 50 = 24.25$ ft per unit cross section. That is,

$$q = 1.64 \times 10^{-6} \text{ ft/yr} = \frac{24.25 \text{ ft}}{t \text{ yrs}} \text{ and}$$

$$t = \frac{24.25}{1.64 \times 10^{-6}} = 14.8 \times 10^6 \text{ yrs}.$$

Under the assumed conditions, it would require approximately 15 million years for escape of the water released in conversion of the gypsum to anhydrite.

Evaporites Along Thrust Zones

Laubscher has given renewed interest to the early interpretation of Buxtorf (1908, p. 103–106, 108, 113–114) that evaporites, notably salt and gypsum, are quite common along and have played an essential role in the sliding of the Jura thrust zone. Laubscher suggested that not only might the salt be very weak at normal pore pressures under long-term loading and hence act as the active gliding element itself, but that λ could be increased by converting gypsum to anhydrite plus water (1961, p. 248–250) and sustained at a high value by trapping the water in the weak impermeable saline sequence, thereby decreasing the shear stress necessary for movement on the thrust sole. Laubscher (1961, p. 246) calculated the maximum permissible

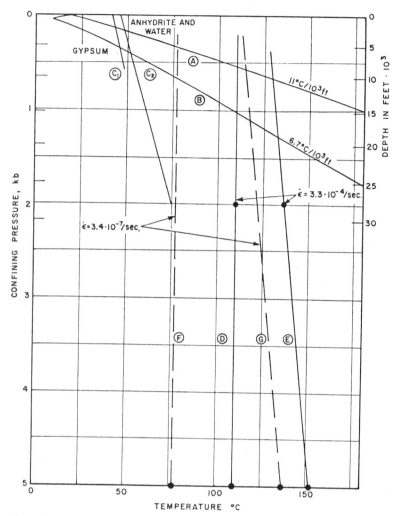

Figure 5.   Confining pressure *vs* temperature showing extremes in Gulf Coast. Curve *A*—36° C/km, onshore Texas. Curve *B*—22° C/km, offshore Louisiana (based on temperature data from Nichols, 1947, p. 46; plot of pressure *vs.* depth *by* Dickinson, 1953, p. 430; and temperature data *from* Moses, 1961). Curve C_1 is an extrapolation of MacDonald's (1953) calculated gypsum-anhydrite boundary; C_2 is Zen's (1965) calculated gypsum-anhydrite boundary. Curves *D* and *E* represent the beginning and completion of the transitional strength at $3.3 \cdot 10^{-4}$/sec. from Figures 1–3. Curves *F* and *G* show the similar strength transition region at $3.4 \cdot 10^{-7}$/sec., assuming the boundaries to be parallel to *D* and *E*.

shear stress of the material at the sole of the Jura thrust at 30–90 bars. The measured shear stress in the anhydrite plus water paste ranges from 50 bars for the fast tests ($3 \cdot 10^{-4}$/sec.) to 35 bars for the slow tests ($3 \cdot 10^{-7}$/sec.); shear stresses at natural strain rates can be expected to be even lower.

In many other European localities besides the Jura Mountains, gypsum or anhydrite and salt occur along the zones of movement of overthrust faults or at the lower boundaries of nappes. In the East Limestone Alps of Austria (Suess, 1909, p. 178, 221); the Bergamasc Alps of Italy (de Sitter and de Sitter-Koomans, 1949, p. 131–132); the Alps of eastern Switzerland (Cadisch, 1941, Pl. 1, profile 1, p. 31; Kläy,

1957, Pl. 1; Schindler, 1959, p. 59–60); the French Alps (Termier, 1927, p. 75–76; Gignoux, 1930, p. 350–358; Schneegans, 1938, p. 43, 44, 304; Gignoux and Moret, 1952, p. 63, 64, 206, 215, 308; Goguel, 1953, p. 11, 95, 97; Trümpy, 1960, p. 851); and the Pyrenees (Ashauer, 1934, Pl. 3; Misch, 1934, Pl. 2; de Sitter, 1956, p. 279) contorted beds of gypsum and anhydrite or the collapse-breccias (cornieules, cargneules, carñiolas, Rauhwacken) of dolomite, limestone, red siltstone, and sandstone left by the solution of such evaporites (Brückner, 1941) crop out, in some areas continuously for many miles, along major tectonic boundaries. The same relationships have also been observed in the Mediterranean Atlas Mountains in north Africa (Termier, 1906; Suess, 1909, p. 221; Gignoux, 1930, p. 343–351; van de Fliert, 1953, p. 71, Pls. 3, 4); in and near the Zagros Mountains of Iran (Busk, 1929, p. 79, 85, 86, 88, 95; de Böckh and others, 1929, p. 81–85; Harrison, 1930, p. 510; Harrison and Falcon, 1936, p. 96, 97; Kent, 1958, p. 2963); in the Salt Range of India (de Böckh and others, 1929, p. 81–85; Wadia, 1947; Krishnan, 1960, p. 235–239); and in Guatemala (de Böckh and others, 1929, p. 161, 164).

It is interesting that a similar association of evaporites and thrust faults appears not to have been noted so widely or at least not reported so frequently in North America. A few somewhat incidental references in the geologic literature suggest, however, that the association may be more common than appears at first glance.

Drilling in the past decade has shown that many thrust faults in the central Appalachians follow bedding surfaces at three general stratigraphic levels—at two predominantly argillaceous zones in the Cambrian and the Ordovician and, very widely in West Virginia and western Pennsylvania, at the level of the Upper Silurian Salina evaporite sequence (Rodgers, 1963, p. 1533–1535; Gwinn, 1964, p. 864–865, 870, 875–891). More recently, wells in West Virginia and southwestern Virginia have shown the presence of anhydrite in the lowermost of the three slide zones—the Cambrian argillaceous sequence (Rodgers, 1964, p. 78).

Longwell (1949) has described numerous blocks, chiefly of Kaibab Limestone, that range from a few tens of feet to 1 mile or more in greatest dimension that have been caught up and dragged along beneath the Glendale thrust fault in the northern part of the Muddy Mountains, Nevada. Movement of the blocks appears to have been localized along beds of gypsum in the lower part of the Moenkopi Formation and within the Kaibab Limestone (Longwell, 1949, p. 931, 949–950).

Mackin (1960, p. 100) has reported a décollement thrust fault in the Iron Springs district in southwestern Utah in which the zone of gliding is a gypsiferous unit in the lower part of the Carmel Formation immediately overlying the Navajo Sandstone.

Some statistics regarding one region of thrust faults may possibly be typical of others. In an area of approximately 2500 square miles in western Wyoming mapped geologically by Rubey, seven major thrust faults crop out over a total trace length of about 255 miles. Although these thrusts tend to approximate closely to bedding-plane faults, the lowest formation exposed above the soles changes from place to place. For a total linear distance of 78 miles, however, the soles of various thrust plates in this area are formed by rocks that are either within the Amsden Formation or at the base of the immediately overlying Wells Formation. From exposures of numerous layers of collapse-breccias within it and from the direct evidence of wells drilled into it, the Amsden Formation is known to contain numerous beds of gypsum or anhydrite. Furthermore, for a total length of 54 miles, the soles of the thrusts in this area are formed by rocks at three other stratigraphic levels—the middle of the Brazer Limestone, the Gypsum Spring Member of the Twin Creek Limestone, and the Preuss Redbeds—that are known from nearby outcrops or from well data to contain evaporites—beds of gypsum, anhydrite, or rock salt. Thus, for approximately half the total length of their outcrop, the thrust faults in this region follow stratigraphic levels that are characterized by deposits of evaporites, chiefly gypsum or anhydrite. It is equally important, however, to note that for the other half of their length these faults follow stratigraphic levels characterized not by evaporites but chiefly by argillaceous sequences.

The rather widespread occurrence of gypsum and, less commonly, of rock salt along zones of tectonic movement has been explained most commonly as the result of these rocks being highly "plastic" (weak and very ductile) and thus serving as lubricating materials that facilitate fault movement. However, abnormally high fluid pressures have been encountered in drilling sequences of anhydrite interbedded with impermeable salt and clay in the Permian of Germany, the Oligocene of Iran, and evaporite sequences elsewhere (Lane, 1949;

Mostofi and Gansser, 1957, p. 81, 82; Hubbert and Rubey, 1959, p. 155–156; Thomeer and Bottema, 1961, p. 1725–1727). It seems permissible, therefore, to suggest an alternative explanation of this widespread occurrence in terms of the concept underlying the present article. Perhaps the common association of evaporites with thrust faults may be interpreted as the result of (1) the low strength of salt and associated red clays which, when the rocks are progressively buried, tends to produce very low porosity and thus very low permeability, and (2) the gypsum-anhydrite transition which releases water that, with low boundary permeability, cannot escape and consequently develops high fluid pressures and a zone of very low strength along which faults readily develop.

Progressive Metamorphism of Serpentinites, Pelitic Rocks, and Carbonate Rocks

The gypsum-anhydrite transition was chosen for this study partly because of the known occurrence of evaporites along thrust zones but more especially because it was considered to be an example, more tractable than others to laboratory investigation, of the hydrous to anhydrous transitions that characterize many metamorphic reactions. Hence, the marked weakening of this rock if the water released on its dehydration cannot readily escape may have tectonic implications that go considerably beyond the occurrence of evaporites *per se* along thrust faults. It is of interest to indicate briefly some of these other possible implications.

The association of evaporites with zones of fault movement has been observed more frequently around the outer margins of folded mountain ranges than in their more intensely deformed or metamorphosed interiors. A somewhat similar association involving a different rock type, however, has been widely noted within the inner and medial chains of many folded mountain ranges. Suess (1909, p. 565) called attention to the tendency for serpentinite, peridotite, and other ultramafic and mafic rocks to occur as ". . . sills in dislocated mountains, which sometimes follow the bedding planes and at others the planes of movement." Benson (1927), in a monograph on the tectonic relationships, of mafic and ultramafic rocks, noted that the alpine type of serpentinites, peridotites, and gabbros occurs in regions that have been intensely disturbed by overthrusting and alpine orogeny and that rocks of this type tend to occur as concordant sills emplaced dur-

ing deformation (p. 6, 69). Bucher (1933, p. 272) suggested that the frequent association of these ultramafic and mafic rocks (or ophiolites as they are often called) with thrust plates may perhaps be explained ". . . by the weakness and lubricating behavior of serpentinized basic rocks within a part of the crust undergoing deformation." Laboratory investigations have shown that serpentinite is not in fact a weak rock but that it has a strength as great as that of most other rocks (Raleigh and Paterson, 1965, p. 3981–3982); the association that Bucher noted must have some other explanation. Hess (1939, p. 263, 264, 269–270), in a review of the serpentinized peridotites of the strongly deformed belts of the world, concluded that these rocks were intruded during compression, perhaps as "cold" intrusions, into fold axes and along bedding or foliation planes. Turner and Verhoogen (1951, p. 241) note that ". . . major intrusions of peridotite and serpentinite tend to be located along zones of strong dislocation or at least to be bounded on one or both sides by faults of great magnitude."

Ultramafic intrusive masses or injections of this type have been recorded along thrust planes and major tectonic boundaries in the Caledonides of Norway (Dietrichson, 1960, p. 65; Strand, 1960, p. 265), the French, Swiss, Italian, and Austrian Alps and the Carpathians (Suess, 1909, p. 564, 566; Benson, 1927, p. 28–29, 68–69; Leupold, 1935, profile; Cadisch and others, 1948, p. 224; Merla, 1952, p. 240–242; Cadisch, 1953, profile; Jaffé, 1955, p. 147; Kläy, 1957, Pl. I; Milovanovic and Karamata, 1960, p. 409; Trümpy, 1960, p. 864, 898; Cadisch and others, 1963), the Balkan Peninsula (Hiessleitner, 1951–1952, p. 33–37, 98–105, 119, 233, 236, 237, 240), and the Pindus Mountains of Greece (Aubouin, 1959, p. 338, 339, 345, 490, Pls. 2, 6) and also in numerous other regions such as Southern Rhodesia and the Taurus Mountains of Turkey (Benson, 1927, p. 31, 48, 76), Iran (de Böckh and others, 1929, p. 120, 130–134; Gray, 1950, p. 189), the Himalayas (Suess, 1909, p. 564–565; de Terra, 1936, p. 864; Krishnan, 1960, p. 73), Timor and Celebes (de Roever, 1940, p. 336–337; 1953, p. 71, 74, 75, 77; Brouwer, 1942, p. 376–377), New Caledonia, Australia, and New Zealand (Benson, 1927, p. 39, 41, 42; Wilkinson, 1953, p. 306), Cuba (Thayer and Guild, 1947), and eastern United States and Canada (Pratt and Lewis, 1905, p. 141–142).

In some regions of thrust faults or nappe structure, serpentinites are closely associated with evaporites at tectonic boundaries. In the Limestone Alps of Austria, the Lower Engadine window in eastern Switzerland, the Pyrenees and nearby, north Africa, the Persian Gulf, and Zagros Mountains, contorted evaporites are associated intimately with layers, sills, or elongate wedges of serpentinite, greenstone, and other ophiolites (Suess, 1909, p. 179–180, 222, 248, 561–567; Gignoux, 1930, p. 346–349; Cadisch, 1941, p. 31–33; Gray, 1950, p. 189).

To note this occurrence of ultramafic rocks at tectonic boundaries is, of course, not even remotely to imply that great faults occur in all areas of serpentinized peridotites. It is merely that this spatial association is sufficiently widespread to suggest that under certain conditions, serpentine may play a part in the development of major faults.

The tectonic setting in which ophiolites are found in many regions early suggested a genetic relationship of some kind between these rocks and thrust faults and that the ophiolites were probably injected at the time of overthrusting along the soles of the thrusts (Suess, 1909, p. 564–567; Kober, 1912, p. 394, 397, 408, 461; Heritsch, 1929, p. 57). This interpretation has been questioned by others who, noting the common association in many mountainous regions of serpentinized periodotite with spilitic pillow lavas and radiolarian cherts, have concluded that ophiolites are extrusive rocks formed in a deep-sea or geosynclinal environment and therefore can have only an accidental or, at most, an incidental association with thrust faults (Steinmann, 1906, p. 45–49, 59, 62; 1927, p. 648–650; Staub, 1922, p. 105, 123, 143; Bailey and McCallien, 1953, p. 404, 421, 422; Trümpy, 1960, p. 861, 866, 898). It may be contended, however, that the difference in age of formation of the injected serpentinite and the radiolarite and pillowy spilite demonstrates that the association of these rocks in, for example, the Arosa Schuppenzone of eastern Switzerland cannot be the result of an identical origin (Grunau, 1946) and thus that the environments in which the radiolarites and the pillow lavas accumulated afford no valid evidence that the serpentinites accumulated under similar conditions. Benson sought to reconcile these two apparently conflicting interpretations. He (1927, p. 70) suggested that mafic magma rising along a plane of shearing in a geosynclinal zone may possibly be pressed out at the front of an advancing overthrust anticline onto the ocean floor and there consolidated as a submarine volcanic rock. As fault move-

ment continues, this rock may be overridden by the advancing sheet, and magma passing along the thrust plane may then be injected into the previously formed volcanic rocks. This or some equally complex hypothesis is perhaps required to explain the observed field relationships of serpentinites and related rocks to radiolarites and pillow lavas on the one hand and to zones of thrust faulting on the other.

It seems possible that the occurrence of ultramafic rocks at tectonic boundaries in some areas may be the result of essentially the same phenomenon that in the earlier part of this paper is proposed to account for the occurrence of gypsum or anhydrite along tectonic boundaries, except that here the high fluid pressures are generated by hydrous-anhydrous reactions in the system $MgO-Al_2O_3-SiO_2-H_2O$ (Bowen and Tuttle, 1949; Yoder, 1952; 1955; Roy and Roy, 1955; Thompson, 1955; Bennington, 1956; Shimazu, 1960; Pistorius, 1963; Robie, 1964; Raleigh and Paterson, 1965).

The following four reactions involve the dehydration of serpentine and are reasonably typical of others in this system:

5 serpentine $\rightarrow$ 6 forsterite + talc + 9 water vapor
Serpentine + brucite $\rightarrow$ 2 forsterite + 3 water vapor.
"Aluminous serpentine" + 2 brucite $\rightarrow$ 3 forsterite + spinel + 6 water vapor.
3 serpentine + 5 anorthite $\rightarrow$ actinolite + epidote + 2 almandine + 2 quartz + 5 water vapor.

The experimental investigations of Raleigh and Paterson (1965, p. 3971–3975) demonstrate that the first two of these reactions run at temperatures of from 300° to 600° C at confining pressures of 1–5 kb. The range in temperature is dependent upon which reaction is involved, the total pressure, what mineral species are present, the amount and composition of minor impurities (Raleigh and Paterson, 1965, p. 3981), and the extent to which equilibrium is attained (Ramberg, 1952, p. 71). In the second of these four reactions, the volume of solids decreases from 107 cm³ for the mole of serpentine and 24 cm³ for the mole of brucite, a total of 131 cm³, to 88 cm³ for the 2 moles of fosterite (a total solid volume decrease of 33 per cent). However, the 3 moles of water released at 325° C and 3.5 kb (approximately the temperature and pressure at which the transition proceeds in the two mesh-texture serpentinites examined by Raleigh and Paterson, 1965, p. 3975) would have a volume of 58 cm³ (Sharp, 1962, p. 20, 21, 22, 25), thus giving a total solid plus vapor volume increase of 11 per cent. If this added volume of water vapor cannot readily escape, its pressure will rise until it is equal to or slightly greater than the overburden pressure. The rock can then expand as new cracks and pores open up along grain boundaries to make room for the increased volume of water. And as the rock begins to expand by this minute fracturing along grain boundaries, it begins also to lose its cohesive or intrinsic strength, just as in the gypsum-anhydrite reaction.

The experiments of Raleigh and Paterson on jacketed specimens of serpentinites under confining pressures of 1–5 kb show clearly that something of this sort must happen although other mechanisms of weakening are probably also operative (Raleigh and Paterson, 1965, p. 3978, 3979). When serpentinites, with an ultimate strength comparable to that of granite, are heated above their dehydration temperatures, their strength drops sharply to approximately one tenth this value. Griggs and others (1953, p. 1333) and Handin (1964) had earlier found marked weakening of jacketed specimens of serpentinite but at temperatures significantly lower than those expected for dehydration reactions of pure minerals.

If hydrous-anhydrous reactions in the MgO-Al_2O_3-SiO_2-H_2O system can have this effect on fluid pressure and rock strength, then by this mechanism, serpentinite and peridotite and associated brucite, talc, and related minerals might be injected along fault boundaries or other zones of weakness or serve as the soles of thrust faults. If this has happened widely, regeneration of serpentine from peridotite must be a common phenomenon because the ultramafic rocks along tectonic boundaries characteristically show evidence of hydration rather than dehydration. Perhaps this evidence of widespread hydration should be viewed as an expected corollary of the suggested mechanism. As the weak olivine-water vapor mush formed by dehydration in a high pressure-high temperature environment below was squeezed upward into cooler rocks above, extensive reserpentinization may be the normal consequence that should be anticipated. It is worth noting, however, that these possible tectonic effects of dehydration could be produced only in environments where temperatures are high enough for these reactions to run and where the boundary permeabilities are low enough to allow λ to approach 1.0.

The association of evaporites and serpentinites with thrust faults and tectonic boundaries prompts the question of whether the even more abundant slates, phyllites, schists, and gneisses resulting from the dehydration and decarbonation of pelitic and carbonate rocks likewise exhibit the same tectonic relationships. Nearly all stages in the progressive metamorphism of common argillaceous and carbonate-bearing rocks (chlorite, biotite, garnet, sillimanite, *etc.*, and their counterparts) depend upon reactions that involve loss of water, carbon dioxide, or both, from the preceding lower-grade stage or facies (Eskola, 1920; Bowen, 1940; Thompson, 1955; Yoder, 1955; Turner, 1958; Turner and Weiss, 1963, p. 459). Thus, if the tectonic relationships of evaporites and serpentinites have been correctly interpreted in the preceding paragraphs, similar tectonic relationships might be expected for the slates, phyllites, schists, and gneisses.

To an extent, but only to a limited extent, this expectation is realized. In many of the great mountain systems of the world, important tectonic boundaries are marked by sharp contrast in the grade of metamorphic rocks at these boundaries. Low-angle thrust faults which involve metamorphic rocks are well known in the Alps, Scotland, the Scandinavian Peninsula, the Himalayas, and eastern and western North America.

The temperatures and pressures at which a hydrate or a mineral assemblage containing carbonates or hydrates becomes unstable and reacts to form other more stable minerals are, in any particular locality, governed by the local geothermal gradient and the weight per unit area of overburden; thus, they depend largely upon the depth below the surface. Hence, during regional metamorphism, the boundary between one metamorphic facies and the next should tend to be approximately horizontal; if the rocks are deeply enough buried or for any other reason are essentially impermeable, a zone of very low rock strength, approximately horizontal or gently inclined, would be expected at the facies boundary. With the maximum principal effective stress subhorizontal, the circumstances would then be favorable for the pulling apart of basement and overlying rocks at small effective stress differences. If the detached plate of overlying rocks then breaks across and rides out over rocks that are relatively impermeable but not yet fully consolidated nor highly metamorphosed, high fluid pressures are likely to be generated in the overridden rocks, either directly by the weight of the superposed riding plate or indirectly by the dehydration or decarbonation of minerals when the overridden rocks are thus buried to greater depths and higher temperatures. High fluid pressures from either cause would establish one of the conditions favorable to the development of further thrust-faulting, either on the same or on a new and deeper fault surface.

The particular circumstances that would favor development of a flat thrust plate, however, may be only a special case in a more general relationship between metamorphic reactions and rock deformation. If the critical vapor pressure-temperature gradient of a metamorphic reaction happened to coincide closely with the local geothermal and lithostatic pressure gradient, and if the rocks were of similar composition through an appreciable thickness, or if the critical zones of several different metamorphic reactions overlapped one another, the zone of low rock strength caused by the increased fluid pressure would extend through a considerable thickness. Under these circumstances, some type of more uniformly distributed deformation such as close folding or crumpling would be expected rather than the pulling apart of basement and overlying rocks and the development of thrust faults.

Although metamorphic rocks mark important tectonic boundaries in many mountain systems, large-scale low-angle thrust faults are not the prevailing tectonic environment of metamorphic rocks generally. Characteristically, these rocks are intensely deformed, but the deformation tends to be more or less uniformly distributed throughout large rock masses. It is possible that the conditions favoring thick rather than thin zones of rock weakness during hydrous to anhydrous metamorphism are much more common, and that, unless the bulk composition or texture of adjacent rock masses differs markedly, it is only under exceptional conditions that thin subhorizontal zones of weakness and thus low-angle thrust faults develop. It may be for some reason such as this that evaporites, which occur in sedimentary beds, and serpentinites, many bodies of which are sheetlike, are frequently found associated with thrust faults and tectonic boundaries; whereas, the rocks resulting from metamorphism of pelitic and carbonate rocks (which are much more common and tend to occur in thick masses) are, for the most part, characterized by isoclinal folding, crumpling, and other types of more uniformly distributed deformation.

Syntectonic Metamorphism, a Requirement of the Proposed Mechanism

If the deformation of some metamorphic rocks (either by low-angle faulting, close folding, or both) is to be explained by the loss of rock strength caused by metamorphic hydrous-anhydrous reactions, then the deformation must be contemporaneous with at least one episode of regional metamorphism, and this fact affords a means by which the hypothesis can be tested. Metamorphic regions of intense deformation in Europe and North America that have been studied in detail by structural, petrological, or geochronological methods have yielded complex records of several distinct episodes of metamorphism; in these regions, at least some of the individual metamorphic episodes appear to have coincided with periods of thrust faulting or intense orogenic deformation (Ovchinnikov and Harris, 1960, p. 44–45; Strand and Holmsen, 1960, p. 11–13; Johnson, 1961, p. 429–430; Bryant and Reed, 1962, p. 175; Davis and others, 1963, p. 227–229; Long and Lambert, 1963, p. 223–224; Hadley, 1964, p. 34, 36–39; Lapham and Bassett, 1964, p. 661; Steiger, 1964, p. 5408). For example, in an area of complex polymetamorphic history, a plate of old crystalline rocks, metamorphosed in late Paleozoic time and now lying in thrust contact with high- to low-grade metamorphic Mesozoic rocks, may have been thrust into its present position during one or more of the stages of metamorphism that affected the Mesozoic rocks.

To find evidence of coincidence in time between metamorphism and thrusting or orogeny will not, of course, prove that metamorphic reactions which released water or carbon dioxide necessarily caused marked decrease in strength of the rocks and thus intense deformation or thrusting. The enclosing rocks may have been relatively permeable, and consequently the fluids released during metamorphism may have leaked away as rapidly as they were formed. But to find after careful examination no such evidence of a wave of metamorphism at the same time as thrusting or nappe movement would prove the hypothesis untenable for that area.

To point out as we do here that syntectonic metamorphism is required if the proposed mechanism is to work does not imply that there needs to have been an episode of thrusting or nappe movement at each stage of metamorphism in a region or that there was necessarily a stage of metamorphism at each episode of thrusting. High fluid pressures are only one of several conditions favorable to thrust faulting, and some thrust faults evidently develop without any help from high fluid pressures. The thesis of this paper is simply that fluid pressures from dehydration reactions deserve careful consideration as one of the possible factors that may help to account for some thrust faults.

High fluid pressures at the time of thrusting or orogeny may in some areas be caused by mechanisms entirely different from dehydration reactions—magmatic (Barth, 1936, p. 829–833; Platt, 1962), tectonic (F. A. F. Berry, 1965, paper presented at Am. Assoc. Petroleum Geologists 1965 Annual Meeting), diagenetic (Zen and Hanshaw, 1965; M. Powers, 1965, paper presented at Am. Assoc. Petroleum Geologists 1965 Annual Meeting), and sedimentary loading. Partial melting of merely the lowest-melting constituents in a rock, such as

the fractional percentages of combined water and potassium-feldspar in a deeply buried peridote, would probably create high fluid pressure and thus weaken the rock greatly and facilitate upward injection. But alongside these and other possible mechanisms to account for thrusting, the hypothesis seems worth testing that episodes of syntectonic metamorphism may have developed high fluid pressures that would help to explain thrust-faulting and rock deformation in some regions. Future structural studies and precise geochronological dating of events in the metamorphic history of these regions will probably show whether or not the hypothesis is tenable.

REFERENCES CITED

Ashauer, Hans, 1934, Die östliche Endigung der Pyrenäen: Ges. Wiss. Göttingen, Abh. math. -phys. Kl., ser. 3, H. 10, p. 1–115 (1285–1397)

Aubouin, Jean, 1959, Contribution à l'étude géologique de la Grèce septentrionale: les confins de l'Epire et de la Thessalie: Athens, Annales géologiques des pays Helléniques, v. 10, 525 p.

Bailey, E. B., and McCallien, W. J., 1953, Serpentine lavas, the Ankara mélange and the Anatolian thrust: Royal Soc. Edinburgh Trans., v. 62, pt. 2, p. 403–442

Barth, T. F. W., 1936, Structural and petrologic studies in Dutchess County, New York: Part 2. Petrology and metamorphism of the Paleozoic rocks: Geol. Soc. America Bull., v. 47, p. 775–850

Bennington, K. O., 1956, Role of shearing stress and pressure in differentiation as illustrated by some mineral reactions in the system MgO-SiO₂-H₂O: Jour. Geology, v. 64, p. 558–577

Benson, W. N., 1927, The tectonic conditions accompanying the intrusion of basic and ultrabasic igneous rocks: Natl. Acad. Sci., v. 19, Memoir 1, p. 1–90

Borg, I. Y., Friedman, Melvin, Handin, John, and Higgs, D. V., 1960, Experimental deformation of St. Peter sand: A study of cataclastic flow, p. 133–191 in Griggs, D. T., and Handin, John, Editors, Rock deformation: Geol. Soc. America Memoir 79, 382 p.

Bowen, N. L., 1940, Progressive metamorphism of siliceous limestone and dolomite: Jour. Geology, v. 48, p. 225–274

Bowen, N. L., and Tuttle, O. F., 1949, The system MgO-SiO₂-H₂O: Geol. Soc. America Bull., v. 60, p. 439–460

Brouwer, H. A., 1942, Summary of the geological results of the expedition: Geol. Exped. Univ. Amsterdam to Lesser Sunda Islands, v. 4, p. 345–402

Brückner, Werner, 1941, Über die Entstehung der Rauhwacken und Zellendolomite: Eclog. geol. Helvet., v. 34, no. 1, p. 117–134

Bryant, Bruce, and Reed, J. C., Jr., 1962, Structural and metamorphic history of the Grandfather Mountain area, North Carolina; a preliminary report: Am. Jour. Sci., v. 260, p. 161–180

Bucher, W. H., 1933, The deformation of the earth's crust: Princeton Univ. Press, 518 p.

Busk, H. G., 1929, Earth flexures: Cambridge Univ. Press, 106 p.

Buxtorf, August, 1908, Geologische Beschreibung des Weissenstein-tunnels und seiner Umgebung, pt. A. Stratigraphie und Tektonik: Beitr. geol. Karte Schweiz, N. F. Lief. 21, p. 1–125

Cadisch, Joos, 1941, Ardez Blatt 420: Geol. Atlas der Schweiz 1:25,000, Erläut., 51 p.

—— 1953, Unterengadin-Samnaun, Geologie, p. 17–23: Bern, Schweiz. Alpenposten, PTT, 104 p.

Cadisch, Joos, Eugster, H., Wenk, Eduard, Toricelli, G., and Burkard, G., 1963, Scuol-Schuls-Tarasp (Atlasblatt 44): Geol. Atlas der Schweiz 1:25,000

Cadisch, Joos, and others, 1948, An account of the long field meeting held in Switzerland, 6th–21st September, 1947. Excursion from Davos: Davos-Weissfluhjoch-Weissfluh: Geol. Assoc. London Proc., v. 59, pt. 4, p. 223–225

Crowell, J. C., 1962, Displacements along the San Andreas fault, California: Geol. Soc. America Special Paper 71, 61 p.

Davis, G. L., Tilton, G. R., Aldrich, L. T., Hart, S. R., Steiger, R. H., and Kouvo, Olavi, 1963, The ages of rocks and minerals, p. 218–229: Carnegie Inst. Wash. Year Book, v. 62, 551 p.

de Böckh, H., Lees, G. M., and Richardson, F. D. S., 1929, Contribution to the stratigraphy and tectonics of the Iranian ranges, p. 58–176 *in* Gregory, J. W., *Editor*, The structure of Asia: London, Methuen & Co., 227 p.

de Roever, W. P., 1940, Geological investigations in the southwestern Moetis region (Netherlands Timor): Geol. Exped. Univ. Amsterdam to Lesser Sunda Islands, v. 2, p. 97–344

—— 1953, Tectonic conclusions from the distribution of the metamorphic facies in the island of Kabaena, near Celebes: Proc. 7th Pacific Sci. Cong. (New Zealand) 1949, v. 2, p. 71–81

de Sitter, L. U., 1956, Structural geology: New York, McGraw-Hill Book Co., 552 p.

de Sitter, L. U., and de Sitter-Koomans, C. M., 1949, The geology of the Bergamasc Alps, Lombardia, Italy: Leidsche Geol. Medellinger, v. 14B, p. 1–257

de Terra, Hellmut, 1936, Himalayan and Alpine orogenies: 16th Internat. Geol. Congr., Washington, Rept., v. 2, p. 859–871

Dickinson, George, 1953, Geological aspects of abnormal reservoir pressures in Gulf Coast Louisiana: Am. Assoc. Petroleum Geologists Bull., v. 37, p. 410–432

Dietrichson, Brynjulf, 1960, Intact recrystallized rocks of ultrabasic to intermediate composition along the movement-zone of the Jotun-norite nappe in the east-Jotunheimen, central Norway: 21st Internat. Geol. Congr., Norden, Rept., pt. 19, p. 64–88

Eskola, Pentti, 1920, The mineral facies of rocks: Norsk. Geol. Tidsskr., v. 6, p. 143–194

Gignoux, Maurice, 1930, La tectonique des terrains salifères; son rôle dans les Alpes françaises: Livre jub. Soc. Géol. France, v. 2, p. 329–360 (Cent. de la Soc. Géol. France)

Gignoux, Maurice, and Moret, Léon, 1952, Géologie dauphinoise. Initiation à la géologie par l'étude des environs de Grenoble, 2d Edition: Paris, Masson et Cie, 391 p.

Goguel, Jean, 1953, Les Alpes de Provence: Géologie régionale de la France, v. 8, 123 p.

Gray, K. W., 1950, A tectonic window in southwestern Iran: Geol. Soc. London Quart. Jour., v. 105, p. 189–223

Griggs, D. T., 1940, Experimental flow of rocks under conditions favoring recrystallization: Geol. Soc. America Bull., v. 51, p. 1001–1022

Griggs, D. T., and Handin, John, 1960, Observations on fracture and a hypothesis of earthquakes, p. 347–364 *in* Griggs, D. T., and Handin, John, *Editors*, Rock deformation: Geol. Soc. America Memoir 79, 382 p.

Griggs, D. T., and Miller, W. B., 1951, Deformation of Yule marble, Part 1: Geol. Soc. America Bull., v. 62, p. 853–862

Griggs, D. T., Turner, F. J., and Heard, H. C., 1960, Deformation in rocks at 500° to 800° C, p. 39–104 *in* Griggs, D. T., and Handin, John, *Editors*, Rock deformation: Geol. Soc. America Memoir 79, 382 p.

Griggs, D. T., Turner, F. J., Borg, Iris, and Sosoka, John, 1953, Deformation of Yule marble: Part V— Effects at 300° C: Geol. Soc. America Bull., v. 64, p. 1327–1342

Grunau, Hans, 1946, Die Vergesellschaftung von Radiolariten und Ophiolithen in den Schweizer Alpen: Eclog. geol. Helvet., v. 39, p. 256–260

Gwinn, V. E., 1964, Thin-skinned tectonics in the Plateau and northwestern Valley and Ridge provinces of the central Appalachians: Geol. Soc. America Bull., v. 75, p. 863–899

Hadley, J. B., 1964, Correlation of isotopic ages, crustal heating and sedimentation in the Appalachian region, p. 33–45 *in* Lowry, W. D., *Editor*, Tectonics of the Southern Appalachians: Va. Polytech. Inst. Dept. Geol. Sci. Memoir 1, 114 p.

Handin, John, 1964, Strength at high confining pressure and temperature of serpentinite from Mayaguez, Puerto Rico, p. 126–131 *in* Burk, C. A., *Editor*, A study of serpentinite: Natl. Acad. Sci., Natl. Res. Council Pub. 1188, 175 p.

Handin, John, and Hager, R. V., Jr., 1957, Experimental deformation of sedimentary rocks under confining pressure: Tests at room temperature on dry samples: Am. Assoc. Petroleum Geologists Bull., v. 41, p. 1–50

Handin, John, Hager, R. V., Jr., Friedman, Melvin, and Feather, J. N., 1963, Experimental deformation of sedimentary rocks under confining pressure: Pore pressure tests: Am. Assoc. Petroleum Geologists Bull., v. 47, p. 717–755

Harrison, J. V., 1930, The geology of some salt-plugs in Laristan (southern Persia): Geol. Soc. London Quart. Jour., v. 86, p. 463–522

Harrison, J. V., and Falcon, N. L., 1936, Gravity collapse structures and mountain ranges, as exemplified in southwestern Iran: Geol. Soc. London Quart. Jour., v. 92, p. 91–102

Heard, H. C., 1960, Transition from brittle fracture to ductile flow in Solenhofen limestone as a function of temperature, confining pressure, and interstitial fluid pressure, p. 193–226 *in* Griggs, D. T., and Handin, John, *Editors*, Rock deformation: Geol. Soc. America Memoir 79, 382 p.

—— 1963, Effect of large changes in strain rate in the experimental deformation of Yule marble: Jour. Geology, v. 71, p. 162–195

Heard, H. C., and Rubey, W. W., 1964, Possible tectonic significance of transformation of gypsum to anhydrite plus water, p. 77–78 *in* The Geological Society of America, Abstracts for 1963: Geol. Soc. America Special Paper 76, 341 p.

Heritsch, Franz, 1929, The nappe theory in the Alps (translated by P. G. H. Boswell): London, Methuen & Co., 228 p.

Hess, H. H., 1939, Island arcs, gravity anomalies and serpentinite intrusions: A contribution to the ophiolite problem: 17th Internat. Geol. Congr., Russia, Rept. Pt. 2, p. 263–283

Hiessleitner, Gustav, 1951–1952, Serpentin- und Chromerz-Geologie der Balkanhalbinsel und eines Teiles von Kleinasien: Jahrb. Geol. Bundesanstalt (Wien) Sonder-bd. 1, pts. 1 and 2, 683 p.

Hubbert, M. K., and Rubey, W. W., 1959, Role of fluid pressure in mechanics of overthrust faulting: I. Mechanics of fluid-filled porous solids and its application to overthrust faulting: Geol. Soc. America Bull., v. 70, p. 115–166

Jaffé, F. C., 1955, Les ophiolites et les roches connexes de la région du Col des Gets (Chablais, Haute Savoie): Suisse Minéralog. Petrogr. Bull., v. 35, pt. 1, p. 1–150

Johnson, M. R. W., 1961, Polymetamorphism in movement zones in the Caledonian thrust belt of North-west Scotland: Jour. Geology, v. 69, p. 417–432

Kelley, K. K., 1960, Contributions to the data on theoretical metallurgy, XIII. High-temperature heat-content, heat-capacity, and entropy data for the elements and inorganic compounds: U. S. Bur. Mines Bull. 584, 232 p.

Kelley, K. K., Southard, J. C., and Anderson, C. T., 1941, Thermodynamic properties of gypsum and its dehydration products: U. S. Bur. Mines Tech. Paper 625, 73 p.

Kent, P. E., 1958, Recent studies of south Persian salt plugs: Am. Assoc. Petroleum Geologists Bull., v. 42, p. 2951–2972

Kläy, Louis, 1957, Geologie der Stammerspitze. Untersuchungen im Gebiete zwischen Val Sinestra, Val Fenga und Samnaun (Unterengadin): Eclog. geol. Helvet., v. 50, no. 2, p. 323–467

Kober, Leopold, 1912, Über Bau und Entstehung der Ostalpen: Wien, Mitt. geol. Ges., v. 5, p. 368–481

Krishnan, M. S., 1960, Geology of India and Burma: Madras, Higginbothams, 604 p.

Lane, H. W., 1949, Drilling practices in Iran: Oil and Gas Jour., v. 48 (Aug. 4), p. 56–58, 61

Lapham, D. M., and Bassett, W. A., 1964, K-Ar dating of rocks and tectonic events in the Piedmont of southeastern Pennsylvania: Geol. Soc. America Bull., v. 75, p. 661–668

Laubscher, H. P., 1961, Die Fernschubhypothese der Jurafaltung: Eclog. geol. Helvet., v. 54, no. 1, p. 221–282

Leupold, Wolfgang [1935?], Flüelapass. Davos und der Flüelapass. Geologische Übersicht, p. 1–10: Bern, Schweiz. Alpenposten, PTT, 32 p.

Long, L. E., and Lambert, R. St. J., 1963, Rb-Sr isotopic ages from the Moine Series, p. 217–247 *in* Johnson, M. R. W., and Stewart, F. H., *Editors*: The British Caledonides: Edinburgh, Oliver & Boyd, 280 p.

Longwell, C. R., 1949, Structure of the northern Muddy Mountain area, Nevada: Geol. Soc. America Bull., v. 60, p. 923–967

MacDonald, G. J. F., 1953, Anhydrite-gypsum equilibrium relations: Am. Jour. Sci., v. 251, p. 884–898

Mackin, J. H., 1960, Structural significance of Tertiary volcanic rocks in southwestern Utah: Am. Jour. Sci., v. 258, p. 81–131

Merla, Giovanni, 1952, Geologia dell' Appennino Settentrionale: Soc. geol. Italiana Boll., v. 70, p. 95–382

Milovanović, Branislav, and Karamata, Stevan, 1960, Über den Diapirismus serpentinischer Massen: 21st Internat. Geol. Congr., Norden, Rept., pt. 18, p. 409–417 (Abstract in English)

Misch, Peter, 1934, Der Bau der mittleren Südpyrenäen: Ges. Wiss. Göttingen, Abh. math.-phys. Kl., ser. 3, H. 12, 168 p. (1597–1764)

Moses, P. L., 1961, Geothermal gradients now known in greater detail: World Oil, v. 152, no. 6, p. 79–82

Mostofi, B., and Gansser, Augusto, 1957, The story behind the 5 Alborz: Oil and Gas Jour., v. 55, no. 3, p. 78–84

Nichols, E. A., 1947, Geothermal gradients in Mid-continent and Gulf Coast oil fields: Petroleum Dev. and Tech. Trans., Am. Inst. Min., Metall., Petroleum Engineers, v. 170, p. 44–50

Ovchinnikov, L. N., and Harris, M. A., 1960, Absolute age of geologic formations of the Urals and the pre-Urals: 21st Internat. Geol. Congr., Norden, Rept., pt. 3, p. 33–45

Palache, Charles, Berman, Harry, and Frondel, Clifford, 1951, Dana's System of mineralogy, Volume 2, 7th Edition: New York, John Wiley and Sons, Inc., 1124 p.

Paterson, M. S., 1958, Experimental deformation and faulting in Wombeyan marble: Geol. Soc. America Bull., v. 69, p. 465–475

Pistorius, C. W. F. T., 1963, Some phase relations in the system MgO-SiO₂-H₂O to high pressures and temperatures: Neues Jahrb. Mineral., Monat., H. 11, p. 283–293

Platt, L. B., 1962, Fluid pressure in thrust faulting, a corollary: Am. Jour. Sci., v. 260, p. 107–114

Posnjak, Eugen, 1938, The system CaSO₄-H₂O: Am. Jour. Sci., 5th ser., v. 35-A, p. 247–272

Pratt, J. H., and Lewis, J. V., 1905, Corundum and the peridotites of western North Carolina: Rept. Geol. Survey N. Carolina, v. 1, 464 p.

Raleigh, C. B., and Paterson, M. S., 1965, Experimental deformation of serpentinite and its tectonic implications: Jour. Geophys. Research, v. 70, p. 3965–3985

Ramberg, Hans, 1952, The origin of metamorphic and metasomatic rocks: Univ. Chicago Press, 317 p.

Riecker, R. E., and Rooney, T. P., 1966, Weakening of dunite by serpentine dehydration: Science, v. 152, no. 3719, p. 196–198

Robie, R. A., 1964, Equilibrium of talc with enstatite and quartz: Science, v. 143, p. 1057

Robinson, L. H., Jr., 1959, The effect of pore and confining pressure on the failure process in sedimentary rock: Colo. School of Mines Quart., v. 54, p. 177–199

Rodgers, John, 1963, Mechanics of Appalachian foreland folding in Pennsylvania and West Virginia: Am. Assoc. Petroleum Geologists Bull., v. 47, p. 1527–1536

—— 1964, Basement and no-basement hypotheses in the Jura and the Appalachian Valley and Ridge, p. 71–80 in Lowry, W. D., Editor, Tectonics of the Southern Appalachians: Va. Polytech. Inst. Dept. Geol. Sci. Memoir 1, 114 p.

Roy, D. M., and Roy, Rustum, 1955, Synthesis and stability of minerals in the system MgO-Al₂O₃-SiO₂-H₂O: Am. Mineralogist, v. 40, p. 147–178

Rubey, W. W., and Hubbert, M. K., 1959, Role of fluid pressure in mechanics of overthrust faulting: II. Overthrust belt in geosynclinal area of western Wyoming in light of fluid-pressure hypothesis: Geol. Soc. America Bull., v. 70, p. 167–206

Schindler, C. M., 1959, Zur Geologie des Glärnisch: Beitr. geol. Karte Schweiz, N.F., v. 107, 135 p.

Schneegans, Daniel, 1938, La géologie des nappes de l'Ubaye Embrunnais entre la Durance et l'Ubaye: Mém. Carte géol. détaillée de France, 339 p.

Sharp, W. E., 1962, The thermodynamic functions for water in the range −10 to 1000° C and 1 to 250,000 bars: Univ. Calif. Radiation Lab. Rept. 7118, 50 p.

Shimazu, Yasuo, 1960, A role of water in metamorphism as illustrated by some reactions in the system MgO-SiO₂-H₂O. A thermodynamical aspect of the earth's interior, Part IV: Jour. Earth Sciences, Nagoya Univ., v. 8, p. 86–92

Staub, Rudolf, 1922, Ueber die Verteilung der Serpentine in den alpinen Ophiolithen: Schweiz. Miner. und Petrol. Mitt., v. 2, H. 1–2, p. 78–149

Steiger, R. H., 1964, Dating of orogenic phases in the central Alps by K-Ar ages of hornblende: Jour. Geophys. Research, v. 69, p. 5407–5421

Steinmann, Gustavo, 1906, Geologische Beobachtungen in den Alpen. II. Die Schardt'sche Ueberfaltungstheorie und die geologische Bedeutung der Tiefseeabsätze und der ophiolitischen Massengesteine: Naturf Gesell. Freiburg i.B., Ber., v. 16, p. 18–66

—— 1927, Die ophiolitischen Zonen in den mediterranen Kettengebirgen: 14th Internat. Geol. Congr., Madrid, Rept., pt. 2, p. 637–667

Strand, Trygve, 1960, The pre-Devonian rocks and structures in the region of Caledonian deformation, p. 170–284 in Geology of Norway: Norges Geologiske Undersökelse, no. 208, 540 p.

Strand, Trygve, and Holmsen, Per, 1960, Stratigraphy, petrology, and Caledonian nappe tectonics of central southern Norway; Caledonized basal gneisses in a north western area (Oppdal-Sunndal): 21st Internat. Geol. Congr., Norden, Guide to excursions A 13 and C 9, 31 p.

Suess, Eduard, 1909, The face of the earth (Sollas translation): Oxford, v. 4, 673 p.

Termier, Pierre 1906, Sur les phénomènes de recouvrement du Djebel Ouenza (Constantine) et sur l'existence de nappes charriées en Tunisie: Paris, Comptes Rendus Acad. Sci., v. 143, p. 137–139

—— 1927, Nouvelle contribution à l'étude du problème de Suzette: Soc. Géol. France Bull., v. 27, p. 57–76

Terzaghi, Karl, 1936, Simple tests determine hydrostatic uplift: Engin. News-Record, v. 116, p. 872–875

Thayer, T. P., and Guild, P. W., 1947, Thrust faults and related structures in eastern Cuba: Am. Geophys. Union Trans., v. 28, p. 919–930

Thomeer, J. H. M. A., and Bottema, J. A., 1961, Increasing occurrence of abnormally high reservoir pressures in boreholes, and drilling problems resulting therefrom: Am. Assoc. Petroleum Geologists Bull., v. 45, p. 1721–1730

Thompson, J. B., Jr., 1955, The thermodynamic basis for the mineral facies concept: Am. Jour. Sci., v. 253, p. 65–103

Trümpy, Rudolf, 1960, Paleotectonic evolution of the central and western Alps: Geol. Soc. America Bull., v. 71, p. 843–907

Turner, F. J., 1958, Mineral assemblages of individual metamorphic facies, p. 199–239 in Fyfe, W. S., Turner, F. J., and Verhoogen, Jean, Metamorphic reactions and metamorphic facies: Geol. Soc. America Memoir 73, 259 p.

Turner, F. J., and Verhoogen, Jean, 1951, Igneous and metamorphic petrology: New York, McGraw-Hill Book Co., 602 p.

Turner, F. J., and Weiss, L. E., 1963, Structural analysis of metamorphic tectonites: New York, Mc-Graw-Hill Book Co., 545 p.

van de Fliert, J. R., 1953, Tectonique d'écoulement et Trias diapir au Chetthaabas, sud-ouest de la ville de Constantine, Algérie: 19th Internat. Geol. Congr., Algeria, Compte Rendu, sec. 3, pt. 3, p. 71–92

Wadia, D. N., 1947, The significance of thrust structure of the Salt Range: India, Proc. Natl. Acad. Sciences, v. 16, pts. 2–4, sec. B, p. 249–252

Whitten, C. A., 1955, Measurements of earth movements in California: Calif. Div. Mines Bull., v. 171, p. 75–80

—— 1956, Crustal movement in California and Nevada: Am. Geophys. Union Trans., v. 37, p. 393–398

Wilkinson, J. F. G., 1953, Some aspects of the alpine-type serpentinites of Queensland: Geol. Mag., v. 90, p. 305–321

Yoder, H. S., Jr., 1952, The MgO-Al$_2$O$_3$-SiO$_2$-H$_2$O system and the related metamorphic facies: Am. Jour. Sci., v. 250 Supplement (Bowen Volume), p. 569–627

—— 1955, The role of water in metamorphism, p. 505–524 in Poldervaart, Arie, Editor, Crust of the earth: Geol. Soc. America Special Paper 62, 762 p.

Young, Allen, Low, P. F., and McLatchie, A. S., 1964, Permeability studies of argillaceous rocks: Jour. Geophys. Research, v. 69, p. 4237–4245

Zen, E-an, 1965, Solubility measurements in the system CaSO$_4$-NaCl-H$_2$O at 35°, 50°, and 70° C and one atmosphere pressure: Jour. Petrology, v. 6, pt. 1, p. 124–164

Zen, E-an, and Hanshaw, B. B., 1965, Osmotic equilibrium and mechanics of overthrust faulting, p. 232–233 in The Geological Society of America, Abstracts for 1964: Geol. Soc. America Special Paper 82, 400 p.

Manuscript Received by the Society January 15, 1965

36

Reprinted from *Geol. Soc. America Bull.*, **80**(6), 927–952 (1969)

Role of Cohesive Strength in the Mechanics of Overthrust Faulting and of Landsliding

K. JINGHWA HSÜ *Geological Institute, Swiss Federal Institute of Technology, Zurich, Switzerland*

Abstract: The Mohr-Coulomb criterion for failure, modified in light of the concept of effective stress, is

$$\tau_c = \tau_0 + (S - p) \tan \phi_i,$$

where τ_c is the critical shear stress at failure, S the normal pressure, and p the pore pressure across the plane of internal slippage at failure, ϕ_i the internal friction angle, and τ_0 an empirical constant, commonly referred to as the cohesive strength. Experiments showed that the τ_0 for sedimentary rocks is about 200 bars.

Hubbert and Rubey (1959) assumed that once a fracture is started, τ_0 is eliminated and further movement results when

$$\tau_c = (S - p) \tan \phi.$$

They proceed, however, to use this formula for the frictional sliding of cohesionless block as the criterion of failure of large thrusts, after they assumed that τ_0 could be eliminated through a concentration of stress. This assumption led to their conclusions that very long overthrust blocks are possible and that such blocks may have moved by gravitational sliding along very gentle slopes.

I present arguments to show that their assumption of zero τ_0 was based upon a faulty argument and to point out that the τ_0 term should not be omitted unless it could be proved the moving block slid along an already existing fracture plane.

The first part of this paper consists mainly of conclusions based on computations. Clearly, an unjustified omission of a 200-bar cohesive strength would lead to erroneous and misleading results; particularly, gravitational sliding cannot be an important mechanism if such a cohesive strength has not been eliminated during overthrust faulting.

The second part presents evidence to distinguish between movements of cohesively bound blocks and cohesionless blocks. The Glarus overthrust, characterized by presence of a ductilely deformed limestone layer within the thrust zone, is considered a typical example of thrusting of cohesively bound blocks. The Heart Mountain thrust, characterized by a shattering of the "upper plate" and absence of a weak layer above the thrust contact, is interpreted as an example of thrusting of cohesionless blocks. The former is compared to slowly creeping slides moving at rates of centimeters or less per year, and the latter with catastrophic landslides (such as the Flims, Goldau, and Vaiont slides) moving at speeds of many meters per second.

Third, the conclusion of Raleigh and Griggs (1963) that large thrusts can only form when a toe of the thrust is continually eroded is also traced to the assumption of zero cohesive strength along thrust plane. Otherwise, the toe effect would produce a zone of imbrication at the front of over-thrust blocks, particularly those sliding downslope under their own weight.

CONTENTS

INTRODUCTION

Hubbert and Rubey (1959) advocate that bodily displacement of large overthrust sheets is made mechanically feasible because of reduction of sliding friction as a result of abnormally high pore pressure at the base of such thrusts. As an illustration of the effect of pore pressure on the sliding friction, they calculate: (1) maximum length of horizontal overthrust, x_1, for various thickness, z_1, and various values of fluid pressure to overburden ratio at the base of the thrust, λ_1; (2) maximum length of a block 6 km thick, which can be pushed down a θ-degree slope for various values of λ_1; (3) angle of a slope, θ_c, which a block will slide down under its own weight for various values of λ_1.

Hubbert and Rubey applied the Mohr-Coulomb Law of failure; they assume that slippage along any internal plane in a rock should occur when the shear stress along that plane reaches the critical value

$$\tau_c = \tau_0 + \sigma \tan \phi \qquad \text{(HR-3)}$$

where σ is the normal stress across the plane of slippage and τ_0 the initial shear strength, or shear strength of the material when σ is zero,

and ϕ the angle of internal friction. However, their calculations neglected the τ_0 term because they believed that τ_0 is eliminated once a fracture is started and further slippage is governed by the law of frictional sliding (Hubbert and Rubey, 1959)

$$\tau_0 = \sigma \tan \phi \qquad \text{(HR-98)}$$

where ϕ is used to designate the angle of sliding friction and is considered identical to that of internal friction.

Birch (1961) criticized this treatment and stated that neglecting τ_0 makes no great difference if the effective normal stress is high. However, neglecting τ_0 can no longer be justified when the value of the internal friction term is small compared to τ_0.

Hubbert and Rubey justify their treatment with an argument which is difficult for me to accept and apparently for other readers also (for example Birch, 1961; Raleigh and Griggs, 1963). Hubbert and Rubey's statement reads (p. 125):

On an area as large as the surface of a regional fault, it appears to be most likely that the fracture will be propagated as a dislocation. If so, the area over which the stress τ_0 would be involved at any one time should comprise but a minute fraction of the total area of the fault surface. We shall accordingly assume that the effect of τ_0 is negligible.

They further amplified this idea in reply to Birch (Hubbert and Rubey, 1961a, p. 1446):

If the fault is initiated and the thrust block is moved by a compressive stress applied along its rearward edge, then the stress must have its greatest intensity in that region. As the stress is built up it reaches the critical value for rupture and a fracture is started. . . . Then as the fracture advances the part of the block to the rear of the leading edge of the fracture suffers only the frictional resistance to sliding . . . with the term τ_0 deleted. . . .

The difference between the total force required to move a block of a given size when the fracture is propagated in this manner than that which would be required if τ_0 were effective simultaneously over the total basal area of the block can accordingly be quite large. Since the force is the product of the stress by the area, if A is the area of a longitudinal strip of the fault surface and F_A the force required to overcome τ_0, assumed to be effective over all of A simultaneously, then

$$F_A = \tau_0 A \, .$$

On the other hand, if τ_0 is effective only over a small ΔA then

$$F_{\Delta A} = \tau_0 \cdot \Delta A$$

and

$$F_{\Delta A}/F_A = \Delta A/A$$

which, as A becomes large, approaches the limit zero.

A good analogy to these two cases is afforded by tearing a piece of cloth. Consider a piece of cloth 1 m wide with a tensile strength, say 5 kgm weight per cm of width. If uniform tensile stress were applied to the whole width of the cloth the force required to tear it would be 500 kgm weight, or half a ton. On the other hand, if the stress is applied initially at one edge only, a tear can be started at that point and then propagated across the cloth by a force that can easily be applied with one's two hands, say 2–5 kgm weight.

The numeration of the equations in this quotation is omitted by me.

I have long puzzled over these paragraphs. It seems to me they say, in effect, that the total force required to propagate a fracture (or to tear a piece of cloth) is reduced if the force can be concentrated to a small area—a self-evident statement. However, neither their equations nor their subsequent illustration demonstrated that we encountered a zero τ_0 during the process of fracturing or tearing. The easy propagation of a tear, once a tear is started, is not related to the elimination of tensile strength of the cloth but rather is

related to the fact that the induced stress can be concentrated at the point of tear and can thus easily exceed the tensile strength of the cloth. Can shear stress also be concentrated at a fracture in the case of faulting? If so, why do not more faults simply follow preexisting fractures of similar orientation in total disregard of cohesive strength of the rocks? Note also that Hubbert and Rubey speak of the concentration of stress in the case of a thrust block being pushed from the rear. Even if they are correct in this case, they still provide no explanation why τ_0 should also be eliminated for the case of gravitational sliding, as they assume in computing the results of their Tables II and III.

Heard and Rubey (1966) further argue that the dehydration of hydrous minerals during thrusting may eliminate cohesion and, therefore, that the assumption of dislocation theory is no longer necessary. Their experiments show a drastic reduction of ultimate strength of dehydrating gypsum but do not prove that cohesive strength of such materials has been reduced to less than 30 bars. Meanwhile, the experiments of Handin and others (1963) clearly show that τ_0 of most sedimentary rocks is not eliminated by abnormal pore pressure.

Is it necessary to neglect τ_0? If so, why? Did we leave out τ_0 because such an assumption would lead to simplification in mathematical computations without greatly affecting the results? Or must τ_0 be eliminated in order to approximate physical reality?

The first question can be easily answered. The mathematics is relatively simple even if the τ_0 term is retained and the calculations can be quickly performed by computers. I have done that and found this term is not numerically negligible. The relevant question is, therefore, whether τ_0 must be omitted in our computations because the thrusting mechanism is governed by the law of frictional sliding of blocks so that an assumption of finite cohesion would lead to wrong conclusions. That Hubbert and Rubey did not prove this theory is the central theme of my paper and I shall explore this theme along two lines:

(1) The implications of omitting τ_0.

(2) The implications of not omitting τ_0.

The implications on the computed lengths of thrust blocks and critical angles of gravitational sliding are clearly shown by numbers presented in the following seven tables. However, the implications on geological thinking are less obvious. The theory of gravitational sliding as a major mechanism of mountain-building, as developed by Schardt (1898), Ampferer (1934), Haarmann (1930), van Bemmelen (1954), and others, has dominated our thinking on tectonics since the publication of the Hubbert and Rubey classic. Their computations showed that blocks must slide on slopes as low as 1 or 2 degrees when λ_1 has a value of 0.94 or 0.95 and indicated that gravitational sliding cannot be avoided when pore pressure becomes equal to overburden pressure. However, it has not been generally recognized that the reduction of the critical angles of sliding to a degree that is geologically realistic is due as much to the omission of τ_0 as to the effect of very high pore pressure; if τ_0 is not omitted but has a value of 2×10^8 dynes/cm², gravitational sliding under normal geological situations is unlikely even if $\lambda_1 = 1$.

Therefore, I include in the first part of the paper three models, assuming (1) a cohesive strength of 200 bars, which has been considered a representative value for average sedimentary rocks, (Hubbert and Rubey, 1959); (2) a cohesive strength of 30 bars, which has been estimated as the maximum of the evaporite layer under the Jura *décollement* (Laubscher, 1961); and (3) a cohesive strength of 2 bars, which is about the order of magnitude of weakened shaly layers near surface (Krsmanovic and Langof, 1964), to compare with the model formulated by Hubbert and Rubey which is equivalent to the sliding of cohesionless blocks.

In the second part of this paper, I argue that Hubbert and Rubey erred when they omitted τ_0 and when they applied the law of frictional sliding to overthrust mechanics. To illuminate my arguments, I use as an example the famous Glarus overthrust. To bring into focus our differences, I discuss the analogous mechanics of certain types of landslides, which underwent (1) a creep phase when the slide block was still cohesively bound to its underground, and (2) a catastrophic slide phase when its movement was governed by frictional sliding.

The toe effect is omitted in the three models presented in Part I, although in the second part of my paper, I discuss the implications of omitting finite cohesive strength on toe effects. The effect of eroding and overriding toes has been quantitatively evaluated by Raleigh and Griggs (1963). They found toe effects reduce the maximum lengths of thrust blocks and increase the angle necessary for gravitational sliding; however, such quantitative evaluations serve mainly as a refinement of the computed results or to obtain a second approximation. In order to simplify the mathematics of my treatment, I omit the toe effect in order to compare the first-order implications of my models with the original Hubbert and Rubey models. The omission of τ_0 led Raleigh and Griggs to overemphasize the role of an eroding toe to overthrust tectonics; they failed to consider the possibility that thrust blocks, especially those sliding gravitationally, tend to create new toes by gliding along new shear surfaces. Interpretations of tectonic styles as a result of the different toe effects under various stressed situations will be included.

Finally, in the third section of my paper I attempt to clear up two seemingly minor points: (1) The relation established by the experiments of Handin and others is, strictly speaking, not the Mohr-Coulomb criterion for the rupture of brittle materials but an empirical criterion for ultimate strength. (2) The role of increased fluid pressure is to reduce internal or external friction but not to eliminate cohesive strength.

In order to facilitate discussion, the notations and expressions used by Hubbert and Rubey are adopted in this paper, except "cohesive strength" is used as a synonym for their expression "initial shear strength," and the symbol ϕ is replaced by ϕ_i where it refers to angle of internal friction and ϕ_s where it refers to angle of sliding friction.

Equations quoted from their 1959 paper are numerated as given with the prefix HR and number of equation in parenthesis. Formulae which did not appear in their paper are given new numbers. I accept as valid all the equations quoted except those which involve an assumption of zero cohesive strength, as this assumption represents my point of departure from their treatment of overthrust mechanics.

ACKNOWLEDGMENTS

This work is a part of a research project I have initiated at Zürich to study mass movements under gravity. It was possible because I was left in the isolated confines of my office where I have the opportunity for meditation. For this I am grateful to my colleagues, particularly A. Gansser and R. Trümpy. I also appreciate the forbearance of my wife Christine who cheerfully ignored my mental absence at home during the weeks of my intense intellectual activities.

R. Trümpy kindly enlightened me on several points pertaining to the geology of Glarus overthrust. I wish to express special thanks to my assistant, C. Siegenthaler, who not only had to be a patient audience when I had the need to think out loud but also programmed the computations.

Profs. King Hubbert, Hans Laubscher, John Handin, and Dr. L. Rybach read a first draft of the manuscript critically and offered many helpful suggestions.

ROLE OF COHESIVE STRENGTH

Effect of a 200-Bar Cohesive Strength

Maximum lengths of a horizontal overthrust block. The revolutionary aspect of Hubbert and Rubey's model of overthrusting mechanics is their theory that friction during overthrusting must have been reduced because of abnormally high pore pressure at the base of the thrust. This theory provides a solution to the paradox of the apparent mechanical impossibility of very large overthrusts, as posed by Smoluchowski (1909) and Oldham (1921). The feasibility of gravitational sliding on slopes of very small inclination is a corollary of the theory. Their approach departed drastically from the approach of their predecessors (for example, Rich, 1934), who sought the solution of the paradox through the assumption of the reduction of coefficients of friction. Although this latter opinion has been held by some, even after the Hubbert and Rubey revolution (for example, Birch, 1961; Carlisle, 1965), I do accept the Hubbert and Rubey model that ϕ is nearly constant as a working hypothesis in order to go deeper into the inquiry on overthrusting mechanics.

The theoretical aspects of the treatment of the mechanics of fluid-filled porous rocks also have been questioned by Laubscher (1960) and Moore (1961); however, Hubbert and Rubey (1960, 1961b) aptly replied to these critics. Furthermore, the effect of pore pressure in reducing the internal friction term of the Mohr-Coulomb law has been elegantly proven by experimental studies (McHenry, 1948; Handin and others, 1963), and I accept the concept of effective normal stress as given by Hubbert and Rubey

$$\sigma = (S - p), \qquad \text{(HR-75)}$$

where p is the pressure of the fluid in a porous rock and S the normal stress.

This leads us to the equation of equilibrium of the x-components of force when a horizontal block is at a state of incipient motion (Figure 1):

$$\int_0^{z_1} S_{xx} dz - \int_0^{x_1} \tau_{zx} dx = 0, \qquad \text{(HR-88)}$$

where S_{xx} is the total stress component applied to the rearward edge of a thrust block, and τ_{zx} the shear stress along its base, and

$$S_{xx} = a + [b + (1 - b)\lambda]\rho_b g z_1, \qquad \text{(HR-93)}$$

where ρ_b is the bulk density of the rock, g the gravitational acceleration and a and b are experimental constants relating maximum and minimum principal stresses at failure.

According to Hubbert and Rubey

$$\tau_{zx} = \sigma_{zz_1} \tan \phi = (1 - \lambda_1)\rho_b g z_1 \tan \phi \qquad \text{(HR-94)}$$

because they assumed a zero τ_0, but

$$\tau_{zx} = \tau_0 + \sigma_{zz_1} \tan \phi_i$$
$$= \tau_0 + (1 - \lambda_1)\rho_b g z_1 \tan \phi_i \qquad (4)$$

if such an assumption is not made.

Substituting the expressions for S_{xx} and τ_{zx} from equation (HR-93) and equation (4) into equation (HR-88) and assuming the value of λ at depth less than z_1 shall not exceed λ_1, we obtain

$$x_1 = \frac{1}{\tau_0 + (1 - \lambda_1) \tan \phi_i \rho_b g z_1}$$
$$\times \left[a z_1 + \frac{[b + (1 - b)\lambda_1]}{2} \rho_b g z_1^2 \right]. \qquad (5)$$

This formula can be compared to equation (HR-96), which can be written

$$x_1 = \frac{1}{(1 - \lambda_1) \tan \phi_s \cdot \rho_b g \cdot z_1}$$
$$\times \left[a z_1 + \frac{[b + (1 - b)\lambda_1]}{2} \rho_b g z_1^2 \right]. \qquad (6)$$

Following Hubbert and Rubey and using values of

$$a = 7 \times 10^8 \text{ dynes/cm}^2$$
$$b = 3$$
$$\rho_b = 2.3 \, g/\text{cm}^3$$

and a value of τ_0 of 2×10^8 dynes/cm², my calculations of x_1 are shown in Table 1 for different λ_1 and z_1 values. Also listed (the lower figure in each square) are the values computed from equation (HR-96) and quoted

TABLE 1. Maximum Length (in km) of Horizontal Overthrust for Various Thicknesses and Values of Pressure-Overburden Ratio, λ_1, Assuming (a) a 200-Bar Cohesive Strength, (b) a Zero Cohesive Strength.

z_1 (km)	λ_1							
	0	0.465	0.5	0.6	0.7	0.8	0.9	1.0
1	3.1	3.5	3.5	3.6	3.7	3.8	3.9	4.1
	8.0	13.4	14.2	17.3	22.5	32.9	64.0	(∞)
2	6.0	6.9	7.0	7.3	7.6	8.1	8.6	9.3
	10.6	16.7	17.6	21.2	27.1	39.0	74.4	(∞)
3	8.7	10	11	11	12	13	14	16
	13.2	20.1	21.1	25.1	31.8	45.1	84.8	(∞)
4	11	14	14	15	16	18	20	23
	15.8	23.5	24.6	29.0	36.4	51.2	95.2	(∞)
5	14	17	17	19	20	23	26	32
	18.4	26.8	28.0	32.9	41.0	57.3	106	(∞)
6	17	20	21	23	25	28	33	42
	21.0	30.2	31.5	36.8	45.6	63.4	116	(∞)
7	19	24	24	26	29	33	40	53
	23.6	33.6	34.9	40.7	50.3	69.5	126	(∞)
8	22	27	28	30	34	39	47	65
	26.2	36.9	38.4	44.6	54.9	75.6	137	(∞)
9	25	30	31	34	38	44	55	78
	..	..	..	..	..	..	..	..
10	27	34	35	38	42	50	63	93
	..	..	..	..	..	..	..	..

The upper figure in each square is computed from my equation (5), the lower figure in each square has been computed from equation (HR-96) by Hubbert and Rubey, (except those in parenthesis).

from Hubbert and Rubey. Clearly in no case is the effect of τ_0 numerically negligible. The difference becomes particularly great in the case of thin blocks and in the cases of very high pore pressures.

If we accept the assumption that τ_0 is not zero but has a value given by Hubbert and Rubey based on experimental work of Handin and others (1963), the following conclusions can be drawn:

(1) Cohesive strength rather than the friction term is a more important factor in determining the lengths of near-surface thin thrust blocks; the effect of the pore pressure is minor: For example if $z_1 = 1$ km, we have $x_1 = 3.2$ km for $\lambda_1 = 0$, and $x_1 = 4.1$ km for $\lambda_1 = 1$.

(2) Very high pore pressure, say λ_1 approaching 1, must be assumed to explain the dimension of the Glarus overthrust, if it was a horizontal block, because the main unbroken thrust plate is more than 30 km long (Trümpy and Herb, 1962, Fig. B-43) and its thickness was about 5 km or 6 km (Trümpy, personal commun.).

(3) Very long horizontal thrust blocks, say longer than 100 km, or blocks with extreme length to thickness ratio (say 64 : 1 for 1 km thick block at $\lambda_1 = 0.9$, as given by Hubbert and Rubey) are not likely to be found in nature. Blocks approaching infinity in length are impossible.

During this comparison, I noted an irony: Hubbert was inspired by the Glarus overthrust to invoke high pore-pressure to explain the length of large overthrusts (Hubbert and Rubey, 1959), but the dimensions of the Glarus overthrust need not be explained in terms of very high pore-pressure if we could neglect the effect of τ_0: Hubbert and Rubey's computations showed that a 6 km thick thrust block with *normal pore pressure at its base* would be 30.2 km long if the block were horizontal and 41.9 km long if it were pushed down a 5 degree incline (*see* Table 2).

On the other hand, the existence of thin and long thrust blocks such as the Jura *décollement*, which has been estimated to be about 1.6 to 2 km thick and 80 km long at the time of thrusting (Laubscher, 1961) presents a problem if the cohesive strength of the *décollement* layer was 200 bars. Four possible alternatives exist:

(1) The Jura thrust was much thicker than that estimated by geologists.

(2) The Jura slid down slope gravitationally.

(3) The cohesive strength at the *décollement* horizon was eliminated by a dislocation mechanism as proposed by Hubbert and Rubey.

(4) The cohesive strength at the *décollement* horizon (Triassic evaporites) was much less than 200 bars-only about 20 or 30 bars.

The first alternative can be ruled out by geological evidence. The second alternative is equally unlikely, as the slope angle required for gravitational sliding for $\tau_0 = 200$ bars would be much too high ($\sim30°$) to be realistic. The third alternative is the point of contention; arguments presented in the second part of this paper are directed mainly against this alternative. I prefer the fourth alternative, particularly in view of the work by Heard and

TABLE 2. Maximum Length (in km) of a Block 6 km Thick, with a Pressure-Overburden Ratio λ_1, which Can Be Pushed Down θ-Degree Slope, Assuming (a) a 200-Bar Cohesive Strength, and (b) a Zero Cohesive Strength.

θ (degrees)	λ_1							
	0	0.465	0.5	0.6	0.7	0.8	0.9	1.0
0	17	20	21	23	25	28	33	42
	21.0	30.2	31.5	36.8	45.6	63.4	116	(∞)
1	17	21	22	24	26	30	36	47
	21.6	31.8	33.4	39.7	50.6	74.4	163	(∞)
2	18	22	23	25	28	32	40	55
	22.3	33.9	35.8	43.3	57.2	91.1	..	..
3	18	23	24	26	30	35	44	65
	23.0	36.2	38.4	47.5	65.3	116	..	..
4	19	24	25	28	32	39	50	80
	23.9	39.0	41.6	52.8	76.7	..	..	..
5	19	25	26	29	34	42	58	*104*
	24.7	41.9	45.0	59.0	..	..	..	..
6	20	27	28	31	37	47	68	*149*
	25.7	45.7	49.5	67.5	..	..	..	..
7	21	28	29	33	40	52	82	v.l.
	26.7	50.2	54.9	..	..	..	..	..
8	21	30	31	36	44	*60*	*105*	v.l.
	27.8	55.5	..	..	..	..	..	..
9	21	31	33	39	49	*70*	*145*	n.a.
	28.9	..	..	..	..	..	..	..
10	22	33	35	42	*55*	*85*	v.l.	n.a.
	30.3	..	..	..	..	..	..	..

The upper figure in each square is computed from my equation (8); the lower figure has been computed from equation (HR-125) and quoted from Hubbert and Rubey (except those in parentheses). Italicized figures represent values for which $x_1 \cdot \sin > 8$ km., v.l. represents very large values (>200 km), and n.a. represents conditions which cannot satisfy my equation (8).

Rubey (1966) who show experimentally that the ultimate strength of dehydrating gypsum can be reduced to 35 to 50 bars. The effect of a 30-bar cohesive strength on overthrust mechanics will, therefore, be discussed in a later section.

Maximum lengths of blocks which can be pushed down slope. The equation of equilibrium of the x-components of force, when a block begins to be pushed down a plane which is parallel to the top of the block and the x-y plane (Figure 1), is, according to Hubbert and Rubey,

$$\int_0^{z_1} s_{xx}dz_1 + \rho_b g z_1 x_1 \sin \theta - \int_0^{x_1} \tau_{zx}dx = 0,$$
(HR-120)

where S_{xx}, the total stress applied at the rearward edge of the block, is

$$s_{xx} = \sigma_{xx} + p = a + [b + (1 - b)\lambda_1]$$
$$\times \tan \phi_i \rho_b g z_1 \cos \theta$$
(HR-122)

and, if we do not omit the τ_0 term,

$$\tau_{zx} = \tau_0 + \sigma_{zz_1} \tan \phi_i$$
$$= \tau_0 + (1 - \lambda_1) \tan \phi_i \rho_b g z_1 \cos \theta .$$
(7)

Substituting equations (HR-122) and (7) into equation (HR-120) and assuming the value of λ at depth less than z_1 shall not exceed λ_1, we have

$$x_1 = \frac{1}{[\tau_0 + (1 - \lambda_1) \tan \phi_i \rho_b g z_1 \cos \theta - \rho_b g z_1 \sin \theta]}$$
$$\times \left[az_1 + \frac{[b + (1 - b)\lambda_1]}{2} \rho_b g z_1{}^2 \cos \theta \right].$$
(8)

This formula can be compared with, for example, (HR-125), which can be written

$$x_1 = \frac{1}{[(1 - \lambda_1) \tan \phi_i \rho_b g z_1 \cos \theta - \rho_b g z_1 \sin \theta]}$$
$$\times \left[az_1 + \frac{[b + (1 - b)\lambda_1]}{2} \rho_b g z_1{}^2 \cos \theta \right].$$
(9)

The equations (8) and (HR-125) should yield negative x_1 values if the quotient is negative. Overthrust blocks with negative lengths are, of course, physically meaningless. Actually, we should note that if there is no push from behind, that is $\sigma_{xx} = 0$, the S_{xx} term in equation (HR-120) is equal to p, then the condition of equilibrium cannot be satisfied if

$$p + \rho_b g z_1 x_1 \sin \theta > \int_0^{x_1} \tau_{zx}dx ,$$
(10)

or

$$\lambda \rho_b g z_1 \cos \theta + \rho_b g z_1 x_1 \sin \theta > \tau_0$$
$$+ (1 - \lambda_1) \tan \phi_i \rho_b g z_1 \cos \theta ,$$
(11)

when the block must glide downslope under its own weight. This analysis shows that the validity of equation (8) is limited. Negative x_1 values would be obtained if the angle of slope is already larger than the critical angle of gravitational sliding. Similarly, the validity of equation (HR-125) is also limited by the condition,

$$\tan \theta \leq (1 - \lambda_1) \tan \phi_s .$$
(12)

I might point out that the x_1 values, omitted in Table 3 of Hubbert and Rubey (1959), include

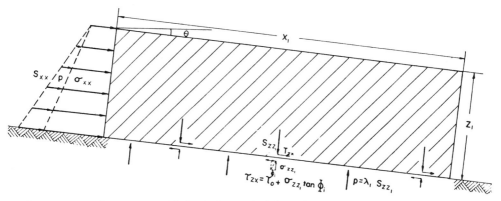

Figure 1. Boundary stress on a block which is pushed down slope θ. ($\theta = 0°$ for a horizontal block.)

not only those which give $x_1 \cdot \sin \theta$ values greater than 8 km, but also negative x_1 values.

Using equation (8), I calculated the values of the maximum length of blocks 1–10 km thick that can be pushed down a θ-degree slope for different values of λ_1 and z_1 using the same values of ρ_b, ϕ_i, a and b as did Hubbert and Rubey. Tables 2 and 3 show the computed values for 6 km and 1 km blocks. Also listed (in the lower figure of each square) are values computed from equation (HR-125) quoted from Hubbert and Rubey (1959) for comparison. The italicized numbers in Table 2 represent x_1 values, which are so large that $x_1 \cdot \sin \theta$, or the elevation difference between the front and the rear of the block, exceeds 8 km, a practical limit set up by Hubbert and Rubey (1959).

Tables 2 and 3 clearly show that the effect of τ_0 is in no case numerically negligible, and the difference becomes very significant when the pore pressure is very high. Therefore, if the coherent strength at the base of the sliding block is 200 bars, the following conclusions can be drawn:

(1) The length of thin, near-surface thrust block, approximately 1 km thick, is limited to several kilometers, regardless of the slope or the pore pressure. Even at the extreme pore pressure of $\lambda_1 = 1$ and a relatively very large 10-degree slope, the length of the 1 km block can only be 5.1 km. Again the cohesive strength, rather than the frictional term, is the most important factor in determining the length of such thin near-surface thrusts.

(2) The Glarus overthrust, 5–6 km thick and 30–35 km long, could have been pushed down an incline of about 10 degrees even if the pore pressure was normal. However, the present 10°-dip of the thrust resulted from steepening after thrusting (Trümpy and Herb, 1962, p. 113; Trümpy, personal commun.). The original incline was probably less than 5 degrees and may have been as low as 2 or 3 degrees. The Glarus thrust could have been pushed down such gentle inclines only if λ_1 was equal to 0.9 or larger. An assumption of very high pore pressure is, therefore, necessary to explain the dimensions of the Glarus thrust, if it was pushed down an incline then existing.

(3) Thrust blocks 6 km thick cannot be more than 100 km long if we limit the elevation difference between the rear and front of the block to 8 km, even if $\lambda_1 = 1$. The Bannock thrust of western Wyoming discussed by Rubey

TABLE 3. Maximum Length (in km) of a Block 1 km Thick with a Pressure-Overburden Ratio λ_1, Which Can Be Pushed Down a θ-Degree Slope, Assuming a 200-Bar Cohesive Strength.

θ (degrees)	λ_1							
	0	0.465	0.5	0.6	0.7	0.8	0.9	1.0
0	3.1	3.5	3.5	3.6	3.7	3.8	3.9	4.1
1	3.2	3.5	3.5	3.6	3.7	3.9	4.0	4.2
2	3.2	3.6	3.6	3.7	3.8	3.9	4.1	4.2
3	3.3	3.6	3.7	3.8	3.9	4.0	4.2	4.3
4	3.3	3.7	3.7	3.8	4.0	4.2	4.2	4.4
5	3.3	3.7	3.8	3.9	4.0	4.2	4.3	4.5
6	3.4	3.8	3.8	4.0	4.1	4.3	4.4	4.6
7	3.4	3.9	3.9	4.0	4.2	4.3	4.5	4.7
8	3.5	3.9	4.0	4.1	4.3	4.4	4.6	4.8
9	3.5	4.0	4.0	4.2	4.3	4.5	4.7	5.0
10	3.6	4.1	4.1	4.3	4.4	4.6	4.8	5.1

All values are computed from my equation (8).

and Hubbert (1959, Fig. 7) has dimensions of 6 km thick and about 100 km long; such a block could be pushed down a 5-degree slope if $\lambda_1 = 1$, but then the elevation difference $x_1 \cdot \sin \theta$ would be about 9 km.

(4) Long thrust blocks must have a minimum thickness: The Roberts Mountain thrust, for example, was about 80 km long (Carlisle, 1965, p. 273). However, the thrust block could not have been as thin as 3 km, as Carlisle estimated. For example, an 80 km block pushed down a 6° slope at the extremely high pore pressure of $\lambda_1 = 1$, must be at least 5 km thick.

It should be pointed out that maximum lengths of the Glarus and the Bannock thrusts would be hundreds of kilometers or infinity if $\tau_0 = 0$ and λ_1 is 0.8, 0.9, or 1.0 (a condition postulated by Hubbert and Rubey). A question could then be raised: Why are such thrusts not actually longer? One can, of course, assume various conditions which could rationalize the existence of a thrust block much shorter than the maximum length. On the other hand, such rationalization would not be necessary if the $\tau_0 = 200$ bars. Table 2 shows that these thrusts were probably almost as long as that permitted by the condition of equilibrium (HR-120).

Critical angle required for gravitational sliding. By following the treatment of Hubbert and Rubey for gravitational sliding of a subaerial block, except that I assume a finite cohesive strength along the glide plans (Figure 2),

$$\rho_b g z_1 \sin \theta_c = \tau_0 + (1 - \lambda_1) \tan \phi_i \rho_b g z_1 \cos \theta_c ,$$
$$(12)$$

where the angle θ_c is the critical angle of a slope which a block will glide down gravitationally.[1] Equation (12) solved for θ_c gives

$$\theta_c = \text{arc cos}$$
$$\times \frac{-mn + \sqrt{(mn)^2 + (1 - m^2)(1 + n^2)}}{(1 + n^2)} ,$$
$$(13)$$

where m and n are parameters, namely

---

[1] In contrast to previous considerations of blocks being pushed, when Hubbert and Rubey assumed the presence of a face free from lateral pressure in the front of the pushed block, they did not assume such a free face in the case of gravitational sliding. Equation (12) differs from equation (11) because of the absence of a free face.

[2] Errors so introduced are smaller than ten percent for θ_c less than 25°.

$$m = \frac{\tau_0}{\rho_b g z_1} , \qquad (14)$$

$$n = (1 - \lambda_1) \tan \phi_i , \qquad (15)$$

and equation (13) yields real values when $m^2 < 1$.

Applying equation (13) to calculate the critical angle θ_c of slope, which a subaerial block (cohesively bound to its underlying layer) will glide down under its own weight, I have obtained the values listed in Table 4, having used the same values of ρ_b, ϕ_i as Hubbert and Rubey and a τ_0 value of 2×10^8 dynes/cm².

If τ_0 is assumed to be zero, as was done in the treatment by Hubbert and Rubey, equation (13) is reduced to

$$\tan \theta_c = (1 - \lambda_1) \tan \phi_s . \quad \text{(HR 117)}$$

This was the formula used by them to compute the critical angle θ_c of a slope which a block will slide down. For the sake of comparison, I might introduce an approximation of equation (15) through an assumption, $\cos \theta_c \cong 1$, for small θ_c values[2] so that

$$\tan \theta_c \cong \sin \theta_c \cong \frac{\tau_0}{\rho_b g z_1} + (1 - \lambda_1) \tan \phi_i . \quad (16)$$

It is clear that the first term in equation (16) represents the major difference between my treatment and that of Hubbert and Rubey. According to them, θ_c is related to fluid pressure only and is independent of the cohesive strength, the thickness (z_1) and the bulk density of the thrust block. Their oversimplified relation actually holds only for cohesionless blocks or material. Note that the maximum angle of a slope which a block will slide down or the maximum angle of subaerial slope, according to Hubbert and Rubey, would be only 30° when $\tau_0 = 0$. This is about the angle of repose of unconsolidated dry sands. The presence of such majestic scenery as the Eiger Northwall or the Matterhorn clearly indicates that the cohesive strength of the rocks cannot be ignored.

The effect of z_1 on the angle θ_c is less obvious, but is clearly brought out by the numerical results shown in Table 4. The effect of bulk density is relatively minor because of the small differences in bulk density of rocks.

A comparison of computed results of θ_c is included in Table 4. An assumption of a $\tau_0 = 200$ bars would lead to the following conclusions:

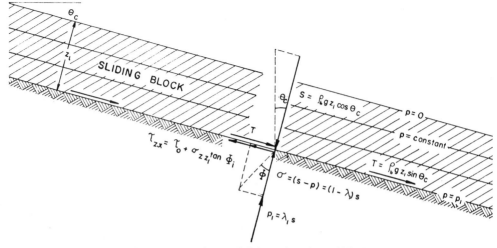

Figure 2. Gravitational sliding of a subaerial block.

(1) High pore pressure and great thickness both favor gravitational sliding. However, gravitational sliding down a gentle slope of near surface blocks less than 1 km thick can only take place if the coherent strength is less than 200 bars; reduction of the friction term alone is not sufficient.

(2) The Glarus overthrust, if it was 6 km thick, which was probably the maximum, could slide down a slope of 12° if $\lambda_1 = 0.9$, and a slope of 8° if $\lambda_1 = 1$, assuming the bulk density of the thrust were 2.3 g/cm³. If, however, the bulk density is 2.65 g/cm³, nearly the maximum possible limit, the θ_c would be 10° for $\lambda_1 = 0.9$ and 7° for $\lambda_1 = 1$. Since the slope at the time of thrusting was probably 5 degrees or less and a push from the rear in the form of advancing Pennine Nappes must have existed (Trümpy, personal commun.), the Glarus overthrust was probably pushed down a slope, rather than sliding down under its own weight.

(3) Gravitational sliding cannot be an important mechanism for the movement of large overthrusts such as those with dimensions comparable to the Bannock or the Robert Mountain thrust, if we set 8 km as a limit of elevation difference between the front and rear of the thrust. A 6 km thick block can glide down an 8-degree slope gravitationally under the most favorable condition $\lambda_1 = 1$, but then $x_1 \sin \theta$ value for an 80 km long thrust would be 11 km and for a 100 km long thrust would be 14 km.

This analysis has shown that large overthrusts could easily have been the result of gravitational sliding if $\tau_0 = 0$, but they must have been mostly pushed from the rear if $\tau_0 = 200$ bars. What if the cohesive strength under a thrust is more than zero but much less than 200 bars?

Effect of a 30-Bar Cohesive Strength

The maximum cohesive strength at the base of a sliding block can be computed on the basis of the dimensions of horizontally thrust blocks. Laubscher (1961) formulated a relation to estimate the maximum permissible shear stress τ_{max} at the base of a horizontal thrust:

$$\tau_{max} = \frac{1}{x_1}\left[az_1 + [b + (1 - \lambda_1)b]\frac{\rho_b g z_1^2}{2}\right]. \quad (17)$$

This relation was derived through a consideration of the equilibrium of x-component of forces at the base of the thrust, equation (HR-88) and the relation of the maximum and minimum principal stresses at failure, equation (HR-7); no assumption was made pertaining to the magnitude of the cohesive strength.

If the shearing resistance at the bottom of a thrust block consists of cohesive resistance and internal friction, equation (7), the cohesive strength can be equal to τ_{max} if $\lambda_1 = 1$, but it cannot be greater than τ_{max}. We can, therefore, utilize our knowledge of the dimensions of known thrusts to estimate the maximum possible cohesive strength at their base.

Substituting estimated dimensions of 35 km long and 6 km thick for the Glarus overthrust in the Laubscher equation, the maximum

TABLE 4. Angle θ_0 of Slope Down Which Block Will Glide Under Its Own Weight as Function of λ_1 and z_1, Assuming (a) a 200-Bar Cohesive Strength, and (b) a Zero Cohesive Strength.

z_1 (km)	λ_1							
	0	0.465	0.5	0.6	0.7	0.8	0.9	1.0
1	79	73	73	71	69	66	64	60
	30.0	..	..	13.0	..	6.6	3.3	(0)
2	52	42	41	38	35	32	29	26
	30.0	..	..	13.0	..	6.6	3.3	(0)
3	45	33	32	29	26	23	20	17
	30.0	..	..	13.0	..	6.6	3.3	(0)
4	41	29	28	25	22	19	16	13
	30.0	..	..	13.0	..	6.6	3.3	(0)
5	39	27	26	23	20	17	13	10
	30.0	..	..	13.0	..	6.6	3.3	(0)
6	37	25	24	21	18	15	12	8
	30.0	..	..	13.0	..	6.6	3.3	(0)
7	36	24	23	20	17	14	10	7
	30.0	..	..	13.0	..	6.6	3.3	(0)
8	35	23	22	19	16	13	10	6
	30.0	..	..	13.0	..	6.6	3.3	(0)
9	35	22	21	18	15	12	9	5.5
	30.0	..	..	13.0	..	6.6	3.3	(0)
10	34	22	21	18	15	12	8	5.0
	30.0	..	..	13.0	..	6.6	3.3	(0)

The upper figure in each square is computed from my equation (14). The lower figure is computed from equation (HR-117), Hubbert and Rubey (1959) (except those figures in parentheses). All θ_0 are in degrees.

permissible shearing stress would be about 360 bars, if the λ value within the block has an average of 0.5. The cohesive strength of 200 bars, assumed for the previous models, does not exceed this τ_{max}.

In contrast, the maximum permissible shearing stress at the base of the Jura computed by Laubscher (1961, p. 244) should be of the order of 30 bars. I have shown previously that the dimensions of the Jura indicate that the cohesive strength of the décollement layer under the Jura could not have been as great as 200 bars. Experiments on dehydration of gypsum by Heard and Rubey (1966, p. 749) revealed that the "measured stress in the anhydrite plus water paste ranges from 50 bars for the fast tests (3×10^{-4}/sec) to 35 bars for the slow tests (3×10^{-7}/sec)" under conditions of $\lambda_1 = 1$. Assuming that the Mohr-Coulomb relation holds, the measured stress of 35 − 50 bars

should be a measure of the cohesive strength of "the anhydrite plus water paste."

Whether the low cohesive strength under the Juras is related to the dehydration of gypsum (Laubscher, 1961; Heard and Rubey, 1966) or to some other cause (Laubscher, 1967), the existence of décollement horizons with coherent strength less than that of the average sedimentary rocks should be considered. I have, therefore, carried out another series of computations to determine the x_1 and θ_c for thrust blocks underlain by a décollement horizon with $\tau_0 = 30$ bars, using the same values of ρ_b, ϕ_i, a, and b as before. The results are shown by Tables 5, 6, and 7. These tables serve to demonstrate that the presence of a layer with a low coherent strength of 30 bars (instead of 200 bars) would (1) increase the maximum permissible length of thrust blocks, (the Jura décollement is comparable to the dimensions of a horizontal thrust block whose basal shear resistance consists of a cohesion of about 25 bars and a zero internal friction); (2) make it unnecessary to invoke high pore pressure to explain the dimensions of thrusts such as the Glarus overthrust, which could have been pushed down a gentle slope of 3 or 4 degrees with a normal pore pressure at base; (3) greatly reduce the gravitational stability of thin slabs under abnormally high pore pressure: 1 km thick slabs could slide down slopes of 7.5° and 2 km thick slabs down 3.7° (instead of 26°) under conditions of $\lambda_1 = 1$.

Therefore, gravity sliding could be an important deformational mechanism for such thrust blocks underlain by a weak layer with a coherent strength of about 30 bars.

Effect of a 2-Bar Cohesive Strength

Table 7 shows that a cohesive strength of 30 bars will prevent a slab 200 m thick or thinner from sliding down slopes less than 56° if the pore pressure is normal and less than 41° if the pore pressure is extreme. However, landslides of thin slabs of sedimentary rocks down slopes 20° or less are not uncommon (Heim, 1932, p. 60). The cohesive strength of the materials at the base of such sliding slabs must, thus, be less than 30 bars.

The initiation of a landslide down a dip slope along a weak sedimentary horizon can be compared to the gravitational sliding discussed in this paper; the x-component of forces at equilibrium has been given by equation (12). Since the initial slope angle of a landslide represents θ_c, the coherent strength at its base can thus be estimated

$$\tau_0 = \rho_b g z_1 [\sin \theta_c - (1 - \lambda_1) \tan \phi_i \cos \theta_c] . \quad (18)$$

Listed in Table 9 are values of maximum τ_0 for blocks of 0.1 km thick which have slid on various θ_c angles, assuming various values of λ_1, and using the same values of ϕ_i and ρ_b as before. (Note also that τ_0 is directly proportional to z_1; if all conditions are equal, a 1 km thick slab could begin to slide along a layer that has a cohesive strength 10 times greater than that given by Table 8).

Heim (1932) described several landslides of about 100 m thick down 20 degree slopes; the cohesive strength under such slabs should be of the order of a few bars for normal or somewhat higher than normal pore pressures. Such an estimate is consistent with experimental data.

Rocha (1964) made in situ tests in weathered granites, for example, and found their cohesive strength reduced or 10^6 to 10^7 dynes/cm². Krsmanovic and Langof (1964) tested the

TABLE 5.   Maximum Length (in km) of Horizontal Overthrust for Various Thicknesses and Values of Pressure-Overburden Ratio, λ_1, Assuming a 30-Bar Cohesive Strength.

z_1 (km)	λ_1							
	0	0.465	0.5	0.6	0.7	0.8	0.9	1.0
1	6.4	9.3	9.7	11	13	15	19	27
2	9.4	14	14	16	19	25	35	62
3	12	17	18	21	25	32	48	105
4	15	21	22	25	30	39	60	155
5	17	25	26	29	35	46	72	v.l.
6	20	28	29	33	40	53	84	v.l.
7	23	31	33	37	45	59	95	v.l.
8	25	35	36	41	50	66	106	v.l.
9	28	38	40	45	55	72	117	v.l.
10	31	42	43	49	59	78	128	v.l.

v.l. represents very large values (>200 km).

TABLE 6.   Maximum Length (in km) of a Block 1 km Thick, With a Pressure-Overburden Ratio λ_1, which Can Be Pushed Down a θ-Degree Slope, Assuming a 30-Bar Cohesive Strength.

θ (degrees)	λ_1							
	0	0.465	0.5	0.6	0.7	0.8	0.9	1.0
0	27	34	35	38	43	50	63	93
1	28	35	36	40	46	54	71	116
2	29	37	38	43	49	60	82	154
3	30	39	40	45	53	67	98	v.l.
4	30	41	43	49	58	75	121	v.l.
5	31	44	45	52	63	87	157	v.l.
6	32	46	48	57	71	103	v.l.	n.a.

v.l. represents very large values (>200 km) and n.a. represents conditions which cannot satisfy my equation (8).

shear strength of large specimens of interbedded sedimentary materials; they found the strength of limestone with clay interbeds is only about a few bars.

Obviously, rocks near the surface can be weakened to such a degree by weathering processes and by the presence of weathered shaly interbeds that the cohesive strength of near-surface sedimentary slabs must be, in general, considerably less than 200 bars. Using a τ_0 value of 2 bars, the gravitational stability of near-surface slabs, as a function of λ_1 and z_1, is computed.

Table 9 shows that gravitational stability of weak dip slopes ($\tau_0 = 2$ bars) is very sensitive to changes in pore pressure and is relatively independent of z_1; furthermore, that dry slabs are stable on slopes 30 degrees or more and wet slabs with normal pore pressure are unstable on slopes 18–20 degrees is the basic principle underlying the well-known engineering practice that a slowly creeping landslide can be stopped through diversion of drainage from the sliding area (Heim, 1932).

SLIDING OF COHESIVELY BOUND AND COHESIONLESS BLOCKS

Mechanisms of Overthrusting and of Catastrophic Sliding

Previous discussions have shown that the τ_0 term can not be disregarded. In this section, I would like to present evidence to show that overthrusts in general did not move as co-

hesionless blocks. For this purpose, the Glarus overthrust will be taken as an example, as this represents the very thrust which inspired Hubbert and Rubey (1959), to develop ideas on the role of fluid pressure in the mechanics of overthrust faulting. I quote the following description by Hubbert and Rubey (1959, p. 122), because it is, on the whole, accurate except perhaps for the one sentence which I have italicized:

Returning to the Alps in Switzerland, one of the most striking nappe complexes, the Mürtschen-Glarus, is bounded by a great fault which for convenience is here referred to as the "Glarus overthrust." This overthrust is easily seen in the field of the area of the Helvetian Alps between Glarus and Elm. In this fault, as described by Oberholzer (1933, . . . 1942), Permian sediments from the south have overridden Tertiary Flysch sediments for an aggregate distance of at least 40 km. . . . The fault zone comprises a thin limestone a few tens of centimeters thick, which is said to be Jurassic or Triassic in age. This occurs beneath and parallel to the bedding of the Permian conglomerate of the upper plate of the fault and rests upon Flysch sediments which are truncated at an angle of about 30 degrees. *The fault surface proper occurs as a discrete polished surface in a clay layer about 1 cm thick, within the fault-zone limestone.* In view of the distance of displacement, the trivial amount of deformation evident a meter or so away from the fault zone is most impressive.

This thin limestone has been called *Lochseitenkalk*, since the time of Arnold Escher von der Linth, and has played a major role in

TABLE 7. Angle θ_c of Slope Down Which Block Will Glide under Its Own Weight, As Function of z_1 and λ_1. Assuming a 30-Bar Cohesive Strength.

z_1 (km)	λ_1							
	0	0.465	0.5	0.6	0.7	0.8	0.9	1.0
0.2	64	56	55	52	50	47	44	41
1.0	37	24	23	20	17	14	11	7.5
2.0	33	21	20	17	14	10	7.0	3.7
3.0	32	20	18	15	12	9.1	5.8	2.5
4.0	32	19	18	15	12	8.4	5.2	1.9
5.0	31	19	18	14	11	8.1	4.8	1.5
6.0	31	18	17	14	11	7.8	4.6	1.3

All θ_c figures are in degrees.

helping unravel the tectonics of the Glarus overthrust (see Bailey, 1935). The *Lochseitenkalk* is not a thin septum that moved *en bloc* between the overlying Permian sediments and the underlying autochthonous Tertiary. This is a smeared-out, partly recrystallized, and mainly Upper Jurassic limestone, a zone of flowage. In the southern areas of the Mürtschen-Glarus thrust near the Peak of Vorab, for example, the *Lochseitenkalk* is absent, and the thrust rests directly on a normal Cretaceous-Jurassic sequence of a parautochthonous fold (Oberholzer, 1933). Elsewhere, the *Lochseitenkalk* can be traced under the Mürtschen-Glarus thrust mass for almost 20 km in a north-south direction and for more than 40 km in an east-west direction (Oberholzer, 1920, 1942). Locally, particularly in the southwestern areas, the limestone may reach 20 m or more in thickness where the upper surface is flat as a table top and the lower surface is crenulated by minor folds (Oberholzer, 1933). More commonly, however, the limestone is less than 1 m thick, having been reduced to a thin shear zone under the Permian sediments. This limestone is, thus, a block of mainly Upper Jurassic limestone caught under the thrust and spread out for about 20 km from its place of origin, while its thickness was reduced to a hundredth or less of its original thickness during the shearing movement. Younger (Cretaceous and Tertiary) and older (Jurassic and Triassic) rocks also are found locally under the thrust as smeared-out tectonic inclusions (Oberholzer, 1933).

This geologic description should suffice in indicating that the main movement of the Glarus overthrust did not slide along the clay lamina mentioned by Hubbert; nor can the thrust mechanism be compared to the sliding of one rigid plate over another with a thin film of compressed liquid between the plates to reduce sliding friction (Birch, 1961). As Griggs and Handin (1960) pointed out, faulting without loss of cohesion is not the same as brittle shear fracture, propagating as a dislocation or otherwise. The Glarus overthrust might best be compared to the movement of one plate over another, with the upper plate being carried by the flow of a thin layer of pseudoviscous material between the plates. This layer (the *Lochseitenkalk*) had a finite cohesive strength and flowed when

$$\tau_c = \tau_0 + (1 - \lambda_1) \tan \phi_i \rho_b g z_1 \cos \theta ,$$

a condition given by our equation (7).

In the Swiss Alps block movements, demonstrably governed by the law of frictional sliding, become the catastrophic slides (*Schlipfsturz*) described by Heim (1932). The best known example is the Rossberg-Goldau (Switzerland) slide of 1806. A slab of Molasse sediments, about 100 m thick and 1.8 km long, slid down an initial slope of 19.5°–20°. A less well known but more impressive slide of this type is the interglacial Flims slide of Grisons (Switzerland). The Flims block was about 700 m thick and 5 km long, and the slide started on a 10°–

15° slope and slid down the 7°–10° northern slope of the Rhine Valley.

The catastrophic phase of the slides is characterized by fantastic speeds (up to 100 meters per second), enormous power outputs, and the explosive shattering effect exhibited by the slid block. The Rossberg-Goldau slide of 1806 came down a distance of 3 km in a matter of a minute or so, according to eyewitnesses (Heim, 1932). The slab was broken during the sliding movement and gave rise to a current of debris (*Trümmerstrom*) consisting of broken blocks, dusts, and air. This dry debris current flowed down an average slope of about 3.5° and was arrested during its upward course (Heim, 1932, Fig. 9). The power output was in the order of 5×10^9 kilowatts (Lanser, 1967). The Flims slide was not witnessed by man, but its enormous energy output is well indicated by the tremendous shattering of the slid block. The slab, with an estimated volume greater than 12 billion cubic meters, has been broken into many small slabs tens of meters thick, countless blocks, fragments, and much dust. All broken slabs were shattered internally and to such an extent that often it is difficult to define the boundary between a slab and the dusts between the slabs. The broken debris flowed in part down the gentle 2° slope of the Rhine Valley, eastward for some 8 km, and in part westward upstream for some 6 km!

The mechanics of this type of mass movement has been treated by Shreve (1966, 1968). He elaborated on the air-layer lubrication hypothesis to account for the gravitational sliding of detached blocks on slopes consider-ably less than 30 degrees. Shreve (1966, p. 1641) further developed that the rock debris of such slides "did not flow like a viscous fluid, but instead slid like a flexible sheet . . . Thus, the 'low coefficient of friction' is due not to low internal friction, but to low sliding friction." The mechanism he envisioned is thus nearly comparable to that illustrated by the beer can or the concrete block experiments of Hubbert and Rubey (1959), which demonstrated the reduction of sliding friction under detached blocks because of high pore pressure.

Rates of Overthrusting and of Catastrophic Sliding

The law of frictional sliding governs the movement of landslide blocks after the blocks were detached, when the sliding friction could be reduced by a compressed layer of trapped air (Shreve, 1966, 1968). However, why should such blocks begin to fall in the first place?

Heim, in his writings on landslides (for example, 1881, 1932), repeatedly emphasized the fact that such slides were not made in one day but were preceded by a long and intensive preparation. I cannot resist the temptation to inject a little tragic humor in this dry discussion on mechanics by quoting the famous last words of one who died in the Goldau slide:

"Drissig Jahr händ mir jez scho druf gwartet, dass de Berg chömi, er wird wol no warte, bis ich mis Pfiffli gstopft ha!"

Translated into English, they read:

"For thirty years have we waited for the mountain to come, well, now it can wait until I finish stuffing my pipe."

But the mountain did not wait.

TABLE 8. Maximum Value of Cohesive Strength (in Bars) At Base of 0.1 km Block Which Begins To Slide Down a θ_e-Degree Slope Under Its Own Weight.

θ_e (degrees)	λ_1							
	0	0.465	0.5	0.6	0.7	0.8	0.9	1.0
1	0	0	0	0	0	0	0	0.4
5	0	0	0	0	0	0	0.7	2.0
10	0	0	0	0	0.1	1.4	2.7	4.0
15	0	0	0	0.8	2.1	3.4	4.7	6.0
20	0	1.2	1.6	2.9	4.1	5.4	6.6	7.9
25	0	3.3	3.7	4.9	6.1	7.3	8.5	9.7
30	0	5.4	5.8	6.9	8.1	9.2	10	12

TABLE 9. Angle θ_c of Slope Down Which Block Will Glide Under Its Own Weight, As Function of λ_1 and z_1 Assuming a 2-Bar Cohesive Strength.

z_1 (km)	λ_1							
	0	0.465	0.5	0.6	0.7	0.8	0.9	1.0
0.1	34	22	21	18	15	12	8.3	5.0
0.2	32	20	18	15	12	9.1	5.8	2.5
0.3	31	19	18	15	11	8.2	5.0	1.7
0.4	31	18	17	14	11	7.8	4.6	1.3
0.5	31	18	17	14	11	7.6	4.3	1.0
0.6	31	18	17	14	11	7.4	4.1	0.8
0.7	31	18	17	14	11	7.3	4.0	0.7
0.8	31	18	17	14	10	7.2	3.9	0.6
0.9	30	18	17	14	10	7.1	3.9	0.6
1.0	30	18	17	13	10	7.1	3.8	0.5

All θ_c figures are in degrees.

Creep movements preceded catastrophic landslides, and the eventual scar was usually localized by a crack or fissure which widened itself gradually. Such occurrences have been described for several famous historic landslides by Heim (1932), and the slow movements before the Vaiont slide of North Italy have been studied in detail (Selli and others, 1964). Nearly constant rates from 1 mm to 2 mm per day were reported for the Vaiont slide during 1961 and 1962 (Selli and others, 1964, Fig. 5) but the final rate was accelerated to about 80 cm/day just before the catastrophe of October 9, 1963 (Nonveiller, 1967, Fig. 5). Although the slide mass crept slowly, that it moved at all under its own weight is sufficient evidence that shearing stress (τ_{zx}) at the base of the sliding block (z_1 km thick) is greater than shearing resistance, or

$$\tau_{zx} > \tau_0 + (1 - \lambda_1) \tan \phi_i \rho_b g z_1 \cos \theta .$$

Slow ground movement was an expression of slow deformation of a creeping layer at the base of the slide block, and the creep rate, u_c, is

$$u_c = \int_0^{z_2} (\tau_{zx}/\eta) dz , \qquad (19)$$

where z_2 is the thickness of this creeping layer, η its equivalent viscosity. That is, the ground movement rate is a function of stress and equivalent viscosity and is independent of coefficients of friction.

The nearly constant creep rate at Vaiont for two years and the final dramatic acceleration of ground movement leading to the catastrophe is indeed similar to that part of strain-time (creep) curve for a rock under constant differential stress when the creep changed from a pseudoviscous (steady) flow to a tertiary (accelerating) flow, leading to rupture (see Handin, 1966, Fig. 11–3).

Once the block is detached, becoming a catastrophic slide, the velocity is no longer related to the equivalent viscosity of a creeping layer. The movement would then be governed by the law of frictional sliding, and the velocity, according to Heim (1932), of blocks sliding down an inclined plane is

$$u_s = u_0 + ct , \qquad (20)$$

where u_0 is the initial velocity (or the final creep rate) and t the time after the catastrophic slide began; c, the acceleration, is

$$c = g(\sin \beta - \tan \phi_s' \cdot \cos \beta) , \qquad (21)$$

where the angle β refers to the initial slope of the slide, and $\tan \phi_s'$ is the equivalent coefficient of friction of the slide and is the ratio between the height of the fall and the horizontal distance traveled by the slide (Heim, 1932; Shreve, 1966, 1968).

Using equation (20), the velocity of frictional slides can be calculated. Heim, helped by his physicist friend Müller, computed the velocities of the Elm Landslide and found its maximum speed 83.5 m/sec. The computed speeds are comparable to the estimates of eyewitnesses (Heim, 1932). Using the same approach, I found the maximum speed of the Flims slide to be about 40 m/sec. Similar calculations gave a 18 m/sec maximum speed for the catastrophic Vaiont slide (Selli and others, 1964).

Rate considerations brought to focus the well known fact that a landslide may pass through two stages of development: (1) a long creep phase when the rate of movement is largely determined by the equivalent viscosity of the creeping material at the base of the slide block, where the shearing stress must overcome the cohesive strength and the internal friction of this creeping material, and (2) a catastrophic sliding phase when the rate of movement is largely determined by the angle of the initial slope and by the "equivalent coefficient of friction," and when the shearing stress at the base of the slide must only overcome the sliding friction (*see* Terzaghi, 1950).

Returning to the Glarus overthrust, the mechanism of the thrust is comparable to the slow creeping before the catastrophic slide: nowhere in the Glarus overthrust do we witness shattering as is impressively displayed by the Flims and other catastrophic landslides. The deformation mechanism is characterized by the flow of creep of the *Lochseitenkalk* at its base.

In theory, equation (19) should permit an estimate of the rate of the Glarus overthrust. Unfortunately, the equivalent viscosity of limestone is not a material constant, but is a function of strain rate, which in turn is sensitive to small changes in differential stress during deformation (Heard, 1963; Hahn, and others, 1967). For example, the equivalent viscosity and strain rates varied four orders of magnitude while the differential stress changed only half an order of magnitude during a series of creep tests on Yule Marble by Heard (*in* Handin, 1966, Tables 11–17). Since we cannot yet accurately determine the stress difference during deformation, we cannot now calculate the strain rate.

If the Glarus overthrust was formed during a fraction of the Miocene Epoch, as suggested by geologic evidence (Trümpy and Herb, 1962; Trümpy, personal commun.), say 10^5 to 10^6 years, the rate of ground movement would be of the order of centimeters or fractions of a centimeter per year, comparable to the Holocene rate of thrusting at Buena Vista, California (Gilluly, and others, 1958). The strain rate of the *Lochseitenkalk* would then be of the order of 10^{-2} or 10^{-3} per year, or 10^{-9} to 10^{-9} per second. The total strain of the *Lochseitenkalk* was estimated on the assumption that it was a packet of limestone originally less than 1 km long and some tens of meters thick, now smeared into a thin lamina some 20 km long and less than 1 m thick. The slow rate of thrusting is thus comparable to the creep of landslide.

Heart Mountain Thrust—Sliding of Cohesionless Thrust Blocks?

Is it possible that some thrusts did slide as detached cohesionless blocks? If so, such thrusts should be characterized by features typical of catastrophic landslides such as the Flims slide. Could the unusual Heart Mountain thrust be such a tectonic slide?

I have not had the opportunity to visit the Heart Mountain thrust. I was, however, much impressed by the following group of phrases by Pierce, (1957, 1963, 1966) describing this thrust:

(1) The mappable part of the thrust plane is very gentle for long distances, of the order of a few degrees for some 50 km . . . (1963).

(2) The absence of solid materials that could have had a low coherent strength along the thrust. . . .

(3) The absence of materials that could have been characterized by a very high pore pressure along the thrust. . . .

(4) The existence of a zone of brecciation up to 50 feet or more thick above the thrust, but no brecciation of the rocks under the thrust, and the brecciation has been described as "shattered and broken" . . . (1957).

(5) The disorderly arrangement of thrust blocks above the thrust, necessitating the assumption of local faulting which does not cut the thrust plane; the presence of numerous blocks of younger limestones directly on the fault with the intervening older formations absent. . . .

(6) The discontinuous distribution of carbonate thrust blocks. . . .

(7) The presence of an "early basic breccia" in the spaces between the carbonate thrust blocks, and directly on the rocks below the thrust, "yet this contact horizon can not logically be considered an erosion surface or a plane of intrusion" . . . (1957).

(8) The thrust "was derived from a source without any known roots," and the frontal part has ridden across a former land surface (1957).

(9) The thrust of movement appears to have been rapid, rather than a slow creep (1963).

(10) The existence of a conglomerate consisting almost exclusively of carbonate debris derived from the thrust blocks, but dipping 30 degrees in a region where the thrust has remained flat—a curious fact which led Pierce (1957, p. 612) to suggest that the conglomerate was "involved to some extent in the thrusting," but "is not shattered and broken" like the thrust block on which it rests.

Those readers who have had a chance to visit the Flims slide of Grisons may have recognized that practically all the preceding descriptions are applicable to the Flims slide, particularly the movement of the broken slabs for many kilometers on very gentle slopes, the absence of low coherent-strength or high pore-pressure materials above the sliding surface, the shattering of the broken slabs, the disorderly arrangement of the broken blocks, the very rapid rate of sliding, and finally, the presence of a conglomerate, dipping up to 30°, deposited on the steep deltaic slope of a lake resulted from the damming of the Rhine by the slide debris (Heim, 1932, p. 127). In fact, one of my American colleagues who visited the Flims slide with me was impressed with its similarity to a chaotic thrust mass and remarked on the difficulty of recognizing such a slide in the geologic record.

Even the scale of the Flims slide is comparable to the Heart Mountain thrust: The Flims slab, some 700 m thick moved a combined distance of some 20 km in the narrow Rhine Valley and partly upstream. With this occurrence in mind, I can well imagine a Heart Mountain slab, some 600 m thick, sliding down the flank of a monocline and moving some 50 km across an open plain (Eocene) of Wyoming.

The postulate of thrusting by catastrophic sliding seems incredible. I was, therefore, somewhat relieved when King Hubbert kindly called my attention to the fact that Bucher may have entertained a similar idea. Unfortunately Bucher published only two short abstracts on this subject. In one he emphasized the "extremely shattered" nature of the Heart Mountain thrust masses and noted the "peculiar" fracturing and the presence of "curious shatter marks" of quartzite pebbles beneath several of the thrust masses (1933, p. 57). Later, he cited the "practically complete absence of signs of compression and the dominance of tensional fractures" throughout the Heart Mountain thrust masses as evidence that they could not have been formed by "normal orogenic processes" (1935, p. 69). Evidence for explosive deformation led Bucher to invoke "volcanic explosive forces" as the chief cause of the thrust. However, all features, noted by him, could be cited to corroborate the hypothesis of catastrophic sliding.

While geological evidence suggests an origin by catastrophic sliding of the Heart Mountain thrust, this interpretation is almost the only reasonable alternative to a consideration of overthrust mechanics. The length-to-thickness ratio alone would indicate that the Heart Mountain thrust was not pushed from behind (see Tables 1, 3, and 5). However, as Pierce (1963, p. 1234) asked, how could such a mass slide down a slope of 2.5° or less? Gravity sliding would only be possible if the thrust was underlain by a layer of nearly zero cohesive strength and characterized by an extremely high pore pressure (Tables 8, 9). Yet all evidence for the presence of such a layer is absent (Pierce, 1966). The only substance of such character, which could have been present under the thrust but is now gone and leaves no trace behind, would be an "air cushion," as Shreve postulated for the type of slides like the Flims.

Although the Heart Mountain slab may have slid along a nearly flat land surface after it was detached, the thrust may have originated on a steeper slope, say 5° to 10° when the slab was still cohesively bound. The thrust, like all landslides, may have started as a creep along a zone of weakness where the cohesive strength of the rocks has been much reduced by near-surface processes.

I would like to emphasize, however, the idea of a catastrophic slide travelling at great speed for 50 km or more is too fantastic to be accepted without a great deal of evidence. It is beyond the scope of this paper to resolve the origin of the Heart Mountain thrust. I only offer the bold idea so that this mystery may be examined from a rather unorthodox point of view. It would be ironic indeed if this thrust, which was cited by Davis (1965) to refute the Hubbert and Rubey theory, may best be interpreted by their model of brittle fracturing and frictional sliding.

Cohesive Strength in Toe Effects

Both Hubbert and Rubey's and my computations, tabulated in Tables 1, 2, 5, and 6, clearly indicate that the maximum length of blocks that can be pushed down a slope increases with increasing slope angle, and, gravitationally slid blocks have a maximum length of infinity. Conversely, the blocks that can be pushed uphill should be shorter and are thus likely to be more broken up than the gravitationally slid blocks.

Paradoxically, thrust blocks in orogenic belts, where evidence of crustal shortening by horizontal compression is clearly demonstrated by geological studies, such as the Alps (Trümpy, 1959, p. 848) and the Canadian Rockies (Bally and others, 1966, p. 359) are far less broken up than the mélange of broken slabs of presumably gravity-sliding terranes, such as the *argille scagliose* of the Apennines (Page, 1963) or the Franciscan of the California Coast Ranges (Hsu, 1965; 1966). A consideration of stress patterns confirms the geological interpretations: The overthrust belts of compressional tectonics are characterized by the absence of synorogenic extensional features, a fact that motivated Hubbert and Rubey (1959) to renounce the hypothesis of purely gravitational sliding to explain such overthrusts. In contrast, the tectonic mélanges show a complex superposition of compressional and extensional shear fractures with the maximum and minimum principal stresses alternatively parallel to the direction of tectonic transport (Hsu, 1965)—a stress pattern characteristic of gravity sliding movements such as glaciers or landslides (Heim, 1932; Shreve, 1966, 1968). The answer of this paradox in tectonic style may well lie in the toe effect!

Gravity sliding of rocky slabs at a creep rate along a bedding plane (Heim's Bergsturz Typus XI) or along a pre-existing tectonic plane (Typus XII) is not uncommon in the Swiss Alps. An interesting observation made by Heim is that the front of such slowly sliding slabs has been broken into many small fragments, which are mixed with the soft materials originally under the slides. In fact, the toe of such gravity-sliding slabs is not a wedge but a heap of slowly creeping debris!

In contrast, the Jura Mountains, which has been interpreted as a slab pushed up a nearly horizontal incline (Laubscher, 1961), passes transitionally into the Jura Plateau; there is no toe. Other thrust belts, such as the Canadian Rockies, are characterized by a zone of imbrication, with many thin thrust slices near the mountain front (Bally and others, 1966, plates 12, 15) where the toe must have been.

In considering the toe effect, Raleigh and Griggs (1963) followed the approach of Hubbert and Rubey and omitted the τ_0 term in their treatment. The movement of the thrust up an inclined slope under the toe was thus a problem of frictional sliding of a detached, flexible plate. I shall not go through, in this paper, the mathematical computations to show the implications on numerical results of omitting τ_0 in their analysis. The far more important fact is that we might have to assume a different model in treating the toe effect if the thrust did not move like a cohesionless block.

Raleigh and Griggs (1963) gave the condition of gravitational sliding of a thrust block whose volume is V_1 with a toe of volume V_2 down a slope θ_1 as in Figure 3a.

$$\rho_b g V_1 \sin \theta_1 > (1 - \lambda_1) \tan \phi_s$$
$$\times (\rho_b g V_1 \cos \theta_1 + F_h \sin \theta_1) + F_h \cos \theta_2 . \quad (22)$$

The first term on the right side of equation (22) is the friction of sliding down a slope θ_1 of the main thrust block, and the second term is the parallel-to-the-slope component of the force, F_h, required to push the toe up a slope θ_2. This force was given as

$$F_h \cos \theta_2 = (1 - \lambda_2) \tan \phi_s$$
$$\times (\rho_b g V_2 \cos \theta_2 + F_h \sin \theta_2)$$
$$+ \rho_b g V_2 \sin \theta_2 , \quad (23)$$

where λ_2 is the average pore pressure-overburden pressure ratio beneath the toe. F_h becomes greater as V_2 increases, as more force will be required to overcome the friction and inertia of an ever-enlarging toe. Eventually dynamic equilibrium will be reached when the driving force will be equal to the resisting force and no more thrusting will be possible. Further movment can only be induced if the size of the toe is reduced by erosion. This appears to me the reason why Raleigh and Griggs (1963, p. 829) came to the conclusion that "large thrusts can only form when the toe is continually eroded." If this model were

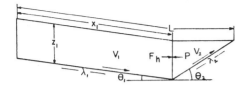

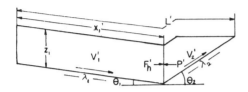

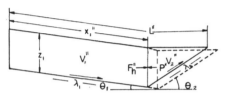

Figure 3. The toe effect in the mechanics of overthrust faulting.

applicable, the gravitationally sliding slabs described by Heim (Bergsturz Typus XI and XII) would not break; they would simply stop and wait until erosion of their toes permitted further downhill movement.

The toe effect will be different if the cohesive strength at the base of the sliding slab is finite and is about the same as that of the slab itself. The equation of forces for sliding equilibrium can then be given by

$$\rho_b g V_1 \sin \theta_1 = \tau_0 A + (1 - \lambda_1) \tan \phi_i$$
$$\times [\rho_b g V_1 \cos \theta_1 + F_h \sin \theta_1]$$
$$+ F_h \cos \theta_2 . \quad (24)$$

According to this model, sliding must not follow any pre-existing sliding plane. If the toe volume V_2 is enlarged to V'_2 (Fig. 3b) so that

$$\rho_b g V_1' \sin \theta_1 < \tau_0 A + (1 - \lambda_1) \tan \phi_i$$
$$\times [\rho g V_1' \cos \theta_1 + F_h \sin \theta_1] + F_h' \cos \theta_2 , \quad (25)$$

sliding will no longer follow the pre-existing slide plane, but will take place along a new surface which underlies a new toe (Fig. 3c) with a volume V''_2 ($< V'_2$) so that

$$\rho g V_1'' \sin \theta_1 > \tau_0 A + (1 - \lambda_1) \tan \phi_i$$
$$\times [\rho g V_1'' \cos \theta_1 + F_h'' \sin \theta_1] + F_h'' \cos \theta_2 , \quad (26)$$

where F_h' and F_h'' represent the resistive forces of the enlarged old toe and the new toe respectively, and V_1' and V_1'' represent the volume of down-sliding segment of the old and new slab.

If all other factors are constant, the driving force of a gravitationally sliding slab is greatest at the front of a slab, where a new toe will be created near the old toe (Fig. 3c).

In theory, erosional removal of old toes would prevent the creation of new toes; in nature, however, the rate of erosion may not be sufficiently rapid. Consider, for example, the rate of Holocene thrusting at Buena Vista or the estimated rate of the Glarus overthrust; a forward movement of 2 mm to 2 cm per year of a downslope sliding slab would correspond to an upward elevation of the toe at 1 mm to 1 cm per year rate, assuming as Raleigh and Griggs, a 30° angle for θ_2. At such a rate, a 10,000 m high plateau would be formed in 1 to 10 million years. As the enlarged toe volume could not be effectively removed by erosion, the sliding slab would tend to seek new toes. The end result might well be the mélange of pervasively sheared slabs and blocks such as the *argille scagliose* of the Appennines, the Franciscan of the California Coast Ranges, or the *Wildflysch* of the Prealps.

The toe effect of a block being pushed horizontally or upslope can be very different. For if a pushed block is at its maximum length, the stress at its base (T_{zz} of Fig. 1) would be the maximum that can be transmitted by the block. As the shearing resistance to thrusting is increased because of the increased toe size, the transmissible stress would become less than that required to move the block. The forward movement of the block would thus stop, unless erosion reduced the toe size or more push is applied at the rear of the block. If the latter is the case, the length of the block that could be pushed by this increased stress would be shorter, and a new sliding surface under a new toe would be initiated some distance behind the old toe, which would be relieved of the stress induced by the push from the rear. The spacing between the original and the new toes of a block being pushed from the rear depends upon the relative magnitude of the three terms on the right side of the formula (26). If the force required to create a new toe (the third term) is very small compared to the shearing resistance at the base of the thrust block (the first two terms), the new thrust block would be nearly as long as the old block, and the new toe would be closely adjacent to the old toe. Otherwise the new thrust block would have to be considerably shorter than the old thrust block, resulting in widely spaced toes. We might thus speculate that the imbricate thrusts on the front of the Canadian Rockies may be related to the fact that thrusts moved along limestones of similar cohesive strengths. In contrast, the absence of imbrication at the boundary of the chain and Table Juras may reflect the large difference in cohesive strength of the rocks within the décollement ($\tau_0 \cong 200$ bars) and of the sliding layer at its base ($\tau_0 \cong 30$ bars).

MOHR-COULOMB LAW AS A CRITERION OF ULTIMATE STRENGTH

The Mohr-Coulomb law forms the common basis of the treatments of overthrust mechanics by Hubbert and Rubey and by me in this paper. The relation

$$\tau_c = \tau_0 + \sigma \tan \phi_i \qquad \text{(HR-3)}$$

was originally proposed by Coulomb in 1776 to express the condition of rupture in simple compression of a brittle material (*see* Nadai, 1950, p. 219). This failure criterion constitutes a special case of the Mohr strength theory when the Mohr envelope consists of two straight lines. A modification, taking into consideration the effect of pore fluid pressure, was applied by Terzaghi (1943; 1950) to soil mechanics. Hubbert with others further developed the effective-stress concept (that is, Hubbert and Willis, 1957; Hubbert and Rubey, 1959), and the Mohr-Coulomb law was written as

$$\tau_c = \tau_0 + (S - p) \tan \phi_i. \qquad \text{(HR-84)}$$

Experiments, mainly by Handin and others (1963), showed that this relation also holds for materials undergoing permanent deformation except in the case where τ_c is the shearing stress at ultimate strength. As pointed out by Handin (written commun., 1968), the significant fact is "a single linear envelope, derived from the principal stresses at ultimate strength, predicts the failure in shear, whether the mechanism is brittle fracture with total loss of cohesion, faulting without loss of cohesion, or cataclastic flow." Such an empirical extension of the Mohr-Coulomb law is particularly important to geologists, because rock deformation

in nature does not always lead to brittle fracture. The *Lochseitenkalk* under the Glarus overthrust evidently flowed. In fact many large thrust sheets are underlain by mylonites, which are products of cataclastic flow (Hsu, 1955) when the milling down of the original rock material occurred under such conditions that the rock retained its coherence (Waters and Campbell, 1935). There is no justification for omitting the τ_0 term in the Mohr-Coulomb law as it is applied to those problems of thrusting by flowage.

ROLE OF PORE PRESSURE IN OVERTHRUSTING AND IN LANDSLIDING

Heard and Rubey (1966) performed dehydration experiments and speculated (p. 744–745):

When a single crystal of a mineral undergoes a reconstructive transformation as is the case for gypsum dehydrating to anhydrite plus fluid, most chemical bonds must be broken, and thus its strength must approach zero. In an aggregate of such crystals, the process would not only affect individual crystal strength but would also disrupt intergranular cohesion, so that in polycrystalline gypsum, the strength could range from the crushing strength of a nearly pure, coherent gypsum aggregate through zero, to that of a secondary anhydrite aggregate of variable grain size and cohesion. This could result in a lower initial shear strength τ_0 for the Mohr failure criterion. . . . Thus we can argue that τ_0 may be small; one does not need necessarily to regard fault propagating as a dislocation and assume therefore, that τ_0 is negligible compared to τ (as did Hubbert and Rubey, 1959, p. 125).

They further cited the experimental studies by Raleigh and Paterson (1965) and Riecker and Rooney (1966) to extend their conclusions for all dehydration reactions. In essence they argued that the effect of pore pressure is not only to reduce friction, but also to reduce or eliminate cohesive strength of dehydrating aggregates.

From their published experimental data it appears that the reduction of the ultimate strength of dehydrating gypsum resulted in part from a reduction of τ_0 and in part from an elimination of internal friction under conditions $\lambda_1 = 1$. However, the estimated experimental value of τ_0 is not zero, but 35 bars or so, and computations have shown that even this small cohesive strength should not be eliminated in the consideration of overthrust mechanics (Tables 5, 6, 7).

Furthermore, the arguments of loss of cohesion during dehydration are not applicable to deformations which do not involve such reactions, such as the flow of the *Lochseitenkalk*. An increase of pore pressure alone reduces only the internal friction but not the cohesive strength as shown clearly by the experiments of Handin and associates (1963) on the Berea sandstone. In fact, their results suggest that the cohesive strength of the muddy shale actually increased for unjacketed specimens ($\lambda_1 = 1$), although the ultimate strength of the specimens has been reduced because of a decrease in internal friction (Handin and others, 1963, Figs. 23 and 24).

In conclusion, I would like to emphasize that as demonstrated by Hubbert and Rubey, the role of high pore pressure is to reduce the frictional resistance to overthrust faulting. This role should not be exaggerated where an assumption of zero cohesive strength is not justified.

SUMMARY

In this analysis of overthrust mechanics, following the principle of Hubbert and Rubey that pore fluid pressure reduces effective confining pressure, I have presented arguments to show:

(1) Hubbert and Rubey have not given convincing proof that the cohesive strength can be omitted in a treatment of the overthrust mechanics; the dislocation "theory" advocated by them is an assumption, but not established, and the example of tearing a piece of cloth may be irrelevant.

(2) The term τ_0 is not numerically negligible as shown by computations herein.

(3) An application of the Mohr-Coulomb law, with an assumption of $\tau_0 = 0$, would lead to such unlikely deductions as thrust plates of infinite length, gliding on infinitesimal slopes when pore pressure is equal to overburden pressure, or instability of all slopes over 30 degrees.

(4) If $\tau_0 = 0$, we need not assume abnormal pore pressure to explain the dimension of such long thrust plates as the Glarus overthrust; the argument that abnormal pore pressure must have played a role in such thrusts would no longer be a physical necessity.

(5) Dimensions of thrust plates, calculated on the basis of finite τ_0 values, more nearly correspond to those in nature, such as the Glarus overthrust and the Jura *décollement*.

(6) If τ_0 is 200 bars (considered representative for sedimentary rocks) and if it cannot be omitted, overthrust by gravity sliding without push from the rear could only have occurred under very unusual geological situations—even if pore pressure is extremely high.

(7) If τ_0 is 30 bars (estimated for the evaporite layer under the Jura *décollement*), overthrust by gravity sliding on gentle slopes of a few degrees could have been an important mechanism of deformation, if pore pressure is extremely high.

(8) Landslide slabs, some 100 m thick, down an initial slope of 20°, suggest that rocks, particularly shales, could have their cohesive strength reduced to the order of a few bars by near-surface processes.

(9) Mechanism of overthrusts, such as that of Glarus overthrust, involved permanent deformation of rocks under the thrust mass; the mechanics of such deformation is not governed by the law of frictional sliding.

(10) Mechanism of catastrophic landslide is governed by the law of sliding friction. Such slides could move down slopes less than 30 degrees because sliding friction has been reduced through the abnormally high pressure of the air trapped under the slide, thus providing a natural analogue to Hubbert and Rubey's "beer can experiment."

(11) An overthrust "creeps" at a slow rate, if it is coherently bound to its underground because of the very high equivalent viscosity of the materials being deformed at the thrust zone. A catastrophic slide "flies" at a fantastic rate, tens of meters per second, and is thus shattered and broken into small blocks and dusts during its movement.

(12) The unusual Heart Mountain thrust may have been a detached thrust block and moved like a catastrophic slide.

(13) If thrusts move along a plane of finite cohesive strength, the toe effect may force the thrusts (particularly those sliding gravitationally) to seek a new surface of sliding, or may stop the thrusts (particularly those being pushed uphill from the rear) movement altogether. Erosional thrust would, however, form if erosion was able to reduce the toe size at a sufficiently rapid rate.

(14) The experiments by Handin and others extended the Mohr-Coulomb Law as a criterion of ultimate strength. This law is thus applicable to natural deformation of rocks which did not involve loss of cohesion.

(15) The effect of pore pressure reduces the internal friction term of the Mohr-Coulomb law. The speculation by Heard and Rubey that τ_0 is completely eliminated at very high pore pressures has not been proven by experimental evidence.

The foregoing analysis indicates that the Hubbert and Rubey treatment of the overthrust mechanics is a special theory, applicable in particular to the thrusting or landsliding of cohesionless blocks. The inclusion of a finite cohesive strength in the consideration of stress equilibrium represents an extension of their theory to a general theory applicable to overthrusts and landslides in general (with $\tau_0 = 0$ being the special case).

REFERENCES CITED

Ampferer, O., 1934, Über die Gleitformung der Glarner Alpen: Sit. Ber. Ak. Wiss. Wien, Math.-Natw. Kl., Abt. I, v. 143, p. 109–121.

Bailey, E. B., 1935, Tectonic essays, mainly Alpine: Oxford Univ. Press, p. 200.

Bally, A. W., Gordy, P. L., and Steward, G. A., 1966, Structure, seismic data, and orogenic evolution of southern Canadian Rocky Mountains: Canada Petroleum Geology Bull., v. 14, p. 337–381.

Birch, Francis, 1961, Role of fluid pressure in mechanics of overthrust faulting; discussion: Geol. Soc. America Bull., v. 72, p. 1441–1443.

Bucher, W. H., 1933, Problem of the Heart Mountain thrust: Geol. Soc. America Proc., 1933, p. 57.

—— 1935, Remarkable local folding, possibly due to gravity, bearing on the Heart Mountain thrust problem: Geol. Soc. America Proc., 1935, p. 69.

Carlisle, Donald, 1965, Sliding friction and overthrust faulting: Jour. Geology, v. 73, p. 271–292.

Davis, G. A., 1965, Role of fluid pressure of overthrust faulting: discussion: Geol. Soc. America Bull., v. 76, p. 463–468.

Gilluly, James, Waters, A. C., and Woodford, A. O., 1958, Principles of geology: Freeman & Co., San Francisco, p. 534.

Griggs, David, and Handin, John, 1960, Observations on fracture and a hypothesis of earthquakes: Geol. Soc. America Mem. 79, p. 347–364.

Haarmann, E., 1930, Die oszillationstheorie: Ferdinand Enke, Stuttgart, 260 p.

Hahn, S. J., Ree, Taikyue, and Eyring, Henry, 1967, Mechanism for the plastic deformation of Yule Marble: Geol. Soc. America Bull., v. 78, p. 773–782.

Handin, John, 1966, Strength and ductility: in Clark, S. P., Jr., Editor, Handbook of physical constants: Geol. Soc. America Mem. 97, pp. 223–289.

Handin, John, Hager, R. V., Friedman, Melvin, and Feather, J. N., 1963, Experimental deformation of sedimentary rocks under confining pressure: pore pressure tests: Am. Assoc. Petroleum Geologists Bull., v. 47, p. 717–755.

Heard, H. C., 1963, Effect of large changes in strain rate in the experimental deformation of Yule Marble: Jour. Geology, v. 71, p. 162–195.

Heard, H. C., and Rubey, W. W., 1966, Tectonic implications of gypsum dehydration: Geol. Soc. America Bull., v. 77, p. 741–760.

Heim, Albert, 1881, Die geologischen Verhältnisse des Bergsturzes von Elm: in Buss and Heim "Der Bergsturz von Elm," Wurster & Cie., Zürich, p. 130–141.

—— 1932, Bergsturz und Menschenleben: Fretz & Wasmuth Verlag, Zürich, 218 p.

Hsu, K. J., 1955, Granulites and mylonites of San Gabriel Mountains, California: Calif. Univ. Pubs. Geol. Sci., v. 30, p. 223–352.

—— 1965, Franciscan rocks of the Santa Lucia Range, California, and the argille scagliose of the Apennines, Italy: a comparison in style of deformation: Geol. Soc. America Abs., 1965.

—— 1966, Mesozoic geology of the California coast ranges—a new working hypothesis: in Etages tectoniques, p. 279–296, Baconnière, Neuchâtel, Suisse.

—— 1968, The principles of mélanges and their bearing to the Franciscan-Knoxville paradox: Geol. Soc. America Bull., v. 79, p. 1063–1074.

Hubbert, M. K., and Rubey, W. W., 1959, Role of fluid pressure in mechanics of overthrust faulting: Geol. Soc. America Bull., v. 70, p. 115–206.

—— 1960, Role of fluid pressure in mechanics of overthrust faulting: reply to discussion by H. P. Laubscher: Geol. Soc. America Bull., v. 71, p. 617–628.

—— 1961a, Role of fluid pressure in mechanics of overthrust faulting: reply to discussion by Francis Birch: Geol. Soc. America Bull., v. 72, p. 1445–1451.

—— 1961b, Role of fluid pressure in mechanics of overthrust faulting: reply to discussion by W. L. Moore: Geol. Soc. America Bull., v. 72, p. 1581–1594.

Hubbert, M. K., and Willis, D. G., 1957, Mechanics of hydraulic fracturing: Trans. A. I. M. E., v. 210, p. 153–168.

Krsmanovic, D., and Langof, Z., 1964, Large scale laboratory tests of the shear strength of rocky material: Felsmechanik u. Ingenieurgeologie, Supplement 1, p. 20–30.

Lanser, O., 1967, Felsstürze und Hangbewegungen in der Sicht des Bauingenieurs: Felsmechanik u. Ingenieurgeologie, v. 5, p. 89–113.

Laubscher, H. P., 1960, Role of fluid pressure in mechanics of overthrust faulting: Discussion: Geol· Soc. America Bull., v. 71, p. 611–615.

—— 1961, Die Fernschubhypothese der Jurafaltung: Eclogae Geol. Helvetiae, v. 54, p. 221–282.

—— 1967, Geologie und paläontologie: Verhandl. Naturf. Ges. Basel, v. 78, p. 24–34.

McHenry, D., 1948, The effect of uplift pressure on the shearing strength of concrete: Internat. Congr. on Large Dams, p. 31.

Moore, W. L., 1961, Role of fluid pressure in mechanics of overthrust faulting: A discussion: Geol· Soc. America Bull., v. 72, p. 1581–1586.

Nadai, A., 1950, Theory of flow and fracture of solids, McGraw Hill Co., N. Y., 572 p.

Nonveiller, E., 1967, Zur Frage der Felsrutschung im Vajont-Tal: Felsmechanik u. Ingenieurgeologie, v. 5, p. 2–9.

Oberholzer, J., 1920, Geologische Karte der Alpen zwischen Linthgebiet und Rhein, 1:50,000: Geol. Kommiss. Schweiz, Spezialkarte No. 63.

—— 1933, Geologie der Glarneralpen: Beitr. Geol. Karte d. Schweiz, N.F., Lieferung 28, 626 p.

—— 1942, Geologische Karte des Kantons Glarus, 1:50,000: Geol. Kommiss. Schweiz, Spezialkarte No. 117.

Oldham, R. D., 1921, Know your faults: Geol. Soc. London Quart. Jour., v. 57, p. lxxvii–xcii.

Page, B., 1963, Gravity tectonics near passo della cisa, northern Apennines, Italy: Geol. Soc. America Bull., v. 74, p. 655–672.

Pierce, W. G., 1957, Heart Mountain and South Fork detachment thrusts of Wyoming: Am. Assoc. Petroleum Geologists Bull., v. 41, p. 591–626.

—— 1963, Reef Creek detachment fault, northwestern Wyoming: Geol. Soc. America Bull., v. 74, p. 1225–1236.

—— 1966, Role of fluid pressure in mechanics of overthrust faulting: Discussion: Geol. Soc. America Bull., v. 77, p. 565–568.

Raleigh, C. B., and Griggs, D. T., 1963, Effect of the toe in the mechanics of overthrust faulting: Geol. Soc. America Bull., v. 74, p. 819–830.

Raleigh, C. B., and Paterson, M. S., 1965, Experimental deformation of serpentinite and its tectonic implications: Jour. Geophys. Research, v. 70, p. 3965–3985.

Rich, J. L., 1934, Mechanics of low-angle overthrust faulting as illustrated by Cumberland thrust block, Virginia, Kentucky, and Tennessee: Am. Assoc. Petroleum Geologists Bull., v. 18, p. 1584–1596.

Riecker, R. E., and Rooney, T. P., 1966, Weakening of dunite by serpentine dehydration: Science, v. 152, p. 196–198.

Rocha, M., 1964, Some problems on failure of rock masses: Felsmechanik u. Ingenieurgeologie, Supplement I, p. 1–9.

Rubey, W. W., and Hubbert, M. K., 1959, Role of fluid pressure in mechanics of overthrust faulting: Geol. Soc. America Bull., v. 70, p. 167–206.

—— 1965, Role of fluid pressure in mechanics of overthrust faulting: Reply: Geol. Soc. America Bull., v. 76, p. 469–474.

Schardt, H., 1898, Les régions exotiques du versant nord des Alpes suisses: Bull. Soc. vaudoise des Sci. Nat. v. 34, p. 114–219.

Selli, R., Trevisan, L., Carloni, G. C., Mazzanti, R., and Ciabatti, M., 1964, La Frana del Vaiont: Geionale di Geologia, Annali del Museo geol. die Bologan, ser. 2, v. 32, p. 1–154.

Shreve, R. L., 1966, Sherman landslide, Alaska: Science, v. 154, p. 1639–1643.

—— 1968, The Blackhawk landslide: Geol. Soc. America Spec. Paper 108, 47 p.

Smoluchowski, M. S., 1909, Some remarks on the mechanics of overthrusts: Geol. Mag., n.s., Dec. V, v. 6, p. 204–205.

Terzaghi, K., 1943, Theoretical soil mechanics: New York, John Wiley & Sons, 510 p.

—— 1950, Mechanism of landslides, p. 83–123 *in* Paige, Sidney, *Editor*, Application of geology to engineering practice (Berkey Volume): Geol. Soc. America, p. 327.

Trümpy, R., 1959, Paleotectonic evolution of the central and western Alps: Geol. Soc. America Bull., v. 71, p. 843–908.

Trümpy, R., and Herb, R. 1962, Eastern and northern Alps: *in* Lombard, A., *Editor*, Guidebook for the international field institute, Alps, 1962, American Geol. Inst. Rept., p. 86–130.

van Bemmeln, R. W., 1954, Mountain building: Nijhoff, Den Haag, p. 177.

Waters, A. C., and Campbell, C. D., 1935, Mylonites from San Andreas Fault zone: Am. Jour. Sci., v. 229, p. 437–503.

MANUSCRIPT RECEIVED BY THE SOCIETY JULY 16, 1968

Reprinted from *Geol. Soc. America Bull.*, 83(10), 3073–3081 (1972)

Stress Distributions and Overthrust Faulting

GEORGE Z. FORRISTALL   *Geological Institute, Swiss Federal Institute of Technology, Zürich, Switzerland*

ABSTRACT

Previous calculations for the maximum possible length of overthrust blocks under various conditions have not taken into account the full state of stress of the overthrust block. We show the significance of this omission in this paper and correct for it by taking an elasticity solution for the stress state. Calculations are presented for various pore pressures, block thicknesses, cohesive strengths, and slope angles which show that the maximum theoretical length of an overthrust block is significantly less than previously supposed. However, it is still possible to explain the long overthrusts observed in nature if the pore-pressure ratio and angle of slope are large enough.

INTRODUCTION

Hubbert and Rubey (1959) revolutionized geologists' thoughts concerning the possibility of large overthrust faults, with their lucid demonstration of the strong influence that fluid pore pressure can have on the stresses involved in such thrusts. Their thesis provided a solution to the paradox of the apparent mechanical impossibility of large overthrusts, as stated by Smoluchowski (1909), among others. Using the concept of effective stress in a fluid-filled porous solid, they were able to show that the stresses possible in a rock mass could push an overthrust block of dimensions much greater than formerly supposed. The mechanics of fluid-filled solids have been the subject of much discussion, such as that by Laubscher (1960), Moore (1961), and Hubbert and Rubey (1960, 1961). However, the recent theoretical demonstration by Nur and Byerlee (1971) that Hubbert and Rubey used the physically important limiting case of a more general law should do much to settle the controversy. Hsü (1969) has demonstrated the importance of the cohesive strength at the slip surface and recalculated the possible lengths of overthrust blocks for various values of the cohesive strength.

The present paper is concerned not with criticism of the role of fluid pore pressure or the cohesive strength at the slip surface, but rather with previous calculations of the maximum stresses permitted in the overthrust block and on its faces. Hubbert and Rubey and their critics have all made the tacit assumption in their work that the maximum and minimum principle stresses in the overthrust block were parallel to the faces of the block. This assumption cannot be true because of the presence of shearing stresses throughout the block which are caused by the shearing action at its base, and the incorrect use of the assumption can be shown to lead to an overestimation of about 50 percent in the possible length of the overthrust block. This correction will exist regardless of whether or not the slip surface has any cohesive strength.

To demonstrate these points, we first consider in some detail a simple example which shows the effect of neglecting the shear stresses in the material. We use the Mohr-Coulomb theory of failure to calculate the maximum horizontal stress possible for given vertical and shear stresses. Using these data and elasticity theory solutions for the state of stress in the block, we then calculate the maximum theoretical lengths of overthrust blocks for various thicknesses, pressure ratios, and angles of slope.

EFFECT OF IGNORING SHEAR STRESS

Let us first consider the simple example of a thrust block with no pore pressure on a horizontal surface, as shown in Figure 1. This block is to be pushed to the right by the forces on its trailing edge, with the resisting force coming from the shear stress at its base due to the internal coefficient of friction of the rock in question. Clearly the longer the block, the greater the total of the resisting forces, and thus the greater the stresses required at the trailing edge of the block. There will come a point at which these stresses are so great that they will cause failure in the block, and this

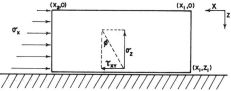

Figure 1. Thrust block on a horizontal surface.

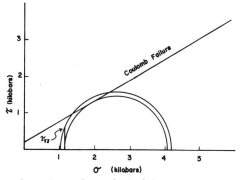

Figure 2. Mohr-Coulomb failure criterion. The small circle shows the state of stress in the example in the text when the shear stress is neglected; the larger circle the state of stress when it is included. The larger circle violates the failure criterion.

determines the maximum length of block that can be overthrust.

To determine the maximum stresses allowed, Hubbert and Rubey (1959) used the Mohr-Coulomb criterion which has been well tested experimentally by Handin and others (1963). This law states that failure will occur when

$$\tau \geq \tau_0 + \sigma \tan \phi, \tag{1}$$

where τ and σ are, respectively, the shear stress along and the normal stress across some plane in the solid, and τ_0 and ϕ are experimentally determined constants known as the cohesive strength and internal angle of friction of the material. In keeping with geological practice, compressive stresses will be taken as positive in this work. The Mohr-Coulomb criterion can be represented graphically as the straight line in the Mohr diagram of Figure 2. The criterion states that failure will occur when the Mohr circle representing the state of stress at any point in the rock crosses or touches the straight line.

For a two-dimensional problem, the state of stress at a point is determined by knowledge of the normal stresses across two perpendicular planes and the shear stress along them. The magnitude of these stresses will depend upon their direction, which is usually taken as parallel to the coordinate directions, giving us σ_x, σ_z, and τ_{xz}. Stresses in other directions can be found by manipulation of Mohr's circle. There exists one direction at each point in the material for which the shear stress vanishes, and the normal stresses taken in this direction are the maximum and minimum principal stresses, σ_1 and σ_3. It is convenient to express equation (1) in terms of maximum and minimum principal stresses by the equation

$$\sigma_1 = a + b\sigma_3, \tag{2}$$

where

$$a = 2\tau_0\sqrt{b}$$

and

$$b = \frac{1 + \sin \phi}{1 - \sin \phi}. \tag{3}$$

Returning to our example, we take as material properties

$$\begin{aligned} a &= 0.7 \text{ kb} \\ \phi &= 30° \\ b &= 3 \\ g &= 980 \text{ dynes/gm} \end{aligned} \tag{4}$$

and

$$\rho = 2.31 \text{ gm/cm}^3.$$

Then, at the base of the block,

$$\sigma_z = \rho g z = 1.13 \text{ kb} \tag{5}$$

and for sliding to occur in the case where cohesive strength at the base of the block is zero,

$$\tau_{xz} = \sigma_z \tan \phi = 0.65 \text{ kb}. \tag{6}$$

If, following Hubbert and Rubey, we take σ_x and σ_z as the principal stresses, then at the ower left corner of the block

$$\sigma_x = a + b\sigma_z = 4.09 \text{ kb}. \tag{7}$$

The stress values from equations (5) and (7) plot as the smaller circle in Figure 2, which is just tangent to the failure line and thus represents the maximum possible stress state.

But now note that we have completely neglected the shear stress which exists all along the base of the block, as given by equation (6). Thus, the true state of stress at the lower left corner of the block is given by the larger circle in Figure 2, which clearly exceeds the Coulomb

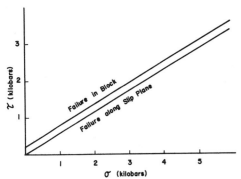

Figure 3. Different failure criteria used in the block and along the plane of sliding.

criterion. The reason for this unhappy state of affairs is that in calculating equation (7), we assumed that σ_x and σ_z were principal stresses, although this cannot be true because the shear stress in this coordinate system is not zero. If we calculate the forces on the rear edge of the block without considering the shear stresses present, many other points in the block will also be overstressed. An elasticity theory solution for the stresses in the interior of the block will give a continuous stress field, and thus, for example, the shear stress on the trailing edge of the block will be nonzero for at least some distance up from the base. If the Mohr circle taking σ_x and σ_z as principal tresses was already tangent to the failure line, as it would be for maximum forces, then the true Mohr circle including the shear stress must be larger and cross the failure line.

Another serious difficulty arises if we take the cohesive strength at the base of the block equal to that in the block. In this case, the normal stress across the base and the shear stress along the base must combine to give a state of failure along the base if sliding is to occur. That is, the Mohr circle for the points along the base must be tangent to the failure line. However, a solution for the stress field at that point will show that σ_x is not zero, which makes the circle larger, showing that failure will occur in some other plane before the block begins to slip at its base.

Thus for sliding along a horizontal plane, material properties must change across that plane. The pore pressure could change rapidly, or the failure criterion for the plane could be weaker than that for the block. This could come, for example, from a weak layer in the bedding or from previous fractures which left

a roughly horizontal surface with no cohesive strength. Many choices could be made for the parameters of the two failure curves required. We have chosen those displayed in Figure 3, which shows that the failure curve for the plane of sliding is parallel to the failure curve in the block, but has lower cohesive strength. This choice facilitates direct comparison with the previous results and there is no compelling reason to believe that such a case is not well represented in nature.

EXTREME VALUES OF HORIZONTAL STRESS

To calculate the maximum possible lengths of thrust blocks subject to our failure criterion, we must first know the extreme values of surface stress which are possible. It is convenient to rewrite the Mohr-Coulomb criterion in the block, equation (2), in terms of stresses in the x-z coordinate system. Referring to the Mohr diagram, it is easily seen that

$$\sigma_1 = \tfrac{1}{2}(\sigma_x + \sigma_z) + r$$
$$\sigma_3 = \tfrac{1}{2}(\sigma_x + \sigma_z) - r \qquad (8)$$

where

$$r^2 = \tfrac{1}{4}(\sigma_x - \sigma_z)^2 + \tau^2_{xz}.$$

At the bottom of the overthrust block, the failure criterion for slipping gives

$$\tau_{xz} = \sigma_z \tan \phi + \tau_a, \qquad (9)$$

where τ_a is the cohesive strength at the base of the block and should be distinguished from τ_0, the cohesive strength of the block. To compare our results with those of Hubbert and Rubey, we use the case $\tau_a = 0$; to compare with Hsü, other values are used. Inserting (9) and (8) into (2) and performing the necessary algebra gives the extreme values of σ_x possible at the base of the block if failure is not to occur.

$$\sigma_x = \left(\frac{1}{2b} + \frac{b}{2}\right)\sigma_z + \frac{a(b-1)}{2b}$$
$$\pm \frac{1}{2b}\{[(b^2 - 1)\sigma_z + a(1 + b)]^2 - 4b(1 + b)^2$$
$$\times (\tau_a + \sigma_z \tan \phi)^2\}^{1/2}. \quad (10)$$

Equation (10) gives two values of σ_x, the maximum and minimum values which are possible for a given value of σ_z at the base. These values are plotted in Figure 4 for the material constants given in equation (4). Note that we do not permit any negative stresses, since we do not allow direct tension in the rock mass. From the stress distribution developed in the next section, we will see that these extreme values for the lower surface of the block are the critical stresses that we must deal with.

MAXIMUM LENGTH OF HORIZONTAL OVERTHRUST BLOCK

Given the boundary stresses on the block, we can use elasticity theory to calculate the stress distribution inside the block. An excellent discussion with geological motivation of the Airy stress functional method is given in Hafner (1951). Briefly, the equations of static equilibrium are satisfied if the stress components are given by

$$\sigma_x = \frac{\partial^2 \phi}{\partial z^2}$$

$$\sigma_z = \frac{\partial^2 \phi}{\partial x^2} + \rho g z \tag{11}$$

$$\tau_{xz} = \frac{\partial^2 \phi}{\partial x \partial z}$$

where Φ is a potential function satisfying the compatibility equation

$$\frac{\partial^4 \phi}{\partial x^4} + 2 \frac{\partial^4 \phi}{\partial x^2 \partial z^2} + \frac{\partial^4 \phi}{\partial z^4} = 0 . \tag{12}$$

In the case where the stress in the vertical direction is due only to gravity, we must have

$$\frac{\partial^2 \phi}{\partial x^2} = 0 \quad \text{for all } z . \tag{13}$$

Integration of equation (13) gives the stress function

$$\phi = c f_1(z) x + a x + b f_2(z) + d . \tag{14}$$

But to satisfy the compatibility equation (12), we must have

$$c x \frac{d^4}{dz^4} f_1(z) + b \frac{d^4}{dz^4} f_2(z) = 0 . \tag{15}$$

Thus the most general stress distribution for a block in which the vertical stress is due only to gravity is given by

$$\sigma_x = c x z + d x + e z$$

$$\sigma_z = \rho g z \tag{16}$$

$$\tau_{xz} = \frac{c}{2} z^2 + d z$$

where c, d, and e are unknown constants.

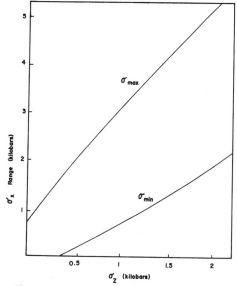

Figure 4. Maximum and minimum horizontal stress permitted for a given vertical stress for $\tau_a = 0$.

The condition that σ_z is due only to gravity is equivalent to the condition that the shear stress is constant for any given depth. We adopt this condition as likely for the large structures we are considering.

Let us refer again to Figure 1. The form of equations (16), along with the shape of the curves in Figure 4 show us that if the block is stressed up to the limit at $(x_1, 0)$, (x_1, z_1), $(x_2, 0)$, and (x_2, z_1), then the state of stress at every other point in the block will be below the failure line. Roughly speaking, the curves for the extreme values of σ_x bow outward, whereas σ_z increases linearly and τ_{xz} quadratically. We thus specify the following boundary stresses:

$$\begin{aligned}
\sigma_x &= 0 \quad \text{at } z = 0, x = x_1 \\
\sigma_x &= \sigma_0 \quad \text{at } z = 0, x = x_2 \\
\sigma_x &= \sigma_{\min} \text{ at } z = z_1, x = x_1 \\
\sigma_x &= \sigma_{\max} \text{ at } z = z_1, x = x_2 .
\end{aligned} \tag{17}$$

Referring to Figure 4, we see that for the material constants we have chosen, $\sigma_0 = 0.7$ kb. For a block of thickness z_1, equation (16) gives the vertical stress σ_z, and entering this value in Figure 4 or equation (10) gives the required values for $\sigma_{\min}$ and $\sigma_{\max}$.

The boundary conditions (17) insure that the block will be stressed as much as possible.

We must also insure that the block will slip along its lower surface by specifying that the failure curve for sliding is met there. That is,

$$\tau_{xz} = \sigma_z \tan \phi + \tau_a \text{ at } z = z_1. \qquad (18)$$

We now insert equation (16) in conditions (17) and (18), and solve for x_1, x_2, c, d, and e.

$$x_1 = 0$$

$$x_2 = \frac{\sigma_{\max} + \sigma_0 - \sigma_{\min}}{2\rho g \tan \phi + 2\tau_a/z_1}$$

$$e = \frac{\sigma_{\min}}{z_1} \qquad (19)$$

$$d = \frac{2\rho g \sigma_0 \tan \phi + 2\sigma_0 \tau_a/z_1}{\sigma_{\max} + \sigma_0 - \sigma_{\min}}$$

$$c = \frac{2}{z_1}(\rho g \tan \phi - d + \tau_a/z_1) .$$

The maximum possible length of the block is of course given by $x_2 - x_1 = x_2$.

The results of some numerical calculations for the length of the block are presented in the next section, but the reader may find it instructive at this point to calculate one example for himself, and draw circles on a Mohr diagram representing the state of stress at various points to see how they compare to the failure criteria. It is interesting to note that equations (16) give non-zero shear and horizontal stresses on the front end of the block. This is more realistic than a free leading edge, and, in any case, it is unavoidable once the vertical stress is specified as due only to gravity. The required forces could, for example, be due to the toe of the thrust block.

EFFECT OF FLUID PORE PRESSURE

Nur and Byerlee (1971) have shown that for a porous material which behaves elastically, the total vertical stress due to the overburden S_z is the sum of an effective pore pressure and the stress in the solid σ_z; that is,

$$S_z = [1 - (K/K_s)]p + \sigma_z \qquad (20)$$

where p is the fluid pressure in the pores, and K and K_s are the bulk moduli of elasticity for the aggregate and the grains, respectively. For many materials of geological interest, $K_s \gg K$, and we are justified in using the limiting case adopted by Hubbert and Rubey (1959), where

$$S_z = p + \sigma_z . \qquad (21)$$

For convenience we will here take

$$p = \lambda S_z \qquad (22)$$

where λ is constant everywhere in the overthrust block. This is admittedly a simplification, but as pointed out by Hubbert and Rubey, it is a conservative assumption. Thus the vertical stress in the solid can be calculated as a function of depth by the formula

$$\sigma_z = (1 - \lambda)\rho g z \qquad (23)$$

where ρ is here the bulk density of the fluid-solid mixture. For numerical calculations we will keep the same material constants given in equation (4).

Since the fluid pressure only varies in the vertical direction, we may take the stress almost as before in equation (16), with only σ_z changed, and the unknown constants c, d, and e.

$$\sigma_x = cxz + dx + ez$$
$$\sigma_z = (1 - \lambda)\rho g z \qquad (24)$$
$$\tau_{xz} = c/2 \, z^2 + dz .$$

Since failure in the block is governed by the properties of the solid, the boundary stresses permitted will again be as given in equation (17), and similarly, slip will occur when

$$\tau_{xz} = \sigma_z \tan \phi + \tau_a \text{ at } z = z_1. \qquad (25)$$

Solving for the unknown constants using (17), (24), and (25) gives

$$x_1 = 0$$

$$x_2 = \frac{\sigma_{\max} + \sigma_0 - \sigma_{\min}}{2\rho g(1 - \lambda) \tan \phi + 2\tau_a/z_1}$$

$$e = \frac{\sigma_{\min}}{z_1} \qquad (26)$$

$$d = \frac{2\rho g \sigma_0(1 - \lambda) \tan \phi + 2\tau_a \sigma_0/z_1}{\sigma_{\max} + \sigma_0 - \sigma_{\min}}$$

$$c = \frac{2}{z_1}[\rho g(1 - \lambda) \tan \phi - d + \tau_a/z_1] .$$

To calculate the maximum length of an overthrust block, one first calculates σ_z for the base of the block from equation (23), then finds the corresponding values of $\sigma_{\min}$ and $\sigma_{\max}$ from Figure 4 or equation (10), and then calculates X_2 from equation (26). This has been done for a variety of thicknesses, pore pressure ratios, and cohesive strengths, and the results are plotted in Figures 5, 6, 7, and 8. Figures 5 and 6 show the diminishing effect of increasing the thickness of the block, which is to be expected,

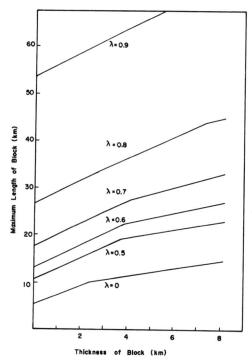

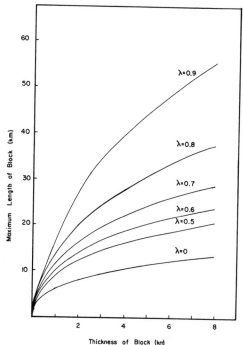

Figure 5. Maximum possible length of horizontal overthrust block, plotted against block thickness for various pore-pressure ratios with $\tau_a = 0$.

Figure 6. Maximum possible length of horizontal overthrust block, plotted against block thickness for various pore-pressure ratios, with $\tau_a = 30$ bars.

since, as shown in Figure 4, the difference between the stresses at opposite ends of the block increases less than linearly with the thickness of the block. Figure 7 shows the very large effect of increasing pore pressure which was noted by Hubbert and Rubey (1959), and Figure 8 shows how the maximum length of the overthrust block decreases rapidly as the cohesive strength at the base increases, as noted by Hsü (1969). In general the lengths calculated for a given set of parameters are considerably smaller than those calculated by previous investigators, who neglected the internal state of stress in the block.

We now consider the situation shown in Figure 9, in which a block with the same properties as before is pushed down a θ — degree slope. It is convenient to keep the coordinate system in line with the surface of the block, and when this is done, the equilibrium conditions in the solid matter of the block can be satisfied by the set of stress functions

$$\sigma_x = \frac{\partial^2 \Phi}{\partial z^2}$$

$$\sigma_z = \frac{\partial^2 \Phi}{\partial x^2} + \rho g x (1 - \lambda) \cos \theta \qquad (27)$$

$$\tau_{xz} = \frac{\partial^2 \Phi}{\partial x \partial z} + \rho g z \sin \theta \,.$$

Note that this choice again requires shear stresses to be maintained on the ends of the block. If this were not true, then σ_z and τ_{xz} could not be constant at a given depth.

Following our previous development, the internal state of stress is

$$\sigma_x = cxz + dx + ez$$

$$\sigma_z = \rho g z (1 - \lambda) \cos \theta \qquad (28)$$

$$\tau_{xz} = \frac{c}{2} z^2 + dz + \rho g z \sin \theta$$

and for boundary conditions we still require equations (17) and (25). A solution for the unknown constants gives

$$x_1 = 0$$

$$x_2 = \frac{\sigma_{max} + \sigma_0 - \sigma_{min}}{2\rho g[(1 - \lambda) \cos \theta \tan \phi - \sin \theta] + 2\tau_a / z_1}$$

345

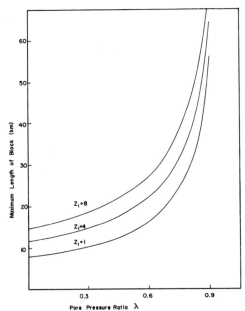

Figure 7. Maximum possible length of horizontal overthrust block, plotted against pore-pressure ratio for various block thicknesses, with $\tau_a = 0$.

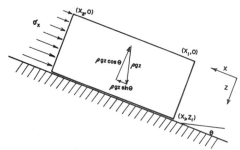

Figure 9. Overthrust block pushed down a sloping surface.

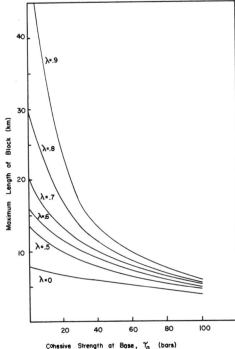

Figure 8. Maximum possible length of a 1-km-thick horizontal overthrust block, plotted against the cohesive strength at the base for various pore-pressure ratios.

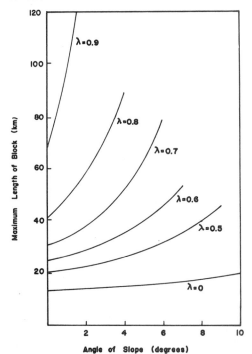

Figure 10. Maximum length of a 6-km-thick overthrust block which can be pushed down a slope of angle theta, for various pore-pressure ratios, with $\tau_a = 0$.

$$e = \frac{\sigma_{min}}{z_1}$$

$$d = \frac{2\rho g \sigma_0[(1 - \lambda)\cos\theta\tan\phi - \sin\theta] + 2\tau_a\sigma_0/z_1}{\sigma_{max} - \sigma_{min} + \sigma_0} \quad (29)$$

$$c = \frac{2}{z_1}\{\rho g[(1 - \lambda)\cos\theta\tan\phi - \sin\theta]$$

$$- d + \tau_a/z_1\} .$$

The maximum overthrust block lengths for

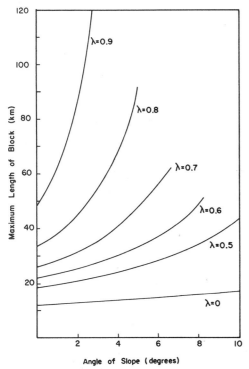

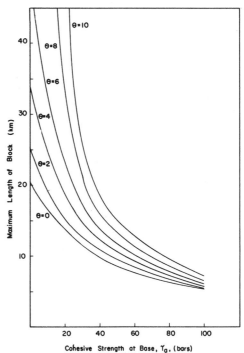

Figure 11. Maximum length of a 6-km-thick over-thrust block which can be pushed down a slope of angle theta, for various pore-pressure ratios, with τ_a = 30 bars.

Figure 12. Maximum length of a 1-km-thick over-thrust block which can be pushed down a slope of angle theta, plotted against the cohesive strength at the base of the block, with λ = 0.7.

given values of thickness, pore pressure ratio, and angle of slope can now be calculated and the results of calculation for various values of thickness, pore pressure, cohesive strength, and angle of slope are presented in Figures 10, 11, and 12. Figures 10 and 11 show that the effect of increasing the slope is more important for larger pore pressures, and Figure 12 shows that increasing the cohesive strength at the base is more important for large slopes. Again, the lengths are considerably less than calculated by previous investigators.

CONCLUSIONS

Previous calculations for the maximum lengths of overthrust blocks have not taken into account the complete elastic state of stress in the block. By using the same assumptions and material properties as have earlier authors, we have clearly shown the analytical and numerical effect of a more complete solution, using elasticity theory to calculate the state of stress in the block. Our results are also

simplifications in the sense that they do not include viscous or plastic effects, or variable pore pressures. However, we have shown the important consequence of a rather insignificant looking omission in previous elastic analyses. The data presented in our graphs show that the maximum lengths calculated by Hubbert and Rubey (1959) and Hsü (1969) are in general about 50 percent too large, quite a significant error. In addition, we see that increasing the thickness of the block has less effect than previously supposed. These results emphasize the importance of assuming abnormally high pore pressures and low cohesive strengths to explain the long overthrusts seen in nature.

ACKNOWLEDGMENT

I would like to thank Professor K. J. Hsü for many helpful discussions and a critical reading of the manuscript.

REFERENCES CITED

Hafner, W., 1951, Stress distributions and faulting: Geol. Soc. America Bull., v. 62, p. 373–398.

Handin, John, Hager, R. V., Friedman, Melvin, and Feather, J. N., 1963, Experimental deformation of sedimentary rocks under confining pressure: Pore pressure tests: Am. Assoc. Petroleum Geologists Bull., v. 47, p. 717–755.

Hubbert, M. K., and Rubey, W. W., 1959, Role of fluid pressure in mechanics of overthrust faulting: Geol. Soc. America Bull., v. 70, p. 115–206.

—— 1960, Role of fluid pressure in mechanics of overthrust faulting: Reply to discussion by H. P. Laubscher: Geol. Soc. America Bull., v. 71, p. 617–628.

—— 1961, Role of fluid pressure in mechanics of overthrust faulting: Reply to discussion by W. L. Moore: Geol. Soc. America Bull., v. 72, p. 1581–1594.

Hsü, K. J., 1969, Role of cohesive strength in the mechanics of overthrust faulting and of landsliding: Geol. Soc. America Bull., v. 80, p. 927–952.

Laubscher, H. P., 1960, Role of fluid pressure in mechanics of overthrust faulting: Discussion: Geol. Soc. America Bull., v. 71, p. 611–615.

Moore, W. L., 1961, Role of fluid pressure in mechanics of overthrust faulting: Discussion: Geol. Soc. America Bull., v. 72, p. 1581–1586.

Nur, A., and Byerlee, J. D., 1971, An exact effective stress law for elastic deformation of rocks with fluids: Jour. Geophys. Research, v. 76, p. 6414–6419.

Smoluchowski, M. S., 1909, Some remarks on the mechanics of overthrusts: Geol. Mag., n.s., Dec. v, v. 6, p. 204–205.

MANUSCRIPT RECEIVED BY THE SOCIETY DECEMBER 27, 1971
REVISED MANUSCRIPT RECEIVED APRIL 25, 1972
AUTHOR'S PRESENT ADDRESS: SHELL DEVELOPMENT CO., EXPLORATION AND PRODUCTION RESEARCH CENTER, P.O. BOX 481, HOUSTON, TEXAS 77001

Editor's Comments
on Paper 38

38 ROBERTS

The Mechanics of Overthrust Faulting: A Critical Review

While venturing to resolve some fundamental questions concerning vertical jointing in rock, an attempt was made to diagrammatically illustrate the stress history of various rocks. An example is given in Figure 1; reference coordinates are effective stresses in the horizontal $(\bar{\sigma}_x, \bar{\sigma}_y)$ and vertical $(\bar{\sigma}_z)$ directions, normalized with respect to the uniaxial compressive strength of the material $(C_0{}^*)$. With burial and compaction, horizontal stresses increase as a function of depth; this increase reflects lateral components due to increase of both vertical load and temperature. Typically, after burial, $\bar{\sigma}_x \cong \bar{\sigma}_y < \bar{\sigma}_z$. It was next assumed that horizontally imposed tectonic forces were applied in the x direction so rapidly that excess fluid pressures could not be dissipated. The change in fluid pressure could then be related to the

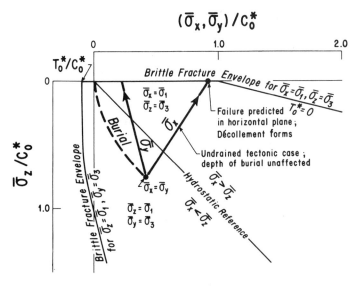

Figure 1 Idealized stress path for an arbitrary material. After burial, $\bar{\sigma}_x \cong \bar{\sigma}_y < \bar{\sigma}_z$. Application of tectonic stress under virtually undrained conditions results in increase of $\bar{\sigma}_x$ and decrease in $\bar{\sigma}_z$; failure (hydraulic fracture parallel to bedding) is predicted when $\lambda \cong 1$, with $\bar{\sigma}_x = \bar{\sigma}_1$ and $\bar{\sigma}_z = \bar{\sigma}_3 = 0$.

change in principal stress components by the expression

$$\Delta p = 1/3(\sigma_1 + \sigma_2 + \sigma_3) + \xi[(\Delta\sigma_1 - \Delta\sigma_2)^2 + (\Delta\sigma_2 - \Delta\sigma_3)^2 + (\Delta\sigma_3 - \Delta\sigma_1)^2],$$

i.e., basically the same as that employed in soil mechanics for problems involving undrained soil strength (Skempton, 1954, 1961). ξ is a fluid-pressure coefficient expressing inelastic dilatation under a given load increment.

The increase of total stress in the x direction produced an increase in the effective stress $\bar{\sigma}_x$ in the same direction, for the example shown a decrease in $\bar{\sigma}_y$, and a marked decrease in $\bar{\sigma}_z$. Obviously, $\bar{\sigma}_z$ decreased even though depth of burial remained constant, because of the marked fluid pressure increase. Failure was thus predicted when the stress path intersected the envelope representing the brittle fracture criterion; this apparently occurred, if $\xi \geq 0$, at very low effective confinement (i.e., $\bar{\sigma}_z \cong 0$). The researcher was, of course, dismayed to discover that the predicted orientation of hydraulically induced jointing was thus approximately horizontal, not vertical! Of what use are horizontal fractures, when the majority of observed joints seem to be steeply inclined? There is another possible application, however: what about progressive propagation of décollement as hydraulically induced quasi-horizontal joints? The tensile strength parallel to bedding must be virtually zero, and the development of a fluid pressure–to–overburden ratio slightly in excess of unity would be sufficient to induce propagation of a "sill" of excess fluid at the leading edge of a décollement. The stress trajectories at the leading edge would be somewhat inclined, not horizontal, but perhaps anistropy or inhomogeneity would be effective in restraining the décollement, for the most part, to specific bedding horizons. What about the effect of lithology? Uncemented argillaceous materials, which can often be regarded as "normally consolidated" clays (cf. Paper 32) seem to meet the requirement $\xi \geq 0$; rock types that are dilatant do not. Fracture or hydraulic sill propagation may thus be preferentially expected in argillaceous materials as a consequence of orogenic loading, the volume of fluid available being a direct function of initial porosity, compressibility, and imposed load. Could not this mechanism lead to the development of surfaces or zones of low residual cohesion and to progressively smaller frictional resistance? This would permit the formation of overthrust blocks of large dimensions. Thus ran the writer's argument; it was paralleled by the work of John Roberts (Paper 38). Through qualitative arguments, Roberts reached essentially the same conclusion as that presented above: dilation hardening affects sandstone or limestone horizons, which thus form overthrust sheets, whereas shale, capable of further compaction, forms the locus of the overthrust faults.

38

Reprinted from *Rept. Sec. 3, 24th Internat. Geol. Congr., Ottawa, 1972,*
pp. 593-598

The Mechanics of Overthrust Faulting:
A Critical Review

JOHN L. ROBERTS,
U.K.

ABSTRACT

The discussion concerning the pore pressure hypothesis of Hubbert and Rubey (1959) is summarized and certain conclusions drawn. Thus, frictional sliding may be governed by an adhesive strength τ_a, of lesser value than the cohesive strength τ_0, in addition to the coefficient of sliding friction (Hsu, 1969; Hubbert and Rubey, 1969). Although the cohesive strength must be taken into account when considering the initiation of the thrust planes underlying thrust sheets (Birch, 1961), it can be neglected or replaced by the adhesive strength τ_a when considering the edgewise propagation of such thrust planes (Hubbert and Rubey, 1959 & 1961b). Overthrusts, therefore, cannot develop unless the thrust sheet is underlain by a weaker layer or unless abnormal pore pressures are restricted to this layer (Birch, 1961). The conditions of failure implied by the Hubbert-Rubey hypothesis indicate that both requirements are met if the ready ingress or egress of pore fluid is prevented during impending shear failure. Under these circumstances, sediments capable of further compaction undergo an increase in pore pressure so that the effective value of λ at failure is unity. Such behaviour is typical of shale horizons which, therefore, act as the locus of overthrust faults, whereas dilatation hardening affects the intervening sandstone or limestone horizons, which, therefore, form the overthrust sheets. Once shear failure is initiated, movement is essentially frictionless as long as the excess pore pressures are maintained. Finally, the development of overthrusts is considered in relation to the tectonic evolution of orogenic belts. Such overthrusts, steepening at depth, are related to step-like displacements developed at the miogeosynclinel-eugeosynclinal boundary. Gravitational sliding is rejected as a mechanism of prime importance.

INTRODUCTION

HUBBERT AND RUBEY (1959) argue that overthrust faulting and gravitational sliding can only occur by virtue of abnormal pore pressures. Thus, according to the concept of effective stress, which they place on a firm theoretical basis, the Coulomb-Navier criterion for shear failure:

$$|\tau| = \tau_0 + \mu\sigma \qquad (1)$$

becomes:

$$|\tau| = \tau_0 + \mu \ (\sigma\text{-}p) = \tau_0 + \mu \ (1\text{-}\lambda)\sigma \qquad (2)$$

and the law of sliding friction:

$$|\tau| = \mu\sigma \qquad (3)$$

becomes:

$$|\tau| = \mu \ (\sigma\text{-}p) = \mu \ (1\text{-}\lambda) \ \sigma \qquad (4)$$

if a pore pressure p is developed, where τ is the shear stress parallel to the plane of shear failure or frictional sliding, σ the stress normal to this plane, τ_0 the cohesive strength of the material, μ the coefficient of internal or sliding friction and λ the ratio of the pore pressure p to the normal stress σ. As a result, the shear stress τ necessary for either shear failure or frictional sliding is reduced as the pore pressure p increases relative to the normal stress σ so that the ratio λ approaches unity.

On this basis, Hubbert and Rubey (1959) analyse the maximum length of a rectangular thrust block which can be pushed across a horizontal surface. For movement to occur, the shear stresses acting along the base must exceed the value given by equation (4), and the stresses developed within the block must not exceed the value given by equation (2). The results show that the length of thrust sheets with a thickness of a few kilometres could exceed 100 km, provided that the ratio λ approached unity. They also consider the case of a rectangular block sliding down an inclined surface under its own weight. The results show that slopes of a few degrees or less were required for gravitational sliding to occur, provided that the ratio λ approached unity. Subsequently, Raleigh and Griggs (1963) extended the analysis to cover thrust blocks with toes as a more realistic shape, without greatly altering the conclusions of Hubbert and Rubey (1959) except that erosion of the toe was required for movement to occur.

This general hypothesis excited considerable discussion. Laubscher (1960) and Moore (1961) made criticisms of the theoretical aspects of the analysis which were answered by Hubbert and Rubey (1960 and 1960a). Davis (1965) and Pierce (1966) considered that the hypothesis could not be applied to several specific overthrusts, although Rubey and Hubbert (1965), in reply to Davis, argued to the contrary. Carlisle (1965) argued for a very much lower coefficient of sliding friction than assumed by Hubbert and Rubey (1959), basing his argument on experimental evidence of the frictional sliding of metal surfaces under high normal stress. The relevance of this evidence to the present problem is uncertain.

Criticisms of more substance were made by Birch (1961) and Hsü (1969). In their analysis, Hubbert and Rubey (1959) neglect the term τ_0 in equation (2) when they consider the development of a sole thrust underlying a thrust sheet, arguing that the fault plane will propagate itself edgewise as a dislocation. Although this argument, reiterated in more detail by Hubbert and Rubey (1961b), appears logically sound *as far as it goes*, it is not accepted by Birch (1961) or Hsü (1969). In addition, Hsü (1969) argues that frictional sliding must overcome an adhesive strength, so that an additional term τ_a, corresponding to the term τ_0 in equation (2), should be incorporated in equation (4). Experimental studies by Byerlee (1966) indicate that the frictional sliding of granite under high normal stress is governed by such a relationship. If so, the adhesive strength τ_a would be less than the cohesive strength τ_0 for shear failure to be restricted to a particular plane of fracture (Hubbert and Rubey, 1969). Thus, it is possible that the propagation of the fault plane underlying a thrust sheet, and the subsequent movements on that plane, are governed by a relationship in the form of equation (2) rather than (4), thus resulting in a decrease in the maximum length of thrust sheets and an increase in the minimum slope required for gravitational sliding (Hsü, 1969; cf. Hubbert and Rubey, 1959).

Birch (1961) also makes a more fundamental criticism which is not satisfactorily answered by Hubbert and Rubey (1961b). Thus, in their analysis, Hubbert and Rubey (1959) only consider the conditions for the edgewise propagation of the fault plane underlying a thrust sheet; they do not consider the conditions for the initiation of the fault plane in the first instance. Evidently, this occurs at that point where the lateral stresses first exceed the shear strength of the rock, as given by equation (2) with the term τ_0 retained. Moreover, unless the fault plane is already in existence, shear failure under these circumstances will result in a normal thrust rather than an overthrust. Birch (1961), therefore, argues that overthrusts cannot develop unless abnormal pore pressures are restricted to the underlying layer or unless this layer is relatively weak. A more detailed consideration of the conditions leading to failure than that given by Hubbert and Rubey (1959) indicates the likelihood of both propositions.

THE CONDITIONS OF FAILURE

Hubbert and Rubey (1959) relate the abnormal pore pressures required by their hypothesis to the compaction of sediment containing pore fluid. In sediments undergoing compaction, the total stress can be divided into two components; namely, the effective stress supported by the framework of solid particles and the hydrostatic pressure developed by the pore fluid. Any increase in the total stress is transferred to the pore pressure, to result in a subsequent increase in the effective stress as the pore fluid escapes and the pore pressure declines. During this increase in effective stress, the void ratio decreases and the sediment undergoes compaction (Hubbert and Rubey, 1959).

The rapid loading of relatively impermeable sediment results in abnormal pore pressures which can approach the total overburden pressure (Hubbert and Rubey, 1959; Rubey and Hubbert, 1959; Bredehoeft and Hanshaw, 1968; Hanshaw and Bredehoeft, 1968). As abnormal pore pressures are directly associated with a high void ratio, the sediment can evidently undergo further compaction. In addition, the shear failure of such material under any stress differences which may develop is controlled by the ease with which the pore fluid can escape. In the present case, however, it is clear that the pore fluid cannot escape readily, because abnormal pore pressures would not otherwise be developed by compaction. Under impending failure, the framework of solid particles tends to collapse progressively under shear, resulting in an increase in the pore pressure relative to the total stress and in a decrease in the effective stress to a *minimum* value which can just be supported by the framework of solid particles. An analog is provided by the undrained test of soil mechanics carried out on sediment capable of further compaction, such as wet clay. The Mohr envelopes obtained in terms of total stress from such tests (Bishop and Henkel, 1962) are practically horizontal, so that the stress difference $(\sigma_1-\sigma_3)/2$ at failure equals the "cohesive strength" and the apparent coefficient of internal friction is zero. In terms of equation (4), the ratio λ increases to an effective value of unity as failure occurs. It is this increase in pore pressure which Hubbert and Rubey (1959) neglect in their analysis. It can be argued, therefore, that the movement of overthrust sheets is essentially frictionless, provided that the excess pore pressure is maintained long enough for movement to occur. Although the "cohesive strength" must be overcome during the initiation of the thrust planes, it is rather unlikely that there will be a finite adhesive strength to retard the subsequent movements. Moreover, the "cohesive strength" will probably be much less than the 200 bars which is taken by Hubbert and Rubey (1959) as the average value of the cohesive strength τ_0 of fully consolidated sedimentary rocks.

Such behaviour can be contrasted with the behaviour of material which undergoes an increase in volume under impending failure. In this case, there is a drop in pore pressure leading to dilatation hardening (Frank, 1965). Such behaviour is typical of granular material with a dense packing, in which the grains have to move over one another, and of solid rock, in which fractures open. Thus, those parts of the sedimentary sequence consisting of sandstone or limestone will be resistant to shear failure during overthrust movements, whereas the intervening shale horizons will be the locus of these movements.

THE DEVELOPMENT OF OVERTHRUSTS

Overthrust faulting typically affects the miogeosynclinal and exogeosynclinal areas marginal to the eugeosyncline. As Price (1971) points out, there has been a recent tendency to interpret such belts of overthrusting in terms of gravitational sliding, despite a clear lack of evidence. There are three major arguments contrary to the hypothesis of gravitational sliding. First, the intensity of deformation within the belts of overthrusting decreases away from the eugeosyncline; gravitational sliding would result in the reverse. Second, the sediments involved thicken rapidly toward the eugeosyncline; if gravitational sliding occurred, it appears to have been uphill. Third, an inspection of the Tectonic Map of North America (King, 1969) suggests that most overthrusts, apart from the Taconic overthrust, steepen in depth toward the eugeosyncline; corroboration is provided by Shackleton (1969), who states that orogenic belts have steep margins where they are exposed at depth within Precambrian shields. It is, therefore, argued that overthrusts steepen at depth into upthrusts marginal to orogenic belts so that gravitational sliding cannot be a mechanism of prime importance in their formation.

Acceptance of such arguments means that the displacements of 100-200 km across belts of overthrusting pose a considerable problem, because displacements of this order evidently cannot be maintained as the overthrusts steepen in depth. However, the margin of an orogenic belt marks a boundary where the deformation affecting the belt dies away. It can be assumed that there is a certain level at depth where this deformation results in neither upward nor downward movement of material. At this level, there will be no displacements at the margin of the belt. At successively higher levels, such displacements will increase progressively to accommodate the deformation occurring at an intervening level, thus giving rise to the major displacements across the belt of overthrusting at the highest level.

The tectonic development of orogenic belts follows a consistent pattern (Kay, 1951; Aubouin, 1965). A eugeosynclinal-miogeosynclinal couple develops as the result of long-continued sedimentation. The eugeosynclinal area is then affected by deformation, metamorphism and intrusion during the orogenic phase. This phase is associated with the influx of coarse clastic sediments, derived from the uplifted parts of the eugeosyncline, into the miogeosyncline, thereby marking a transition from miogeosynclinal to exogeosynclinal conditions. Toward the end of this phase, overthrusting affects the miogeosynclinal area (Armstrong and Oriel, 1965; Bally, Gordy and Stewart, 1966).

Prior to overthrusting, continued subsidence of the miogeosynclinal area, coupled with uplift of the adjacent parts of the eugeosynclinal area, must have resulted in a step-like displacement of the crust located along the miogeosynclinal-eugeosynclinal boundary. According to Sanford (1959), the corresponding stress distribution in terms of elastic behaviour results in horizontal compression within the area of subsidence and horizontal extension within the area of uplift. If failure occurs, shear fractures formed at a low angle within the area of subsidence steepen in depth as they are traced toward the zone of differential movement. Such a model, therefore, can be applied to the development of upthrusts, passing into overthrusts at a higher level, at the margin of an orogenic belt.

Although the overthrusts developed within the area of subsidence may follow bedding for long distances for the reasons advanced in the previous section, there is an additional factor to be considered. During the orogenic phase, various reactions leading to the evolution of water, carbon dioxide and hydrothermal fluids will occur at depth within the eugeosynclinal area (Platt, 1962; Hanshaw and Zen, 1965; Raleigh and Paterson, 1965; Heard and Rubey, 1966; Scarfe and Wyllie, 1967). As these pore fluids escape toward the surface, excess pore pressures will be developed at higher levels. Moreover, because the excess pore pressure at depth will approach the overburden pressure, and because the gradient of the pore pressure is less than the gradient of the overburden pressure under equilibrium conditions, the pore pressure can exceed the overburden pressure above a certain level, particularly if permeable rocks are capped by impermeable rocks. Under these circumstances, shear failure would give way to extension failure as the pore pressure came to exceed the minimum principal stress by an amount equal to the tensile strength of the rock (Roberts, 1970). Thus, the thrusts at the margins of orogenic belts may be formed by shear failure at depth, but by extension failure near the surface. Moreover, it can be argued that the opening of shear fractures at depth will provide a pathway for the escape of pore fluids, thereby facilitating extension failure at a higher level. Once initiated in this manner, overthrust movements on these extension fractures would be essentially frictionless because the ratio λ would have an effective value of unity, thus forming the extensive bedding-plane thrusts typical of belts of overthrusting.

REFERENCES

Armstrong, F. C., and Oriel, S. S., 1965. Tectonic development of Idaho-Wyoming thrust belt. Bull. Am. Assoc. Pet. Geol. 49, p. 1847-1886.
Aubouin, J., 1965. Geosynclines. Elsevier, Amsterdam.
Bally, A. W., Gordy, P. L., and Stewart, G. A., 1966. Structure, seismic data and orogenic evolution of southern Canadian Rocky Mountains. Bull. Can. Pet. Geol. 14, p. 337-381.
Birch, F., 1961. Role of fluid pressure in mechanics of overthrust faulting: discussion. Bull. Geol. Soc. Am. 72, p. 1441-1444.
Bishop, A. W., and Henkel, D. J., 1962. The measurement of soil properties in the triaxial test. Arnold, London.
Bredehoeft, J. D., and Hanshaw, B. B., 1968. On the maintenance of anomalous fluid pressures: I, thick sedimentary sequences. Bull. Geol. Soc. Am. 79, p. 1097-1106.
Byerlee, J. D., 1966. Frictional characteristics of granite under high confining pressure. J. Geophys. Res. 72, p. 3639-3648.
Carlisle, D., 1965. Sliding friction and overthrust faulting. J. Geol. 73, p. 271-292.
Davis, G. A., 1965. Role of fluid pressure in mechanics of overthrust faulting: discussion. Bull. Geol. Soc. Am. 76, p. 463-468.
Frank, F. C., 1965. On dilatancy in relation to seismic sources. Rev. Geophys. 3, p. 485-503.
Hanshaw, B. B., and Bredehoeft, J. D., 1968. On the maintenance of anomalous fluid pressures: II, source layer at depth. Bull. Geol. Soc. Am. 79, p. 1107-1122.
Hanshaw, B.B., and Zen E-an, 1965. Osmotic equilibrium and overthrust faulting. Bull. Geol. Soc. Am. 76, p. 1379-1386.
Heard, H. C., and Rubey, W. W., 1966. Tectonic implications of gypsum dehydration. Bull. Geol. Soc. Am. 77, p. 741-760.
Hsü, J. K., 1969. Role of cohesive strength in the mechanics of overthrust faulting and landsliding. Bull. Geol. Soc. Am. 80, p. 927-952.
Hubbert, M. K., and Rubey, W. W., 1959. Role of fluid pressure in mechanics of overthrust faulting: part I. Bull. Geol. Soc. Am. 70, p. 115-166.
———, 1960. Role of fluid pressure in mechanics of overthrust faulting: a reply to discussion by Hans P. Laubscher. Bull. Geol. Soc. Am. 71, p. 617-628.
———, 1961a. Role of fluid presure in mechanics of overthrust faulting: a reply to discussion by Walter L. Moone. Bull. Geol. Soc. Am. 72, p. 1587-1594.
———, 1961b. Role of fluid pressure in mechanics of overthrust faulting: a reply to discussion by Francis Birch. Bull. Geol. Soc. Am. 72, p. 1445-1452.

————, 1969. Role of cohesive strength in the mechanism of overthrust faulting and of landsliding: a discussion. Bull. Geol. Soc. Am. 80, p. 953-954.

Kay, M., 1951. North American geosynclines. Mem. Geol. Soc. Am. 48, p. 1-143.

King, P. B., 1969. Tectonic map of North America. U.S. Geol. Surv., Washington.

Laubscher, H. P., 1960. Role of fluid pressure in mechanics of overthrust faulting: a discussion. Bull. Geol. Soc. Am. 71, p. 611-615.

Moore, W. L., 1961. Role of fluid pressure in overthrust faulting: a discussion. Bull. Geol. Soc. Am. 72, p. 1581-1586.

Pierce, W. G., 1966. Role of fluid pressure in mechanics of overthrust faulting: a discussion. Bull. Geol. Soc. Am. 77, p. 565-568.

Platt, L. B., 1962. Fluid pressure in thrust faulting, a corollary. Am. J. Sci. 260, p. 107-114.

Price, R. A., 1971. Gravitational sliding and the foreland thrust and fold belt of the North America Cordillera: discussion. Bull. Geol. Soc. Am. 82, p. 1133-8.

Raleigh, C. B., and Griggs, D. T., 1963. Effect of the toe in the mechanics of overthrust faulting. Bull. Geol. Soc. Am. 74, p. 819-830.

Raleigh, C. B., and Paterson, M. S., 1965. Experimental deformation of serpentinite and its tectonic implications. J. Geophys. Res. 70, p. 3965-3985.

Roberts, J. L., 1970. The intrusion of magma into brittle rocks. In Newall, G., and Rast, N. (Editors), Mechanism of igneous intrusion. Geol. J., Spec. Issue, No. 2.

Rubey, W. W., and Hubbert, M. K., 1959. Role of fluid pressure in mechanics of overthrust faulting: part II. Bull. Geol. Soc. Am. 70, p. 167-206.

————, 1965. Role of fluid pressure in mechanics of overthrust faulting: reply to discussion by Gregory A. Davis. Bull. Geol. Soc. Am. 76, p. 469-474.

Sanford, A. R., 1959. Analytical and experimental study of simple geologic structures. Bull. Geol. Soc. Am. 70, p. 19-52.

Scarfe, C. M., and Wyllie, P. J., 1967. Serpentine dehydration curves and their bearing on serpentinite deformation in orogenesis. Nature, Lond., 215, p. 945-946.

Shackleton, R. M., 1969. Displacements with continents. In Kent, P. E. et al. (Editors). Time and place in orogeny. London (Geol. Soc.), 1-8.

Editor's Comments
on Papers 39 and 40

39   VOIGHT
*Clastic Fluidization Phenomena and the Role of Fluid Pressure
in Mechanics of Natural Rock Deformation*

40   LAUBSCHER
Abstract from *Die Mobillsierung klastischer Massen*

In 1965, Gregory Davis, in a discussion of the Hubbert–Rubey papers, expressed his concern about the general geological applicability of the fluid-pressure mechanism. He cited four areas of low-angle faulting in which he believed that high fluid pressures could have played no important role in the development and movement of thrust plates: (1) thrusting during regional metamorphism in the Klamath Mountains in California; (2) certain (Middle Penninic to Upper Austroalpine) Alpine nappes; and two cover faults: (3) the Heart Mountain décollement of northwestern Wyoming, and (4) the Muddy Mountain thrust of Nevada (Longwell, Paper 20, 1949). In their reply, Rubey and Hubbert (1965, p. 469) remark that "it is impossible, as we see it, to prove rigorously that high fluid pressures have been the major factor permitting large-scale horizontal movements in any area of overthrust faulting." They seek to "find out not whether our theory can be proven to account for the thrust faults of some particular area, but rather if it in any way helps to make these improbable-looking faults seem any more believable."

Perhaps ironically, in that same year the writer was, in fact, attempting to prove in the field that a condition had existed whereby fluid pressure equaled overburden pressure. The critical evidence sought involved clastic dikes along a major "overthrust" plane; the location was on the land-surface segment of the Heart Mountain detachment.

The monumental work of William G. Pierce of the U.S. Geological Survey in elucidating the geometry of the Heart Mountain problem has already been mentioned in this volume. That structure basically consists of more than four dozen discrete plates, some as large as 1 km thick and 8 km across, scattered widely over an area on the order of 2000 km^2; before movement, these plates continuously covered an area of about 1300 km² (Pierce, 1973). Three segments of the décollement have been identified; a bedding-plane segment (ca. 55 × 20 km), a "toe" or "transgressive fault" segment (ca. 20 × 3 km), and a former (Eocene) land-surface segment, now widely dissected (Fig. 2). Heart Mountain itself, which lent its name to the overall structure (Dake, 1918), is one of the larger plates sitting on the former land surface, far out in the Bighorn Basin. There are some astounding features, e.g., probably at no time did the overall average dip on the décollement exceed 2°! Furthermore, the décollement follows, almost everywhere, the same stratigraphic horizon: a thin (2.5-m) bed of dolomite. Inasmuch as the base of the moving plate is also dolomite, the mechanical implications seem incredible, for a coefficient of friction between two smooth dolomite slabs *cannot* have been less than 0.6. Pierce (1963, p. 1234; 1966) had, in fact, considered the Hubbert–Rubey fluid-pressure mechanism as a possible means of reducing the required forces, but had rejected it principally because he did not see how high fluid pressure could be developed at shallow depth, nor, if somehow developed, how it could be maintained under a fragmenting fault mass. He was thus led to suggest, following Bucher (1947, p. 196) that the

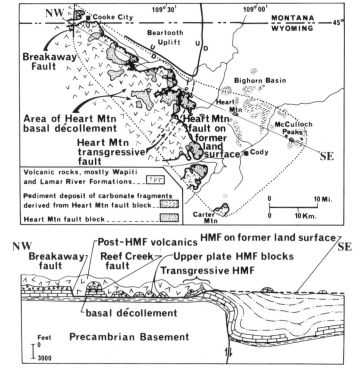

Figure 2 Schematic map and section of the Heart Mountain area, northwestern Wyoming.

"shaking motion of innumerable earthquakes," combined with the action of gravity, was responsible for fault motion (Pierce, 1963, p. 1234; 1973, p. 468–470; cf. Reeves, Paper 19). This mechanism, aptly described as a "vibrating conveyor," apparently seemed unconvincing; one difficulty that it left unexplained, for example, is the virtually complete absence of deformation below the Heart Mountain décollement (Voight, 1973, p. 118).

In any event, as a prominant North American "showpiece" of gravity tectonics (Scholten and De Jong, 1973, p. 4), the Heart Mountain structure after the middle 1960s attracted the international attention of numerous geologists, each of whom, dissatisfied with existing mechanical interpretations of that structure, independently began to search for new, sometimes novel, solutions. We have here, perhaps, a minor example of a "scientific crisis" in Kuhn's terminology. Thus in 1969, Goguel, from Paris, proposed frictional (creep)-induced vaporization of water, and Hsü from the Geological Institute at Zürich, suggested (Paper 36) a comparison with the catastrophic Flims landslide, in association with an "air-cushion" mechanism (cf. Shreve, 1968). Neither author had visited the slide. Both had confidence, on the one hand, in Pierce's (1957, 1960, 1963, 1966) lucid descriptions of field relationships; yet neither, from the vantage point of overthrust mechanics, could readily accept Pierce's outright rejection of a fluid-pressure mechanism. "The problem," as Goguel was moved to diplomatically remark upon examination of the Mineral Industries Art Collection at The Pennsylvania State University, "has not yet been solved!"

In the following year, Kehle suggested (Paper 37) a viscous décollement *zone* at Heart Mountain, and he interpreted the structure as a "high speed slide," i.e., a slide with a velocity of a few meters a year; Charles Hughes (1970), from Newfoundland, coined the term "Hovercraft Tectonics" and postulated the lateral voluminous intrusion of volcanic gas along the décollement as a mechanism for lubrication. Hughes' proposal, curiously enough, appears to be a revival of the gas lubrication mechanisms previously suggested by Walter Bucher (Paper 44, p. 1311), for the "Amargosa chaos" structure, and is akin to Bucher's early (1933a) hypothesis for Heart Mountain, which involved the "horizontal component of the force of a large volcanic explosion." Hughes' evidence was based on the "examination of only one exposure and one hand specimen," but he at least had the benefit of a brief 1968 field reconnaissance of the Heart Mountain structure, led by Pierce and Willard Parsons in association with a National Science Foundation summer program on "Structures and Origin of Volcanic Rocks."

Pierce (1973), however, vigorously opposed the new mechanism hypotheses. He argued, I believe successfully, that Hsü's suggestion of Shreve's air-cushion mechanism is incompatible with a slope of less than 2°, and that Kehle's suggestion is incompatible with the geological evidence: no viscous zone (which presumably would extensively involve Cambrian shales) seems to exist in the bedding-plane décollement segment. Pierce's adverse criticism concerning the merits of Hughes' hypothesis, however, seems less convincing to me, inasmuch as my own investigations had led to a conclusion similar to that of Hughes. The search for clastic dikes, which began in 1965, was successful; clastic injection dikes were found [the first by Pierce (1968)] in association with a thin layer of crushed carbonate fault breccia along bedding-plane and transgressive fault segments of the Heart Mountain décollement. The tectonic significance of clastic dikes is given in Paper 39, and details concerning their significance at Heart Mountain appear in Voight (1973, p. 118–120, 1974; cf. Pierce, 1968); in Voight (1973) the Heart Mountain structure is compared to the 1964 Turnagain Heights landslide of Anchorage, Alaska. During the same period of investigation a more intimate association of volcanism with Heart Mountain deformation was discovered, chiefly by Harold Prostka of the Geological Survey, who had extended his detailed mapping of Absaroka Range volcanics east of the Yellowstone Park boundary. The present writer has recently prepared a detailed paper exclusively concerned with the mechanics of Heart Mountain deformation, in collaboration with Barton Jenks, a lubrication engineer at The Pennsylvania State University; in this paper the dynamics of earthquake oscillation "vibrating conveyor" and "fluid pressurization" hypotheses are compared.

Finally, in the preparation of this volume it was discovered, by now without surprise, that the use of clastic dikes as a criterion of fluid pressurization along thrust faults had previously been suggested in an *Eclogae* article by Laubscher; the English-language abstract of that article is reprinted as Paper 40 (cf. Maxwell, 1962).

Davis' selection of North American cover thrusts in his critique of the Hubbert and Rubey concept appears to have been an ironic one; soon after Paper 39 appeared, John Dennis wrote (personal communication, 1973) that similar "clastic dikes" of fault-zone material had been found along the Muddy Mountain thrust of Nevada, a view confirmed in the next year by Stanley and Morse (1974). Thus the two overthrusts cited in 1965 to refute the fluid-pressure mechanism now appear to be among the few overthrust locations at which the former existence of high fluid pressures can in point of fact be demonstrated.

39

Reprinted from *Geol. Soc. America Northeastern Section 8th Annual
Meeting Abst., Allentown, Pa., 1973*, p. 233

CLASTIC FLUIDIZATION PHENOMENA AND
THE ROLE OF FLUID PRESSURE IN MECHANICS
OF NATURAL ROCK DEFORMATION

Barry Voight

Department of Geosciences, Pennsylvania State University

The Hubbert–Rubey resolution of the paradox of overthrust faulting arose from recognition of the influence of fluid pressure upon effective (intergranular) stresses, as given in the Terzaghi equation $\bar{\sigma}_i = \sigma_i - u$, $i = 1$–3, where $\bar{\sigma}_i$ are effective principal stresses, σ_i are total principal stresses, and u is pore fluid pressure; e.g., if $u = \lambda\sigma_i$, the effective weight per unit area of a horizontal block is diminished as λ increases, vanishing when $\lambda \geqslant 1$. The effective stress principle seems valid and applicable to the mechanics of a variety of structural elements. Unfortunately, apart from direct measurement it has thus far seemed impossible to prove that high fluid pressures have been an essential factor in individual examples of natural rock deformation. The contention here is that geological proof of the $\lambda = 1$ condition is possible to obtain. Intrusive bodies of detrital materials, e.g., clastic injection plugs, dikes and sills, and extrusive features, e.g. detrital ridges, vents, and sheets, develop as a direct consequence of hydrodynamic or pneumodynamic processes; some pore fluid is required to produce the "liquefied" or "fluidized" condition, and fluid pressure must be great enough at time of injection to totally neutralize the external stress field even under conditions of large strain, i.e., $\lambda = 1$. Clastic intrusives occupy either pre-existing void space or pneumatic/hydraulic fractures that develop as a direct consequence of intrusion. In the latter instance, in the absence of significant anisotropy, fracture orientation directly gives information on principal stress orientation. Obviously, field evidence must be examined with scrutiny both to establish contemporaneity and to demonstrate an intimate relation between fluidization phenomena and associated deformational features.

361

40

Reprinted from pp. 283, 284 of *Eclogae Geol. Helv.*, 54(2), 283–334 (1961)

Die Mobilisierung klastischer Massen[1])

I. Teil: Die Sandsteingänge in der San Antonio-Formation (Senon) des Rio Querecual, Ostvenezuela

II. Teil: Die Mobilisierung klastischer Massen und ihre geologische Dokumentation

Von **Hans Peter Laubscher** (Basel)

Mit 25 Textfiguren

* * * * * *

ABSTRACT

In the San Antonio formation (Senonian) of eastern Venezuela there are numerous sandstone dikes in sapropelitic shaly sediments. They occur both as dikes and sills and frequently form anastomosing networks. They often are complexly folded and sheared. From these peculiarities it is concluded that they were injected as high pressure gas sands into the cracks of an embryonic submarine slide mass, their deformation being due to concomitant and subsequent sliding movements. Internal structure reveals the intruding mass to have been violently turbulent at first, reworking large amounts of wall material. Subsequently, internal friction increased because of decreasing pore pressure. As a result, later deformation was essentially confined to numerous shear fractures. It is believed that sliding was facilitated if not initiated by the high pore pressures built up by bacterial activity.

Mobilization of clastic aggregates by high pore pressures has occurred in a variety of geological settings. Of particular interest are the phreatic effects of magmatic activity. Here, magmatic heat may raise pressure tremendously by evaporation of pore water in permeable clastic sediments. These may temporarily become suspended in the pore fluid so as to intrude into any crack opening in the surrounding rock. At the same time metasomatism may convert the intrusive mass into a crystalline rock of apparently magmatic origin. Other examples of mobilized clastic masses include the mylonite dikes (pseudotachylites) and the crystal mushes of deep-seated deformation.

* * * * * *

[1]) Gedruckt mit Unterstützung des Kollegiengeldfonds der Universität Basel.

[*Editor's Note:* A row of asterisks indicates that material has been deleted.]

Editor's Comments
on Paper 41

Orogenic theories are diverse; detailed information from nature must be sought in order to examine the relative merits of competing theories. The application of recently developed direct stress-measurement procedures is discussed in Paper 41 from that vantage point.

Of interest is the fact that Hubbert–Rubey fluid-pressure or Smoluchowski viscosity overthrust mechanisms need merely to be reversed to effect the locking-in of orogenic stress. The writer is under few illusions concerning practical difficulties in the interpretation of stress measurements, but the potential of this approach appears great enough to encourage its investigation.

Reprinted from *Amer. Jour. Sci.*, 274, 662–665 (June 1974)

A MECHANISM FOR "LOCKING-IN" OROGENIC STRESS

BARRY VOIGHT

Department of Geosciences, The Pennsylvania State University,
University Park, Pennsylvania 16802

ABSTRACT. The dissipation of fluid pressure is shown to provide a mechanism for "locking in" orogenic stresses in superjacent rock. These stresses can be measured presumably by existing in-situ stress measurement procedures; the measurements may thus be used presumably to evaluate alternative models of orogenesis.

To cite merely one example, Kehle (1970, p. 1650) has shown that for a given constitutive idealization, model geometry, and assumed velocity, an estimate of orogenic stress may be obtained. The equations may be difficult to evaluate if deformable plates, buckling instabilities, décollement zone "shifts", fluid pressure discontinuities, and other complexities are assumed to have been involved, but in principle results can be obtained by employment of numerical methods. *In any case, it seems clear that measurement of orogenic stress, if this were possible, could be used to provide tests of model idealization* (compare Voight and Taylor, 1969).

Significant Appalachian orogenic stresses are known to have extended well into Ohio, for example, as based on evaluation of regional joint patterns (Nickelsen and Van Hough, 1967). The long-distance transmission of such stresses was undoubtedly related to the development of décollement as suggested by Rich (1934), Woodward (1959), Rodgers (1963), Gwinn (1964), and others. If these décollement were characterized by a condition whereby the interstitial fluid pressure was of a magnitude approximately equal to overburden pressure, as suggested by the Hubbert-Rubey (1959) "resolution" of the overthrust "paradox",[1] then a suitable mechanism would be available for "locking in" such stresses; *once fluid pressure dissipated, the frictional strength imposed in the décollement zone by effective overburden pressure would prevent regional relaxation of the orogenic stress existing at that point in space and time* (fig. 1). A conversion is thus implied within the décollement zone, from a low effective stress Coulomb state or a quasi-viscous fluid state (depending on choice of idealization) to a Coulomb state involving moderate to high effective stresses normal to the décollement[2]; this conversion "freezes" lateral stresses existing in superjacent rock. Décollement should thus be expected to represent a discontinuity with respect to existing stress fields. Certain changes in stress would be expected as a consequence of denu-

---

[1] The existence of clastic dike-like bodies projecting from certain Appalachian décollement argues in favor of an enhanced fluid pressure mechanism (Voight, 1973).

[2] A conversion from a low to high viscosity state would provide an analogous mechanism for "freezing-in" orogenic stresses; the viscosity change could be related to a change in fluid pressure, temperature, or other deformation factors; this phenomenon need not be restricted to décollement but could occur generally within large volumes in orogenic belts.

dation and/or subsequent orogenic events[3]; under numerous conditions, however, the remnant tectonic components would remain significant.

The local and regional program of stress measurement by rock mechanics specialists thus could take on added meaning to tectonic geologists. A suitable and perhaps typical example is provided by stress measurements by the U.S. Bureau of Mines in eastern Ohio (Obert, 1962; Voight, 1967, p. 341-342; compare Hooker and Johnson, 1967, 1969; Voight, 1969, 1971; Rough and Lambert, 1971; Sbar and Sykes, 1973); these measurements, in limestone at a depth of about 700 m, suggest an orogenic stress component on the order of 300 bars. That stress is the pressure in excess of the horizontal pressure accounted for by existing overburden. The direction is in the east-west plane, and although it cannot be presumed *a priori* that existing stress should bear a symmetrical geometric relationship to previously developed (Paleozoic) structures, that assumption seems plausible in this instance.[4] If this stress could be assumed to be uniformly distributed in a thrust plate (this seems unlikely in detail), assuming, for purposes of example only, the adequacy of the simple model treated by Kehle (1970, p. 1649-1650), and further assuming a velocity of 10 km/m.y. (equivalent to the reported sea-floor spreading rate in the North Atlantic), a décollement zone 0.1 km thick should possess a (Newtonian) viscosity of about 4×10^{-2} m.y.-b[5]. For a décollement zone 0.01 km thick the predicted viscosity is one order of magnitude less; both values fall into the range ($10^{-1}-10^{-3}$ m.y.-b) considered reasonable by Kehle for shale under orogenic conditions.

Whether this model is even crudely appropriate remains to be tested adequately, for indeed, within that portion of the Appalachians, movement may have been largely restricted to a thin décollement horizon; the Hubbert-Rubey idealization (with inclusion of a residual cohesive term) or some alternative may be in fact more appropos at that locality. However, the proposed "locking" mechanism seems sufficiently

[3] It should be noted that the subsequent imposition of regional strains may change the entire stress field, that is, changes may occur in stresses both above and below existing décollement. In such an instance the actual value of stresses at the conclusion of the overthrust event would not be preserved; however, consideration of the *difference* in stress fields determined above and below décollement horizons, as well as the absolute magnitude of stresses within individual thrust plates, may lead to useful interpretations of the mechanics of deformation.

[4] *Note added in press*: The observed westward inclination of trajectories of maximum compression does seem in accord with inferred Paleozoic stress patterns in contrast to a previously expressed opinion (Voight, 1967, p. 342) now considered to be erroneous. If the measured stresses do in fact reflect, in part, Paleozoic deformation, these stress components can be regarded as *residual*, of type II (Voight, 1967, p. 343), that is, stresses that reflect structural inhomogeneity.

For a recent review of the complex subject of residual stress on a smaller scale the reader is referred to a paper by M. Friedman (1972); the concept of *locked-in versus locking strains* (or corresponding stresses) as discussed by Friedman seems particularly relevant to the fluid pressure mechanism presented herein.

[5] The viscosity unit "million year-bars" (m.y.-b) is used herein to permit direct comparison with Kehle's (1970) paper. Kehle has argued that this unit is more convenient than poise for geologic problems, inasmuch as time is typically measured in millions of years rather than seconds, and stress is measured in bars rather than dynes per square centimeter; 1 m.y.-b is approximately 10^{19} poise.

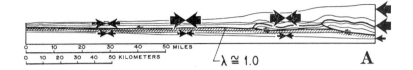

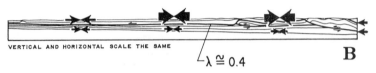

Fig. 1. Cross section of Appalachian Plateau Province in West Virginia (modified from Rodgers, 1963) to illustrate the "locking-in" mechanism for orogenic stress fields. The term λ refers to the ratio of fluid pressure to total overburden pressure. Allegheny orogenic forces are transferred within a plate lying on a low friction décollement characterized by high λ (A). The décollement represent discontinuities in regional stress fields. Dissipation of λ occurs prior to release of significant orogenic boundary forces, "locking-in" upper plate stresses; subsequent denudation produces changes in surface topography and some change in subsurface stress field, but significant stresses remain to the present day (B).

plausible to warrant further examination of the question of locked-in orogenic stress fields and its possible relationship to mechanics of décollement evolution and underlines the tectonophysical potential of rock stress measurements.

ACKNOWLEDGMENT

I am particularly grateful to Professor Jean Goguel, who pointed out the significance of the *difference* in stress fields between a tectonic "basement" and its superjacent thrust sheet.

REFERENCES

Friedman, Melvin, 1972, Residual elastic strain in rocks: Tectonophysics, v. 15, p. 297-330.

Gwinn, V. E., 1964, Thin-skinned tectonics in the Plateau and Northwestern Valley and Ridge provinces of the Central Appalachians: Geol. Soc. America Bull., v. 75, p. 863-900.

Hooker, V. E., and Johnson, D. E., 1967, *In situ* stresses along the Appalachian Piedmont, *in* 4th Symposium on rock mechanics Proc.: Ottawa, Ontario, Canada, Dept. Energy, Mines and Resources, p. 137-154.

———— 1969, Near-surface horizontal stresses including the effects of rock anisotropy. U.S. Bureau Mines Rept. Inv. 7224, 29 p.

Hubbert, M. K., and Rubey, W. W., 1959, Role of pore pressure in the mechanics of overthrust faulting: Geol. Soc. America Bull., v. 70, p. 115-166.

Kehle, R. O., 1970, Analysis of gravity sliding and orogenic translation: Geol. Soc. America Bull., v. 81, p. 1641-1664.

Nickelsen, R. P., and Hough, V. N. D., 1967, Jointing in the Appalachian Plateau of Pennsylvania: Geol. Soc. America Bull., v. 78, p. 609-630.

Obert, Leonard, 1962, In situ determination of stress in rock: Mining Eng., August, p. 51-58.

Rich, J. L., 1934, Mechanics of low-angle overthrust faulting illustrated by Cumberland thrust block, Virginia, Kentucky, and Tennessee: Am. Assoc. Petroleum Geologists Bull., v. 18, p. 1584-1596.

Rodgers, John, 1963, Mechanics of foreland folding in Pennsylvania and West Virginia: Am. Assoc. Petroleum Geologists Bull., v. 47, p. 1527-1536.

Rough, R. L., and Lambert, W. G., 1971, In situ strain orientations: A comparison of three measuring techniques: U.S. Bur. Mines Rept. Inv. 7575, 17 p.

Sbar, M. L., and Sykes, L. R., 1973, Contemporary compressive stress and seismicity in Eastern North America: An example of intraplate tectonics: Geol. Soc. America Bull., v. 84, p. 1861-1882.

Voight, Barry, 1967, Interpretation of in situ stress measurements: Internat. Soc. Rock Mechanics Cong., Panel Rept. Theme 4, Proc. 1, v. 3, p. 332-348.

————— 1969, Evolution of North Atlantic Ocean: Relevance of rock-pressure measurements, *in* North Atlantic-Geology and continental drift: Am. Assoc. Petroleum Geologists Mem. 12, p. 955-962.

————— 1971, Prediction of in situ stress patterns in the earth's crust: Internat. Soc. Rock Mechanics, Determination of Stresses in Rock Masses Symposium, Lisbon, 1966, p. 111-131.

————— 1973, Mechanics of retrogressive block-gliding, with emphasis on the evolution of the Turnagain Heights landslide, Anchorage, Alaska, *in* DeJong, K., and Scholten, R., eds., Gravity and Tectonics: New York, John Wiley & Sons, p. 97-121.

Voight, Barry, and Taylor, J. W., 1969, Tectonophysical implications of rock stress determinations: Geol. Rundschau, v. 58, no. 3, p. 655-676.

Woodward, H. P., 1959, The Appalachian region: World Petroleum Cong., 5th, Proc., Sec. 1, p. 1061-1079.

Editor's Comments
on Papers 42 and 43

Little is apparently known concerning the velocity of moving overthrust sheets. Stratigraphic information, in general, provides only minimum estimates: e.g., the *Médianes Plastiques, Médianes Rigides,* and *Brèche* nappes of the Prealps were set into motion after the middle Eocene and arrived at their present position during the Oligocene, which allows a few million years for a displacement of 100–150 km and a (minimum mean) velocity of 10^{-8}–10^{-9} knots* (Lemoine, 1973, p. 212; cf. Trümpy, 1973, p. 247–248). Nevertheless, as Lemoine observes, the velocity could have been much higher; in reality, we know little about it.

Similarly, at Heart Mountain some fault blocks rest on early Eocene rocks, and yet the fault is overlain by volcanic rocks that are either late-early or early-middle Eocene. Thus, the main movement occurred during a rather brief time interval, as emphasized by the conspicuous absence of erosion on the tectonically denuded bedding-plane fault surface subsequently buried under younger volcanic rocks (e.g., Pierce, 1973, p. 462). The fault movements are consequently known to have been very rapid; yet sufficient latitude perhaps existed (dependent upon interpretation of available evidence) such that two mathematical calculations of velocity cover a surprisingly wide range; cf. Kehle (Paper 47, p. 1649), 10^{-5}–10^{-6} knots, and Voight (1973, p. 119; 1974), 10^2 knots!

One of the first quantitative approaches appropriate for overthrust velocity calculations is due to Albert Heim, whose detailed interest in Alpine landslide phenomena spanned over a half-century (Heim, 1881; 1932). In the latter work Heim cites a letter from his physicist colleague E. Müller, in which an attempt is made to calculate the mean and maximum velocities of the classic, catastrophic Elm slide. This

---

*One knot $\cong 1.62 \times 10^4$ cm/yr.

approach, based on energy conservation, leads to the calculation of the "apparent" coefficient of friction of a gravity slide as the ratio of the height of fall and the horizontal distance traveled (Heim, 1932; Shreve, 1968; Hsü, Paper 36, p. 943; 1975; in press).

This approach is, however, restricted to catastrophic gravity sliding; it thus seems severely limited as a general approach to overthrust kinetics. The suggestion of Harold Jeffreys (Paper 42) is thus of importance. Jeffreys drew on his broad experience in seismology and physics of deformation to suggest upper and lower velocity limits. The most important arguments, from the point of view of overthrust mechanisms, involve his suggestions concerning kinetics and frictional heating: if velocities are sufficiently slow, sufficient heat can be conducted from the fault zone, and local fusion would be prevented (cf. Oxburgh and Turcotte, 1970). The principal observation from which Jeffreys begins is, however, rather old, as the following quotation from Mellard Reade (1886, p. 3–4) indicates:

> The enormous forces which have (as so clearly shown by Dr. Calloway, Professor Lapworth, and the officers of the Geological Survey) folded the strata of the Northern Highlands in such a way as to superimpose Archaean Gneiss upon Silurian rock, so as to deceive even the practised eye of a Murchison as to their succession, have not developed heat sufficient to melt any of the beds in question. Enormous ruptures of the strata have taken place, beds being driven over beds for miles; rocks have been ground to powder, yet none of these, so far as we know, have been thereby fused. . . .

Reade, in fact, was enormously preoccupied with the thermal question, which he regarded as fundamental to the origin of mountain ranges; indeed, he viewed mountains as, primarily, structural instabilities arising from temperature changes in sedimentary basins. Jeffreys began from an observation similar to that quoted above (there is no direct evidence that he derived it directly from Reade), and then proceeded, as did Smoluchowski before him, to formulate it in terms of a paradox which he then resolved.

A similar argument was subsequently employed by William H. Pierce; Pierce also makes the interesting suggestion that the potential energy released in gravity-driven overthrusts is transferred, almost entirely, into heat. Although I do not accept the author's argument (Paper 43, p. 228), that his analysis, without further modification to take into account fluid transport, is applicable for fluid-saturated porous rocks, the overall approach seems most promising. Of historical interest is that William H. Pierce, an engineer, became chiefly motivated for this work by his father's lasting contributions to the Heart Mountain fault problem. Not surprisingly, the numerical example given on his p. 230 employs data that seem, for the most part, entirely appropriate for examination of the Heart Mountain question, although it is not identified as such by the author. Finally, the essence of Pierce's conclusion is worthy of repetition: *Velocity is not merely a geologic statistic—it is an important piece of significant evidence relevant to the mechanical analysis of thrust faults and décollement.*

Reprinted from pp. 291–293 of *Geol. Mag., 79*, 291–295 (1942)

On the Mechanics of Faulting

By HAROLD JEFFREYS

1. GEOLOGICAL observation provides abundant evidence of fractures in the upper crust, the relative displacements of the sides of a fracture being anything from a few centimetres to thousands of metres. It is generally believed that earthquakes are produced in these displacements, but little attention seems to have been given to the quantitative relations involved. The geologist sees the record of the total displacement, but has no direct information about how long it took to occur, or whether it took place all at once or by several stages. Except in cases where an earthquake has been associated with fracture of the outer surface, and we can say how much displacement occurred up to the time of inspection, he has no means of knowing whether it took seconds, hours, years, or geological periods to form. Even in these rather rare cases the displacement is not more than a few metres, and extrapolation to the larger known faults is impossible. Seismology on the other hand is well adapted to the study of sudden shifts, but the instruments do not record large movements spread over a long time ; thus the geological and seismological data are largely complementary. It is known also that a large earthquake is often followed by a swarm of small ones, called aftershocks, apparently from the same focus and continuing for, possibly, several months ; there may be a thousand in a single series. The geological displacement observed long afterwards will include those due to all the aftershocks, supposing, as is generally done, that they all are due to shifts on the same fault.

Two lines of argument suggest upper and lower limits to the time required for a single displacement. A sudden change of shearing stress, as in a fracture, generates a transverse elastic wave, and a sudden change of velocity is associated with it. This is calculable from the strength of the material and the elastic properties of the material. If we divide it into the total displacement occurring in one stage of the formation of the fault we get an estimate of the time needed ; this will be a minimum because the velocity will be gradually annulled by friction and by the reduction of the stress due to the spreading out of the pulse.

Again, along a belt of overthrusting fused and partly fused rocks may have a wide distribution, and such pseudotachylytes are held to have been made amorphous by heat generated by friction.[1] If the motion was too slow this could not happen, for the heat would be conducted away before a high enough temperature was reached. Consideration of the temperature attainable should therefore give a maximum time of formation for any dislocation containing pseudotachylyte.

2. From the Appendix (1) we see that the two sides of a fault will slip in opposite directions with initial velocities $= \frac{1}{2}S\beta/\mu$, where S is the limiting stress difference, β the velocity of transverse waves, and μ the rigidity. The relative velocity of the two faces is therefore $S\beta/\mu$. For compact rocks we have approximately $S = 10^9$ dynes/cm.^{2}, $\mu = 3 \times 10^{11}$

---

[1] T. J. Jehu and R. M. Craig, *Trans. Roy. Soc. Edin.*, 53, 430, 1923.

dynes/cm.², $\beta = 3$ km./sec. The relative slip velocity is therefore about 1,000 cm./sec. and an apparent throw of 100 metres will occur in ten seconds.

If the mean depth of the fault is 1 km. the frictional resistance will be about 2×10^8 dynes/cm.², and for this and smaller depths may be regarded as a small correction. Possibly the absorption of energy in extending the fracture itself is more important, but little appears to be known about this factor. The distant parts of the body may determine a new position of equilibrium, about which the fault may perform a few oscillations. We may therefore say that the first passage of the faulted rocks through the new position of equilibrium, if the final relative displacement is 100 metres, will be more than 10 seconds after the first fracture ; allowance for the corrections mentioned may double this estimate, but hardly more. For other displacements the times will be in proportion. The data refer to granite and compact rocks such as quartzite. For sediments S, β, and μ will all be smaller, the variation of S being the most important, and the time for the same displacement may be much longer—indeed for thoroughly loose sands the elastic effect may be so small that gravity is the chief controlling factor.

3. Suppose now that friction generates heat at a rate Q per unit area per unit time over the broken face. The temperature attained on the fault is

$$V = \frac{hQ}{k} \left(\frac{t}{\pi} \right)^{\frac{1}{2}}$$

where k is the conductivity, h^2 the thermometric conductivity, and t the time taken. Suppose that the fracture takes place at a depth of 1 km. Then the normal pressure is about $2 \cdot 5 \times 10^8$ dynes/cm.² and the frictional stress about half this. The heat energy per unit area per centimetre displacement is therefore about 10^8 ergs/cm.² or 2 cal./cm.². If w is the relative slip velocity in cm./sec., then $Q = 2w$. But if H is the total slip and T the time taken, $w = H/T$, and the final temperature is $2hH/k(\pi T)^{\frac{1}{2}}$. Taking that temperature as 1,000°, $h = 0 \cdot 08$, $k = 0 \cdot 006$, $H = 10^4$ cm., all in c.g.s. centigrade units, we have

$$T = 24,000 \text{ secs.} = 7 \text{ hours roughly.}$$

Hence fusion will occur in faults of this magnitude at a depth of 1 km. if the displacement takes less than about 7 hours to develop. The time suggested is very sensitive to the data used, since all of them enter through their squares ; if we used a mean depth of 200 metres the time would be divided by 25. But 1 km. of cover is perhaps an acceptable value at the time of formation of thrusts that contain pseudotachylyte, and the other data are much less variable as between different compact rocks. The same time limit would apply if the displacement is not simple but intermittent, that is, if it occurs in stages spread over a few hours ; the essential factor is the average rate of slip during about 7 hours.

4. The elastic theory has given a time of the order of 20 secs., and it does not appear that this can be much increased. Thus even for conditions rather far from those contemplated in the last section the formation of pseudotachylyte along thrusts of this magnitude should be a normal

feature if they are produced in a single stage. Such a fused or partly fused layer would lubricate the surfaces and reduce the dissipation of energy ; oscillations would last longer. We may attempt to evaluate the thickness of the pseudotachylyte layer, nevertheless, on the hypothesis that the energy dissipated by friction is used in heating the rocks to the melting point and supplying the latent heat. With the values used above this energy in a displacement of 100 metres is 2×10^4 cal./cm.². About 500 cal./gm. would suffice to melt most rocks, and the mass of the fused layer would be about 40 gm./cm.², corresponding to a thickness of about 16 cm. This is a little too large because some of the heat would be taken up by rock that never reached the melting point. It is, however, of the order of magnitude of the thicknesses of pseudotachylyte that are found, and we may say that such thicknesses are explicable on the hypothesis that the thrust was formed in one stage, and that the heat produced by friction was used in metamorphosing the rocks.

Now though large thrusts often contain pseudotachylyte, many do not, and small ones seldom do. The larger faults often contain instead a crush breccia that has clearly never been fused. The only explanation seems to be that the whole displacement did not occur at one time, but proceeded by stages, between which the heat had time to be conducted away. This suggests a further question : if the relative slip velocity is 1,000 cm./sec. as found in section 2, how far could slip take place without developing a local temperature of 1,000° ? We have then to put $H/T = 1,000$ cm./sec. in the formula for the final temperature, taken as 1,000°, and solve for H. The result is about 4 cm. A fault with a displacement along it of 100 metres would then have to be the result of about 2,000 separate movements if it is not to contain pseudotachylyte. This may be too high, because the shear stress needed to produce new slip on an imperfectly welded crack will be less than is needed to break fresh rock ; the slip velocity will therefore be less and heat conduction will be better able to remove the heat in the time available for any given displacement. A displacement of 4 cm. with a slip velocity of 1,000 cm./sec. would be indistinguishable from an instantaneous one by a seismograph.

This result is in better agreement with data from known earthquakes. A surface fracture 6 metres in extent is highly exceptional, and most earthquakes are not associated in time with any visible surface fracture at all ; and since the frequency of aftershocks diminishes rapidly with time it is not certain that even the large displacements found in a survey after a few days have been produced as single movements. The rough agreement between the number of stages needed to produce a large fault and the number of aftershocks following a strong earthquake is also suggestive. It may be inferred that the commoner kind of earthquakes are associated with the commoner kind of fault, those that do not contain pseudotachylyte. But we also see that there is a possibility of larger single movements, which would be associated, perhaps, with more violent earthquakes than any we know, and these would produce pseudotachylyte to about the amount found in some of the larger thrusts.

[*Editor's Note:* Material has been omitted at this point.]

43

Reprinted from *Geol. Soc. America Bull.*, **81**(1), 227-231 (1970)

A Thermal Speedometer for Overthrust Faults

WILLIAM H. PIERCE   *Electrical Engineering Department, University of Louisville, Louisville, Kentucky,*
40208

ABSTRACT

The energy released in gravity-driven overthrust faults goes almost entirely into heat. The peak temperatures attained depend upon the rate of movement, since slow movement gives the heat time to diffuse away and rapid movement does not. An exact analysis for the movement duration gives a complicated equation, which is reduced to accurate and simple upper and lower bounds. Mineralization evidence of peak temperature appears as an observed physical parameter in these equations. Actual fault movement durations are typically lower bounded by several years, implying velocities so low as to require a re-examination of some theories of fault mechanics.

INTRODUCTION

The purpose of this paper is to study the rate aspect of fault movement, specifically the problem of inferring movement rate of ancient fault movements by using mineralization evidence to determine peak temperatures. The point of departure can be taken to be the following excerpt from Scheidegger's geodynamics text (1963):

Friction [at faults] also causes much heat to be produced which, in turn, might help to soften the material adjacent to the fault surface, thus again lowering the resistance to sliding. The absence of pronounced chemical evidence regarding this point, however, shows that heating cannot be too great. Unfortunately, no further quantitative corroborations of the above qualitative arguments are as yet available.

An analysis will be developed which will use frictional heating effects to determine the velocities of overthrust faults. The greater the observed heating, the more rapid must have been the movement, as large observed heating implies that the heat energy did not have time to diffuse away during the fault movement.

Previous work on overthrust faults has had a mechanical, not rate-inferring, orientation. Hubbert and Rubey (1959) studied strength aspects of fault masses. They, and Raleigh and Griggs (1963), also show how water pore pressure can help float the overburden, thereby reducing the frictional pressure at the fault. Ode (1961) presents a plastic flow theory for overthrust faults. Bostrom (1968) studied large-scale mountain movement in the fjord region of British Columbia, where static rock strength may have been weakened by a phase transition.

THE MODEL

The purpose of this section is to develop a simple but realistic model for the movement of fault blocks under the influence of gravity. Assuming the gravitational potential energy is all converted into heat, a subsequent analysis will develop equations which give rate of movement from thermal evidence data, specifically mineralization conditions near the fault.

The model assumes the following:

(1) A fault block of vertical height H moves downhill on a fault plane of angle Θ from the horizontal. Gravity is the only driving force; the gravitational acceleration is g.

(2) The total horizontal distance traveled is L.

(3) The density above and below the fault block is ρ. The ratio of rock density ρ to the density of water is $\rho_{specific}$.

(4) The rocks above and below the fault have identical thermal properties, namely diffusivity k and specific heat (per mass) of C. (If σ is the thermal conductivity, then $k = \sigma / C \rho$). The latent heat of any phase changes is negligible.

(5) The speed of travel is assumed to be a constant, say R, which is not known. The duration of travel is thus the unknown time $T = L/R$.

(6) The gravitational potential energy is completely converted into heat uniformly throughout a zone of vertical height $(h/\cos \Theta)$ at the fault zone. Thus, heat is generated frictionally in a zone of rock flour at the fault.

(7) The grinding energy is considered negligible compared to the heat energy.

(8) All the rocks are initially at ambient

temperature U_A. Mineralogical studies show that at some known distance x from the center of the fault zone, the peak temperature was U_0. (The analysis also gives a bound upon R when the temperature at x is known to be upper or lower bounded by U_0.)

(9) The height H is so large and the rocks are such poor conductors of heat that the heat energy dissipated at the surface is negligible.

(10) The block is horizontally so large that edge effects are negligible, and the thermal data came from an interior point. No gaps exist within the fault block through which heat could escape to the surface. No fluid circulation cools the fault, and no heat energy is absorbed in phase changes.

(11) The temperature is U, time t, distance normal to the center of the fault zone is z, which is positive in the upward direction. Fault movement ends at $t = 0$ and begins at the unknown $t = -T$.

In a later section, quantitative statements will be developed which can verify the accuracy of assumptions (7) and (9), the latent heat aspect of assumption (4), and the edge aspects of assumption (10). Conservation of energy of course always is valid, so the significance of assumptions (6) and (7) is that there is no sizeable energy form (except heat) into which the lost gravitational energy is converted. By conservation of energy, the magnitude of the frictional force does not affect the amount of heat generated. The analysis will in fact be based upon conservation of energy, and not upon solving Newton's laws of motion. As a consequence, the analysis applies in either the presence or absence of hydrostatic pore pressure.

THE ANALYSIS

The assumed symmetry normal to the fault plane makes the problem one dimensional in the sense that the temperature depends only upon t and z. Let δS be an area in a plane parallel to the plane of the fault. Let the heat energy generated in this area during time $t_0 \leq t \leq (t_0 + dt)$ in the zone $z_0 \leq z \leq (z_0 + dz)$ be the function $F(z_0, t_0)dt\, dz\, \delta S$. Because heat flows only in the z direction, the basic differential equation is the one-dimensional heat flow equation:

$$\frac{\partial U}{\partial t} = k \frac{\partial^2 U}{\partial z^2} + \frac{F(z, t)}{C\rho}. \quad (1)$$

When $F(z,t) = C\rho\, \delta(z - z_0)\, \delta(t - t_0)$, this equation has solution

$$U(z, t) = U_A + \frac{1}{\sqrt{4\pi k(t - t_0)}}$$
$$\times \exp\{-(z - z_0)^2/4k(t - t_0)\}$$
$$\text{for } t \geq t_0. \quad (2)$$

When $F(z,t)/(C\rho)$ is the constant K, when $-h/2 \leq z \leq h/2$ and $-T \leq t \leq 0$, and zero elsewhere, then the linearity of the differential equation shows that U will be the following superposition integral:

$$U(z, t) = U_A + K \int_{-T}^{0} \int_{-h/2}^{h/2} \sqrt{\frac{1}{4\pi k(t - \tau)}}$$
$$\times \exp\{-(z - \alpha)^2/4k(t - \tau)\}d\alpha d\tau$$
$$t \geq 0. \quad (3)$$

The evaluation of K is based upon the fact that the total energy per horizontal area released in the fault is $(L\rho\, gH \tan\Theta)$. The total energy per area in the fault plane therefore will be $L\rho\, gH \tan\Theta \cos\Theta$, which equals $F(0, 0^-)hT$, which equals $KC\rho\, hT$. Thus,

$$K = \frac{LgH \tan\theta \cos\theta}{ChT}. \quad (4)$$

If the peak temperature at x is known to be U_0, the maximum of $U(x, t)$ will occur for nonnegative t. Therefore, the equation follows that

$$U_0 = U_A + \max_{0 \leq t \leq \infty} \left\{ \frac{LgH \tan\theta \cos\theta}{ChT} \right.$$
$$\times \int_{-T}^{0} \int_{-h/2}^{h/2} \sqrt{\frac{1}{4\pi k(t - \tau)}}$$
$$\times \left. \exp[-(x - \alpha)^2/4k(t - \tau)]d\alpha d\tau \right\}. \quad (5)$$

This equation implicitly defines T, and thus serves as the basic equation for deducing T given U_0 and x. Since $RT = L$, this also implicitly defines R, thereby being an equation for a thermal speedometer.

In the special case $x = 0$, the maximization over t occurs at $t = 0$, since heat will only flow outward from the center, but the heat at $x = 0$ increases up to the time the motion stops and then decreases. Let U_0^* denote the $x = 0$ special case. Then

$$U_0^* = U_A + \frac{LgH \tan\theta \cos\theta}{ChT} \int_{-T}^{0} \int_{-h/2}^{h/2} \sqrt{\frac{-1}{4\pi k\tau}}$$
$$\times \exp(\alpha^2/4k\tau)d\alpha d\tau. \quad (6)$$

An upper bound on U_0^* follows from replacing the α^2 in the exponent by zero, while a lower bound results from replacing α^2 by $(h/2)^2$. The resulting integration is independent of α, so the α integration merely introduces a factor of h. Letting $\gamma = -T/\tau$ in the remaining integration, transposing the U_A, taking constants outside the integral, dividing by the resulting constant multiplying the integral gives the following bounds:

$$\int_1^\infty \gamma^{-3/2} \exp(-h^2\gamma/16kT)d\gamma$$

$$\leq \frac{2(U_0^* - U_A)C\sqrt{\pi kT}}{LgH \tan\theta \cos\theta}$$

$$\leq \int_1^\infty \gamma^{-3/2}d\gamma. \tag{7}$$

The upper bound is 2. The lower bound can be expanded in a Taylor series in $(h^2/4kT)$, which is dimensionless. The expansion diverges around $(h^2/4kT) = 0$, so an expansion around $h^2/4hT = 0.4$ is selected. The remainder term for the Taylor series expansion shows that the first two terms serve as a lower bound. Numerical evaluation of the resulting integrals gives

$$1.79 - .0798 \frac{(h^2}{4Tk} - 0.4)$$

$$\leq \frac{2C(U_0^* - U_A)}{LgH \tan\theta \cos\theta}\sqrt{\pi Tk} \leq 2. \tag{8}$$

In situations where $(h^2/4kT) \leq 1.2$, these bounds will be close to each other, which is likely to be the case. In such situations, equation (8) can be used to estimate T as accurately as the model can be expected to apply.

Now suppose that at some x the temperature U_3 was known to exceed some crystallization temperature U_2. Then $(U_2 - U_A) \leq (U_3 - U_A) \leq (U_0^* - U_A)$ so that the upper bound in (8) can be solved for T to give

$$\frac{L}{R} = T \leq \frac{1}{k\pi}\left[\frac{LgH \tan\theta \cos\theta}{(U_2 - U_A)C}\right]^2. \tag{9}$$

On the other hand, if at $x = 0$, it was known that the temperature was some value U_0^* less than or equal to a crystallization temperature U_1, then $(U_1 - U_A) \geq (U_0^* - U_A)$, and substituting this in the lower bound of (8), squaring and multiplying both sides by 0.25 times the right side of (9) gives

$$T \geq \left\{\frac{1}{k\pi}\left[\frac{LgH \tan\theta \cos\theta}{(U_1 - U_A)C}\right]^2\right\}\left\{.91 - \frac{.01h^2}{Tk}\right\}^2 \tag{10}$$

A lower bound on T can be obtained from (10) as follows: Let $T_{1.b.(present)}$ be an initial guess for a lower bound on T, such as half the right side of (9). Substitute $T_{1.b.(present)}$ into the right side of (10), and calculate the right side of (10), calling it $T_{1.b.(new)}$. If $T_{1.b.(new)} \geq T_{1.b.(present)}$, then $T_{1.b.(new)}$ is a valid lower bound on T, because the right side of (10) can be lower bounded by itself and evaluated at any lower bound for T. If $T_{1.b.(new)} < T_{1.b.(present)}$, then reiterate with an appropriately decreased revised estimate of $T_{1.b.(present)}$.

The terms of the type $[LgH \tan\theta \cos\theta]/[\delta UC]$ appearing in (9) and (10) have the dimension of length. This length is the width of rock which could be heated to a temperature change δU by the amount of gravitational energy released in the fault movement. This length will be called the "energy-equivalent length."

Equations (9), valid for any x, and (10), valid for $x = 0$, give explicit bounds for solving the general implicit equation (5) for the unknown T or R.

VERIFICATION AND MODIFICATION OF ASSUMPTIONS

The purpose of this section is to present quantitative criteria to evaluate the accuracy of some of the model's assumptions. Provision will also be made for different rock properties above and below the fault.

A surface at finite z will in actuality always introduce additional cooling. Thus the actual temperature at any x and t will be less than or equal to the temperature at calculated x and t assuming an infinite extension of material in the z direction. Since the time T decreases as the temperature change increases, it follows that equations (5) and (10) give valid lower bounds on T when the rock has only a finite height above the fault plane.

In order for an equation or an upper bound on T to be reasonably accurate, it is necessary and sufficient that the height H be at least three standard deviations of the distance heat will diffuse in time T. Since this standard deviation is $\sqrt{2Tk}$, it follows that even though the assumed infiniteness in the z direction is violated, equation (5) will be reasonably accurate, and equation (9) will be reasonably accurate, if the T so computed obeys $T \leq H^2/(18k)$. Further manipulations of this, combined with (9), show that equation (9) will be reasonably accurate if H is 2.4 or more times the energy equivalent length.

The accuracy in ignoring the grinding energy can be accounted for by assuming that, at the very worst, the grinding energy is 10 percent of the total, and that the actual grinding energy is not more than the total energy per volume necessary to grind the raw materials for portland cement. Thus the appropriate criterion is that the grinding energy (at the cement raw materials rate) not exceed one-tenth the gravitational energy. The total grinding energy for the raw materials of cement is about 7 kilowatt hours per barrel (Witt, 1947). Much of this undoubtedly appears as heat, so this gives a conservative upper bound on the non-heat energy absorbed in grinding. By manipulations of units and formulas for the gravitational energy, it follows that for rocks such as limestones and shales, the assumption that the grinding energy can be neglected is valid if $h \leq 4.5 \times 10^{-6}$ m^{-1} $\rho_{specific} HL \tan \Theta$. For example, if $\rho_{specific} = 2.5$, $H = 500$ m, $L = 5000$ m, $\Theta = 2°$, then the grinding energy can be neglected if $h \leq 1$ m.

If the diffusivities above and below the fault are k_{hi} and k_{low} respectively, then the above results need to be corrected as follows: Return to (1) and let $y = z$ for $z \geq 0$, and $y = k_{hi}/k_{low} z$ for $z < 0$. A constant coefficient equation in y of the form of (1) results.

The ratio energy in $(z < 0)$ to the energy in $(z < 0)$ is

$$\frac{C_{hi}}{C_{low}} \frac{\rho_{hi}}{\rho_{low}} \sqrt{\frac{k_{hi}}{k_{low}}} = \sqrt{\frac{\sigma_{hi}}{\sigma_{low}} \frac{C_{hi}}{C_{low}} \frac{\rho_{hi}}{\rho_{low}}}.$$

To account for the unequal division of energy, L should be replaced by L times this ratio when z is positive, and L divided by this ratio when z is negative, whenever equations (5), (9), or (10) are used, where the other constants in these formulas are taken for the material of the appropriate z.

A comment on the accuracy of neglecting edge effects, assumption 10, is that this is reasonable at distances from the original edges which are several times the energy-equivalent length.

Neglecting the phase change energy will be valid if and only if the latent heat energy released per area in the fault plane is small compared with $L\rho gH \tan \Theta$.

EXAMPLE AND CONCLUSIONS

An example will now be worked to indicate the type of conclusions which the thermal speedometer analysis can give. The data assumed are the following:

$H = 700$ m
$L = 15$ km
$\Theta = 2°$
$C = 1$ joule/(gm° C) (typical of limestone, Birch and others, 1941)
$\rho_{specific} = 2.5$
$g = 9.8$ m/sec^2
$h = 0.5$ m
$\sigma = 1.8 \times 10^{-3}$ watts/(Cm °C) (typical of stone, Esbach, 1952)
U_1 or $U_2 = 450$ °C (This is the approximate marbleization temperature of limestone.)

Calculations show that the energy-equivalent length is 8.6m, k is 7.2×10^{-8} m^2/sec, and assumption (9) is valid. If marbleization was observed, then $U^2 = 450$ °C and (9) shows that $T \leq 10.3$ years. On the other hand, if no marbleization was observed, then $U_1 = 450$ °C and the suggested method of solving (10) gives $T \geq 8.5$ years.

The following are the conclusions about the thermal information obtainable on ancient fault movement rates:

(1) If no heating effects are observed at the fault, only a lower bound on movement duration can be obtained. The order of this duration will be a decade.

(2) If heating effects are observed, then equation (9) can be used to calculate an upper bound upon the movement duration. A lower bound in this case could be obtained either by (a) applying equation (10) if a second and hotter chemical change were not observed, or (b) solving equation (5) numerically.

(3) Instantaneous fault movements would leave a thickness of recrystallized rock approximately equal to the energy-equivalent length. The solution of equation (5) for T may become degenerate as the observed crystallization band approaches the energy-equivalent length.

(4) The usual absence of crystallizations over a width anywhere comparable to the energy-equivalent length indicates that overthrust fault movement durations are seldom less than a year.

(5) In the absence of observed crystallization, fault movement duration could be lower

bounded by the theory of this paper, and in at least some cases (*see* Pierce, 1966, 1968) erosional effects might supply an upper bound within an order of magnitude of the thermal lower bound.

The above conclusions call for a re-examination of earlier theories on the mechanics of overthrust faults, for two reasons:

(1) All mechanical theories must be affirmed to apply at the velocities involved. (There is some doubt as to whether the concept of static coulomb friction can pass this requirement.)

(2) Beginning from a knowledge of typical fault movement velocities, it may be possible to develop more accurate theories. (The new theories are likely to be more closely related to creep and plastic flow than to concepts developed from conditions for bricks to slide on inclined planes.)

The speed of overthrust fault movement is not just a geologic statistic—it is an important piece of experimental evidence relevant to geologic movement mechanisms.

ACKNOWLEDGMENTS

The original stimulus for finding fault movement rates came from discussions with William G. Pierce, my father, concerning the Heart Mountain fault. Arthur Lachenbruch was consulted early in the work and sketched an order of magnitude calculation indicating the potential feasibility of the approach used.

REFERENCES CITED

Birch, Francis, 1941, Elasticity (except compressibility), *in* Birch, F., Schairer, J. F., Spicer, H. C., *Editors*, Handbook of Physical Constants: Geol. Soc. America. Spec Paper 36, p. 82.

Bostrom, R. C., 1968, Oceanward spreading of the continental crust in the fjord region of Northwest America, *in* The Trend in Engineering: Univ. Washington, v. 20, no. 2, p. 4–9.

Esbach, O. W., 1952, *Handbook of Engineering Fundamentals*· New York, John Wiley and Sons, p. 13–39.

Hubbert, M. K., and Rubey, W. W., 1959, Role of fluid pressure in mechanisms of overthrust faulting: Geol. Soc. America Bull., v. 70, p. 115–166.

Ode, H., 1960, Faulting as a velocity discontinuity in plastic deformation, *in* Griggs, D., and Handin, J., *Editors*, Rock Deformation: Geol. Soc. America Mem. 79, p. 293–322.

Pierce, W. G., 1966, Role of Fluid Pressure in Mechanics of Overthrust Faulting: Discussion: Geol. Soc. America Bull., v. 77, p. 565–568.

—— 1968, Tectonic denudation as exemplified by the Heart Mountain fault, Wyoming: 23rd Internat. Geol. Cong. Proc., v. 3, p. 197.

Raleigh, C. B., and Griggs, D. T., 1963, Effects of the toe in the mechanics of overthrust faulting, Geol. Soc. America Bull., v. 74, p. 819–830.

Scheidegger, A. E., 1963, Principles of geodynamics, New York, Academic Press, section 7.24.

Witt, J. C., 1947, Portland cement technology, Brooklyn, Chemical Publishing Co., p. 105.

Editor's Comments
on Papers 44 Through 46

Consideration of time is more explicitly treated in viscous idealizations of rock behavior. The viscous idealization, long dormant, came into focus again with a growing recognition of the utilitarian significance of Gulf Coast salt domes and the publication of M. K. Hubbert's (1937) paper, "Theory of Scale Models as Applied to the Study of Geologic Structures."

In 1954, S. Warren Carey, an Australian, published an influential paper, "The Rheid Concept in Geotectonics," in which he contrasted the "earth of seismology" with the "earth of geotectonics." *Rheidity* (Greek ρεω, flow) was defined as a time parameter which determines whether a substance will behave as a fluid or as a solid for a given loading condition. Essentially employing a Maxwell rheologic idealization, rheidity was then derived as the ratio of "viscosity" to "rigidity" multiplied by an arbitrary number, say 10^3, which expresses the insignificance of elastic strain in comparison with viscous strain components.

For loading over a duration substantially greater than the rheidity (e.g., several weeks for glacier ice, 10 years for rock salt, 10^6 years for schist) Carey assumed that elastic and transient strain terms could be wholly neglected; hence, the deformation could be discussed mathematically in terms of fluid mechanics. A cross section of the Alps was thus compared with morainal contortions in the Malaspina glacier and with flowing scum in a river eddy.

Walter Bucher, meanwhile, had been deeply involved in field studies in overthrust areas, most notably in early work in the Heart

Mountain and Beartooth uplift region. In his European collegiate studies (Ph.D. Heidelberg, 1911), and in his subsequent attempt to "assemble all essential geological facts of a general nature that bear on the problem of crustal deformation and to derive from them inductively a hypothetical picture of the mechanics of diastrophism" (which study led to the epoch-making volume, *Deformation of the Earth's Crust*), Bucher became well aware of the varying modes of tectonic deformation displayed in overthrust belts. Familiar with the recumbent fold nappes of the Alps, and impressed by similar, more recent (1939) interpretations by Arnold Heim and Gansser in the Himalayas, he became (as Professor of Structural Geology at Columbia University) intensely involved in experimental work on thrust faults. His experimental procedures include consideration of questions of model similitude; indeed, he was familiar with that subject long before Hubbert's presentation (Bucher, 1933b, p. 144).

The role of experimentation in Bucher's work was to examine concepts and to discover unanticipated effects of model parameter variations; experiments typically have served to stimulate the imagination of the experimenter, but it might be noted that lack of creative imagination was never one of Bucher's defects. In Paper 44 herein, emphasis is given to the importance of viscous components in elastico-viscous materials, and to the role of gravity in effecting major deformation of such materials, given the benefit of geologic time intervals. This paper was an influential one, and indeed, modern concepts of "gravitational spreading" as applied to the southern Canadian Rockies (Price, 1973, p. 496–501) seem to owe their stimulus directly to Bucher's work. The earlier experiments of Sollas (Paper 7; cf. Bucher, Paper 44, Fig. 1, Pl. 1) have been long forgotten.

A more extensive program of experiments, with a similar purpose to examine theories of orogenesis, has been more recently carried out by Hans Ramberg and his colleagues at Uppsala. Ramberg's centrifuge approach is similar to that of the American mining engineer Bucky, in the 1930s, but his experiments are far more sophisticated in the analysis of complex and finely laminated composite geologic models.

Similarities between Pennine-type nappes and salt diapirs, mantled gneiss domes, and sialic plutons make it tempting for Ramberg to believe that orogenesis is a process powered by gravitational instability. In the excerpts included herein (Paper 45) Ramberg and Sjöström demonstrate the advantage of centrifugal modeling and produce significant insights into the question of "crustal shortening" in orogenesis (cf. however, Mellard Reade, 1886, p. 170–172).

In soil mechanics, as well as in tectonics, the role of viscous deformation had been much neglected. There appear to be good reasons for this neglect. Rigid-plastic idealizations lead generally to simpler mathematical approaches for engineering work, and the dominantly "plastic" behavior of sands at failure appear to be precisely mirrored by clay deformation if analyses are carried out in terms of effective stress. Furthermore, the engineer is typically interested in material behavior over relatively short time spans, say 50–100 years, and thus the time factor is seldom evaluated in material property investigations.

Even so, engineers confronted with serious problems of landslide control have certainly appreciated the reality of creep in slopes (see, e.g., Zaruba and Mencl, 1969). Virtually imperceptible in early stages, such slope behavior occasionally leads to catastrophic failure. More often slope creep leads to less spectacular problems of seasonal concern, e.g., retaining-wall damage, which, however, still involve expensive remedial work on an annual basis; the problem often does not simply "disappear," and the specter of potential disaster lingers in slowly moving masses of earth and rock.

Close attention has therefore been devoted to the problem, particularly in the last few decades. Notable studies have been conducted by, for example, Ter-Stepanian and others in Russia, Suklje in Jugoslavia, and Saito in Japan. A considerable advance of knowledge has resulted and a useful presentation of the state-of-the-art is summarized in Paper 46 by Bing Yen.

Of great importance in future work on the mechanics of thrust faulting, it would appear, will be the adaption of existing mechanical theory to specific field occurrences of faults. In general, particularly with the availability of such modern numerical approaches as finite-element methods (Dieterich, 1969; Voight and Dahl, 1969; cf. Goodman et al., 1968; Wang and Voight, 1969; Zienkiewicz, 1967, 1970; Desai and Abel, 1972), *the major difficulty will no longer lie in available mechanical theory* but in the adequate determination of boundary and environmental conditions, load paths, and material (rheologic) properties, for rock at the time of deformation. The general approach may follow directly from Rubey and Hubbert's classical analysis of the Western Wyoming Thrust Belt, but modern approaches may be expected to be more sophisticated, inasmuch as consideration of elastic–plastic–viscous interactions, internal discontinuities, inhomogeneities, local finite-amplitude buckling, energy expenditure in flexure at toe and ramp structures, and changing boundary conditions all seem to be numerically achievable. A burden is thus placed squarely on the shoulders of both the field geologist and the experimentalist, who must now provide data of equivalent refinement in order to provide realistic ground rules for mathematical modeling of particular structures.

44

Reprinted from pp. 1295–1302, 1305, 1306, 1311, 1314, 1316, 1317, 1318 of
Geol. Soc. America Bull., **67**(10), 1295–1318 (1956)

ROLE OF GRAVITY IN OROGENESIS

By Walter H. Bucher

(*Address as Retiring President of The Geological Society of America*)

Abstract

In elastico-viscous materials, the concept of "viscosity" makes model experiments possible, since time appears in its definition as an independent dimension when gravity is the only force acting on the material in both nature and the laboratory. Stitching wax ("shoemaker's pitch") is discussed as a useful material for tectonic experiments. The limit of stress below which solid flow does not take place is very low in this material, permitting slopes of only a few degrees to be maintained. Material raised to greater relative elevations flattens under its own weight, as does ice in an ice cap. Laboratory experiments using stitching wax are described which show that such flattening produces superficial folding in stratified layers of the same material if zones of lower viscosity, such as petrolatum, are present among the layers. Good examples of disharmonious folding are figured.

In a second series of experiments, bars of layered stitching wax (without weak zones) were subjected to slow compression. When one part was kept a little warmer than the rest, a recumbent anticline was formed at the interface; the cooler "foreland" pushed under the uparching "welt," and it in turn pushed the former down, flattening it. In this way, the features which in alpine mountains have been thought to call for either an actively overriding master thrust sheet ("*traineau écraseur*") or an active downsucking ("engulfment") were produced simultaneously. This leads to the conjecture that orogenic belts, at least of the alpine type, arise because, at the time of their formation, they are weaker than the normal crust. This thought is pursued further in the last chapter. The structural details of the recumbent folds produced in the experiments are instructive.

In the course of such close folding of the basement surface, the weaker materials of sedimentary mantle must be forced out onto the slope toward the foreland. The structural consequences are discussed hypothetically. The stronger of such detached sedimentary units pile up as "peel thrusts." The weaker ones, such as sediments of flysch type and serpentine, must flatten out greatly to form "thrust flow sheets." The weakest may continue to spread still farther as the foreland is warped ("tilt flow"). "Chaotic" structures are briefly mentioned. They seem to be special cases involving gaseous volcanic activity.

Role of Gravity in Marginal Deformation

Problem of Shallow Deformation

The problem of surficial deformation shall serve to introduce the subject of this lecture.

Belts of long, shallow folds in sedimentary rocks, such as those of the Swiss Jura, and extensive thin thrust sheets, such as the Cumberland thrust sheet in Kentucky, Tennessee, and Virginia, have this in common: they have moved differentially over the formations beneath them. They do not represent a shortening of the earth's crust, but merely a sloughing off of its outermost layers.

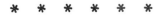

The movements of such thin sheets cannot be due to external stress applied to them endwise.

The sediments of which they consist are too weak, and the frictional resistance of such large sheets is far too great to make such deformation possible. This was recognized long ago (*e.g.*, Lawson, 1922, p. 342–343).

In such cases it would seem natural to seek the explanation in the action of gravity. One thinks of loose soil wrinkling as it creeps down a slope of bedrock, or of the snow cover on a sloping roof thrown into folds as it slumps down (de Sitter, 1954, p. 322; Seibold, 1955, p. 295–296).[1] But the efficacy of such action depends upon the existence of a sufficient slope.[2] A gentle slope did indeed exist in both cases,

---

[1] For references to early examples of this thinking *see* Haarmann, 1930, p. 101–106.
[2] For the Jura, Goguel computes, on reasonable assumptions, that a slope of 8° would suffice to cause gliding; he does not suggest, however, that this actually took place (Goguel, 1943, p. 396).

[*Editor's Note:* A row of asterisks indicates that material has been deleted.]

the Swiss Jura and the western edge of the Appalachian folded belt, but it was in the wrong direction. The folds of the Jura and the Cumberland thrust sheet moved up the gently sloping side of a trough of sedimentation.

The first step toward the understanding of such folding must be, then, to learn how elastico-viscous materials behave and search for materials suitable for scale-model experimentation.

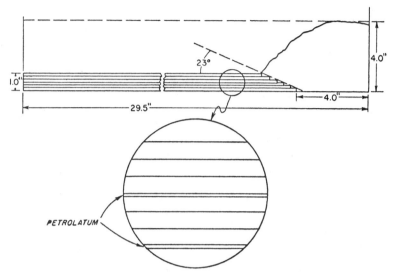

FIGURE 1.—DIAGRAMMATIC SECTION OF EXPERIMENTAL SET-UP

The high body of clear stitching wax meets layers of the same material (including grease layers) along an inclined plane representing a thrust plane.

This leaves us with the puzzling problem: How can orogenic pressure which must ultimately reside in the crust as a whole be translated into skin-deep action at the outer edge of its reach?

The writer has suspected for some time that the answer does indeed lie in the action of gravity on the rocks of the outer crust, but in a more fundamental sense than has been generally contemplated. He asked himself this question: Suppose mountain folding affected layers of glacier ice instead of rocks. Its folds would not stand up under their own weight, that is certain. Are we sure that rocks such as limestones, gneisses, and granites stand up under their own weight to any possible height? Glacier ice is elastico-viscous—*i.e.*, it behaves as an elastic and brittle material under rapidly applied stress and yet flows like a viscous substance under prolonged stress. The smooth folds into which quartzites and gneisses as well as limestones and shales can be formed show that these rocks behave in a similar way.

A brief reference in the first edition of Jeffrey's *The Earth* called the writer's attention to shoemaker's pitch—or stitching wax[3]—as a promising material (Jeffreys, 1924, p. 113). Experimentation with this substance has become possible under the sponsorship of the Humble Oil and Refining Company. The experiments reported in this paper were designed by the writer and set up under his supervision in the company's Research Laboratory at Houston. For the time-consuming preparation and execution of the experiments themselves, including skillful photography, the writer is indebted to his collaborator, Mr. Peter Masson, a geologist. Systematic studies of the physical properties of stitching wax and other materials are being made by a second able collaborator, Dr. Charles Arnold, a chemical engineer. Some of the things we learned in the course of these and related studies underlie much of what follows.

---

[3] Essentially a resin with a plasticiser.

In a first experiment, layers of three kinds of stitching wax (clear, white, and brown) and two thin layers of grease (petrolatum) were built up between well-lubricated walls of plexiglass to simulate a stratigraphic sequence (Fig. 1). At one end, the layers were stepped back to form a slope of about 23°. A lump of clear wax was laid on top of that slope, rising to four times the height of the layered sections, to represent some massive basement rock brought up along a low-angle thrust. Then the material was left alone for several days. The wax of the high block, too weak to support its own weight, flattened out under the action of gravity. In so doing, it threw the upper layers of the section into a series of folds. When the folding ceased, the whole block was chilled with crushed ice and then sawed through lengthwise.

The structure thus made visible is shown in the overall picture in Figure 1 of Plate 1, and in detail for the proximal part in Figure 2 of Plate 1. Note especially:

(1) The deformation decreases both upward and downward from a maximum attained below the middle of the stratigraphic column.

(2) The structure varies from strongly overturned folds near the source of pressure to almost symmetrical open folds away from it. This decrease in the intensity of folding with distance is characteristic of all "marginal folding" (Bucher, 1933, p. 157).

(3) The inclined base of the high block, the "thrust plane," has flattened to a nearly horizontal position in its lowest part and steepened to over 60° near the surface. This reproduces a peculiarity of many thrust planes associated with shallow folding which, near the surface, are high-angle reverse faults, some nearly vertical, and bedding-plane thrusts at depth.

Usefulness of Stitching Wax for Tectonic Model Experiments

The similarity between the folds produced in this experiment and the surficial folds in nature is obvious. But does it mean anything? Can we take seriously the implication that solid rock bodies uplifted sufficiently will flatten out under their own weight and produce results comparable to those shown in the experiment?

Note that much higher stresses are required to produce fracturing by rapid "forced deformation" than are needed for the slow deformation (creep) under the action of gravity. All available evidence suggests that the same is true of most rocks. Yet many geophysicists base their thinking concerning orogenesis on the assumption that the rocks of the earth's crust will not deform permanently until the "crushing strength" is reached, forgetting that under the unhurried action of gravity flow must begin long before that limit is attained.

Except for Newtonian liquids, viscosity is as yet an ill-defined property that is measured in industry by a multitude of empirical devices. These provide data that are useful for comparisons among like materials, but for different types of materials give only a very crude approximation to a measure of relative behavior.

The concept of viscosity makes possible model experiments that introduce the time factor in correct perspective. In his important paper on model theory for geologists, Hubbert has shown that, to insure dynamic as well as geometric similitude in model experiments in which gravity is the only significant force, the material must be chosen such that the viscosity of the rock in nature (η_n) is to that of the substance used in the model (η_m) as the product of the model ratios of · the three independent variables density (δ), length (λ), and time (τ) (Hubbert, 1937, p. 1489–1490).[8]

To see how useful stitching wax is in tectonic model experiments, we solve the equation for the average "apparent viscosity" of the sediments involved in such folding as that of the Swiss Jura, inserting the scale ratios for the dimensional units of viscosity used in the ex-

---

[8] Geophysicists will be interested in the chapter on Dimensional Analysis and Similarity by Corrsin (1953).

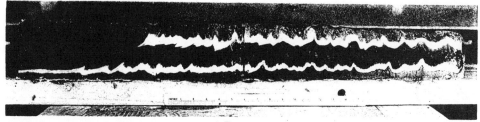

FIGURE 1

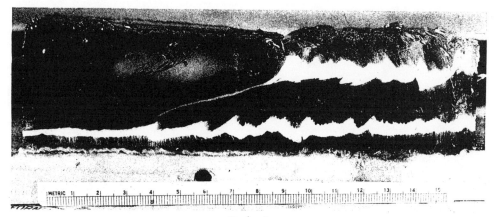

FIGURE 2

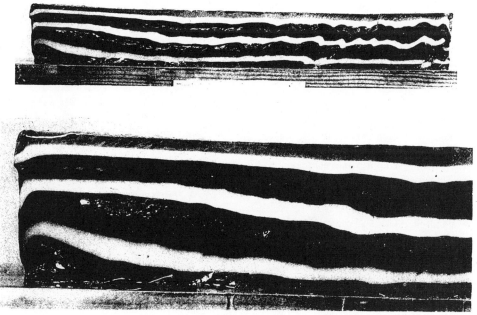

FIGURE 3

FOLDING PRODUCED BY SOLID FLOW IN LAYERS OF STITCHING WAX
(WITH LAYERS OF GREASE)

periment described below. These ratios are:

$$\frac{\text{Density in nature}}{\text{Density in model}} = \delta = \frac{2.5}{1} = 2.5$$

$$\frac{\text{Length in nature}}{\text{Length in model}} = \lambda = \frac{3 \times 10^6 \text{ cm}}{30 \text{ cm}} = 10^5$$

$$\frac{\text{Time in nature}}{\text{Time in model}} = \tau = \frac{10^6 \text{ years}}{2.7 \times 10^{-3} \text{ years}} = 4 \times 10^8$$

$$\frac{\text{Viscosity in nature}}{\text{Viscosity in model}} = \frac{x \text{ poises}}{10^6 \text{ poises}}$$

$\eta_n = 10^6 \times 2.5 \times 10^5 \times 4 \times 10^8 = 10^{20}$ poises
= "apparent viscosity" of folded sediments.

Is this figure, 10^{20} poises, reasonable?

Viscosity as a Function of Stress, Temperature, and Chemical Environment

The rocks involved in the folding of the Swiss Jura are chiefly limestones and shales. No figures are available for shale. Griggs has determined the "equivalent viscosity" of dry Solenhofen limestone at 2×10^{22} poises, under a differential pressure of 1,400 kg/cm² (Griggs, 1939, p. 235; 1940, p. 1016).

The viscosity of non-Newtonian liquids and of plastic solids is, of course, not a constant but is itself a function of stress. It is also a function of temperature and, above all, of chemical factors.

Let us look at these factors separately and see how they affect the value of viscosity. Changes in differential stress seem to have the smallest effect. The viscosity of the stitching wax is merely reduced to half, from 8×10^6 to 4.7×10^6 poises, when in the falling-ball method the weight of the ball is increased 500 times. The best approximation to a figure for the "apparent viscosity" of glacier ice is that derived from observations on the velocity of glacier movement in relation to the thickness of the glacier. The best values for the viscosity of glacier ice computed in that way range through three orders of magnitude, from 10^{14} to 10^{12} poises, with increasing velocity (Deeley, 1908, p. 251–253). The range of these values is probably too great, since it includes the effect of temperature which is large for ice as for all elastico-viscous substances.

According to Höppler (1941, p. 157) "quasi-viscosity" of polycrystalline ice subject to deformation, at right angles to the optic axes of coplanar crystals, rises from 10^9 to 10^{13} poises, as the temperature drops from $-1°$ to $-30°C$. The viscosity of stitching wax rises similarly, from 10^2 to 10^6, when the temperature drops $45°$, from $69.5°C$ to $25°C$.

Still another factor enters to complicate the picture. The rise in temperature probably changes the mutual solubility of the constituents of stitching wax. Changes in the chemical environment of the particles whose movement produces solid flow is probably the most potent factor to influence viscosity.

This is indicated by Griggs' observations on alabaster. Dry alabaster is very viscous. Under a differential stress of 420 kg/cm² (5174 psi), its "equivalent viscosity" is of the order of magnitude of 10^{18} poises. But in the presence of its own saturated aqueous solution, under a lower differential stress, the viscosity drops to 10^{17} poises. Under greater differential stress and a confining pressure of 1000 atmospheres, it drops to 10^{14} poises (Griggs, 1940, p. 1017–1020).

Limestone must behave in the same manner. Griggs found the "equivalent viscosity" of the very fine-grained, homogeneous Solenhofen limestone very high at 1 atmosphere confining pressure and a differential pressure of 1400 kg/cm², viz., 2×10^{22} poises (Griggs, 1939, p. 235). Under very high confining and differential pressures the "equivalent viscosity" of the dry rock could be reduced to as low as 10^{14} poises (Griggs, 1939, p. 245). It is to be expected that the presence of its own saturated solution will also reduce the viscosity of limestone by several orders of magnitude.

* * * *. * *

A comparison of the results of the model experiments in the light of the considerations just presented encourages us in the belief that stitching wax comes near being adequate for experiments that give an essentially correct picture of the kinematics and dynamics involved in crustal deformation. For the overall deformation of the complex sequence of limestones and shales of which the Jura folds consist, the figure 10^{20} poises seems entirely reasonable.

No figures are available to the writer for the apparent viscosity of shales. It must be materially lower than that of limestones and sandstones. Shales are above all inhomogeneous, with a strong tendency to part more easily parallel to the bedding planes than at an angle to them. That property combined with their affinity for moisture must tend to lower the viscosity of shales by several orders of magnitude below that of the stronger sediments. If we use stitching wax, with a viscosity of 10^6 poises for the limestones, we must use something like grease to represent shale formations.

Details of the Folding in Sedimentary Series

That layers of stitching wax and grease come close to reproducing to scale the typical folding in sedimentary series becomes evident when such a sequence is subjected to simple compression. In the next two experiments, alternating layers of different kinds of stitching wax with intercalated layers of grease were compressed in an electrically controlled device that allowed the compression to proceed very slowly, $\frac{1}{2}$ cm per hour and less. This rate is comparable to that at which the elevated wax spread in the first experiment and produces folding of the same kind as seen on Plate 1, but on a larger scale. Artificially applied force is thus here substituted for the force exerted by spreading wax to facilitate control over the experiment and to make possible experiments on a larger scale.

In the first of these experiments (Pl. 1, fig. 3), the active pressure was applied at the left end of the column, which was thickened and thereby produced a slope. Figure 4 of Plate 1 and Figure 1 of Plate 3 show that the two thin layers of grease, being weaker than the white layers of stitching wax, have induced folding with much smaller wave length, thereby furnishing a fine example of disharmonious folding.

* * * * * *

ROLE OF GRAVITY IN AXIAL FOLDING

Details of the Formation of Recumbent Folds

In further experiments, the chilled, stiffer part of the bar of stitching wax was replaced by a wooden bar, to save time and effort. As the wooden bar with its high strength and rectangular shape has no counterpart in nature, the value of such experiments is limited to the study of the formation of a recumbent anticline out of a column of stitching wax which undergoes horizontal compression, and the effect of the advancing front of the "nappe" on the layers of the foreland. Plate 5 shows five successive steps in the development of the recumbent nappe in one such experiment. Note that the foreland beds have thickened somewhat, and that there is no foreland folding, because the stratigraphic column contained no weaker layers that facilitate *décollement*.

When grease layers are included among the wax layers on top of the wooden ("foreland") bar, surficial folding of the Jura type develops in front of the advancing recumbent fold. This is but one of the instructive features of the experiment to be described next. Figure 1 on Plate 6 shows the general set-up: (1) the motor-driven piston on the left, pushing the wooden bar toward the weak zone on the right; (2) the strata on top of the "foreland" bar, 3 cm thick, which include layers of grease; (3) the 8-cm thick zone without grease layers, which represents the "metamorphic core." The piston moved forward 2.5 cm/hr—*i.e.*, several times faster than the wax can advance by creep under the action of gravity. As a result, during the day the wax piled up at the weak end somewhat higher than the plexiglass walls and had to be held between raised walls. During the night, the material spread out again. The experiment lasted about 4 days.

Figure 1 of Plate 6 shows the set-up on the first day, when the piston had advanced about 8 cm. Compression was stopped when the piston had advanced more than twice as far. After spreading of the material had practically come to an end, the bar was removed and sawed open after having been cooled in ice. Figure 2 of Plate 6 shows the face of the bar sawed open, broken into five pieces which were propped up on lumps of modelling clay for photographing.

The recumbent anticline is the most striking feature, with its vertical "root" on the right-hand side and its apex not far from the center of the picture. Note the greatly thinned and drawn-out forelimb of the anticline (the

FIGURE 1

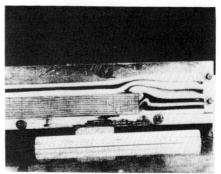

FIGURE 2

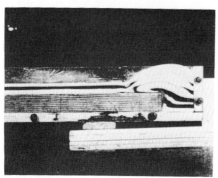

FIGURE 3

FIGURE 4

FIGURE 5

RECUMBENT ANTICLINE PRODUCED BY COMPRESSING A SHORT, THICK COLUMN OF STITCH-ING WAX BENEATH AN OVER-ALL COVER OF THE SAME MATERIAL (WITHOUT GREASE LAYERS)

FIGURES 1–4.—Four intermediate stages of the experiment seen through the plexiglass walls.

FIGURE 5.—Internal structure at final stage (Specimen sawed through; "root zone" broken off). Note the final slope of surface. No foreland folding developed, because of absence of grease layers from "foreland" beds. (Total length of ruler: 12.6 inches = 32.0 cm)

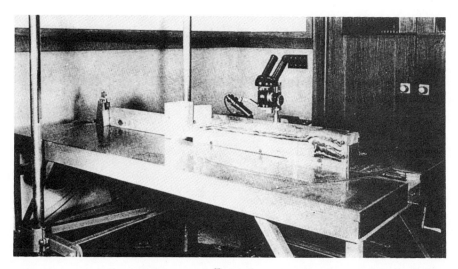

FIGURE 1

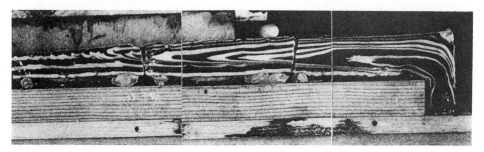

FIGURE 2

FIGURE 3

RECUMBENT ANTICLINE PRODUCED BY COMPRESSING A SHORT, THICK COLUMN OF
STITCHING WAX BENEATH AN OVER-ALL COVER OF THE SAME MATERIAL
(WITH GREASE LAYERS)

FIGURE 1.—The experimental set-up. The "foreland" with its wooden base has advanced about 8 cm from the left to the right. The experiment was eventually stopped when the "foreland" had advanced twice as far. (Length of upper edge of ruler: 15.8 cm = 5.90 inches)

FIGURE 2.—Internal structure of the recumbent anticline at end of experiment (sawed open and the pieces assembled resting on lumps of modeling clay). Note the "root zone"; the digitations on apices in the core; the greatly thinned "middle limb"; and the surficial folding of the foreland, dying off downward.

FIGURE 3.—Detail of Figure 2 of this plate, showing front of the recumbent anticline and the folding in the foreland.

389

"middle limb" of European nomenclature).

Returning to the experiment, note the surficial folds in front of the nappe (Pl. 6, fig. 3). Nearest the apex, these surficial folds are themselves overturned into a near-horizontal position. Farther from the apex of the advancing front, the folds are rather open and little overturned, if at all. The nature of the deformation changes downward and is almost absent in the bottom layers. The second white layer from the top seems not to be folded, but cut by a number of thrust planes. This is, however, a deception. What appear to be thrust planes are in reality the isoclinally drawn-out cores of recumbent folds, which involve the whole layer.

It is of interest to compare the apparent reduction of the distance from end to end of the experimental block with the actual shortening. The distance along the surface of the recumbent anticline, measured at right angles to the strike from one side of the "root" to the apex and back to the other side, is more than 80 cm. The actual shortening produced by the compression is about 20 cm. But in front of the apex, the "foreland" sediments (on the wooden bar) have been thrown into surficial folds. This part of the model is shown in detail in Figure 3 of Plate 6. These folds suggest another shortening of at least 20 cm. Its counterpart in nature would still be attributed by some geologists to "crustal shortening." In other words, the traditional method of estimating the "shortening of the crust" would suggest anywhere from 4 to 5 times the amount actually involved in the experiment. This, then, is an experimental demonstration of the process postulated recently by the writer (Bucher, 1955, p. 357–358). It goes far to explain the conflict between the estimates of crustal shortening made by geologists and the amount deemed possible by geophysicists (Jeffreys, 1952, p. 306–309).

Looking back we thus see gravity as a primary and secondary agency directing the development of orogenic structures, from the great primary recumbent folds of the active cores of orogenic belts to the marginal surficial folds and thrust sheets, including at the surface the "thrust-flow" sheets of serpentine and of allochthonous argillaceous formations, the "tilt-flow" sheets developed from mobile clays, and "chaotic" structures, with decreasing dimensions, down to the "rock glaciers," "mud flows", and "heaving shales" of mountain sides and valleys.

References Cited

Bucher, W. H., 1924, The pattern of the earth's mobile belts: Jour. Geology, v. 32, p. 265–290
—— 1933, The deformation of the earth's crust: Princeton, N. J., Princeton Univ. Press, 518 p.
—— 1955, Deformation in orogenic belts, p. 343-368 in Poldervaart, Arie, Editor, Crust of the earth: Geol. Soc. America Special Paper 62, 762 p.
Goguel, Jean, 1943, Introduction à l'étude mécanique des déformations de l'écorce terrestre: Mém. Carte Géol. de France, Paris (Imprim. Nationale), 514 p.
Griggs, David, 1939, Creep of rocks: Jour. Geology, v. 47, p. 225–251
—— 1940, Experimental flow of rocks under conditions favoring recrystallization: Geol. Soc. America Bull., v. 51, p. 1001–1034
Haarmann, Erich, 1930, Die Oszillations-Theorie: Stuttgart, Ferd. Enke, 260 p.
Höppler, F., 1941, Die Plastizität des Eises: Kolloid Zeitschr., v. 97, p. 154–160
Hubbert, M. K., 1937, Theory of scale models as applied to the study of geologic structures: Geol. Soc. America Bull., v. 48, p. 1459–1520
Jeffreys, Harold, 1924, The Earth: 1st ed., Cambridge, Univ. Press, 278 p.
—— 1952, The Earth: 3rd ed., Cambridge, Univ. Press, 392 p.
Lawson, A. C., 1922, Isostatic compensation considered as a cause of thrusting: Geol. Soc. America Bull., v. 33, p. 337–352
Seibold, E., 1955, Ein Hangrutsch als tektonisches Modell: Neues Jahrb. Geol., Monatsh., p. 278–297
Sitter, L. U., de, 1954, Gravitational gliding tectonics. An essay in comparative structural geology: Am. Jour. Sci., v. 252, p. 321–344

45

Reprinted from pp. 105, 107–108, 122–128, 130, 131, 132 of
Tectonophysics, **19,** 105–132 (1973)

EXPERIMENTAL GEODYNAMICAL MODELS RELATING TO CONTINENTAL DRIFT AND OROGENESIS

HANS RAMBERG and HÅKAN SJÖSTRÖM

University of Uppsala, Uppsala (Sweden) and University of Connecticut, Storrs, Conn. (U.S.A.)
University of Uppsala, Uppsala (Sweden)

(Accepted for publication February 9, 1973)

ABSTRACT

Ramberg, H. and Sjöström, H., 1973. Experimental geodynamical models relating to continental drift
and orogenesis. In: E. Irving (Editor), *Mechanisms of Plate Tectonics. Tectonophysics,* 19(2): 105–132.

After introductory comments on scale-model theory, descriptions and discussions of experimental
models of continental drift and orogenesis are presented. The driving mechanism of the models is the
spontaneous overturn in layered systems with gravitationally unstable density stratification. The pattern of the overturn movement is similar to thermal convection currents. A model Pangaea breaks in
tension above the upwelling branches of the overturn cells in the model mantle, and the fragments are
set adrift by the horizontal limbs of the current. Though buckling occurs along the leading edges of the
wandering continental fragments, this fold structure is not of the kind encountered in natural orogens.
The latter kind of structures — including basement upheaval, batholith intrusion, nappe formation
etc. — is generated in models with no net horizontal shortening across the orogen.

"Though the primary direction of the force which thus elevated them must have been from below upwards, yet it has been so combined with the gravity and resistance of the mass to which it was applied, as to create a lateral and oblique thrust, and to produce those contortions of the strata, which, when on the great scale, are among the most striking and instructive phenomena of geology."
John Playfair: Illustration of the Huttonian Theory of the Earth, 1802.

[*Editor's Note:* A row of asterisks indicates that material has been deleted.]

391

THE EXPERIMENTAL METHOD

To overcome the experimental obstacles caused by the small strength and low viscosity of the materials as required for models exposed to gravity as the only body force, we run our models in a high-capacity centrifuge (for description of the apparatus see Ramberg, 1967, 1971; Stephansson, 1972). The reaction against the centripetal acceleration (i.e., the negative of the centripetal-acceleration vector) plays the same role in the models as does the acceleration due to gravity in nature. In the centrifuged models the two above-mentioned force ratios take the forms $\rho la/p$ and $\rho l^2 a/\mu v$ where a is the centripetal acceleration affecting the model while it is whirled in the machine. At dynamic similarity between model and nature the conditions:

$$\frac{\rho_n l_n g_n}{s_n} = \frac{\rho_m l_m a_m}{s_m} \quad \text{and} \quad \frac{\rho_n l_n^2 g_n}{\mu_n v_n} = \frac{\rho_m l_m^2 a_m}{\mu_m v_m}$$

must be fulfilled. Here subscript 'n' refers to the natural geologic structure, 'm' to the model. Since the acceleration a_m in our machine is up to $4000\,g_n$ it follows that the strength, s_m, and the viscosity, μ_m, of the model materials can be up to 4000 times larger than these properties of the materials in an otherwise identical model not run in the centrifuge. This circumstance enables one to study scale models of intricate geologic structure not susceptible to experimental tests by other means. It is of particular value that the method permits us to study model structures containing any number of unlike rocks or earth materials with different densities, viscosities or strengths. Since there is practically no sagging of the rather strong materials during the construction of the initial structure of the models, the initial pattern — which of course has to be potentially unstable in a body-force field — can be made as complicated as we wish in order to conform to the natural situation. The dynamic evolution of the model is confined to the period of centrifugation. Thus the finite structure can be scrutinized to any desired degree after the run.

*   *   *   *   *   *

Models simulating orogenic belts

The European Alps, the Scandinavian Caledonides, the Appalachians in America and oth-er orogens exhibit convincing field evidence of large horizontal relative movements of the rock masses. At some places rocks are displaced horizontally over their substratum for dis-tances of the order 100 km. There are also abundant signs of lateral compression in the form of buckle folds in stratified rock complexes. To many the obvious conclusion from such findings is that an orogen represents a large lateral shortening across the belt from the one edge to the other. This is the "giant vise" hypothesis of orogenesis.

This idea of overall shortening has for a long time been a prevailing basic assumption in the geologists' models of orogenesis, an assumption which has reached the status of an ax-iom for many theoreticians concerned with continental drift and plate tectonics.

However, our model studies warn that orogenic belts should not unconditionally be taken as evidence of crustal shortening. A remarkable result of the experimental work is that structures strikingly similar to the very orogenic structures which are regarded as proof of overall lateral compression form in fact in the models without a net shortening across the orogens. Moreover, these model structures form spontaneously from a gravita-tionally unstable mass distribution of a kind which, we believe, may well have existed in "geosynclines" and along continental edges from where orogens rise.

The chief field observations taken as indicative of an overall lateral shortening across orogenic belts are in part the large recumbent folds (nappes) and thrust sheets of rock masses which override their substratum of both younger and older rocks, in part the abun-dant buckle-folded strata with steep axial plane. For observations and reasoning behind the large-scale horizontal movement hypothesis see, e.g., Rodgers (1970) for the Appalachians, and Kautsky (1953) and Magnusson et al. (1962) for the Scandinavian Caledonides.

The idea that a thrust sheet is evidence of shortening across an orogen appears to rest on the assumption that the sheet which overrides one edge of the orogen is rigidly connected with the craton (whether the latter be oceanic or continental) on the opposite side of the fold belt. This, of course, need not be so, and the present authors know of no place where

393

such a situation is sustained by recorded field observations. For the Scandinavian Caledonides, for example, there is convincing evidence that the frontal part of the thrust sheets in
Sweden has moved southeastward over the old Baltic Shield, but there is no sign of significant relative movement between the sheets and their substratum in the interior of the
Caledonides in Norway (see e.g., Nicholson and Rutland, 1969).

There is no physical objection against a process which permits the front of a sheet to
move laterally relative to its substratum at the edge of the orogen while in the interior of
the orogen the same sheet remains fixed in relation to the substratum even if the latter has
not been shortened. In other words, the movement is in the form of vertical flattening and
compensating lateral spreading of the thrust sheet, much like the movement in a giant ice
cap. It is also possible that the sheet simply glides downhill from above elevated basement
culminations in the central parts of the orogen.

A spreading sheet of sufficient thickness may even move up the slope of its substratum
provided that the surface of the sheet slopes in the direction of movement. These kinds of
movements are demonstrated in many of our models as exemplified in Fig. 13, 14 and 15.

These are gravity driven, or at least gravity controlled, tectonic processes as visualized
in one version or another by Haarmann (1930), Van Bemmelen (1933, 1960), Ramberg
(1945, 1963, 1971), Beloussov (1961) and others.

A condition which strongly supports the spreading and/or gravity slide hypothesis for
the Caledonian thrust sheets in Norway and Sweden is that the individual sheets thin out
toward the interior of the Caledonides in Norway (Nicholson and Rutland, 1969; Zachrisson,
1969). The thinning is not due to erosion, and it is not believed to be primary sedimentary.
The picture fits well the hypothesis of spreading from a center above the rising basement
geanticline along the coast of Norway. Obviously, the general westward thinning of the
sheets does not speak for a forceful thrust from the west. This would give a stress distribution which compresses the thrust sheets laterally more in the west than in the east (because
of the resistance along the thrust plane), and consequently should make the sheets thicker
toward the west.

That recumbent fold-type nappes, such as the well known Pennine nappes in the Alps
and the less known Iltay nappe in Scotland (Craig, 1965) be taken as proof of lateral shortening is also contradicted by the experiments. In the tests Pennine-type nappes form readily
when giant upheavels of low-density gneiss—granitic basement approach the surface and
there spread laterally to form huge folds with quasi-horizontal axial plane and inverted

Fig.13. Section through model of orogen run for 8 min. at acceleration increasing from 1300 g to 2200
g. Initial layered structure shown in Fig.17. The final structure shows domes and nappes of grey, white
and black silicone having risen through putty overburden (inclined hatching) and thin "sedimentary"
layers of modelling clay and silicone putty under cover of soft wax. (From Ramberg, 1967.)

Fig.14. Section through model initially quite similar to $S 114$ in Fig.13; see, however, Fig.17. Run for
15 min. at 2400 g – 2900 g. (From Ramberg, 1967.)

Fig.15. Section through orogen-imitation model $Ny 9$. Run for 290 sec at 2200 g, 10 sec at 2600 g and
632 sec at 3000 g. Initial structure shown in Fig.18.

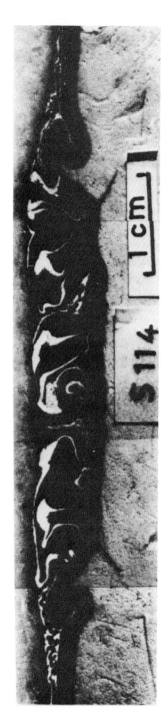

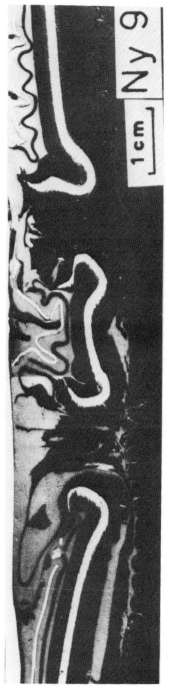

Fig.13.

Fig.14.

Fig.15.

Fig.16. Section through orogen-imitation model *Ny* 7. Run for 95 sec at 2000 g. Initial structure shown in Fig.18.

stratification in their lower limb. Examples are presented in Fig.13, 14 and 15. See also Fig. 16–18. In these experiments there is no shortening of the distance from one craton to the other.

Sometimes the model nappes even exhibit a so-called imbricate structure which is typical in many orogens. This structure is well developed in model *Dv* 7, Fig.19, 20 and 21. Here thin surficial sheets of white and black plasticene broke to small flakes when the recumbent fold came "rolling" over its lower limb. The broken flakes became partly stacked on top of one another in the process, much like the imbricate structure one encounters along the edge of orogens.

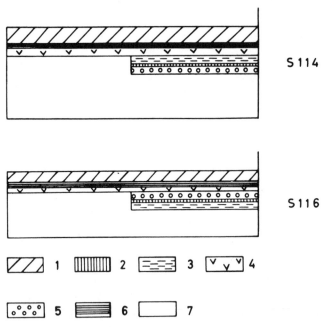

Fig.17. Cross-sections of initial structures of model *S 114* and *S 116* Fig.13 and 14. Only left half of symmetrical models shown.
1 = soft oil-wax mixture, ρ = 0.9 g/cm^3; *2* = white silicone putty, ρ = 1.14 g/cm^3; *3* = grey silicone putty, ρ = 1.25 g/cm^3; *4* = green painter's putty, ρ = 1.87 g/cm^3; *5* = black silicone–magnetite powder mixture, ρ = 1.35 g/cm^3; *6* = silicone putty with thin layers of modelling clay; *7* - painter's putty, ρ = 1.87 g/cm^3.

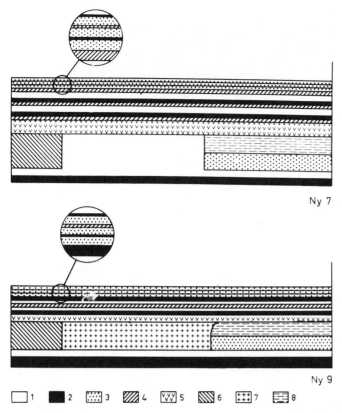

Ny 7

Ny 9

Fig.18. Cross-sections of initial structures of models *Ny 7* and *Ny 9*, Fig.15 and 16. Only left half of symmetrical models shown.

1 = grey painter's putty, ρ = 1.85 g/cm^3; *2* = black modelling clay, ρ = 1.68 g/cm^3; *3* = silicone putty with tungspar powder, ρ = 1.28 g/cm^3; *4* = white modelling clay, ρ = 1.78 g/cm^3; *5* = green painter's putty, ρ = 1.85 g/cm^3; *6* = red modelling clay, ρ = 1.71 g/cm^3; *7* = yellow painter's putty, ρ = 1.85 g/cm^3; *8* = silicone putty with magnetite and tungspar powder, ρ = 1.58 g/cm^3 in model *Ny 7*, 1.33 g/cm^3 in *Ny 9*.

Fig.19. Cross-section of centrifuged model showing spreading surficial nappe or thrust sheet with well developed imbricate structure along sole. Nappe is spreading lobe of silicone-putty diapir which has risen from underneath a layered overburden of painter's putty and modelling clay. Arrows indicate flow directions. Small dome on bottom a little to the right of center consists of stiffer silicone which has risen more slowly. See also Fig.20 and 21.

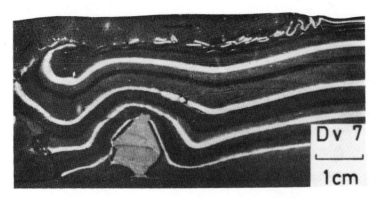

Fig.20. Enlarged part of Fig.19, model *Dv* 7, showing details of imbricate structure along sole of spreading nappe (dotted). Surficial thin sheet of dark modelling clay outlined in ink.

For the understanding of natural imbricate structures along the sole of nappes some details in Fig.20 and 21 are illuminating. The recumbent fold nappe has moved over a surficial multilayer consisting of three thin sheets of competent plasticene embedded in incompetent silicone, the whole complex resting on the "craton". We think that it is tectonically significant that the three sheets have responded very differently when being overridden by the nappe. The uppermost black sheet, which is directly subjacent to the spreading nappe, has broken to small equalsize planar flakes, the one stacked on top of the other. The white plasticene sheet occurring a little below the contact has also broken but some of the flakes have the form of an isoclinal syncline tipped over in the direction of the nappe propagation. None of the flakes form an anticlinal pattern. The lowermost black plasticene sheet has not broken at all but has been thrown into a continuous series of gentle folds with uniform wavelength.

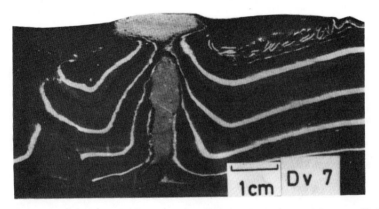

Fig.21. Part of cross-section of model *Dv* 7 taken some distance away from profile in Fig.19 and 20. In this section the small dome shown in Fig.19 and 20 has pierced the nappe of the early-formed spreading diapir. Note the tilted small syncline of the white modelling-clay sheet underneath the nappe.

*   *   *   *   *   *

The evidence of "intense shortening" used in the Appalachians is the same as applied in other orogens, namely folds and thrust sheets of a kind geometrically similar to those produced in the experimented models, but there without overall shortening.

On the other hand the orogen-imitation models do of course not exclude the possibility of shortening across the belts, but they do show that most of the structures in orogenic belts can form without a net shortening and thus indicate that a narrowing of the space occupied by orogens must be proven by other means such as, e.g., paleomagnetic studies of the cratons adjacent to the fold belts.

We feel it appropriate to conclude this account by emphasizing the view on the Alps expressed by Van Bemmelen based upon his extensive field work. Van Bemmelen (1960): "In contrast to the current opinion that the Alps are the result of crustal-shortening in the mobile Tethys belt, the author's field studies in recent years have led to the conclusion that this shortening is not needed to explain the structural overlap of the East Alpine and Pennine nappes . . .".

REFERENCES

Beloussov, V.V., 1961. The origin of folding in the earth's crust. *J. Geophys. Res.*, 66: 2241–2260.

Craig, G.Y. (Editor), 1965. *The Geology of Scotland.* Oliver and Boyd, Edinburgh, 556 pp.

Haarmann, E., 1930. *"Die Oszillationstheorie".* Enke, Stuttgart, 260 pp.

Kautsky, G., 1953. Der geologische Bau des Sulitelma–Salojauregebietes in den nordskandinavischen Kaledoniden. *Sver. Geol. Unders.*, C 5-28: 1–228.

Magnusson, N.H., Thorslund, P., Brotzen, F., Asklund, B. and Kulling, O., 1962. The Pre-quaternary rocks of Sweden, description of map. *Sver. Geol. Unders.*, Ba (16): 1–290.

Nicholson, R. and Rutland, R.W.R., 1969. A section across the Norwegian Caledonides; Bodö to Sulitelma. *Nor. Geol. Unders.*, 260: 1–86.

Ramberg, H., 1945. The thermodynamics of the earth's crust II. *Nor. Geol. Tidskr.*, 25: 307–326.

Ramberg, H., 1963. Experimental study of gravity tectonics by means of centrifuged models. *Bull. Geol. Inst. Univ. Upps.*, XLII: 1–97.

Ramberg, H., 1967. *Gravity, Deformation and the Earth's Crust.* Academic Press, New York, N.Y., 214 pp.

Ramberg, H., 1968a. Instability of layered systems in the field of gravity, I. *Phys. Earth Planet. Inter.*, 1: 427–447.

Ramberg, H., 1968b. Instability of layered systems in the field of gravity, II. *Phys. Earth Planet. Inter.*, 1: 448–474.

Ramberg, H., 1971. Dynamic models simulating rift valleys and continental drift. *Lithos*, 4: 259–276.

Ramberg, H., 1972a. Mantle diapirism and its tectonic and magmagenetic consequences. *Phys. Earth Planet. Inter.*, 5: 45–60.

Ramberg, H., 1972b. Theoretical models of density stratification and diapirism in the earth. *J. Geophys. Res.*, 77: 877–889.

Rodgers, J., 1970. *The Tectonics of the Appalachians.* Wiley-Interscience, New York, N.Y., 271 pp.

Stephansson, O., 1972. Theoretical and experimental studies of diapiric structures on Öland. *Bull. Geol. Instn. Upps., N.S.*, 3: 181–200.

Van Bemmelen, R.W., 1933. The undation theory and the development of the earth's crust. *Int. Geol. Congr. Rend.*, 16 (2): 965–981.

Van Bemmelen, R.W., 1960. New view on the Alpine orogenesis. *Int. Geol. Congr., 21st, Copenhagen, 1960. Rept. Session Norden,* 18: 99–116.

Zachrisson, E., 1969. Caledonian Geology of Northern Jämtland-Southern Västerbotten. *Sver. Geol. Unders., Ser. C,* 644: 1–33.

46

Reprinted from pp. 1075-1084, 1094-1095 of *Proc. Amer. Soc. Civil. Eng.*, SM4, 1075-1096 (July 1969)

STABILITY OF SLOPES UNDERGOING CREEP DEFORMATION

Bing C. Yen

INTRODUCTION

The problem of long-term slope stability in connection with the phenomenon of soil creep has been of interest for many years. "Creep in slopes" is a broad term utilized in engineering geology to describe a down hill movement which occurs at an imperceptible rate, (19,31)[2]. Slow movements of this kind, though imperceptible in their early stage, have resulted in many landslides and damaging slope failures in many countries [e.g., in Japan, Fukuoka (5), in England, Henkel and Skemptom (9), in Yugoslavia, Suklje (26), and in Russia, Ter-Stepanian (29)]. In this country, creep is observed on slopes in many areas such as in Pennsylvania, on the California Coast, and along the Columbia River Valley of Northeastern Washington [Sharpe and Dosh (28), Gould (8), and Jones, Embody et al. (11)]. These observed slope failures all seem to have occurred on rather flat natural slopes or on cut slopes of various geological histories. They were usually stable for many years until soil creep and cracks were observed. The rate of soil creep seems to vary with the magnitude and changes of the landslide-producing agents. Among the case histories previously referred to, the soil creep had been actuated by ground water due to percolation of rain or melting snow, Fukuoka (5), under-cutting of the toe of a slope due to erosion; Henkel (9), earthquakes, Suklje (26), and/ or the works of man, Gould (8). Usually, although not always, the rate of soil creep accelerates to landsliding.

Studies of this type of failure are essential for developing a general understanding of land movement, in an attempt to evaluate the safety of hillside slopes and the design of cut slopes. However, for the long-term stability of clay slopes, especially when the clay is fissured, the error of analysis has

Note.—Discussion open until December 1, 1969. To extend the closing date one month, a written request must be filed with the Executive Secretary, ASCE. This paper is part of the copyrighted Journal of the Soil Mechanics and Foundations Division, Proceedings of the American Society of Civil Engineers, Vol. 95, No. SM4, July, 1969. Manuscript was submitted for review for possible publication on July 10, 1968.

[1] Assoc. Prof. of Civ. Engrg., California State Coll., Long Beach, Calif.

[2] Numerals in parentheses refer to corresponding items in the Appendix II.—References.

been generally large. In some cases, the error could be from 300% to 500%, depending on the methods used [Skempton (23)]. Certain empirical approaches, together with some laboratory data, have been proposed to forecast this type of failure [Goldstein and Ter-Stepanian (7), Saito and Uezawa (18), and Saito (17)]. However, no constitutive equations of the soils involved were formulated because of the empirical nature of the approaches. Culling (3) developed a mathematical theory of erosion on a soil-covered slope that concerned creep phenomenon. It was based on random forces on individual soil particles. This physical picture may represent the action of many small forces, such as weathering and the activities of rodents. Another mathematical approach to the creep process, due to cycles of freezing and thawing, was given by Kirkby (12). He showed that seasonal freezing and thawing, together with a downhill component of the soil weight, are causes of a zig zag path of creep. However, no general failure of a slope was assumed in his equations.

In an attempt to analyze the stability of slopes undergoing creep deformation, an understanding of the mechanism of creep and the long-term soil strength, especially the stress-strain-time relationship of soil, seems to be of paramount importance. Significant progress has been made in recent investigations in the understanding of the mechanism of creep in slopes such as summarized by Skempton (22) and Bjerrum (2). Additional considerations in this regard have been given by Rowe (16), and Ter-Stepanian (28), among others. There is also progress in understanding the effects of stress-strain-time in connection with soil strength as exemplified by soil creep phenomenon [Trollope and Chan (32), Wu, et al. (33), Mitchell and Campanella (13), Mitchell, Seed and Paduana (14), and Singh and Mitchell (21)].

In light of the understanding of the mechanism of long-term slope failure and creep phenomena, the writer presents a quantitative stability analysis in terms of the creep velocity of an infinite slope. Although practical problems usually have a slope of different geometry and are initially stable, it is apparent that the creep of a long slope is of primary concern. The infinite slope also has the advantage of easier mathematical treatment than other type of boundary conditions. It is this phase of the problem that is examined herein. The results are then utilized and compared with published field surveying data of a slope undergoing creep deformation in the Caucasus Coast of the Black Sea area [Ter-Stepanian (29)] and a slope on the Southern California Coast, which involves a preconsolidated tertiary marine clay [Moran et al. (15), Gould (8) and Soil Conservation Service (24)].

THEORETICAL CONSIDERATIONS

Prior to a slope failure, some slow deformation occurs within the slope and on its surface. Although this does not always mean the slope is bound to fail, a seasonal creep may develop into a slide; the slide may also be followed by further creep [Terzaghi (31)]. Continuous creep of a slope is a case in which the down-slope component of the weight of soil causes a slow, continuous yield. The state of slow yielding represents a delicate balance of forces. The soil constitutive equations, slope geometry, and pore water pressure are the primary factors that explain the behavior and stability of the slope. Three basic assumptions are made to account for the previously mentioned parameters for a theoretical analysis. None of these assumptions are new. They have been proposed and/or verified by various laboratory investigations, as indicated in the references. An evaluation of the applicability of the three basic assumptions when combined is presented herein.

BASIC ASSUMPTIONS

Soil behaves as a rigid visco-plastic solid during the shearing deformation [Stroganov (25), Ter-Stepanian (28), Terzaghi (30)]. The relationship between shear stress and the velocity gradient of shearing deformation is shown in Fig. 1 and in the following:

$$\tau = \tau_o + \eta\,G \dotfill (1)$$

in which τ = shear strength; τ_o = yield stress; η = coefficient of viscosity; and G = velocity gradient of shearing deformation.

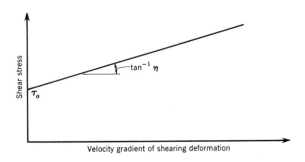

FIG. 1—SHEAR STRESS VERSUS VELOCITY GRADIENT OF SHEARING DEFORMATION

Soil creep along a slope begins when the shearing stress is equal to or greater than the residual strength of the soil (2):

$$\tau_o \geqq C_\gamma + \sigma \tan \phi_\gamma \dotfill (2)$$

in which C_γ, σ, and ϕ_γ are residual cohesion, effective stress and the residual internal angle of friction at the potential plane of failure respectively, and as shown in Fig. 2.

The direction of creep of an infinitely long slope is in the direction of maximum shear stress, as shown in Fig. 3; i.e., a plane failure which is equivalent to minimum work. Consolidation or swelling of the soil during creeping is assumed to be negligible (12,28).

EVALUATION OF ASSUMPTIONS

The first assumption of rigid visco-plastic behavior, as used by previous investigators, is probably an approximate description of soil properties in connection with a generally large deformation such as creep. The term "rigid" is used only to indicate the phase preceding creep. This quasi-definition is not exact, since it suggests that no flow takes place under stress which is smaller than shearing resistance. However, in comparing "rigidity" with large deformation, such as is evidenced in soil creep, the initial deformation of the soil is generally small and may be neglected.

The second assumption is the recognition of the applicability and its associated limitations of the role of residual strength in the long-term slope stability analysis [Bishop et al. (1), Skempton (22) and others]. As pointed out by Bjerrum (2), creep which ultimately leads to sliding must occur on a slope

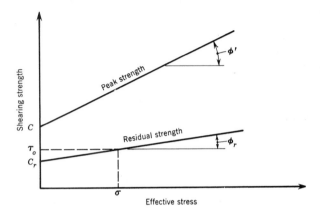

FIG. 2.—SHEAR STRENGTH CHARACTERISTICS [AFTER SKEMPTON (20)]

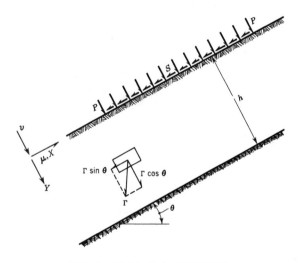

FIG. 3.—COORDINATE SYSTEM

that is so steep that the shearing force is greater than the residual strength. The second assumption used here is a formulation of this statement. Hence, the results derived from this formulated assumption are limited to such slopes.

The third assumption implies isotropic material and constant volume, i.e., incompressibility.

When these assumptions are adopted simultaneously, it can be inferred that during the phase of creep, the soil resistance against deformation consists of the residual strength and viscous resistance of the soil. During the process of the creep movement, there exists no further consolidation or swelling of the soil.

The physical picture of this type of problem may be generally categorized as follows:

1. Along a slope there exists a discontinuity of some type (e.g., a cut) such that the concentration of stress is greater than the peak strength of the soil, thus actuating a progressive failure along a failure plane. The soil strength along the failure zone is already at its residual value and the soil creeps viscously.

2. Along a slope in which there was a prehistoric slide or a geological displacement, the diagenetic bond is broken, and the strength that actually is mobilized along the present plane of movement is now at its residual strength.

STABILITY ANALYSIS

Basic Equations.—There are four basic equations involved in the stability analysis: the constitutive equation of soil behavior as a result of assumptions 1 and 2; the equation of continuity as a result of assumption 3; the equations of state; and finally, the equation of equilibrium as dictated by the slope geometry and kinematics. They are listed as follows with a brief discussion:

Constitutive Equation.—Combining Eqs. 1 and 2, the constitutive equation is

$$\tau = C_r + \sigma \tan \phi_r + \eta \, G \quad \dots\dots\dots\dots\dots\dots\dots\dots\dots\dots\dots \quad (3)$$

Equation of Continuity.—Assuming no volume change, as stated in assumption 3, the equation of continuity in plane strain as shown in Fig. 3 is

$$\frac{\partial u}{\partial x} + \frac{\partial v}{\partial y} = 0 \quad \dots\dots\dots\dots\dots\dots\dots\dots\dots\dots\dots\dots\dots \quad (4a)$$

in which u and v = velocities in the x and y directions respectively. In terms of strain rate

$$\dot{\epsilon}_x + \dot{\epsilon}_y = 0 \quad \dots\dots\dots\dots\dots\dots\dots\dots\dots\dots\dots\dots\dots \quad (4b)$$

in which $\dot{\epsilon}_x$ and $\dot{\epsilon}_y$ = rate in the x and y directions respectively. Since all variables are independent of x for an infinitely long slope,

$$\frac{\partial u}{\partial x} = 0, \quad \text{or} \quad \dot{\epsilon}_x = 0 \quad \dots\dots\dots\dots\dots\dots\dots\dots\dots\dots\dots \quad (5a)$$

Therefore, Eqs. 4a and 4b yield

$$\frac{\partial v}{\partial y} = 0, \quad \text{or} \quad \dot{\epsilon}_y = 0 \quad \dots\dots\dots\dots\dots\dots\dots\dots\dots\dots\dots \quad (5b)$$

Equations of State.—The equations of state resulting from Eqs. 3, 5a, and 5b may be expressed as

$$\sigma_x - \sigma = 2\left(\eta + \frac{C_\gamma + \sigma \tan \phi_\gamma}{G}\right)\dot{\epsilon}_x \quad \dots\dots\dots\dots\dots\dots \quad (6a)$$

$$\sigma_y - \sigma = 2\left(\eta + \frac{C_\gamma + \sigma \tan \phi_\gamma}{G}\right)\dot{\epsilon}_y \quad \dots\dots\dots\dots\dots\dots \quad (6b)$$

$$\tau_{xy} = \left(\eta + \frac{C_\gamma + \sigma \tan \phi_\gamma}{G}\right)\dot{\epsilon}_{xy} \quad \dots\dots\dots\dots\dots\dots\dots \quad (6c)$$

Equations of Equilibrium.—The equations of equilibrium for a long slope, which is independent of the x axis, as shown in Fig. 3, is:

$$\frac{d\sigma_y}{dy} = \Gamma \cos \theta \quad \dots\dots\dots\dots\dots\dots\dots\dots\dots \quad (7a)$$

$$\frac{d\pi_{xy}}{dy} = \Gamma \sin \theta \quad \dots\dots\dots\dots\dots\dots\dots\dots\dots \quad (7b)$$

in which Γ = unit weight of soil with boundary conditions at $y = 0$, $\sigma_y = p$ and $\tau_{xy} = s$. It is necessary to assume that $s < pf$ if f is the friction factor between surcharge and the surface of the slope. Integrating Eqs. 7a and 7b and utilizing the boundary conditions, the equations of equilibrium

$$\sigma_y = \Gamma y \cos \theta + p \quad \dots\dots\dots\dots\dots\dots\dots\dots\dots \quad (8a)$$

$$\tau_{xy} = \Gamma y \sin \theta + s \quad \dots\dots\dots\dots\dots\dots\dots\dots\dots \quad (8b)$$

are obtained.

Slope Stability Analysis.—Substituting Eqs. 5a and 5b into Eqs. 6a and 6b respectively,

$$\sigma_x = \sigma_y = \sigma \quad \dots\dots\dots\dots\dots\dots\dots\dots\dots\dots\dots \quad (9a)$$

is obtained. From the gradient of shear strain, $\dot{\epsilon}_{xy}$,

$$\dot{\epsilon}_{xy} = \frac{\partial u}{\partial y} + \frac{\partial v}{\partial x} = \frac{\partial u}{\partial y}$$

can be written as

$$\dot{\epsilon}_{xy} = \frac{du}{dy} \quad \dots\dots\dots\dots\dots\dots\dots\dots\dots\dots\dots \quad (9b)$$

because the long slope is independent of the x axis, and $\partial v/\partial x = 0$. Taking into account Eqs. 5a, 5b, and 9b,

$$G = \sqrt{(\dot{\epsilon}_x - \dot{\epsilon}_y)^2 + \dot{\epsilon}_{xy}^2} = \dot{\epsilon}_{xy} = \frac{du}{dy} \quad \dots\dots\dots\dots\dots \quad (10)$$

Substituting into the Eq. 6c Eqs. 8a, 8b, 9a, 9b, and 10, the differential equation of a creeping slope at equilibrium

$$-\eta\left(\frac{du}{dy}\right) = (p \tan \phi_\gamma + C_\gamma - s) - \Gamma y \cos \theta (\tan \theta - \tan \phi_\gamma) \quad \dots \quad (11)$$

is obtained. A similar differential equation has been obtained by Stroganov (25) for steady state creep.

Assuming the bedrock below the creep zone is stationary, the boundary condition is $u = 0$ when $y = h$. Integrating Eq. 11, and taking into account the boundary condition, the equation of the velocity field

$$u = -\frac{(h - y)}{\eta} (s - p \tan \phi_r - C_r) - \frac{\Gamma(h^2 - y^2)}{2\eta} \cos \theta (\tan \theta - \tan \phi_r) \quad (12)$$

is obtained. The down-slope velocity, u reaches its maximum u_m, when du/dy = 0 at $y = y_m$ (Eqs. 11 and 12)

$$y_m = \frac{p \tan \phi_r + C_r - s}{\Gamma \cos \theta (\tan \theta - \tan \phi_r)} \quad\dotsnullnull \quad (13a)$$

$$u_m = \frac{-(p \tan \phi_r + C_r - s)}{2 \eta \, y_m} (h - y_m)^2 \dotsnull \quad (13b)$$

The negative sign of Eqs. 11, 12, and 13b indicates a down-slope velocity. If Eqs. 13a and 13b are substituted into Eq 12 and simplified, the velocity pro-

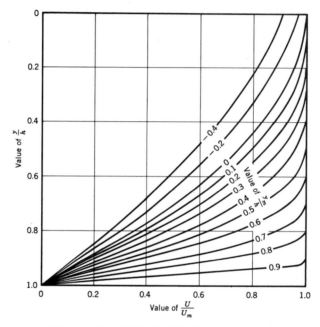

FIG. 4.—VELOCITY PROFILES AT FAILURE

file of the creeping slope is then obtained as the following:

$$\frac{u}{u_m} = 1 - \frac{(y - y_m)^2}{(h - y_m)^2} \quad\dotsnull \quad (14)$$

Eq. 14 can be plotted in a dimensionless coordinate, as shown in Fig. 4. The meaning involved in Eq. 12 through 14 will become apparent after we consider the following three special cases of a slope at equilibrium undergoing creep deformation. The interpretation of the equations follows.

Case I.—Slope Without Surcharge and $C_r = 0$.— For a slope without sur-charge, i.e., $p = 0$ and $s = 0$ and considering that the residual cohesion is small and can be neglected, Skempton (22), Case I may simulate a natural slope of no surcharge, where the diagenetic bond within the creep zone has been broken due to previous slide or other geologic displacement. Eqs. 12 through 14 then reduce to

$$y_m = 0 \quad \dots \dots \dots \dots \dots \dots \dots \dots \dots \dots \dots \dots \dots \quad (15a)$$

$$u_m = -\frac{h^2 \Gamma}{2\eta} \cos \theta \ (\tan \theta \ - \ \tan \phi_r) \quad \dots \dots \dots \dots \dots \quad (15b)$$

$$u = -\frac{\Gamma (h^2 \ - \ y^2)}{2\eta} \cos \theta \ (\tan \theta \ - \ \tan \phi_r) \quad \dots \dots \dots \dots \quad (15c)$$

$$\text{or } u = u_m \left(1 \ - \ \frac{y^2}{h^2}\right) \quad \dots \dots \dots \dots \dots \dots \dots \dots \dots \dots \quad (15d)$$

Case II.—Slope With Surcharge, but $C_r = 0$.—For a slope with $C_r = 0$, but $p \neq 0$ and $s \neq 0$, the corresponding equations of y_m, u_m and u, as reduced from Eq. 12 through 14, can be further reduced. Case II may simulate a slope with surcharge; e.g., if a layer of shale or mudstone along a slope is resting on a clayey layer in which creep occurs and in which the strength mobilized is at its residual strength, then

$$y_m = \frac{p \tan \phi_r \ - \ s}{\Gamma \cos \theta \ (\tan \theta \ - \ \tan \phi_r)} \quad \dots \dots \dots \dots \dots \dots \dots \quad (16a)$$

$$u_m = -\frac{(p \tan \phi_r \ - \ s)}{2\eta \ y_m} \ (h \ - \ y_m)^2 \quad \dots \dots \dots \dots \dots \quad (16b)$$

$$u = \frac{(h \ - \ y)}{\eta} \ (p \tan \phi_r \ - \ s) \ - \ \frac{\Gamma (h^2 \ - \ y^2)}{2\eta} \cos \theta \ (\tan \theta \ - \ \tan \phi_r) \quad (16c)$$

$$\text{or } u = u_m \left[1 \ - \ \left(\frac{y \ - \ y_m}{h \ - \ y_m}\right)^2\right] \quad \dots \dots \dots \dots \dots \dots \dots \dots \quad (16d)$$

Case III.—Slope Without Surcharge, but $C_r \neq 0$.—For a slope without surcharge, but on which the soil has cohesion, and assuming that the cohesion will not decrease due to creep, Eqs. 12 through 14 can be reduced accordingly. Case III may simulate a natural slope where a stress concentration actuates a progressive type of failure (e.g., actuated by a cut or tectonic crack) and the soil creeps viscously as

$$y_m = \frac{C_r}{\Gamma \cos \theta \ (\tan \theta \ - \ \tan \phi_r)} \quad \dots \dots \dots \dots \dots \dots \dots \quad (17a)$$

$$u_m = -\frac{C_r}{2\eta \ y_m} \ (h \ - \ y_m)^2 \quad \dots \dots \dots \dots \dots \dots \dots \dots \quad (17b)$$

$$u = \frac{C_r (h \ - \ y)}{\eta} \ - \ \frac{\Gamma (h^2 \ - \ y^2)}{2\eta} \cos \theta \ (\tan \theta \ - \ \tan \phi_r) \quad \dots \dots \dots \quad (17c)$$

$$\text{or } u = u_m \left[1 \ - \ \left(\frac{y \ - \ y_m}{h \ - \ y_m}\right)2\right] \quad \dots \dots \dots \dots \dots \dots \dots \dots \quad (17d)$$

From a comparison between the general Eqs. 12 through 14 and the equations of the three special cases, the following observations can be made.

General Velocity Profile.—The down-slope velocity profile of a creep zone is a parabola. At a given depth, the velocity is directly proportional to the unit weight of the soil and inversely proportional to the soil coefficient of viscosity. An increase of surcharge shearing load in the down-slope direction will increase the velocity, while an increase of normal surcharge load will decrease the velocity. The velocity at a given depth increases with the sine function of the natural slope. The velocity at a given depth decreases with the tangent of the residual angle of internal friction.

The Location of Maximum Velocity.—The location of the maximum velocity is inversely proportional to the unit weight of the soil and the cosine function of the slope. The increase of normal surcharge load along the slope tends to increase the depth of the maximum velocity, while the increase of shearing surcharge load tends to decrease the depth of maximum velocity. For a slope without surcharge and in which the residual cohesion of the soil is negligible, the location of the maximum velocity is at the surface of the slope. A dimensionless plotting of u/u_m versus y/h is shown in Fig. 4, which describes the creeping velocity distribution as a function of the location of maximum velocity.

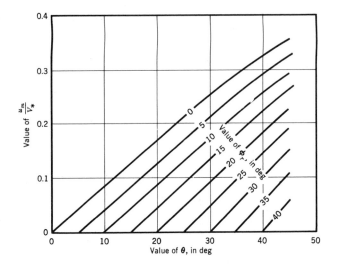

FIG. 5.— u_m/v_* VERSUS θ AND ϕ_r

The Magnitude of Maximum Velocity.—The functional relationship between the maximum velocity and its corresponding depth is given by Eq. 13b. For a special case of general interest of a slope with no surcharge and $C_r = 0$, the maximum velocity is proportional to the square of the depth of creep, the unit weight of soil and the value of (tan θ - tan ϕ_r). A dimensionless plotting of u_m/v_* versus slope angle θ, is shown in Fig. 5, which describes the maximum velocity as a function of ϕ_r, where v_* is equal to $h^2 \Gamma/\eta$. This is a parameter that concerns the depth of the creeping zone and soil properties.

To illustrate the applications of Figs. 4 and 5, consider a natural slope 2 horizontal to 1 vertical ($\theta = 26.6°$) of no surcharge load along the slope, assuming that the soil is a 20-ft layer of weathered clay shale overlying bedrock. The soil has a wet unit weight of 100 pcf and the bedrock dips parallel to the surface of the slope. The residual angle of internal friction of the soil is $\phi_r = 15°$, assuming $C_r = 0$. Assume that no groundwater is found and that the slope has been undergoing creep for some time. From the data of the installed slope indicators, assume the calculated average coefficient of viscosity to be $\eta = 5 \times 10^8$ lb-sec per sq ft.

From the previously mentioned field and laboratory data, Fig. 5 may be utilized to calculate the maximum velocity u_m at failure. Using $\theta = 26.6°$ and $\phi_r = 15°$, a value of $u_m/v_* = 0.105$ can be read off from the ordinate of Fig. 5. Because $v_* = \Gamma h^2/\eta = 100(20)^2/(5 \times 10^8) = 8 \times 10^{-5}$ fps, u_m can be calculated by multiplying 0.105 by 8×10^{-5} to obtain 0.84×10^{-5} fps. Because there is no surcharge load on the surface of the slope, the u_m is located at the surface, according to Eq. 15a. This information may be of interest because generally it is not too difficult to observe the surface displacement along a creeping slope over a period of time. If the observed total displacement is divided by the time elapsed between observations, an average surface movement velocity can be obtained. To compare this observed velocity with the theoretical steady state velocity provides a relative measure of stability at the time of observation.

The creep velocity profile through the 20-ft layer of weathered clay shale can be found from Fig. 4. Since $y_m = 0$, the velocity profile assumes the configuration of $y_m/h = 0$ in Fig. 4. For example, the creep velocity at depth 10 ft from the surface is $0.75 \times u_m$, or 0.63×10^{-5} fps.

It should be emphasized that there are limitations when the theoretical results are applied to field problems. This is partly due to the assumptions used in the derivation of theoretical results and partly from the heterogenous nature of the field problems. First, the theoretical results will be compared with two field case histories and then the limitations will be treated in detail.

✱   ✱   ✱   ✱   ✱   ✱

APPENDIX II.—REFERENCES

1. Bishop, A. W., Bjerrum, L., "The Relevance of the Triaxial Test to the Solution of Stability Problems," ASCE, *Manual of Shear Strength of Cohesive Soils*, Colorado, 1960, p. 437.
2. Bjerrum, L., "The Third Terzaghi Lecture: Progressive Failure in Slopes of Overconsolidated Plastic Clay and Clay Shales," *Journal of the Soil Mechanics and Foundations Division*, ASCE, Vol. 93, Part I, No. SM5, Proc. Paper 5456, Sept., 1967, p. 3–49.
3. Culling, W. E. H., "Soil Creep and the Development of Hillside Slopes," *Journal of Geology*, Vol. 71, No. 2, March, 1963, p. 127.
4. Finn, W. D. L., "Earthquake Stability of Cohesive Slopes," *Journal of Soil Mechanics and Foundations Division*, ASCE, Vol. 92, No. SM1, Proc. Paper 4602, Jan., 1966, p. 1–11.
5. Fukuoka, M., "Landslides in Japan," *Proceedings* 3rd International Conference on Soil Mechanics and Foundation Engineering, Vol. 2, 1953, p. 234.
6. Geuze, E. C. W. A., and Tan, T. K., "Rheological Properties of Clays," 3rd International Conference of Soil Mechanics and Foundation Engineering, 1953.
7. Goldstein, M., Ter-Stepanian, G., "The Long Term Strength of Clays and Depth Creep of Slopes," *Proceedings*, 4th International Conference on Soil Mechanics and Foundation Engineering, Vol. 1, 1957, p. 311–314.
8. Gould, J. P., "A Study of Shear Failure in Certain Tertiary Marine Sediments": *ASCE Manual of Shear Strength of Cohesive Soils*, Colorado, 1960, p. 614.
9. Henkel, D. J., Skempton, A. W., "A Landslide at Jackfield, Shropshire," *Proceedings* European Conference of Stability of Earth Slopes, Stockholm, Vol. 1, 1954, p. 90.

[*Editor's Note:* A row of asterisks indicates that material has been deleted.]

10. Jaeger, J. C., *Elasticity, Fracture and Flow*, John Wiley and Sons, Inc., 1964, p. 72.

11. Jones, F. O., Embody, D. R., and Peterson, W. L., "Landslide Along Columbia River Valley, Northeastern Washington," *USGS Professional Paper 367*, 1961.

12. Kirkby, M. J., "Measurement and Theory of Soil Creep," *Journal of Geology*, Vol. 75, No. 4, July, 1967, p. 359.

13. Mitchell, J. K., Campanella, R. G., "Creep Studies on Saturated Clays," ASTM, *Special Technical Publication. No. 361*, 1963, p. 90–104.

14. Mitchell, J. K., Seed, H. B., Paduana, J. A., "Creep Deformation and Strength Characteristics of Soil Under the Action of Sustained Stresses," *Report No. TE 65-8* to the Bureau of Reclamation, Soil Mechanics and Tituminous Laboratory, University of California, Berkeley.

15. Moran, Proctor, Muser and Rutledge, Consulting Engineers, N.Y., "Final Report of Pacific Palisades Landslide Study," July, 1959.

16. Rowe, P. W., "C = 0, Hypothesis for Normally Loaded Clays at Equilibrium," *Proceedings*, 4th International Conference on Soil Mechanics and Foundation Engineering, Vol. 1, 1957, p. 189.

17. Saito, M., "Forecasting the Time of Occurrence of a Slope Failure," *Proceedings*, 6th International Conference on Soil Mechanics and Foundations Engineering, 1965, p. 537–541.

18. Saito, M., Uezawa, H., "Failure of Soil Due to Creep," *Proceedings*, 5th International Conference on Soil Mechanics and Foundation Engineering, Vol. 1, 1961, p. 315–318.

19. Sharpe, C. F. S., "Landslide and Related Phenomenon," Columbia University Press, 1938, p. 137.

20. Sharpe, C. F. S., Dosch, E. F., "Relation of Soil Creep to Earthflow in the Appalchian Plateaus," *Journal of Geomorphology*, Vol. 5, 1942, p. 132.

21. Singh, A., Mitchell, J. K., "General Stress-Strain-Time Function for Soils," *Journal of Soil Mech. and Foundations Division*, ASCE, Vol. 94, No. SM1, Proc. Paper 5728, Jan., 1968, p. 21–46.

22. Skempton, A. W., "Long Term Slope Stability of Slopes," 4th Rankine Lecture, *Geotechnique*, Vol. 14, No. 2, 1964, p. 77–102.

23. Skempton, A. W., "Opening Address," European Conference of Stability of Earth Slopes, *Geotechnique*, Vol. 4, 1954, p. 6.

24. Soil Conservation Service, USDA, "Soils of the Malibu Area, California," M7-L-*18854*, 1967.

25. Stroganov, A. S., "Viscous-plastic Flow of Soils," *Proceedings*, 5th International Conference on Soil Mechanics and Foundation Engineering, Vol. 2, 1961, p. 721–726.

26. Suklje, L., "A Landslide due to Long Term Creep," *Proceedings*, 5th International Conference on Soil Mechanics and Foundation Engineering, 1961, p. 727.

27. Tan, T. K., "Determination of the Rheological Parameters and the Hardening Coefficients of Clays," *Proc.* International Symposium on Rheology and Soil Mechanics, Grenoble, 1964, pp. 256–272.

28. Ter-Stepanian, G., "On the Long Term Slope Stability of Slopes," *Norwegian Geotechnical Institute*, Publication No. 52, 1963, p. 1–13.

29. Ter-Stepanian, G., "In-Situ Determination of the Rheological Characteristics of Soils on Slope," *Proceedings*, 6th International Conference on Soil Mechanics and Foundation Engineering, 1965, p. 575–577.

30. Terzaghi, C., "Static Rigidity of Plastic Clays," *Journal of Rheology*, Vol. 2, No. 3, 1931, pp. 253–262.

31. Terzaghi, K., "Mechanism of Landslides," *Application of Geology to Engineering Practice*, The Geological Society of America (Berkey Volume), p. 83–123, 1950.

32. Trollope, D., and Chan, C. K., "Soil Structure and Step-Strain Phenomenon," *Journal of Soil Mechanics and Foundations Division*, ASCE, Vol. 86, No. SM11, Proc. Paper, April, 1960, p. 1–39.

33. Wu, T. H., Douglas, A. G., and Goughnour, R. D., "Friction and Cohesion of Saturated Clays," *Journal of Soil Mechanics and Foundations Division*, ASCE, Vol. 88, No. SM3, Proc. Paper 3158, June, 1962, p. 1–31.

34. Yen, Bing C., "Viscosity of Sands Near Liquefaction," *Proceedings*, International Symposium on Wave Propagation and Dynamic Properties of Earth Material, Albuquerque, New Mexico, 1967, p. 877–888.

✳ ✳ ✳ ✳ ✳ ✳

Editor's Comments
on Paper 47

47   KEHLE
Analysis of Gravity Sliding and Orogenic Translation

Existing theories on the mechanics of overthrusting, Kehle emphasizes in Paper 47, impose severe limitations except where abnormal fluid pressures exist (i.e., the Hubbert–Rubey hypothesis); yet Kehle is doubtful that such abnormally pressurized zones occur frequently enough in nature.

What then, is the causative mechanism? The "mechanical paradox of overthrust faulting" is again postulated, and Kehle's answer follows: "deformation occurs in a manner best described as viscous deformation and that almost all such deformation concentrates in the lowest viscosity strata . . ." (p. 1642).

Thus we have apparently come full circle. The mechanical paradox remains vivid. Smoluchowski, however, seems all but forgotten: "Suppose a layer of plastic material, say pitch, interposed between the (overthrust) block and the underlying bed; or suppose the bed to be composed of such material; *then the law of viscous friction will come into play,* instead of the friction of solids; therefore any force, however small, will succeed in moving the block. Its velocity may be small if the plasticity is small, but in geology we have plenty of time; there is no hurry."

47

Reprinted from *Geol. Soc. America Bull.,* **81**(6), 1641–1663 (1970)

Analysis of Gravity Sliding
and Orogenic Translation

RALPH O. KEHLE *Department of Geological Sciences, The University of Texas at Austin, Austin, Texas 78712*

ABSTRACT

Gravity sliding, major thrusting, and the displacement of large crustal plates are accomplished because strata with the lowest viscosity in a rock sequence readily deform by simple (rectilinear) shear flow when subjected to loading. These deforming strata are called *décollement* zones. Strata above these zones are transported without being deformed, provided their viscosity is one or more orders of magnitude larger than that of the décollement zone. Shear stresses acting at the base of these strata, that is, "basal drag," are commonly much lower than stresses accompanying friction across a sole fault (*compare with* Hubbert and Rubey, 1959). Thus, shear flow in décollement zones most likely represents the operative mechanism in natural settings. The speed of gravity sliding and the ease of tectonic transport increase with increasing décollement zone thickness, dip, and depth of burial, and decrease with increasing décollement zone viscosity. End effects, including (in two dimensions) updip attachment of gravity slides and frontal buttressing of both slides and tectonic plates, are as important as basal drag in determining the speed of slides and the resistance to transport of crustal plates.

Structures in the strata above a décollement zone develop independently of those within the zone. In the upper plate, pull-apart grabens form to overcome updip attachment; step thrusts and folds form to overcome frontal buttressing. Structures within décollement zones are proposed to include intensely developed passive disharmonic folds, bedding plane faults at contacts between high and low viscosity strata and through high viscosity masses within the zone, chaotic zones comprised of blocks of high viscosity strata, infolds of adjacent high viscosity strata, and penetrative cleavage, phylonitization, and other features of deformation-induced recrystallization.

Velocity profiles for simple shear portray the general mode of deformation and permit study of transport in specific geologic settings. Composite velocity profiles are constructed by first drawing individual profiles for each rock in the sequence under study. Segments from these curves proportional to the thickness of each succeeding stratum are then spliced together from the bottom up.

INTRODUCTION

General Remarks

Existing theories on the mechanics of gravity sliding and orogenic translation impose severe limitations on the size of thrust plates, except in special settings where they are underlain by abnormally pressured sediments (Hubbert and Rubey, 1959). Frictional drag along the base of all but the smallest thrusts is computed to be so large that stresses in excess of rock strength are needed to sustain motion. For many known thrusts the total frictional resistance must have been smaller than the values computed using standard assumptions. Hubbert and Rubey (1959) show that abnormally high pore fluid pressure in sediments underlying large thrusts reduce the effective normal load on the lower boundary, which in turn lowers the drag. Most abnormal pressures have been measured in thick marine shales (Dickinson, 1951; Thomeer and Bottema, 1961, p. 1723). Yet major tectonic transport commonly occurs over evaporites, thin shales and even limestones. Abnormal fluid pressure is an unlikely assistant in these settings.[1]

---

[1] Abnormal fluid pressure is known to occur *beneath* some thick evaporite sequences (Rubey and Hubbert, 1959, p. 170; Hubbert and Rubey, 1959, p. 155–156; Lane, 1949). Transport of rocks *above* an evaporite sequence is unaided by abnormal pressure *below* the sequence. Influxes of high pressure salt water have been recorded while drilling through a few evaporite sequences, but such occurrences are considered rare (Thomeer and Bottema, 1961, p. 1727, 1723). The frequency of recorded occurrence of abnormal pressure in evaporite sequences appears low compared to the frequency with which they appear to act as décollement zones in thrusting and gravity slides (Heard and Rubey, 1966, p. 749–752).

What then is the deformation mechanism that results in tectonic transport without generating basal drag so large that the size of translating masses is limited to unacceptably small dimensions? A possible answer is that deformation occurs in a manner best described as viscous deformation and that almost all such deformation concentrates in the lowest viscosity strata. The deformation is distributed across these low viscosity zones of décollement rather than being concentrated across infinitely narrow faults. Such a mechanism should be widely applicable because it operates to some extent in all rock sequences. The present study shows that distributed shear is a natural consequence of loading processes thought to be responsible for gravity sliding and sub-horizontal orogenic transport. Large translation of rock masses results from the accumulation of large shear strains in incompetent rock layers rather than from discrete slip along sole faults. Distributing the shear markedly reduces the stresses associated with moving thrusts and the work expended in the process. No limit to the size of thrust sheets is imposed by this theory, provided sufficiently mobile rocks are available to serve as zones of décollement. Mobility is enhanced by the presence of halite in evaporite sequences, by increased water content in shales, and by increased temperature in limestones. The distributed shear theory emphasizes the importance of deformation rate in the formation of thrusts.

The following assumptions are made in the analyses: (1) rock creep is the dominant mechanical process; (2) this creep is accurately modeled as fluid flow, provided an adequate ideal fluid model is employed; and (3) rock layers in the stratified bodies used as examples are perfectly uniform layers of either ideally viscous incompressible fluids or non-linearly viscous incompressible fluids that exhibit power-law behavior. The studied sequences are stylized so as to contain a few thick, distinctly different rock units. Though somewhat artificial, this representation permits visualization of important mechanisms without the complication of unnecessary detail.

The deformation of a three-layer rock sequence containing a low viscosity layer is studied first. The flow analysis is presented in the appendix. Theoretically predicted velocity profiles for gravity sliding of this sequence shows that décollement occurs because of large viscosity contrasts between rock layers. High

viscosity rocks overlying a low viscosity décollement zone can be transported over large distances without suffering internal distortion. Motion of an upper thrust plate does not cause deformation of soft layers within the underlying rock sequence; rather, the plate motion is seen to result from the deformation of these layers. Sliding velocity varies widely, increasing with décollement zone thickness, dip, and depth of burial. Sliding velocity decreases with increasing viscosity of a décollement zone. For simplicity of presentation, the quantitative analysis neglects end effects. These include updip attachment of gravity slides, frictional drag along the base of the front of an advancing thrust sheet, and the energy dissipated if folding of the advancing edge of the thrust provides the space into which the sheet advances. Their neglect leads to overestimating the velocity of gravity slides and underestimating the resistance to orogenic translation. Décollement occurs above basement, because crystalline rock viscosities exceed typical sediment viscosities by orders of magnitude. In the processes studied, basement appears rigid in comparison to sediments. In most geologic settings the present theory requires less work to achieve thrusting than previous theories and thus more likely represents the operative mechanism.

Speculation on structures associated with décollement leads to the suggestion that bedding plane faults can develop only intermittently and over limited areas, unless they transect locally stiff portions of a décollement zone. Here they form because the strain rate imposed on stiffer rocks elevates the shear stress to critical levels. Disharmonic structures adjacent to a lithologic boundary are not necessarily indicators if faulting implies either loss of cohesion or consistent displacement across a plane of discontinuity; rather, they may delineate a boundary of a décollement zone. Such zones may be intensely folded. The folds are passive because they arise from small perturbations of the velocity field in the zone. Erratic blocks and certain tectonic breccias are attributed to shear-induced break-up of stiff layers that are infolded into décollement zones. It is further concluded that several décollement zones could be active simultaneously.

The simple two step method used to construct velocity profiles is discussed in the Appendix. Integration of the strain field is accomplished by a graphic procedure that builds the composite velocity profile from the bottom up as if one were reconstructing the rock sequence under study. A detailed explanation of the construction of Figure 1 serves as an example of the application of the method.

TRANSLATION IN A THREE-LAYER ROCK MASS

When a three-layer stratified rock mass is placed in a setting conducive to gravity sliding, a décollement forms and rigid transport of an upper plate occurs, if a soft layer is in the middle. Rapid creep also occurs if a soft layer is on top but the motion is more like hillside creep than thrusting.

The rock masses used as examples in Figures 1 through 4 are 3 km thick and of infinite lateral extent. The masses are comprised of layers possessing perfectly planar boundaries that dip 20 m/km (105.6 ft/mi, 1.2°). Each layer is homogeneous and isotropic. The bottoms of the sequences lie permanently fixed to rigid basement planes with the same dip. The lowermost layer in these examples is 1.0 km thick and has a viscosity of 10^3 million-year-bars (m.y.-b).[2] This is about 10^{22} poise, a commonly cited viscosity for the crust (Gordon, 1965, p. 2415; Crittenden, 1963, p. 5526; Takeuchi, 1963, p. 2457). The layer behaves mechanically as crystalline basement and is essentially rigid in all examples studied. The other layers have lower viscosities which are varied from example to example. Table 1 lists the Newtonian viscosities used in this paper and suggests rock types that they most reasonably represent under the conditions of temperature and pressure expected for the phenomena studied. A rock sequence is schematically drawn on each of the example figures so that the reader need not refer back to the table.

Actual rock viscosities vary widely and there is considerable overlap of viscosities from one rock type to another (*see*, for example, Handin, 1966, p. 289). Rock viscosities are also strongly dependent on temperature and shear stress (Heard, 1963, p. 178). The assignment of specific viscosities to certain rock types in this

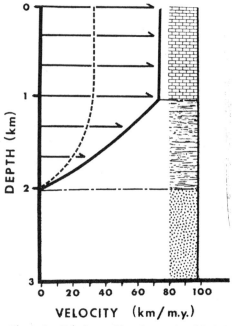

Figure 1. Velocity profiles of a gravity slide in a three layer rock mass. The slide is on a 2 percent grade. Solid line profile for Newtonian viscosity; dashed line profile for power law fluid.

paper is done only to simplify the presentation. Viscosity contrasts of several orders of magnitude between a décollement zone and other rocks in a deforming sequence are used in the examples, but are not essential to the theory. Calculations by Sherwin and Chapple (1968) of viscosity contrasts from the geometry of some single layer folds yield viscosity contrasts of only 14 to 30 for quartz layers in slate. Where viscosity contrasts are this low, an upper thrust plate may deform a noticeable amount during the transport process, but comparison of Figures 1 and 2 shows this is not generally true. Thus, I conclude the results apply to actual rock sequences.

Gravity Sliding in a Stiff-Soft-Stiff Stratified Rock Sequence

When a soft layer (low viscosity) is sandwiched between two stiffer layers, the velocity profile depicting long term creep of the rocks shows that the upper stiff layer slides downhill much faster than the lower stiff layer (Fig. 1) whose creep rate is insignificant. The uppermost kilometer of the section becomes the upper plate of a major thrust. It is able to move many kilometers without undergoing

[2] The viscosity units million-year-bars (m.y.-b) are more convenient than poise in many geologic problems. Time is measured in millions of years rather than seconds, and stress is measured in bars rather than dynes per square centimeter. If a substance one unit of length thick with a viscosity of 1.0 m.y.-b is subjected to a shear stress of 1.0 bar, the upper surface will move, relative to the lower surface, at the rate of 1.0 unit of length per m.y.

TABLE 1. ASSIGNED VISCOSITIES FOR SOME ROCK TYPES

Rock Type	Assigned Newtonian Viscosity (m.y.-b)
Salt	10^{-5}
Gulf Coast shale	10^{-3}
Shale	10^{-1}
Interbedded limestone and shale	10^{0}
Limestone	10^{1}
Crystalline rocks	$>10^{3}$

recognizable internal deformation. The shale sequence below is a décollement zone. No actual bedding plane fault exists; instead, the "detachment" is spread throughout the zone. Clearly, this kilometer of soft rock will be severely deformed during the displacement of the upper block.

The predicted velocity of the upper kilometer of stiff rock is almost 80 km/m.y. (8 cm per year) for Newtonian viscosities, and 35 km/m.y. for non-Newtonian viscosities. Such high velocities would permit substantial thrust fault displacement in very short periods of geologic time. These high speeds are due to two factors: (1) the great thickness of the soft layer (1 km), and (2) the absence of restrictions on either the updip or downdip end of the gravity slide. Compared to the other rocks in the section, the lowermost stiff rock layer is obviously not deforming at a significant rate. The reason for this is the enormous contrast in viscosity (three orders of magnitude) between the sediments and the crystalline basement.

Where velocity remains constant over a range of depth, as it does throughout the upper kilometer, the rocks are not deforming internally. They are simply being carted along on the moving underlying rocks. On the other hand, the faster the velocity decreases with depth (the shallower the velocity profile), the greater the rate of internal deformation (shear-strain-rate of the rock involved). Thus, rocks buried between 1 and 2 km deep are strongly sheared. While the velocity decreases with depth, the shear-strain-rate increases with depth.

The dashed line profile in Figure 1 was constructed under the assumption that the rocks possess non-linear viscosities rather than linear (Newtonian) viscosities used in computing the solid line profiles. The purpose was to show that the conclusions are the same, regardless of whether linear or non-linear viscosity best

depicts the steady creep of rocks. Any flow law that predicts viscosity reduction with increasing shear stress yields similar results. The dashed line profile plots to the left of the solid line because I selected non-linear viscosities that are slightly higher than equivalent linear viscosities. This choice has the advantage that the lines for Newtonian and non-Newtonian fluids do not plot on top of one another.

Although minimum viscosity contrasts of three orders of magnitude (Fig. 1) are probably exceeded in most sedimentary sequences, there is also need to consider rock sequences where viscosity contrasts are not so great. In Figure 2, the original soft layer is replaced by a rock whose viscosity equals 10^{0} m.y.-b. This

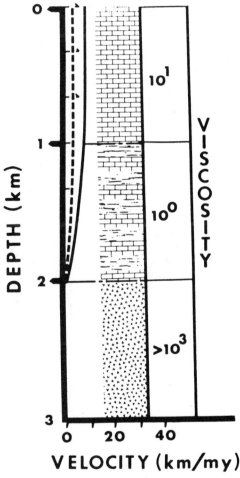

Figure 2. Velocity profile of a gravity slide in a three layer rock mass. The slope is a 2 percent grade. Solid line profile for Newtonian viscosity; dashed line profile for power law fluid.

reduces the viscosity contrast to a factor of 10, yet a décollement still forms. The upper layer moves at a predicted velocity of 10 km per m. y. (4 km/m.y. for non-linear model) without internal deformation. The middle layer undergoes substantial rectilinear shear. The velocity of motion of the upper plate is not nearly so high as in the example of Figure 1, but it is certainly high enough to yield substantial displacements over relatively short periods of geologic time.

The influence of thickness is illustrated in Figure 3, where the shale of Figure 1 has been reduced to one-third its original thickness (to 333 m). The remainder of what was shale is now considered to be limestone. The velocity of the upper plate has correspondingly been reduced to one-third the value in Figure 1. This one-for-one reduction was accomplished only because the *average depths of burial are*

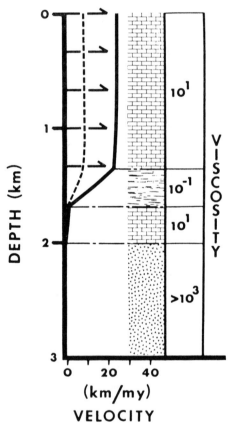

Figure 3. Velocity profile of a gravity slide of a rock sequence with four layers. Slide is down a 2 percent grade. Solid line profile for Newtonian viscosity; dashed line profile for power law fluid.

the same in both examples. If other depths are used, the relationship of the upper plate speed to décollement zone thickness is not as obvious.

From these results one might suspect that a décollement will form in virtually all stratified sequences provided they are inclined. This, of course, is not so. In many natural settings no motion takes place because downdip buttressing which prohibits motion is never overcome. If motion commences but the soft layer is either too thin or too shallow, the shear rate is very small and the motion insignificant.

Internal Deformation Causes Thrusting

In the example above, movement of the upper plate does not cause deformation in the décollement zone. Rather, the deformation in the décollement zone causes the movement of the thrust plate. The zone deforms because it is buried beneath a kilometer of overburden and is inclined at an angle (2 percent grade or 1.2 degree dip). A similar deformation occurs if the rock sequence is pushed by an orogenic compression. The fact that the thrust moves because of internal deformation of the sequence is very important. It severely limits the possible geometry of thrusts and the nature of fault-strata intersections. For example, a low angle thrust cannot decapitate an existing fold (that is, penetrate both limbs) as suggested by the diagram in Billings (1954, p. 182, Fig. 154D). Such fault geometry is mechanically impossible. The remaining structures illustrated by Billings (1954, p. 182, Fig. 154, A-C, E, F) are compatible with the writer's theory.

In the vicinity of the front edge of a translating plate, the motion does cause deformation. If buckling of the upper plate occurs, folding pushes soft décollement zone rocks from beneath growing synclines into the space provided by growing anticlines. Clearly, here thrusting causes the deformation of both the upper plate and the décollement zone. Similarly, if step thrusting occurs, the faulting is caused by actual translation of the upper plate. Additional deformation of both the advancing plate and the underlying lower plate will also be caused by the translation. The mechanical setting at the front edge of a major thrust plate is commonly extremely complex.

This is the portion of large overthrusts that can be demonstrated to exist geologically. Here, older rocks lie on younger, or rocks of the same age, but different facies are superposed. This portion can be of large horizontal extent— more than 50 mi for the Roberts Mountain thrust in Nevada (Gilluly, 1957; I thank W. M. Chapple for suggesting this example). Usually it is much less, and many imbricate step thrusts absorb the total transport (as in the Canadian Rockies, for example, Price,

1962; Bally and others, 1966). In the toes of some overthrusts my mechanism may dominate, but in many it is subordinate. Superposition of relatively stiff rocks with high friction coefficients along a thrust toe leads to the development of both fault gouge and breccia.[3] Once formed, these cataclastic zones probably deform according to the suggested mechanism, but the formation of the gouge and the enlargement of the gouge zone are more important to the mechanics of the process than the deformation of previously formed gouge. Description of a model for gouge formation is beyond the scope of this paper.

Gravity Sliding in a Soft-Stiff-Stiffer Rock Sequence

Figure 4 shows interbedded limestone and shale overlying limestone. If the dip is 20 m per km (2 percent grade, 1.2 degree dip), only the upper kilometer of rock moves at a measurable rate. Although the rate is substantial (up to 4 km per m. y.), the entire motion is accomplished through deformation of the upper layer. It becomes highly deformed. The relative displacement and deformation of various strata within the section is difficult to ascertain. Over-all, the motion is much more akin to soil creep than to thrusting. This may cause mélanges (Hsü, 1968).

END EFFECTS

In the previous examples only basal drag was considered, yet it is certain that end effects are equally important in determining whether a gravity slide will occur and what its speed will be. The end effects in two dimensions are (1) the original updip continuity of gravity slides, and (2) the buttressing afforded by the continuation of a potential thrust unit ahead of the edge of the moving plate. Where crustal shortening is responsible for the development of structures, the resistance due to buttressing is important in determining whether décollement and thrusts occur, or whether some other style of deformation prevails.

Updip Attachment

Prior to formation of a gravity slide the rock mass is continuous with updip rock equivalents not involved in the subsequent slide. When a slide moves, either (1) a transition zone between the slide and its stationary

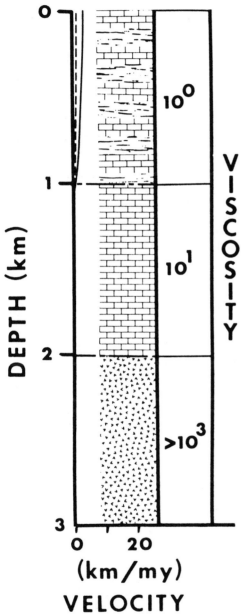

Figure 4. Velocity profile of downhill creep in a three layer rock mass. Slope is a 2 percent grade. Solid line profile for Newtonian viscosity; dashed line profile for power law fluid.

---

[3] The mechanisms discussed by Carlisle (1965) may be important in modeling this phenomena properly.

updip counterparts elongates substantially, or (2) the slide completely detaches itself. Where stretching occurs, thinning and elongation are accomplished through the formation of pull-apart grabens (Fig. 5). Their appearance and mode of formation are thought to be similar to the grabens created experimentally in clay by Cloos (1968, p. 426–427).

The thinning of a transition zone with or without the formation of pull-apart grabens is comparable to necking in a specimen in a tensile test. Considerable energy is expended in establishing the elongation. Consequently, motion of a gravity slide is retarded not only by basal drag, but by viscous retardation accompanying stretching. I know of no satisfactory theoretical model for stretching of rock sequences. Thus, it is not possible for me to estimate the drag associated with this process. It is only possible to say that predicted velocities will be too high because this retarding effect is neglected.

Where complete detachment occurs, the slide is free to move without updip retardation. The equivalent of a bergschrund forms behind the slide. Although complete pull-aparts are probably rare, the Heart Mountain Thrust in Wyoming provides an example. Exposures in the canyon of the Clarks Fork of Yellowstone

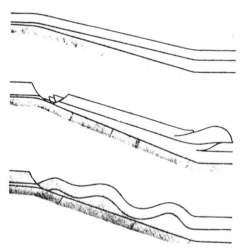

Figure 5. Overcovering updip attachment and downdip buttressing in a gravity slide. Top shows rock sequence before slide. Middle unit is assumed to be mobile shale or salt. Middle diagram shows formation of pull-apart grabens updip and imbricate thrusts downdip. Lower diagram shows single pull-away with reverse drag updip and folding downdip.

River show the bergschrund is filled with volcanic ejecta (Pierce, 1957, p. 604).

Buttressing. Rocks comprising the upper plate of a thrust are originally continuous across the boundary that later becomes the forward edge of the thrust. This continuity of the rocks creates a buttress that prohibits motion of the thrust plate. If the plate begins to move, either a step fault forms at its forward edge, or the plate crumples against the buttress (Fig. 5). All possible variations, from folding without faulting, to faulting without folding, are possible.

Where folding is dominant, the rate of advancement of the sliding plate is limited by the rate at which folding provides space for the sliding plate to advance. In order to compute the retardation of thrust motion, it is necessary to be able to compute the rate of finite amplitude folding as a function of the applied stress. This can be done only on specific examples with the expenditure of great labor using numerical methods (Chapple, 1968). General results are not possible. Nevertheless, it is theoretically possible in specific cases to exactly balance the basal drag and rate of folding with the effective component of gravity. Thus, the "exact" rate of sliding could be computed.

If a step fault forms, considerable frictional drag will accompany the formation of the toe of the thrust. This drag increases in magnitude as displacement of the thrust plate continues. The magnitude of these forces has been computed by Raleigh and Griggs (1963) and has been reconsidered by Hsü (1969, p. 946–948). It is likely that the toe could grow so large that the frictional resistance to motion equals the push causing the thrust. If this happens, a new step fault must form before motion of the upper plate continues. This leads to imbricate thrusting (Fig. 5).

MOTION OF THRUST PLATES NEGLECTING END EFFECTS

Basal drag, which slows the motion of gravity slides and requires high stress levels to sustain motion in orogenic thrusts, increases with increasing décollement zone viscosity, and decreases with increasing décollement zone thickness. Because both the viscosity of rocks and the thickness of potential décollement zones vary over several orders of magnitude, in these deformations they are considered the most important variables, outside of end effects. In gravity sliding, the next most important consideration is the driving force, gravity. The contribution of gravity increases with: (1) the dip of a potential slide mass because the component of gravity in the direction of motion increases with dip; and (2) the depth of burial of the potential décollement zone, because the shear stress in a dipping bed increases linearly with depth of burial. In orogenic thrusting, as explained in more detail below, the dip of the décollement zone aids the motion if it is downdip, or retards the motion if it is updip. The influence of depth of burial on orogenic thrusts is more complicated. If the motion is downdip, thrusting is facilitated by increased depth of burial. If the motion is updip, the reverse is true. For a given thrust plate length, the magnitude of the orogenic stress required to perpetuate the motion decreases as the thickness of the upper plate increases.

Because the end effects have been neglected, neither the predicted velocities of gravity slides or the computed orogenic stresses can be taken literally. Rather, a large calculated velocity corresponds to a high propensity for slide initiation. Whether or not high velocities are achieved depends on end effects. The same reasoning suggests that slides for which the predicted velocity is low are unlikely to form.

The calculated stresses, although they do not include an amount necessary to overcome end effects, do accurately reflect the stresses associated with overcoming basal drag. The theory presented in this paper shows that these stresses are not so large that they limit the size of thrusts.

Velocity of Gravity Sliding

Assuming Newtonian behavior, the equation relating the velocity to the controlling factors is

$$\dot{u} = \{\rho g \sin \theta / \mu\} \{t(2d + t)\} \qquad (1)$$

where ρ is the density of the material, g is the gravitational constant, d the depth of burial to the top of the décollement zone, t the thickness of the décollement zone, θ the angle of dip of bedding, and μ the viscosity of the décollement zone. This equation derives from equation $11a$, which is based on steady state flow of linearly viscous fluids, by assuming that the viscosity of the upper plate is so much higher than that of the décollement zone that the upper plate does not deform significantly.

The influence of dip and décollement zone viscosity on the velocity of gravity sliding is

illustrated in Figure 6. Predicted velocities over the range of dips shown vary from over 10^5 km/m.y. for a slide on a salt bed, to less than 10^{-4} km/m.y. if the décollement must occur in crystalline rocks. Figure 7 plots the predicted velocity of gravity sliding versus depth (d) to the top of the décollement zone (that is, thickness of the upper plate) for a 1.0 percent grade (0.6 degree dip) and viscosity of décollement zone equal to 10^{-1} m.y.-b. Predicted velocities range from zero to 5×10^2 km/m.y. for the range of thicknesses considered.

It appears from Figure 6 that for slopes comparable to those of coastal plains (0.1 percent, 0.1 degree dip) only salt layers or "gumbo" shales could serve as décollement zones. On steeper slopes comparable to moderately steep continental slopes (10 percent, 5.8 degree dip), even limestones could become décollement zones, provided the ambient temperature was high. (At 400° C, Yule marble exhibits a viscosity of about 10^0 m.y.-b; Heard, 1963, p. 182.) Although Figure 7 shows velocities for depths of burial up to 10 km, it is unlikely that depths of greater than 4 or 5 km are meaningful in gravity slides, due to the increasing downdip restriction of motion. In constructing Figure 7,

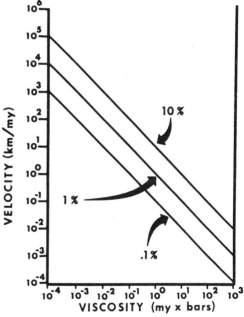

Figure 6. Velocity of a gravity slide versus décollement zone viscosity for slopes of 0.1 percent, 1.0 percent and 10 percent. The sliding plate is 1 km thick and is assumed to be rigid.

the décollement zone viscosity was assumed to be 10^{-1} m.y.-b, even though rocks with lower viscosities are probably found in most rock sequences. A high value was used to illustrate that even if only stiff rocks occur in a sequence, high slide velocities are still possible if the potential décollement zone is thick enough.

The Heart Mountain Thrust in northwestern Wyoming may represent such a high speed slide. The occurrence of Tertiary volcanics on an exposed, uneroded décollement zone surface of this thrust indicates that the time between thrusting and volcanic outpouring was minimal (Pierce, 1957, p. 603). If the viscosity of the décollement zone was between 10^{-3} and 10^{-2} m.y.-b, a velocity of 10^1 to 10^2 m per year could have been attained by the Heart Mountain Thrust. At this rate, the total maximum displacement of some 20 km could have been accomplished in 200 to 2000 years. I know of no way to judge whether any significant erosion of the exposed portions of the décollement surface in the updip pull-away zones would have occurred in such a short time.

Hsü (1969, p. 944–945) suggests an even more catastrophic failure of the type described by Shreve (1966, 1968), preceded by a slower creep stage similar (I presume) to the model invisioned herein. This also appears a likely alternative to me.

Stresses Associated With Orogenic Thrusting

The décollement zone model requires remarkably low stresses to sustain thrusting rates comparable to reported rates for sea-floor spreading (for example, Dickson and others, 1968; Heirtzler and others, 1968; Le Pichon, 1968; Le Pichon and Heirtzler, 1968). In fact, major thrust plates may be moved uphill by orogenic stresses far lower than the brittle strength of most *unconfined* sedimentary rocks. For instance, a thrust plate 10^3 km wide and 10^1 km thick can be easily moved at a rate of 10^0 cm per year by an orogenic stress of 10^0 to 10^1 bars if the mass is underlain by, say, 10 m of soft shale or salt.

The model suggested for orogenic thrusting is very simple (Fig. 8) and is patterned after Hafner's (1951). An almost rigid rectangular upper plate overlies a deformable layer of relatively low viscosity. The low viscosity layer lies on a rigid basement to which it is permanently fixed. Thrusting is accomplished by either of two methods. The first envisions a stationary basement and an orogenic push acting on the upper plate. The low viscosity layer permits the upper plate to move by undergoing rectilinear shear. In the second method, the basement moves and the upper plate is buttressed. The movement causes rectilinear shear in the low viscosity layer, which in turn transmits a shear stress to the base of the upper plate. This shear stress forces the upper plate against its buttress. The resulting stress state in the upper plate is the same for either mechanism (Fig. 8).

The stresses at the right end of the model for the upper plate described above (right end,

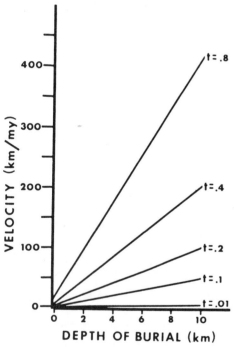

Figure 7. Velocity of gravity slides versus depth to the top of the décollement zone (that is, upper plate thickness). Décollement zone thicknesses are 0.01 km, 0.1 km, 0.2 km, 0.4 km, and 0.8 km. The slope is 1.0 percent, and the décollement zone viscosity 10^{-1} m.y.-b.

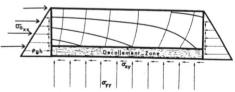

Figure 8. Schematic thrust block showing stress trajectories (*after* Hafner, 1951). σ_{zz} is the overburden, σ_{oxx} is the orogenic stress, and σ_{xx} is the basal shear. (The other lateral stress is the "lithostatic pressure.")

Fig. 8) approach a relaxed stress state (lithostatic "pressure") that represents a far-field stress condition free from the influence of the near field diastrophism (left side, Fig. 8). Thus, the model dissipates the orogenic push entirely through the action of shear stresses (basal viscous drag) acting on the base of the upper plate. To model real thrusts before movement begins, the far end stress field (right end, Fig. 8) should represent the stresses at the buttress. After movement begins, the far field stress state should model the stresses resulting from friction drag of a thrust toe or the complicated stress field accompanying folding of the upper plate, whichever phenomena occurs in the example under study. Because the model doesn't account for the far end properly, the importance of basal drag is overemphasized in determining the feasibility and ease of tectonic transport of the upper plate. Nevertheless, the model shows that the stress required to move a thrust increases with increasing speed of motion and increasing décollement zone viscosity. It decreases with increasing décollement zone thickness. If thrust motion is downdip, the required orogenic stress decreases with increasing dip and increasing plate length. If

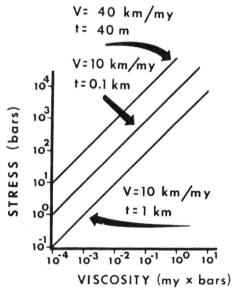

Figure 9. Orogenic stress required to move a thrust plate versus viscosity of the zone of décollement. Plots are for two displacement rates (10 km/m.y. and 40 km/m.y.) and for décollement zone thickness of 1 km, 0.1 km and 0.04 km. End effects have been neglected. The shaded area corresponds to potentially excessive stress levels.

thrust motion is upslope, as it appears to be in many orogenic belts (for example, the Appalachian Mountains: King, 1950; Rodgers, 1950; the Canadian Rockies: Bally and others, 1966; and the Jura: Laubscher, 1961), the required stress increases with increasing dip and plate length. Assuming that the orogenic stress (σ_0) is uniformly distributed over one edge of the thrust plate (Fig. 8), the equation relating this stress to the above factors is

$$\sigma_0 = \mu v l / t d + \rho g l \sin \theta \qquad (2)$$

where μ is the décollement zone viscosity and t its thickness, v the velocity of plate motion, l the length and d the thickness of the thrust plate, ρ the density of the thrust plate, and g the gravitational constant. The first term in equation 2 represents the total viscous drag along the base of the plate, assuming rectilinear shear throughout the décollement zone, whereas the second term represents the component of the upper plate's weight that either helps or hinders the motion. The angle of dip (θ) is positive if the thrust moves updip, and negative if it moves downdip.

Figure 9 plots orogenic stress versus viscosity for three ratios of velocity to décollement zone thickness, assuming the thrust plate length to thickness ratio is 100. The 10 km/m.y. velocity of plate movement approximates the reported rate for sea-floor spreading in the North Atlantic (LePichon, 1968, p. 3664; Heirtzler and others, 1968, p. 2131). Comparison of the curves for zone thicknesses of 1.0 km and for 0.1 km shows that even if only stiff rocks are available for décollement shear (for example, limestone at 300° C, straining at a rate of 10 m.y.$^{-1}$ exhibits an apparent viscosity of 10^0 m.y.-b; Heard, 1963, p. 182), the orogenic stress need not be large if sufficient thicknesses of the rock are incorporated in the deformation. Zone thicknesses of about 0.1 km are thought to be more typical of common décollement zones. If low viscosity rocks, like salt, are available, the required stresses are negligibly small. This reemphasizes the discussion of the last section where it was surmised that end effects contribute markedly to the total drag of a thrust plate. The displacement velocity of Figure 9 (V = 40 km per m.y.) approximates the rate of spreading attributed to parts of the East Pacific Rise (Pitman and others, 1968, p. 2084). With a décollement zone thickness of 40 m, and viscosity no higher than that expected for Gulf Coast shales, the average orogenic stress remains less than 1 kb, a stress level sustainable by most rock sequences. However, average orogenic stresses much above 1 kb are probably excessive. The shaded area of Figure 9 corresponds to these potentially excessive stress levels.

A COMPARISON WITH EXISTING THEORY

The existence of a sole fault below the upper plate of major thrusts is implied in most discussions of the subject (Hubbert and Rubey, 1959; Badgley, 1965, p. 187–250), but more work is commonly required to move a thrust over a sole fault than over a décollement zone that distributes the shear. This is demonstrated by comparing the orogenic stress required to move a thrust across a sole fault with the stress required by décollement (equation 2).

Simple Friction Theory

If the upper plate of a thrust moves as a rigid block along a sole fault, the orogenic stress need only overcome the friction at the base of the plate plus the uphill component of the weight of the thrust plate acting along the plane of motion. The weight of a column of sedimentary rock is approximately 1.0 psi per ft of height. Using this, the equation describing the balance of forces in thrust plate is written

$(\sigma_0, psi)(d, in)(w, in)$
$= (1.0, psi/ft)(d, ft)(l, in)(w, in)(\nu + \sin \theta)$

where it is assumed that the orogenic stress is uniformly distributed along one edge of the plate and d is the thickness of the plate, l its length, w its width, ν the coefficient of friction and θ the angle of dip of the thrust. The term multiplying ν represents frictional resistance to motion, the $\sin \theta$ term the component of the plate's weight along the direction of thrusting. Dividing out the width and thickness yields

$(\sigma_0, psi) = (1/12, psi/ft)(l, in)(\nu + \sin \theta)$.

In order to compare this equation with the results of previous sections, the units must be converted to bars and kilometers. This yields a multiplicative constant of 226.2 bars per km of burial, which is rounded to 230 because of a lack of significant figures in the original constants. The above equation then becomes

$(\sigma_0, bars) = (230, bars/km)(l, km)(\nu + \sin \theta)$. (3)

If the motion is horizontal, the ratio (R) between the stresses predicted by equations 3 and 2 is:

$$R = 230 \, td\nu/\mu\nu . \tag{4}$$

If a sole fault is to form, R needs to be less than 1. Small ratios are obtained for thin thrust sheets with even thinner décollement zones of high viscosity. High thrust velocities also reduce the size of the ratio and thus increase the tendency for faulting. Even so, equation 4 suggests that sole faults ought to be rare. This, perhaps, is best demonstrated with a specific example.

Assume a velocity of 10 km/m.y. for a thrust plate 5 km thick and 100 km long that overlies 100 m of 10^{-1} m.y.-b shale. According to the present theory, the required orogenic stress is 200 bars. The sole fault model requires an orogenic push of 5600 bars if a very low value (0.25) is assumed for the coefficient of friction along the fault. Assuming the same thrust displacement occurs whichever mechanism applies, the ratio of orogenic stresses required by the two models is equal to the ratio of the work done in moving the thrust. This is because work is orogenic stress times plate thickness times thrust displacement, the latter two being the same for any given thrust. Thus, in the example, 26 times more work is required by the sole fault model. The difference between the stresses, and therefore the work or energy expended by the two models, can be reduced by varying the parameters. Assuming that at least 5 m of 10^0 m.y.-b shale will be available in any stratigraphic section (this is very stiff shale), that the maximum velocity is 100 km/m.y. (10 cm/yr), and that the coefficient of friction cannot possibly be less than 0.15, then sole faulting will be preferable only if the 100-km-long plate is less than 600 m thick. If 10 m of shale are available, the plate must be less than 300 m thick, and so forth. Consequently, sole faults are expected only if all potential décollement zones are both thin and highly viscous, or where they are absent altogether.

Reduced Friction Theory

In a classic paper, Hubbert and Rubey (1959) recognized the inadequacies of the simple theory of friction across a sole fault for both orogenic thrusting and gravity sliding. Dip angles needed to initiate gravity sliding were predicted as unrealistically high. Orogenic stresses theoretically needed to move known thrust sheets far exceeded the strength of the rocks involved. Hence, the predicted stresses could not have existed. Previous attempts at resolving this paradox had centered on ways to reduce the coefficient of friction in equation 3. No reasonable reduction was sufficient, however. Instead, Hubbert and Rubey (1959) thought of reducing the effective normal load (the coefficient in equation 3) by supporting a portion of the overburden weight with excess pore fluid pressure. Some recorded pressures from the Gulf Coast Tertiary exceed 90 percent of the overburden. Excess pore pressure reduces the effective overburden weight; in this example by a factor of 10. A further reduction occurs with higher pressure. Such reductions in effective load reduce the friction sufficiently to permit movement on very large thrusts.

A direct comparison of the Hubbert and Rubey theory with mine is not possible. By calling on ever higher abnormal pressure, it may be possible to reduce the orogenic stress required in the Hubbert and Rubey (1959) theory to as low a value as desired. Hsü (1969) claims that it is not proper to neglect the cohesive strength of rocks when computing the stresses required for thrusting as Hubbert and Rubey (1959) do in their analysis. If Hsü[4] is correct, then the stresses required by my present theory are less than those required by the sliding friction theory in all but the most exceptional settings. If Hsü (1969) is not correct, as Hubbert and Rubey (1969) claim in a discussion of Hsü's paper, then by using the reduced friction theory, it is always possible to predict lower values for the required orogenic stresses than the present theory does. But in doing so, one must call on exceptionally high pore pressure—commonly in excess of 99 percent overburden weight. It is questionable whether such excessively pressured sediments occur frequently enough in nature.

Most investigators define abnormal pressure as interstitial fluid pressure that exceeds by an

---

[4] Although Hsü (1969) dwells principally on the question of overcoming cohesive strength of rocks prior to thrusting, his description of actual thrusts (1969, for example, p. 941) emphasizes "flow of a thin layer of pseudo-viscous material," which is in complete accord with my theory of décollement. Hsü's (1969, p. 941) claim that flow of the limestone layer (the *Lochseitenkalk*) occurs when the cohesive strength of the limestone is overcome is not valid. Loss of cohesion is not involved in the transition from elastic behavior to viscoelastic or viscous behavior and the post transition model must be in accord with the post transition behavior. The stress levels at which transitions to visco-elastic or viscous behavior occur depend principally on the strain rate and ambient temperature and can vary significantly (Heard and Rubey, 1966, compare Fig. 1 and Fig. 4). For an expanded discussion of this argument, *see* Kehle and Reid (in prep.).

arbitrary amount the hydrostatic pressure expected at that depth. Rubey and Hubbert (1959, p. 170), following this convention, define abnormal pressure as those pressures exceeding normal pressure by at least 10 percent. This minimum abnormal pressure equals somewhat less than 50 percent of the overburden weight. Rubey and Hubbert (1959, p. 170) state that only 10 percent of the known abnormal pressures exceed 90 percent of the overburden weight—and only five localities are cited where pressures exceed 0.95 psi/ft, that is, only five localities possess pressures presumably representing more than 95 percent of the overburden. This assumption, that .95 psi/ft represents 95 percent of the overburden pressure, is based on the well-known relationship between lithostatic pressure and depth of burial in the Gulf Coast Tertiary (1.0 psi per ft of burial; Dickinson, 1951). This figure, 1.0 psi/ft, is slightly high for depths less than about 5000 ft, and somewhat low at greater depths. Gulf Coast sediments are relatively undercompacted—carbonate sequences and even older sand-shale sequences possess lithostatic pressure gradients in excess of 1.1 psi/ft. Both the North German and the Iran localities cited by Rubey and Hubbert (1959, p. 170) probably exhibit these higher lithostatic gradients. Thus, only three localites remain as possibly possessing pore pressures high enough to favor a reduced friction mode of displacement rather than décollement.

The apparent rarity of exceedingly high pore pressure, the special circumstances under which it occurs, and the possible need to consider cohesive strength, are not consistent with the widespread occurrence of thrusting and gravity sliding, whether in orogenic belts or less active areas. Both large and small thrusts occur over thin marine shales, non-marine red mudstones, shallow water (?) carbonates, and evaporite sequences. But abnormal fluid pressure is only rarely associated with evaporites and is unknown in the other settings. A different mechanism is required, one that works independently of the occurrence of abnormal pressure.

The theory of fluid-like shear of a décollement zone meets this criterion. Interestingly, under the new theory, abnormally pressured sediments may be the preferred sites for décollement—not because the overburden is supported by the fluid pressure, but rather because abnormally pressured sediments are probably less viscous than their normally pressured counterparts.

STRUCTURES ASSOCIATED WITH DECOLLEMENT

If décollement takes place according to the model suggested, structures associated with major crustal translation should be reconcilable with the operative deformation mechanisms. The direct consequences of the model have now been presented, but the model itself is too stylized to tell us much about the nature of structural elements arising from these processes. This deficiency is partially alleviated by drawing on analogies with alpine glacier deformation. These are particularly helpful in discussing bedding plane faults and the internal deformation of décollement zones. In addition, qualitative support for some arguments has been found in the quantitative theories of folding reported by Biot (for example, 1965). The remainder is simply speculation on my part.

Large Bedding Plane Faults

In many actual geologic settings, large bedding plane faults probably do form in association with décollement, but no documented examples are known to me. Two possible types are: (1) faults through stiffer rock masses within décollement zones, and (2) pull-offs at a contact between rapidly shearing rock and a "rigid" stratum (for example, at the base of a décollement zone).

The first type of fault forms because the mechanical properties of a décollement zone probably vary by at least an order of magnitude from place to place. The shear stress in the zone's higher viscosity rocks will be proportionately higher—by an order of magnitude if this is the difference in viscosities. Although average stresses in a décollement zone are predicted to be low, multiplication by a factor of 10 or more within high viscosity zones may elevate these stresses to unstable values. If so, a fault will form. The orientation of such a fault is uncertain. Experiments on rock specimens show shear fractures forming at about 15 degrees to the direction of maximum shear if the samples are free to fail in any direction. This orientation within a décollement zone is 15 degrees from bedding. The resultant fault is a low angle step thrust. On the other hand, if test specimens are constrained by a jig to fail along specified angles other than 15 degrees

off max shear, they will do so. In some settings, the presence of the "rigid" upper thrust plate may act as a constraint, forcing a fault in the décollement zone to form sub-parallel to bedding. Such a fault would be mapped as a bedding plane fault.

The step thrust-like faults mentioned above are not real step thrusts. In fact, I think it unlikely that they would develop into step thrusts. True step thrusts form to overcome buttressing in the upper plate of a thrust sequence. They are entirely an upper plate phenomenon—originating there to satisfy boundary demands on the upper plate and essentially remaining confined to the upper plate because there is no mechanical reason for them to extend backward into the underlying décollement zone.

A second type of bedding plane fault involves discontinuous slip at the base of a rapidly shearing layer. Similar slip occurs at the base of temperate glaciers (see Kamb and Le-Chapelle, 1964; Theakstone, 1967) but the mechanism cannot be the same. In glacial slip, ice melting and refreezing, and the existence of a layer of free water, are important contributing factors. In some glaciers the slip is sometimes accomplished as the glacier locally lifts off its base (for example, the Blue Glacier, Washington; Shreve, 1969, oral commun.). This phenomenon may be responsible for plucking boulders from the beds of glaciers. The same process may be responsible for the incorporation of exotic blocks into a gravity slide such as those reported by de Sitter (1964, p. 236). The occurrence of these blocks and the demonstrated occurrence of flexural slip in some folds argues strongly that some local slip accompanies décollement.

Regardless of the potential existence of bedding plane faults in décollement zones, their actual occurrence is difficult to demonstrate. Being parallel to bedding, no stratigraphic separation is expected. Disharmonic structures within décollement zones adjacent to boundaries are not adequate evidence; they simply indicate local perturbations in the deformation of the décollement zone. Nor do tectonic breccias demonstrate a fault's existence, for breccias may form where interbedded stiff rock becomes involved in the same deformation as the soft rock of the décollement zone. Consequently, two time-honored criteria for recognizing bedding faults may in fact only delineate boundaries of décollement zones or boundaries within a zone. Faulting need not occur on such boundaries, and, in general, ought not to occur there.

Décollement Zone Structures

Development of multiple criteria for the field recognition of décollement zones is desirable to test the present hypothesis. Verification of translation usually depends on the recognition of displaced sedimentary facies. In orogenic regions, rapid facies changes complicate interpretation, and proof of displacement is commonly difficult, if not impossible. The question of the existence of décollement in some regions (for example, the Southern Appalachians) has remained a subject of heated controversy for decades. Criteria for the recognition of such zones because of their large strain or from their internal structures should aid in resolving such question.

Total Shear Within Décollement Zones. Large tectonic transport by décollement produces large strain in the décollement zone. Assuming deformation in the zone to be homogeneous simple shear—an acceptable assumption for thin zones of constant lithology (say, less than 200 to 300 m thick)—it is possible to calculate the finite strain accompanying a given displacement of the upper plate. If λ represents the quadratic elongation $(l_1/l_0)^2$ where l_0 is the original length and l_1 the final length, the principal strains in terms of quadratic elongations are given by

$$\lambda_1, \lambda_2 = \{D^2 + 2 \pm D(D^2 + 4)^{1/2}\}/2 \quad (5)$$

where D is the ratio of the distance of tectonic transport to the décollement zone thickness. (For a derivation of equation 5, *see* Ramsay, 1967, p. 83–85.) A 5 km transport over a 1-km-thick décollement zone (perhaps the Cretaceous carbonates of the Sierra Madre Oriental, Mexico, over the underlying early Mesozoic evaporites) would result in $\sqrt{\lambda_1} = 5.2$ and $\sqrt{\lambda_2} = 0.2$, a rather large strain. Even more impressive are the strains resulting from, say, a 50 km transport over a 50-m-thick shale décollement zone. Here, the maximum strains are $\sqrt{\lambda_1} = 1000$, $\sqrt{\lambda_2} = 0.001$.

How does one measure such large strains and how does the strained material look? In evaporites, recrystallization likely accompanies the entire deformation process, just as it does in ice deformation in glaciers. If only pure, clean glacier ice is studied, evidence of *large* deformation is lacking. Even the presence of dirt layers and other contaminants do not provide quantitative evidence of the enormous deformation sustained by ice in its total travel down a glacier. The commonly contorted, "sheared" aspect of such foreign material does indicate strong deformation, but its magnitude remains unknown. Perhaps no more can be expected in evaporites. Consider, for example, the structures in salt reported by Muehlberger (1959). Here, measurement of large total strains is not amenable to conventional techniques, nor is it expected to be anywhere where recrystallization accompanies deformation.

The situation expected in shales is not much better. In simple shear experiments, preferred orientation of clay minerals parallel to the direction of maximum elongation forms early in the deformation process (B. C. Clark, 1969, written commun.). As deformation continues, the degree of orientation increases. This ultimately results in the formation of cleavage parallel to the direction of maximum elongation. For large simple shear in décollement, the cleavage is subparallel to original bedding. Very likely, strain of the magnitude suggested above will so deform included objects that they would no longer be recognizable. (Imagine trying to recognize an originally equidimensional fossil, say, 1 cm across that is located in the 50-m-thick décollement zone mentioned above that transports the overlying plate 50 km. After deformation the fossil would be 10 m long and only 0.01 m wide.) An exception occurs if the inclusion is stiffer than the zone material, but then its deformation, if any, would not be representative of the décollement zone. Such an object would likely be recognized as a phacoid. Again, the problem of recognizing the large strain and then measuring it appears outside the scope of developed strain analysis.

The fascinating work of Rosenfeld (1968) on rotated garnets suggests that large strain can be measured in metamorphic terranes. His estimates based on rotated garnets apply only to the deformation accrued during and after the period of garnet formation and represent a minimum strain value. Even so, Rosenfeld (1968, p. 192) records strains of $\sqrt{\lambda_1} = 21.87$, $\sqrt{\lambda_2} = 0.0457$ in the Waits River sequence on the flanks of the Chester Dome, but concludes that the deformation is more nearly extending flow than simple shear. If décollement had occurred, a tectonic-transport-distance-to-décollement-zone-thickness-ratio of 15 would have resulted from the measured strains.

Simple Shear in Nature. The velocity

profiles for the décollement zones in Figures 1–3 are very similar to the profiles reported for temperate glacier flow (Reid, 1896; Shumskii, 1964, p. 312). In glacial flow, as in décollement, the primary mode of deformation is simple rectilinear shear, but slight variations of this flow pattern occur in time and space. Such distributed velocity variations result in large internal distortion of the deforming body. In glaciers, one result is infolding of annual firn layers. I surmise that "pure" simple shear does not exist in real décollement zones even though this type of flow predominates. The result is contorted décollement zones. Yet, if a décollement zone occurs in a relatively homogeneous shale, a pervasive cleavage or fissility could form parallel to planes of maximum elongation. These approach original bedding as deformation continues. This mechanically induced fissility would obscure original bedding. Even where real bedding in the zone is severely contorted, this fissility may completely obscure the structure. The fissility may be planar, and thus would not reflect the passive folding in the zone, or it could demonstrate well-developed kink banding. Proof of either substantial shear or of internal contortion could be very difficult under the first of these circumstances. Recognition of "impure" simple shear probably requires the presence of interbeds of contrasting lithology, kink bands, or some other telltale feature that would not be obscured by structurally induced fissility in surrounding shale.

Folding Within Décollement Zones. Annual firn layers in the Blue Glacier, Washington (Raymond and Kamb, 1968), and in Austerdals Glacier, Norway (Shreve and Kamb, 1964), are isoclinally infolded into the glaciers, even though the major flow pattern is that of rectilinear shear. I expect the same structures to develop in décollement zones. Color bands, thin interbedded strata, and other planar and linear features within a décollement zone become folded. Rather severe folding is not only possible but likely. Fold axes are commonly oriented in directions other than perpendicular to the principal displacement direction. Their orientation reflects systematic and random variations in the flow field caused by regional convexity or concavity of both the dip and strike of the décollement zone, facies changes, thickness variation in both the upper plate and the décollement zone, local obstructions to motion, and so forth. This means a single deformation event could contribute several structural patterns. Rocks folded in one locale could

be transported to areas where different patterns originate. New structures would be superimposed on old. This could easily be interpreted as the result of two or more distinct stages of deformation, when in reality the complex structures result from a single continuing process.

Because this type of folding results from intermittent perturbations of an otherwise steady flow field in a décollement zone, these structures will not exist outside the zone. Folds that form in this manner are called passive by Donath and Parker (1964, p. 49–50), whether or not large viscosity contrasts exist between individual beds involved in the folding. The mode of formation is different from buckling folds in competent layers, which form in response to general shortening of a rock sequence. No uncompensated shortening need accompany the formation of folds within a décollement zone. Simple shear parallel to décollement zone boundaries (that is, parallel to original bedding) involves no shortening of the zone. Consequently, folding in a décollement zone is of a "shear" fold type accomplished by stretching and thinning of individual beds in the zone. Estimates of the original length of folded décollement zone beds cannot be made by simply straightening the beds. This unfolding procedure—commonly used to estimate crustal shortening—will erroneously attribute considerable shortening to a décollement zone when, in fact, there is none.

An excellent example of the structural style formed in décollement zones is found in the Lincoln fold system, New Mexico (Craddock, 1964). A relatively undeformed upper plate of San Andres Limestone has moved eastward, propelled by the gravity-induced internal deformation of the underlying Yeso Formation. Primarily a sequence of poorly bedded siltstones and mudstones with lenses of interbedded gypsum, the Yeso also contains thin limestone interbeds that reveal the severe internal contortion of the formation (Fig. 10). Shortening and displacement of the upper plate is reliably estimated by Craddock (1964, p. 132) at 3 km, yet unfolding the Yeso folds yields an unrealistically higher minimum value of 12 km. Although large contrasts in viscosity are suggested by the occurrence of differing lithologies in the Yeso, Craddock's (1964, p. 129–130) analysis of fold wavelengths in the Yeso suggests that no important viscosity contrasts exist. He concludes the folds must have formed by a passive mechanism.

426

Chaotic Blocks

If isolated interbeds of limestone or sandstone in a décollement zone are folded, the folds become distended by the prevailing simple shear. If fold amplitudes are large and viscosity contrast great (for example, limestone or silica-cemented sandstone versus shale), the strain rates induced in the stiffer beds raise the stress in these beds to critical levels. Local faults form here, dissecting the competent beds. Where very large displacement occurs, the limestone or sandstone beds become totally dismembered. Individual blocks move farther and farther from one another, eventually appearing as erratic boulders within the zone. Mélanges (Hsü, 1968) and some wild flysch may have formed in this manner. Some mélanges very likely are décollement zones.

Disturbed Zones. Zones of tectonic brecciation, and so forth, may represent the entire extent of a décollement zone, or only part of it. Away from thrust fault toes, these zones never belong exclusively to either the upper or lower plate. Rather, they form part of the separating décollement zone. No problem exists where the disturbed zone involves all the low viscosity rock that would be expected to form a décollement zone. A problem is encountered where the disturbed zone overlies or underlies a rock interval more suitable for décollement. The presence of a tectonic limestone breccia adjacent to an apparently less deformed shale may lead one to suppose that only the breccia marks the fault zone across which all differential motion has occurred. This is mechanically unlikely. The following examples demonstrate how the breccia can form even though it is not the zone of maximum shear strain or the zone across which the maximum displacement occurs.

Let a stratigraphic sequence contain two adjacent rock units whose viscosities are orders of magnitude lower than the remaining rocks in the section, but let the viscosities of these two units differ by a factor of 2. If the units are of equal thickness, about two-thirds of any thrust displacement will be accomplished through deformation of the softer zone; the rest arises through deformation of the stiffer unit. The resulting deformation process is illustrated in Figure 11, where the stiffer unit is above the softer unit, and in Figure 12, where

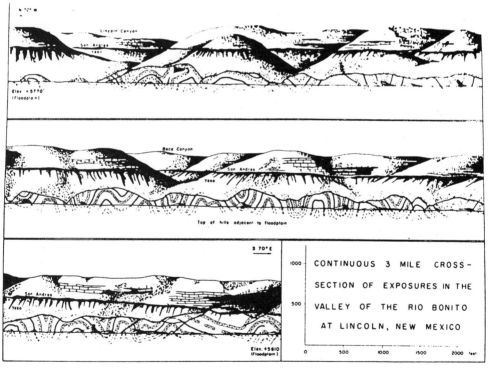

Figure 10.   View to NNE. near Lincoln, New Mexico, showing strongly folded Yeso Formations and conformable overlying San Andres limestone (*from* Craddock, 1964).

427

the stiffer unit is below. The stiffer zone is identified as a stratigraphic transition zone comprised of interbeds of the soft unit and the adjacent "rigid" rock (for example, sandstone or limestone). The softer zone might be a uniform shale or an evaporite-clay sequence and, though severely deformed, may not reveal its state of deformation. Fissility induced by deformation could obscure structure in a shale, and the outcrop quality of evaporite sequences, even in road cuts and quarries, is generally too poor to permit differentiation between major and minor deformation. Although the interbedded sequences in Figures 11 and 12 will have deformed only half as much as the softer zone, the presence of grossly different lithologies would make the deformation visible. If the

Yeso contained no limestone interbeds, its intense deformation in the Lincoln fold system (Craddock, 1964) could have been easily overlooked (Fig. 10).

If the softer rock comprises only a small portion of the total rock volume in the transition zone, and if infolding of stiffer rock takes place, continued deformation breaks up the folds. The resulting rock unit becomes a breccia. Large shear strain rates and relatively thin beds of highly contrasting viscosities favor the formation of infolds (Biot, 1961, p. 1606–1607). A breccia formed in this manner cannot be monolithologic because infolding will not occur without viscosity contrasts that arise from interbedding of different lithologies. Monolithologic breccias, such as are commonly

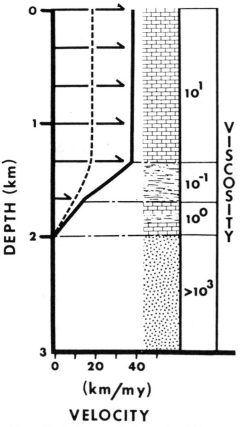

Figure 11. Velocity profile for gravity sliding of a limestone upper plate over a composite décollement zone. The décollement zone includes a pure shale unit ($\mu = 10^{-1}$ m.y.-b) that underlies an interbedded limestone shale sequence ($\mu = 10^0$ m.y.-b) sloping at 1 percent. Solid line profile for Newtonian viscosities; dashed line for non-Newtonian viscosities.

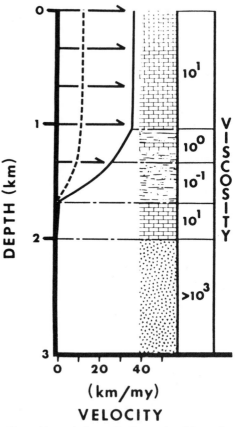

Figure 12. Velocity profile for gravity sliding of a limestone upper plate over a composite décollement zone. The décollement zone includes a pure shale unit ($\mu = 10^{-1}$ m.y.-b) that overlies an interbedded limestone shale sequence ($\mu = 10^0$ m.y.-b) sloping at 1 percent. Solid line profile for Newtonian viscosities; dashed line for non-Newtonian viscosities.

found in limestones, must arise in a different manner.

MULTIPLE DÉCOLLEMENT

Several décollements may be active in a rock sequence at the same time (de Sitter, 1964, p. 236). All that is necessary is for two or more separated softer zones to occur in the sequence. A dual décollement is illustrated in Figure 13. The uppermost limestone is moving 20 km per m.y. faster than the intermediate limestone, which is moving 20 km per m.y. faster than the lowermost limestone. Both the uppermost and the intermediate limestones are thrust plates. Both shales are independent décollement zones. The structures in each plate can develop independently even though both thrusts are moving at the same time. A step thrust could easily connect the lower and upper décollement zones somewhere to the right of the picture.

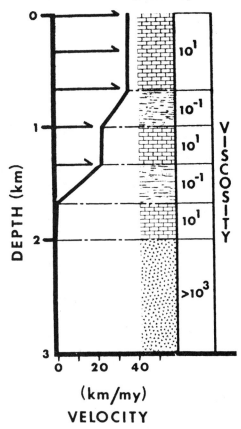

Figure 13. Gravity sliding velocity profile for a double décollement. Both the upper and middle limestone units are "thrust" plates independently undergoing major transport.

Simultaneous Folding

If the middle stiffer unit in Figure 13 was considerably thinner than the surrounding low viscosity units, it would be mechanically isolated from both the rigid basement and the rigid upper plate. This permits the unit to fold, regardless of any continuing deformation of the lower viscosity rocks. Folding in this stiffer unit is due to buckling, and the analyses of Biot (1961) apply. Fold formation would be very likely wherever tectonic transport takes place, provided the viscosity contrasts exceed two orders of magnitude.

If folding occurs, major changes occur in the developing structural style of the region. As the folds grow they protrude into the softer surrounding units. This interferes with the deformation of these units. The rectilinear shear of the décollement zone is transmitted to the folds. They are sheared and overturned as the deformation continues. The shallow limbs of anticlines act as ramps for the formation of strut thrusts through the upper plate. Similarly, break thrusts or stretch thrusts form through the overturned limbs of the folds, thereby connecting the lower and upper décollement zones with a step thrust. What might have been a relatively simple plate translation with attendant frontal thrusts or folds becomes instead a complicated regional structural style not uncommon in orogenic belts.

CONCLUSIONS

Gravity sliding and orogenic translation are accomplished by the deformation of the least competent (lowest viscosity) strata within a rock sequence. The mode of deformation is primarily simple (rectilinear) shear. This deformation may be studied and illustrated with velocity profiles erected under the assumption that rock masses are adequately modeled as either linear or non-linear fluids. This process, dominated by the fluid-like properties of rocks, requires less work to achieve a given displacement than does translation across a sole fault. It ought to be the phenomenon most encountered in nature. Thus, rather than bedding plane faults, entire zones of rock undergo massive shear to accomplish transport of overlying rock masses. Motion of an overlying plate does not cause deformation of the décollement zone. Rather, it is the zone's deformation that results in transport of overlying rocks.

Competent rocks are little affected by the same boundary conditions that result in the severe deformation of rocks within a décollement zone. The rate at which they deform is very slow compared to that of the lower viscosity rocks. Major transport is possible before these thrust plates suffer even minor internal deformation. The drag exerted by the décollement zone on the base of a translating rock mass is commonly so small that no limitations on the size of thrust masses arise from this source. This drag decreases with decreasing zone viscosity and increasing zone thickness. Greater resistance to motion is thought to arise from end effects. They are (1) updip attachment which is overcome by the formation of pull-apart grabens, and (2) downdip buttressing which is overcome by the formation of step thrusts or folds at the leading edge of the translating mass. Because the analysis neglects end effects, velocities of gravity sliding are overestimated and the stresses required for orogenic transport are underestimated.

Structures within a décollement zone are thought to arise because of both inhomogeneities of the zone and unsteady flow within the zone. Bedding plane faults arise in relatively high viscosity masses within a zone and adjacent to undeforming strata where local lift-off may result in plucking of underlying strata. Passive folding, a common feature of décollement zones, is an expression of unsteady flow. Some tectonic breccias are formed by strain-induced break-up of competent beds folded into a décollement zone. All décollement zone structures are disharmonic at zone boundaries. Thus, two classic criteria for recognizing bedding plane faults are actually only indicative of the presence of a décollement zone.

The presence of multiple low viscosity zones within a rock sequence leads to multiple décollement. Step thrusts form to connect these zones. Folding of any of the translating plates can be accomplished while they are being translated. Step or stretch thrusts, or both, form across the overturned limbs of these folds. Strut thrusts originate above their back limbs. The resulting structural style is common to orogenic belts.

ACKNOWLEDGMENTS

I wish to thank W. M. Chapple, S. E. De-Long, P. T. Flawn, J. H. Howard, and W. R. Muehlberger for their many helpful suggestions, and Eddie Williamson for the preparation of the illustrations.

APPENDIX I: METHODOLOGY

VELOCITY PROFILE CONSTRUCTION

Velocity profiles are constructed by a simple two-step method. First velocity profiles are derived for each rock type as if the entire sequence were monolithologic. The second step is to construct a composite velocity profile by stacking appropriate segments of the individual rock curves in the same sequence as the stratigraphic sequence. The resulting profile depicts the over-all displacement rate of the stratified column under the action of either gravity loading or orogenic compression.

Individual Rock Velocity Profiles

The procedure for constructing individual velocity profiles is the same for all rocks. First, the stress state is determined as a function of depth within the rock mass assumed to be comprised entirely of a single rock type. The second step is to decide on an appropriate flow model for the rock (Newtonian or non-linear). By substituting the flow law into the stress equations, the strain rates are determined. The strain rate is then integrated over the column. This yields the velocity as a function of position (depth of burial). The velocity profile is then plotted as velocity versus depth on a scale appropriate for graphical presentation.

Stresses in Gravity Sliding. Because it is assumed that rocks involved in gravity sliding exhibit fluid-like behavior, it is possible to use the classical solution for the stress state in a tilted fluid layer.

$$\sigma_{xx} = \sigma_{yy} = -\rho g \cos \theta (h - y) \quad (6a)$$
$$\sigma_{xy} = -\rho g \sin \theta (h - y) \quad (6b)$$

where ρ is the density, g the acceleration of gravity, θ the dip of the beds, h the thickness of the total moving section, and y the coordinate direction normal to bedding. The origin is taken at the base of the sequence (Fig. 14). (For a derivation of this

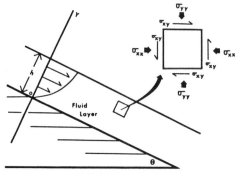

Figure 14. Fluid layer on a plane inclined at an angle showing velocity profile and stress state of an element of fluid.

result, *see* Longwell and others, 1969, p. 259–261.) Knowing the stress state within the fluid permits determination of the strain rate within the fluid if we are able to specify a flow law for the rocks in question.

Flow Equations. At present, it is not known what physical model(s) most accurately depicts the mechanical behavior of rocks. Classical linear viscosity is adequate to solve many problems (for example, Chapple, 1968), but others may require non-linear descriptions (Orowan, 1964). Differing models have been employed by various writers (compare Heard, 1963; Ree and others, 1960; Gordon, 1965). Interestingly, the qualitative results obtained herein are relatively insensitive to the flow law assumed for rock. To illustrate this insensitivity, velocity profiles are developed using two different flow laws: (1) classical linear viscosity, and (2) a shear thinning power law. The latter is used by glaciologists as a flow law for ice (Glen, 1955; Nye, 1952). Both Kehle (1964) and Griggs (1969 A.G.U. address) suggest the applicability of power laws to rocks. The flow equations for Newtonian and power law fluids are, respectively,

$$\sigma_{xy} = \mu \dot{e}_{xy} = (\mu/2)\partial \dot{u}/\partial y \qquad (7a)$$

$$(\sigma_{xy})^n = \mu_0 \dot{e}_{xy} = (\mu_0/2)\partial \dot{u}/\partial y \qquad (7b)$$

where $\dot{u}$ is the velocity in the x-direction and $n \geq 1$ for shear thinning fluids. A "Bingham" plastic corresponds to $n = \infty$, a Newtonian fluid to $n = 1$. Substituting the shear stress given by equation 6b yields

$$\partial \dot{u}/\partial y = (2/\mu)\rho g \sin \theta (h - y) \qquad (8a)$$

and

$$\partial \dot{u}/\partial y = (2/\mu_0)(\rho g \sin \theta)^n (h - y)^n . \qquad (8b)$$

Integrating these equations gives

$$\dot{u} = -(\rho g \sin \theta/\mu)(h - y)^2 + C_a \qquad (9a)$$

$$\dot{u} = -(\rho g \sin \theta)^n (2/\mu_0\{n + 1\})(h - y)^{n+1} + C_b . \qquad (9b)$$

The value of the constant C is determined by the condition that the velocity adjacent to its lower boundary is zero. Thus, at $y = 0$, equations 11a and 11b should be zero. This gives the values

$$C_a = (\rho g \sin \theta/\mu)h^2 \qquad (10a)$$

and

$$C_b = (\rho g \sin \theta)^n (2/\mu_0\{n + 1\})h^{n+1} , \qquad (10b)$$

so equations 11a, and 11b become

$$u = h^2\{\rho g \sin \theta/\mu\} \{1 - (1 - y/h)^2\} \qquad (11a)$$

and

$$u = h^{h+1}\{\rho g \sin \theta\}^n\{2/\mu_0(n + 1)\}$$
$$\times \{1 - (1 - y/h)^{n+1}\} . \qquad (11b)$$

These equations describe the laminar flow of a fluid

layer down an inclined plane; the first for a Newtonian (linear) fluid, and the second for a power law fluid. If $n = 1$, the two equations are identical.

Assigning Rock Viscosities. Published data on the viscosity of rocks are very scarce. They are summarized by Handin (1966, p. 289), and include information on gypsum, halite, limestone, marble, natural rock salt, and artificial rock salt. Estimates of crystalline rock viscosities have been made from studies of glacial rebound (Crittenden, 1963). These meager data serve as a rough guide to assigning rock viscosity. Under the relatively low temperature conditions existing in the first 2 km of the earth's crust, it is likely that crystalline rocks would exhibit viscosities of 10^3 to 10^5 m.y.-b, silica cemented sandstone from 10^{-1} m.y.-b for some sandstones, to 10^5 or more for quartzites, and clay cemented and carbonate cemented sandstones from 10^{-2} to 10^2 m.y.-b. No data have been published on the viscosity of shales, but a comparison, based on personal experience, of its behavior in boreholes with that of other rocks suggests that viscosity varies from 10^{-5} for Gulf Coast gumbo shales to perhaps 10^1 m.y.-b for deeply buried, siliceous lower Paleozoic shales. The viscosity of carbonates is very temperature sensitive (Heard, 1963), varying from 10^2 m.y.-b at room temperature, to 10^{-3} m.y.-b at $400°$ C. Bedded salts probably exhibit viscosities of from 10^{-1} m.y.-b at room temperature, to 10^{-5} m.y.-b at $200°$ C.

Using this set of rough estimates, the investigator must now attempt to assign viscosities to the actual rocks in the sequence under study. If an attempt is being made to model an actual geologic province, these data are probably insufficient. On the other hand, they serve perfectly well for general studies of the type represented by this paper.

Selecting Thickness. When studying an actual geologic setting, the thickness used to construct individual velocity profiles should exceed the actual *total* stratified column thickness by 50 percent. This is done to insure that what has been interpreted petrographically or stratigraphically as basement actually acts as mechanical basement. In most settings of interest, the deformation of the basement is insignificant compared to that of the overlying rocks. In the core of orogenic belts where temperatures are high, careful estimates of sediment and basement viscosities are required. If these do not differ by two orders of magnitude (and they need not), then crystalline basement may not act as mechanical basement. In such cases, the present model will be inadequate and should not be used. For suitable application, a rigid basement is required. This does not mean that the basement need be stationary. A portion of the oceanic crust moving beneath a continental margin can act as a rigid body and would serve as the mechanical basement below the overlying section of continental rise sediments.

Example Profiles. Individual rock velocity profiles are obtained from equations 12a and 12c by entering a viscosity for each rock type and a thickness into the equation. For the examples used in this paper, the stratified sequence thickness was assumed to be 2 km. A thickness of 3 km was used in constructing all the graphs to insure the presence of mechanical basement. Viscosities for shale, interbedded shale and limestone, and limestone (low ambient temperature) were taken to be 10^{-1} m.y.-b, 10^0 m.y.-b, and 10^1 m.y.-b, respectively. Substitution into equations 11a and 11b permitted plotting the profiles for each of these rock types (Fig. 15). The viscosity for crystalline basement was assumed to be 10^3 m.y.-b, but its velocity profile is so low that it cannot be plotted on the figure.

The individual rock velocity plots in Figure 16 are for the non-linear flow laws used in obtaining equation 11. The term $n^{n+2}\{\rho g \sin \theta\}^n\{2/\mu_0(n + 1)\}$ was adjusted so that the maximum velocity for a given rock type was the same for both the linear (Fig. 15) and non-linear (Fig. 16) models.

Composite Profiles

Velocity profiles for a section composed of strata of varying rock type is obtained by splicing together appropriate segments of individual rock profiles. This is done in exact correspondence with the stratigraphic section, beginning from the bottom. The lower one-third of the graph is for basement whose velocity is so low that it does not plot. If it does, it is not acting as mechanical basement and the thickness of the section needs to be increased or the model deemed inapplicable to the problem at hand. Assuming the basement layer plots with a zero profile, the next segment added to the graph should correspond to the lithology of the first layer in the sedimentary column. The segment is taken from the individual rock profile for that rock type *from the appropriate depth level*, not from the bottom of the curve. The profile segment is attached to the top of the existing profile. For the first layer it is attached to the zero profile of basement. The next segment is taken from the next higher depth interval from the curve for the rock type repre-

sented in that stratigraphic interval. This process is continued to the top of the column.

This procedure is based on three assumptions: (1) that the average density of the stratified sequence to any depth is relatively independent of the actual lithologies present to that depth; (2) that the tangential and normal velocities are continuous across all bed boundaries; and (3) that the stresses are continuous across all bed boundaries. The second and third assumptions are standard in fluid mechanics. Only in severe and unusual settings are they not applicable (*see* Langlois, 1964, p. 59–62, for a discussion). Geologists are well acquainted with a situation in which the no-slip condition fails—flexural slip folding. Where bedding slip does occur, the process requires less energy than continuous deformation. We have shown previously that in décollement the slip conditions require more work than continuous deformation. In this light, no-slip conditions appear quite reasonable. Regarding the first assumption, some- variation occurs in every rock sequence but it is negligibly small compared to the variations in viscosity and will not influence the velocity profiles a visually detectable amount.

The process of splicing profiles from the bottom up amounts to summing (integrating) the shear strain rate for each infinitesimal layer from bottom to top. This strain rate depends only on two things once the physical setting has been established: the shear stress (depth of burial in a gravity slide) and the viscosity (rock type). Hence, the reason for selecting that segment for the appropriate depth interval from the profile for the rock type found in that depth interval.

The following example is included to demonstrate the use of the technique.

Constructing Figure 1. The section used in Figure 1 included a basal 1-km-thick layer of crystalline basement, an intermediate 1-km-thick layer of 10^{-1} m.y.-b viscosity shale, and an upper layer of 10^1 m.y.-b viscosity limestone. These lithologic boundaries are indicated on Figure 1. The indi-

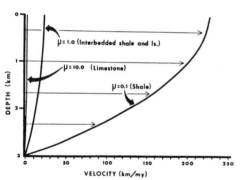

Figure 15. Velocity profiles for gravity creep in three pure rock sections assuming linear viscosities. The viscosities used are 10^1, 10^0 and 10^{-1} m.y.-b

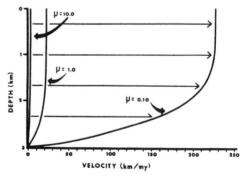

Figure 16. Velocity profiles for gravity creep in three pure rock sections assuming non-linear viscosities. The viscosity parameters were adjusted so that the velocity at $y = h$ for each rock type is the same in this figure as in Figure 15.

vidual rock velocity profiles are illustrated in Figure 15. Starting from the bottom up, the first velocity profile segment is for crystalline basement, which is a zero profile. This accounts for the absence of a visible velocity profile for the lower 1 km of the section in Figure 1. The next layer is shale. Consequently, that part of the shale profile in Figure 15 from the depth interval 1 to 2 km is transferred directly to Figure 1. The lower end is

attached to the top of the existing profile, which is zero. The top portion is obtained by taking the upper kilometer of the limestone profile from Figure 15 and attaching it to the top end of the existing (shale) profile of Figure 1. This completes the construction of the profile for the Newtonian viscosity model. The other profile (dashed line) is constructed in the identical manner using Figure 16.

REFERENCES CITED

Badgley, P. C., 1965, Structural and Tectonic Principles: Harper and Row, New York, 521 p.

Bally, A. W., Gordy, P. L., and Stewart, G. A., 1966, Structure, seismic data, and orogenic evolution of southern Canadian Rocky Mountains: Canadian Petroleum Geology Bull., v. 14, p. 337–381.

Billings, M. P., 1954, Structural Geology: Prentice Hall, Englewood Cliffs, New Jersey, 514 p.

Biot, M. A., 1961, Theory of folding of stratified viscoelastic media: Geol. Soc. America Bull., v. 72, p. 1595–1620.

—— 1965, Theory of viscous buckling and gravity instability of multi-layers with large deformation: Geol. Soc. America Bull., v. 76, p. 371–378.

Carlisle, D., 1965, Sliding friction and overthrust faulting: Jour. Geology, v. 73, p. 271–292.

Chapple, W. M., 1968, A mathematical theory of finite-amplitude rock folding: Geol. Soc. America Bull., v. 79, p. 47–68.

Cloos, E., 1968, Experimental analysis of Gulf Coast fracture patterns: Am. Assoc. Petroleum Geologists Bull., v. 52, p. 420–445.

Craddock, C., 1964, The Lincoln fold system, in Guidebook to the Ruidoso Country: New Mexico Geol. Soc. 15th Field Conf.

Crittenden, M. D., 1963, Effective viscosity of the earth derived from isostatic loading of Pleistocene Lake Bonneville: Jour. Geophys. Research, v. 68, p. 5517–5530.

de Sitter, L. U., 1964, Structural Geology: 2d ed., McGraw-Hill Book Co., New York, San Francisco, Toronto, London, 551 p.

Dickinson, G., 1951, Geological aspects of abnormal reservoir pressures in the Gulf Coast region of Louisiana: Third World Petroleum Congress Proc., The Hague, p. 1–17.

Dickson, G. O., Pitman, W. C., and Heirtzler, J. R., 1968, Magnetic anomalies in the South Atlantic and ocean floor spreading; Jour. Geophys. Research, v. 73, no. 6, p. 2087–2100.

Donath, F. A., and Parker, R. B., 1964, Folds and folding: Geol. Soc. America Bull., v. 75, p. 45–62.

Gilluly, J., 1957, Transcurrent fault and overturned thrust, Shoshone Range, Nevada (abs.): Geol. Soc. America Bull., v. 68, p. 1735.

Glen, J. W., 1955, The creep of polycrystalline ice:

Royal Soc. London Proc., Ser. A, v. 228, p. 519–538.

Goguel, J., 1962, Tectonics (English translation of 1952 French edition by Hans E. Thalmann): W. H. Freeman, San Francisco, 384 p.

Gordon, R. B., 1965, Diffusion creep in the earth's mantle: Jour. Geophys. Research, v. 70, p. 2413–2418.

Hafner, W., 1951, Stress distributions and faulting: Geol. Soc. America Bull., v. 62, p. 373–398.

Handin, J., 1966, Strength and ductility, in Handbook of Physical Constants: Geol. Soc. America Mem. 97, p. 223–291.

Handin, J., and Hager, R. V., 1957, Experimental deformation of sedimentary rocks under confining pressure: Tests at room temperature on dry samples: Am. Assoc. Petroleum Geologists Bull., v. 41, p. 1–50.

Heard, H. C., 1963, Effect of large changes in strain rate in the experimental deformation of Yule marble: Jour. Geology, v. 71, p. 162–195.

Heard, H. C., and Rubey, W. W., 1966, Tectonic implications of gypsum dehydration: Geol. Soc. America Bull., v. 77, p. 741–760.

Heirtzler, J. R., Dickson, G. O., Herron, E. M., Pitman, W. C., III, and LePichon, X., 1968, Marine magnetic anomalies, geomagnetic field reversals, and motions of the ocean floor and continents: Jour. Geophys. Research, v. 73, no. 6, p. 2119–2136.

Hsü, K. J., 1968, Principles of mélanges and their bearing on the Franciscan-Knoxville paradox: Geol. Soc. America Bull., v. 79, no. 8, p. 1063–1074.

—— 1969, Role of cohesive strength in the mechanics of overthrust faulting and of landsliding: Geol. Soc. America Bull., v. 80, p. 927–952.

Hubbert, M. K., and Rubey, W. W., 1959, Role of fluid pressure in mechanics of overthrust faulting: Geol. Soc. America Bull., v. 70, p. 115–206.

—— 1969, Role of cohesive strength in the mechanics of overthrust faulting and of landsliding: Discussion: Geol. Soc. America Bull., v. 80, p. 953–954.

Kamb, B., and LaChapelle, E., 1964, Direct observation of the mechanism of glacier sliding over bedrock: Jour. Glaciology, v. 5, p. 159–173.

Kehle, R. O., 1964, Deformation of the Ross Ice

Shelf, Antarctica: Geol. Soc. America Bull., v. 75, p. 259–286.

King, P. B., 1950, Tectonic framework of southeastern United States: Am. Assoc. Petroleum Geologists Bull., v. 34, no. 4, p. 635–671.

Lane, H. W., 1949, Drilling practices in Iran: Oil and Gas Jour., v. 48 (Aug. 4), p. 56–58, 61.

Langlois, W. E., 1964, Slow Viscous Flow: The Macmillan Company, New York, 222 p.

Laubscher, H. P., 1961, Die Fernschubhypothese der Jurafalting: Eclogae Geol. Helvetiae, v. 54, no. 1, p. 221–282.

LePichon, X., 1968, Sea-floor spreading and continental drift: Jour. Geophys. Research, v. 73, no. 12, p. 3661–3697.

LePichon, X., and Heirtzler, J. R., 1968, Magnetic anomalies in the Indian Ocean and sea-floor spreading: Jour. Geophys. Research, v. 73, no. 6, p. 2101–2117.

Longwell, C. R., Flint, R. F., and Sanders, J. E., 1969, Physical Geology: John Wiley & Sons, Inc., New York, London, Sydney, 685 p.

Muehlberger, W. R., 1959, Internal structure of the Grand Saline salt dome, Van Zandt Co., Texas: Texas Univ. Bur. Econ. Geology Rept. Inv. 38, 18 p.

Nye, J. F., 1952, The flow law of ice from movements in glacier tunnels, laboratory experiments, and the Jungfrau borehole experiments: Royal Soc. London Proc., Ser. A, v. 219, p. 477–489.

Oliver, J., and Isacks, B. L., 1967, Deep earthquake zones, anomalous structures in the upper mantle and the lithosphere: Jour. Geophys. Research, v. 72, p. 4259.

Orowan, E., 1964, Continental drift and the origin of mountains: Science, v. 146, p. 1003–1010.

Pierce, W. G., 1957, Heart Mountain and South Fork detachment thrusts of Wyoming: Am. Assoc. Petroleum Geologists Bull., v. 41, p. 591–626.

Pitman, W. C., Herron, E. M., and Heirtzler, J. R., 1968, Magnetic anomalies in the Pacific and sea floor spreading: Jour. Geophys. Research, v. 73, no. 6, p. 2069–2085.

Price, R., 1962, Fernie map area, east half, Alberta and British Columbia: Canada Geol. Survey, Paper 61–24.

Ramsay, J. G., 1967, Folding and Fracturing of Rocks: McGraw Hill Book Co., New York, San Francisco, Toronto, London, 568 p.

Raleigh, C. B., and Griggs, D. T., 1963, Effect of the toe in the mechanics of overthrust faulting: Geol. Soc. America Bull., v. 74, p. 819–830.

Raymond, C. F., and Kamb, W. B., 1968, Annually cyclic folding in glacier ice (abs.): Geol. Soc.

America Spec. Paper 115, p. 182.

Ree, F. H., Ree, T., and Eyring, H., 1960, Relaxation theory of creep of metals: Am. Soc. Civil Engineers Proc., Jour. Engin. Mech. Div., v. 86, p. 41–59.

Reid, H. F., 1896, The mechanics of glaciers: Jour. Geology, v. 4, no. 8, p. 912–928.

Rodgers, J., 1950, The folds and faults of the Appalachian valley and ridge province: Kentucky Geol. Survey Spec. Pub. 1, Ser. 9, p. 150–166.

Rosenfeld, J. L., 1968, Garnet rotations due to the major Paleozoic deformations in Southeast Vermont, p. 185–202 in Studies of Appalachian Geology: John Wiley & Sons, Inc., New York, London, Sydney, Toronto, 475 p.

Rubey, W. W., and Hubbert, M. K., 1959, Role of fluid pressure in mechanics of overthrust faulting: Geol. Soc. America Bull., v. 70, p. 167–206.

Sanford, A. R., 1959, Analytical and experimental study of simple geologic structures: Geol. Soc. America Bull., v. 70, p. 19–52.

Sherwin, J., and Chapple, W. M., 1968, Wavelengths of single layer folds: a comparison between theory and observation: Am. Jour. Sci., v. 266, p. 167–179.

Shreve, R. L., 1966, Sherman landslide, Alaska: Science, v. 154, p. 1639–1643.

—— 1968, The Blackhawk Landslide: Geol. Soc. America Spec. Paper 108, 47 p.

Shreve, R. L., and Kamb, W. B., 1964, Structure of Austerdals Glacier, Norway: Geol. Soc. America Spec. Paper 76, p. 150.

Shumskii, P. A., 1964, Principles of structural glaciology: Dover Publications, Inc., New York, 497 p.

Sykes, L. R., 1966, The seismicity of deep structure of island arcs: Jour. Geophys. Research, v. 71, p. 2981–3006.

Takeuchi, H., 1963, Time scales of isostatic compensations: Jour. Geophys. Research, v. 68, no. 8, p. 2357.

Theakstone, W. H., 1967, Basal sliding and movement near the margin of the glacier Osterdalsisen, Norway: Jour. Glaciology, v. 6, p. 805–816.

Thomeer, J. H. M. A., and Bottema, J. A., 1961, Increasing occurrence of abnormally high reservoir pressure in boreholes, and drilling problems resulting therefrom: Am. Assoc. Petroleum Geologists Bull., v. 45, p. 1721–1730.

Willis, B., 1893, Mechanics of Appalachian structure: U. S. Geol. Survey, 13th Ann. Rept., Pt. 2.

Manuscript Received by The Society June 4, 1969

Revised Manuscript Received January 7, 1970

REFERENCES

Adams, F. D., 1938, The birth and development of the geological sciences: Dover, New York, 1954, 506 p.

——, and Bancroft, J. A., 1917, Internal friction during deformation and relative plasticity of rocks: Jour. Geology, v. 25, p. 597-637.

Ampferer, O., 1906, Über das Bewegungsbild von Faltengebirge: Jahrbuch der k.k. Reichsanstalt (translation by J. Gilluly on file with U.S. Geol. Survey).

Anderson, E. M., 1942, The dynamics of faulting: Oliver & Boyd, London (2nd ed., 1951, 306 p.).

Bailey, E. B., 1935, Tectonic essays, mainly Alpine: Clarendon Press, Oxford, 200 p.

——, and Mackin, J. H., 1937, Recumbent folding in the Pennsylvania Piedmont—preliminary statement: Amer. Jour. Sci., 5th ser., v. 33, p. 187-190.

Bertrand, M., 1884, Rapports de structure des Alpes de Glaris et du Bassin houiller du Nord: Bull. Soc. Geol. France, 3rd ser., v. 12 (1883-1884), p. 318.

Billingsley, P., and Locke, A., 1933, Tectonic position of ore districts in the Rocky Mountain region: Amer. Inst. Min. Metall. Eng. Tech. Publ. 501, 12 p.

——, and Locke, A., 1939, Structure of ore districts in the continental framework: Amer. Inst. Min. Metall. Eng., New York, 51 p.

Birch, F., 1961, Role of fluid pressure in mechanics of overthrust faulting: discussion: Geol. Soc. America Bull., v. 72, p. 1441-1444.

Bjerrum, L., Casagrande, A., Peck, R. B., and Skempton, A. W., eds., 1960, From theory to practice in soil mechanics—selections from the writings of Karl Terzaghi, Wiley, New York, 425 p.

Blackwelder, E., 1910, New light on the geology of the Wasatch Mountains, Utah: Geol. Soc. America Bull., v. 21, p. 517-542; Discussion, p. 767.

Böker, R., 1915, Die Mechanik der bleibenden Förmanderungen in kristallinisch aufgebauten Körpern: Vereines Deut. Ing. Mitt. Forsch., v. 175, p. 1-51.

Bouasse, M. H., 1901, Sur la théorie des déformations permanentes de Coulomb, son application à la traction, la torsion et le passage à la filière: Ann. Chim. Phys., 7th ser., v. 23, p. 198-240.

Bowden, F. P., and Tabor, D., 1964, The friction and lubrication of solids, part II: Clarendon Press, Oxford.

Bruce, C., 1973, Pressured shale and related sediment deformation: mechanism for development of regional contemporaneous faults: Amer. Assoc. Petrol. Geol. Bull., v. 57, p. 878–886.

Bryant, B., and Reed, J. C., Jr., 1962, Structural and metamorphic history of the Grandfather Mountain area, North Carolina: Amer. Jour. Sci., v. 260, p. 161–180.

Bucher, W. H., 1933a, Volcanic explosions and overthrusts: Amer. Geophys. Union Trans., 14th Ann. Meeting, p. 238–242.

——, 1933b, The deformation of the earth's crust: Princeton University Press, Princeton, N.J., 518 p.

——, 1947, Heart Mountain problem: Wyoming Geol. Assoc. Guidebook, 2nd Ann. Field Conf., p. 189–197.

Burst, J. F., 1969, Diagenesis of Gulf Coast clayey sediments and its possible relation to petroleum migration: Amer. Assoc. Petrol. Geol. Bull., v. 53, no. 1, p. 73–93.

Butts, C., 1927, Fensters in the Cumberland overthrust block in southwestern Virginia: Va. Geol. Survey Bull. 28, 12 p.

Buxtorf, A., 1908, Geologische Beschreibung des Weissensteintunnels und seiner Umgebung: Beitr. Geol. Karte Schweiz, N.F. Lief 21, p. 1–125.

——, 1915, Prognosen und Befunden beim Hauenstein Basis – und Grenchenbergtunnel und die Bedeutung der Letztern für die Geologie des Juragebirges: Verhandl. Naturforsch. Gesell., Basel, v. 27.

Cady, W. M., 1945, Stratigraphy and structure of west-central Vermont: Geol. Soc. America Bull., v. 56, p. 515–587.

——, 1968, Tectonic setting and mechanism of the Taconic slide: Amer. Jour. Sci., v. 266, p. 563–578.

——, 1969, Regional tectonic synthesis of northwestern New England and adjacent Quebec: Geol. Soc. America Mem. 120, 181 p.

Callaway, C., 1883, The Highland problem: Geol. Mag., v. 10, p. 139–140.

Carey, S. W., 1954, The Rheid concept in geotectonics: J. Geol. Soc. Austr., v. 1, p. 67–117.

Christie, J. M., 1958, Dynamic interpretation of the fabric of a dolomite from the Moine thrust-zone in north-west Scotland: Amer. Jour. Sci., v. 256, p. 159–170.

——, 1963, The Moine thrust zone in the Assynt Region, northwest Scotland: Univ. Calif. Publ. Geol. Sci., v. 40, p. 345–440.

Cloos, E., 1947, Oölite deformation in the South Mountain fold, Maryland: Geol. Soc. America Bull., v. 58, p. 843–917.

——, and Heitanen, A., 1941, Geology of the "Martic overthrust" and the Glenarm series in Pennsylvania and Maryland: Geol. Soc. America Spec. Paper 35, 207 p.

Coulomb, C. A., 1776, Essai sur une application des règles de maximis et minimis à quelques problèmes de statique, relatifs à l'architecture: Mémoires de Mathématique et de Physique, Présentés à l'Académie Royale des Sciences par divers Savans, et Lûs dans ses Assemblées, Paris, v. 7, p. 343–382.

Dake, C. L., 1918, The Hart Mountain overthrust and associated structures in Park County, Wyoming: Jour. Geol., v. 26, p. 45–55.

Davis, G. A., 1965, Role of fluid pressure in mechanics of overthrust faulting: discussion: Geol. Soc. America Bull., v. 76, p. 463–468.

de Böckh, H., Lees, G. M., and Richardson, F., 1929, Contribution to the stratigraphy and tectonics of the Iranian ranges, p. 58–176 *in* Gregory, J., ed., The structure of Asia: Methuen, London, 227 p.

De Jong, K. A., and Scholten, R., 1973, Gravity and tectonics: preface, Wiley-Interscience, New York, 502 p.

Desai, C. S., and Abel, J. F., 1972, Introduction to the finite element method: Van Nostrand Reinhold, New York, 477 p.

Dieterich, J. H., 1969, Origin of cleavage in folded rocks: Amer. Jour. Sci., v. 267, p. 155-165.

Dumont, A. H., 1832, La Constitution géologique de la province de Liège: Mém. couronnées Acad. Royale Sci. et Belles-lettres, Bruxelles, v. 8.

Engelder, J. T., 1974, Cataclasis and the generation of fault gouge: Geol. Soc. America Bull., v. 85, p. 1515-1522.

Escher von der Linth, A., 1841, Verhandl. Schweiz. Gesell., Zürich, p. 54, 58.

Friedman, M., 1963, Petrofabric analysis of experimentally deformed calcite-cemented sandstones: Jour. Geol., v. 71, p. 12-37.

——, 1964, Petrofabric techniques for the determination of principal stress directions in rocks, p. 451-550 *in* Judd, W. R., ed., State of stress in the earth's crust: Elsevier, New York, 732 p.

Garg, S. K., and Nur, A., 1973, Effective stress laws for fluid-saturated rocks: Jour. Geophys. Res., v. 78, p. 5911-5921.

Gauthey, E. M., 1813, Traité de la construction des ponts, published by L. M. H. Navier; new edition, 1843.

Geertsma, J., 1957, The effect of fluid pressure decline on volumetric changes of porous rocks: Trans. Amer. Inst. Min. Eng., p. 331.

Geikie, A., 1884, The crystalline rocks of the Scottish Highlands: Nature, Nov. 13, p. 29-35.

——, 1905, The founders of geology, 2nd ed.: Macmillan, London, 486 p.

——, 1913, Mountains: their origin, growth, and decay: Oliver and Boyd, London, 311 p.

——, 1924, A long life's work: Howard, London.

Gilluly, J., 1957, Transcurrent fault and overturned thrust, Shoshone range, Nevada (abstract): Geol. Soc. America Bull., v. 68, p. 1735.

Goguel, J., 1962, Tectonics: W. H. Freeman, San Francisco (English translation of French edition of 1952, by H. E. Thalmann), 384 p.

——, 1969, Le Rôle de l'eau et de la chaleur dans les phénomènes tectoniques: Rev. Geogr. Phys. Geol. Dynam., v. 11, p. 153-164.

Goodman, R., Taylor, R., and Brekke, T., 1968, A model for the mechanics of jointed rock: Proc. Jour. Soil Mech. Found. Div., Amer. Soc. Civil Eng., v. 94, no. SM 3, p. 637-659.

Griggs, D. T., 1936, Deformation of rocks under high confining pressures: Jour. Geol., v. 44, p. 541-577.

——, and Handin, J., eds., 1960, Rock deformation (a symposium): Geol. Soc. America Mem. 79, 382 p.

Haarmann, E., 1930, Die Oszillationstheorie, eine Erklärung der Krustenbewegungen von Erde und Mond: Stuttgart, F. Enke Verlag, 260 p.

Handin, J., 1966, Strength and ductility, sec. 11, p. 223-289, *in* Clark, S. P., ed., Handbook of physical constants: Geol. Soc. America Mem. 97, 587 p.

Hanshaw, B. B., and Zen, E-an, 1965, Osmotic equilibrium and overthrust faulting: Geol. Soc. America Bull., v. 76, p. 1379-1386.

Heard, H. C., 1963, Effect of large changes in strain rate in the experimental deformation of Yule marble: Jour. Geol., v. 71, p. 162-195.

Heim, Albert, 1878, Untersuchungen über den Mechanismus der Gebirgsbildung, Schwabe, Basel.

——, 1881, Die geologischen Verhältnisse des Bergsturzes von Elm, in Buss, E., and Heim, A., Der Bergsturze on Elm, Wurster, Zürich, p. 130-141.

——, 1919-1922, Geologie der Schweiz: Leipzig.

——, 1932, Bergsturz und Menschenleben: Fretz und Wasmuth Verlag, Zürich, 218 p.

Heim, Arnold, and Gansser, A., 1939, Central Himalaya: geological observations of the Swiss Expedition, 1936: Schweiz. Naturforsch. Gesell., Denkschrift, v. 73, Mem. 1, 245 p.

Hewett, D. F., 1920, The Heart Mountain overthrust, Wyoming: Jour. Geol., v. 28, p. 536-557.

Heyman, J., 1972, Coulomb's memoir on statics: Cambridge University Press, London, 212 p.

Hicks, H., 1878, On the metamorphic and overlying rocks in the neighborhood of Loch Maree, Ross-shire: Quart. Jour. Geol. Soc. London, v. 34, p. 811-818.

Hsü, K J., 1969a, Role of cohesive strength in the mechanics of overthrust faulting and of landsliding: reply: Geol. Soc. America Bull., v. 80, pp. 955-960.

——, 1969b, A preliminary analysis of the statics and kinetics of the Glarus overthrust: Eclogae Geol. Helv., v. 62, no. 1, p. 143-154.

——, 1975, Catastrophic debris streams generated by rock falls: Geol. Soc. America Bull., v. 86, p. 129-140.

——, in press, Albert Heim: contributions to mechanics of landslides, *in* Voight, B., ed., Rockslides and Avalanches, Elsevier, Amsterdam.

Hubbert, M. K., 1937, Theory of scale models as applied to the study of geologic structures: Geol. Soc. America Bull., v. 48, p. 1459-1520.

——, 1940, The theory of ground-water motion: Jour. Geol., v. 43, p. 785-944.

——, 1945, Strength of the earth: Amer. Assoc. Petrol. Geol. Bull., v. 29, p. 1630-1653.

——, 1956, Darcy's law and the field equations of the flow of underground fluids: Amer. Inst. Min. Metall. Eng. Trans., v. 207, p. 222-239.

——, 1972, Structural geology: Hafner, New York, 329 p.

——, and Rubey, W. W., 1960, Role of fluid pressure in mechanics of overthrust faulting: reply to discussion by H. P. Laubscher: Geol. Soc. America Bull., v. 71, p. 617-628.

——, and Rubey, W. W., 1961a, Role of fluid pressure in mechanics of overthrust faulting: reply to discussion by F. Birch: Geol. Soc. America Bull., v. 72, p. 1445-1451.

——, and Rubey, W. W., 1961b, Role of fluid pressure in mechanics of overthrust faulting: reply to discussion by W. L. Moore: Geol. Soc. America Bull., v. 72, p. 1587-1594.

——, and Rubey, W. W., 1969, Role of cohesive strength in the mechanics of overthrust faulting and of landsliding: discussion: Geol. Soc. America Bull., v. 80, pp. 953-954.

——, and Willis, D. G., 1957, Mechanics of hydraulic fracturing: Amer. Inst. Min. Metall. Eng. Trans., v. 210, p. 153-166; Discussion, p. 167-168.

Hughes, C. J., 1970, The Heart Mountain detachment fault—a volcanic phenomenon?: Jour. Geol., v. 78, p. 107-116.

Jacobeen, F., and Kanes, W. H., 1974, Structure of Broadtop synclinorium and its implications for Appalachian structural style: Amer. Assoc. Petrol. Geol. Bull., v. 58, p. 362-375.

Jaeger, J. C., 1962, Elasticity, fracture and flow, 2nd ed.: Methuen, London, 212 p. (3rd ed., 1969, 268 p.).

Keith, A., 1912, New evidence on the Taconic question (abstract): Geol. Soc. America Bull., v. 23, p. 720-721.

——, 1932, Stratigraphy and structure of northwestern Vermont: Wash. Acad. Sci. Jour., v. 22, p. 357-379, 393-406.

Kick, F., 1892, Die Principien der mechanischen Technologie und die Festigkeit-slehre: Z. Vereines Deut. Ing., v. 36, p. 278, 919.

King, P., Ferguson, H. W., Craig, L., and Rodgers, J., 1944, Geology and manganese deposits of northeastern Tennessee: Tenn. Div. Geol. Bull. 52, 283 p.

Knopf, E. B., and Ingerson, E., 1938, Structural petrology: Geol. Soc. America Mem. 6, 270 p.

Kuhn, T. S., 1962, The structure of scientific revolutions: University of Chicago Press, Chicago, 172 p.

Lapworth, C., 1883, The secret of the Highlands: Geol. Mag., v. 10, p. 120-128; 193-199; 337-344.

——, 1885, The Highland controversy in British geology: Nature, v. 32, p. 558-559.

Laubscher, H. P., 1960, Role of fluid pressure in mechanics of overthrust faulting: discussion: Geol. Soc. America Bull., v. 71, p. 611-615.

Lemoine, M., 1973, About gravity gliding tectonics in the Western Alps, p. 201-216 *in* De Jong, K. A., and Scholten, R., Gravity and tectonics, Wiley-Interscience, New York, 502 p.

Logan, W. E., 1861, Remarks on the fauna of the Quebec group of rocks and the primordial zone of Canada: Am. Jour. Sci., 2nd ser., v. 31, p. 216-220; Can. Naturalist, v. 5, p. 472-477 (1860).

Longwell, C. R., 1922, The Muddy Mountain overthrust in southeastern Nevada: Jour. Geol., v. 30, p. 63-72.

——, 1949, Structure of the northern Muddy Mountain area, Nevada: Geol. Soc. America Bull., v. 60, p. 923-968.

——, Pampayan, E., Bowyer, B., and Roberts, R. J., 1965, Geology and mineral deposits of Clark County, Nevada: U.S. Bur. Mines Bull., v. 62, 218 p.

Lugeon, M., 1902a, Les Grandes nappes de recouvrement des Alpes du Chablais et de la Suisse: Bull. Soc. Geol. France, 4th ser., v. i, p. 723.

——, 1902b, Sur la coupe géologique du massif du Simplon: Compt. Rend. Acad. Sci. Paris, v. 134, p. 726.

——, and Argand, E., 1905, Sur les grandes nappes de recouvrement de la zone du Piedmont: Compt. Rend. Acad. Sci. Paris, v. 111, p. 1364.

Mackin, J. H., 1950, The down-structure method of viewing geologic maps: Jour. Geol., v. 58, p. 55-72.

——, 1962, Structure of the Glenarm Series in Chester County, Pennsylvania: Geol. Soc. America Bull., v. 72, p. 557-577.

Mansfield, G. R., 1927, Geography, geology, and mineral resources of part of southeastern Idaho: U.S. Geol. Survey Prof. Paper 152, 453 p.

Maxwell, J. C., 1962, Origin of slaty and fracture cleavage in the Delaware Water Gap area, N.J. and Pa., p. 281-311 *in* Engel, A. J., et al., eds., Petrologic studies: a volume in honor of A. F. Buddington: Geol. Soc. America, 660 p.

McIntyre, D. B., 1954, The Moine Thrust—its discovery, age, and tectonic significance: Proc. Geol. Assoc., v. 65, p. 203-223.

Merriam, C. W., and Anderson, C. A., 1942, Reconnaissance survey of the Roberts Mountains, Nevada: Geol. Soc. America Bull., v. 53, p. 1675-1728.

Miller, R. L., and Brosgé, W. P., 1954, Geology and oil resources of the Jonesville district, Virginia: U.S. Geol. Survey Bull. 990, 240 p.

——, and Fuller, J. O., 1947, Geologic and structure contour maps of the Rose Hill oil field, Lee County, Virginia: U.S. Geol. Survey Oil and Gas Invest. Prelim. Map 76.

Milnes, A. G., 1973, Structural reinterpretation of the classic Simplon Tunnel section of the Central Alps: Geol. Soc. America Bull., v. 84, p. 269-274.

References

Misch, P. 1957, Magnitude and interpretation of some thrusts in northeast Nevada (abstract): Geol. Soc. America Bull., v. 68, p. 1854.

——, 1960, Regional structural reconnaissance in central-northeast Nevada and some adjacent areas: Guidebook to Geology of East-Central Nevada, Intermountain Assoc. Petrol. Geol. 11th Ann. Field Conf., p. 17-42.

Miser, H. D., 1924, Structure of the Ouachita Mountains of Oklahoma and Arkansas: Okla. Geol. Survey Bull., v. 50, 30 p.

Mohr, O. C., 1882, Über die Darstellung des Spannungszustandes und des Deformations-zustandes eines Körperelementes und über die Anwendung derselben in der Festigkeitslehre: Der Zivilingenieur, v. XXVIII, p. 113-156.

Mohr, O. C., 1914, Abhandlungen aus dem Gebiete den Technischen Mechanik, 2nd ed.: W. Ernst, Berlin, p. 192-235.

Moore, W. L., 1961, Role of fluid pressure in mechanics of overthrust faulting: discussion: Geol. Soc. America Bull., v. 72, p. 1581-1586.

Murchison, R. I., 1849, On the geological structure of the Alps, Appenines, and Carpathians: Quart. Jour. Geol. Soc., v. 5, p. 157-312.

——, 1859, On the succession of the older rocks in the northernmost counties of Scotland: Quart. Jour. Geol. Soc. London, v. 15, p. 353-418.

——, and Geikie, A., 1861, On the altered rocks of the western islands of Scotland, and the north-western and central Highlands: Quart. Jour. Geol. Soc. London, v. 17, p. 171-228.

Nádai, A., 1927, Der bildsame Zustand der Werkstoffe: J. Springer, Berlin, 171 p.

Navier, L. M. H., 1839, Résumé des Lecons données à l'Ecole des Ponts et Chaussées sur l'Application de la Mécanique à l'Etablissement des Constructions et des Machines (new ed.), Bruxelles.

Nicol, J., 1857, On the red sandstone and conglomerate and the superposed quartz-rocks, limestones, and gneiss of the north-west coast of Scotland: Quart. Jour. Geol. Soc. London, v. 13, p. 17-39.

Nolan, T. B., 1935, The Gold Hill mining district, Utah: U.S. Geol. Survey Prof. Paper 177, 172 p.

Nur, A., and Byerlee, J. D., 1971, An exact effective stress law for elastic deformation of rock with fluids: Jour. Geophys. Res., v. 76, p. 6414-6419.

Odenstad, S., 1951, The landslide at Sköttorp on the Lidan River: Roy. Swedish Geotech. Inst. Proc., no. 4, 40 p.

Oldham, R. D., 1893, A manual of the geology of India. Stratigraphic and structural geology: Calcutta, 2nd ed. rev., 543 p.

Oxburgh, E. R., and Turcotte, D. L., 1970, Thermal structure of island arcs: Geol. Soc. America Bull., v. 81, p. 1665-1688.

Paterson, M. S., 1970, Experimental deformation of minerals and rocks under pressure, p. 191-233 in Pugh, H. L. D., ed., Mechanical behavior of materials under pressure: Elsevier, Amsterdam.

Peach, B. N., Horne, J., Gunn, W., Clough, C. T., Hinxman, L. W., and Cadell, H. M., 1888, Report on the recent work of the Geological Survey in the North-West Highlands of Scotland: Quart. Jour. Geol. Soc. London, v. 44, p. 378-441.

Perry, E. A., and Hower, J., 1971, Late-state dehydration in deeply buried pelitic sediments: Amer. Assoc. Petrol. Geol. Bull., v. 56, no. 10, p. 2013-2021.

Phillips, J., 1843, On certain movements in the parts of stratified rocks: Rept. Brit. Assoc., Cork, p. 60.

Pierce, W. G., 1941, Heart Mountain and South Fork thrusts, Park County, Wyoming: Amer. Assoc. Petrol. Geol. Bull., v. 25, p. 2021-2045.

——, 1957, Heart Mountain and South Fork detachment thrusts of Wyoming: Amer. Assoc. Petrol. Geol. Bull., v. 41, p. 591-626.

440

——, 1960, The "break-away" point of the Heart Mountain detachment fault in northwestern Wyoming, p. B236-B237 *in* Geological Survey research 1960: U.S. Geol. Survey Prof. Paper 400-B.

——, 1963, Reef Creek detachment fault, northwestern Wyoming: Geol. Soc. America Bull., v. 74, p. 1225-1236.

——, 1966, Role of fluid pressure in mechanics of overthrust faulting: discussion: Geol. Soc. America Bull., v. 77, p. 565-568.

——, 1968, Tectonic denudation as exemplified by the Heart Mountain fault, Wyoming: 23rd Int. Geol. Congr., Prague, sec. 3, p. 191-197.

——, 1973, Principal features of the Heart Mountain fault and the mechanism problem, p. 457-471 *in* de Jong, K., and Scholten, R., eds., 1973, Gravity and tectonics: Wiley-Interscience, New York, 502 p.

Price, R. A., 1973, Large-scale gravitational flow of supracrustal rocks, southern Canadian Rockies, p. 491-502 *in* De Jong, K. A., and Scholten, R., eds., Gravity and tectonics, Wiley-Interscience, New York, 502 p.

Platt, L. B., 1962, Fluid pressure in thrust faulting: a corollary: Amer. Jour. Sci., v. 260, p. 107-114.

Powers, M. C., 1967, Fluid release mechanisms in compacting marine mudrock and their importance in oil exploration: Amer. Assoc. Petrol. Geol. Bull., v. 51, p. 1240-1252.

Reade, T. M., 1886, The origin of the mountain ranges: London.

Reeves, F., 1925, Shallow folding and faulting around the Bearpaw Mountains: Amer. Jour. Sci., 5th ser., v. 10, p. 187-200.

Reyer, E., 1888, Theoretische Geologie: Stuttgart, 868 p.

——, 1892, Geologische und geographische Experiments: I. Heft: Deformation und Gebirgsbildung: Engelmann, Leipzig, 52 p.

Richards, R. W., and Mansfield, G. R., 1912, The Bannock overthrust, a major fault in southeastern Idaho and northeastern Utah: Jour. Geol., v. 20, p. 681-709.

Rodgers, J., 1970, The Tectonics of the Appalachians: Wiley-Interscience, New York, 271 p.

Rogers, W. B., and Rogers, H. D., 1843, On the physical structure of the Appalachian chain, as exemplifying the laws which have regulated the elevation of great mountain chains, generally: Assoc. Amer. Geologists and Naturalists Rept., p. 474-531.

Ruedemann, R., 1909, Types of inliers observed in New York: N.Y. State Mus. Bull. 133, p. 164-193.

Rosenfeld, J., 1968, Garnet rotations due to the major Paleozoic deformations in southeast Vermont, Chap. 14 *in* Zen, E-an, et al., eds., Studies of Appalachian geology, Wiley-Interscience, New York, 475 p.

Rubey, W. W., and Hubbert, M. K., 1965, Role of fluid pressure in mechanics of overthrust faulting: reply to discussion by G. A. Davis: Geol. Soc. America Bull., v. 76, p. 469-474.

Safford, J. M., 1869, Geology of Tennessee: Nashville, 550 p.

St. Julien, P., and Hubert, C., 1975. Evolution of the Taconian Orogen in the Quebec Appalachians: Amer. Jour. Sci., v. 275A, p. 337-362.

Saussure, H. B., de, 1779-1796, Voyages des Alpes, Neuchatel.

Schardt, H., 1893, Sur l'origine des Préalpes romandes: Arch. Sci. Phys. Nat., Genève, 3rd ser., v, 30, p. 570.

——, 1898, Les Régions exotiques du versant nord des Alpes suisses: Bull. Soc. Vaudoise Sci. Nat., v. 34, p. 114-219.

Schultz, A. R., 1914, Geology and geography of a portion of Lincoln County, Wyoming: U.S. Geol. Survey Bull. 543, 141 p.

Scholten, R., and K. A. De Jong, 1973, Introduction to Part 1, p. 3-5 *in* De Jong,

References

K. A., and Scholten, R., eds., Gravity and tectonics: Wiley-Interscience, New York, 502 p.

——, Keenmon, K. A., and Kupsch, W. O., 1955, Geology of the Lima region, southwestern Montana and adjacent Idaho: Geol. Soc. America Bull., v. 66, p. 345–404.

Shreve, R., 1968, The Blackhawk landslide: Geol. Soc. America Spec. Paper 108, 47 p.

Sharpe, D., 1847, On slaty cleavage: Quart. Jour. Geol. Soc. London, v. 5, p. 111.

Skempton, A. W., 1954, The pore-pressure coefficients A and B: Geotechnique, v. 4, p. 143–147.

——, 1960, Effective stress in soils, concrete and rock, in Pore pressure and suction in soils, Butterworth, London, p. 4–16.

——, 1961, Horizontal stresses in an over-consolidated Eocene clay: Proc. 5th Int. Conf. Soil Mech. Found. Eng., Paper 1/61.

Smoluchowski, M. S., 1909, Versuch über Faltungserscheinungen Schwimmender elastisichen Platten: Anz. Akad. Wiss., Krakau, Math. Phys. Kl., p. 727–734.

Sollas, W. J., 1895, Experiments with pitch glaciers: Quart. Jour. Geol. Soc. London, v. 51, p. 361.

Sorby, H. C., 1853, On the origin of slaty cleavage: Edinburgh New Phil. Jour., v. 55, p. 137.

Smith, E. A., 1893, Underthrust folds and faults: Amer. Jour. Sci., 3rd ser., v. 45, p. 305–306.

Stanley, R. A., and Morse, J. D., 1974, Fault zone characteristics of two well exposed overthrusts: the Muddy Mountain thrust, Nevada, and the Champlain thrust at Burlington, Vermont: Geol. Soc. America Abstracts with Programs, v. 5, no. 1, p. 78–79.

Stearns, R. G., 1954, The Cumberland Plateau overthrust and geology of the Crab Orchard Mountains area, Tennessee: Tenn. Div. Geol. Bull. 60, 47 p.

Straub, H., 1949, Die Geschichte der Bauingenieurkunst: Verlag Birkhauser, Basle; English ed., 1964, MIT Press, Cambridge, Mass., 258 p.

Suess, E., 1883–1885, Das Antlitz der Erde, Prague and Leipzig.

——, 1892, Conversations with Albert Heim at Zürich.

——, 1909, The face of the earth, v. 4 (Sollas translation): Oxford, 673 p.

Swinnerton, A. C., 1922, Geology of a portion of the Castleton, Vermont, quadrangle: Harvard Univ. thesis, 262 p.

Termier, P., 1906, Sur les phénomènes de recouvrement du Djebel Ouenza (Constantine) et sur l'existence de nappes charriées en Tunisie: Compt. Rend. Acad. Sci. Paris, v. 143, p. 137–139.

Terzaghi, K., 1936, Presidential Address, Proc. 1st Int. Conf. Soil Mech. Found. Eng., Cambridge, Mass., v. 3, p. 22–23.

Thompson, J. B., Jr., 1956, Skitchewauge nappe, a major recumbent fold in the area near Claremont, New Hampshire (abstract): Geol. Soc. America Bull., v. 67, p. 1826–1827.

——, Robinson, P., Clifford, T., and Trask, N., 1968, Nappes and gneiss domes in west-central New England, p. 203–218 in Zen, E-an, et al., eds., Studies of Appalachian geology, Wiley-Interscience, New York, 475 p.

Törnebohm, A. E., 1888, Om Fjäll problemet: Geol. Fören. Stockholm Förhandl., v. 10, p. 328–336.

Trümpy, R., 1973, The timing of orogenic events in the Central Alps, p. 229–251, in De Jong, K. A., and Scholten, R., eds., Gravity and tectonics, Wiley-Interscience, New York, 502 p.

Turner, F. J., 1953, Nature and dynamic interpretation of deformation lamellae in calcite of three marbles: Amer. Jour. Sci., v. 251, p. 276–298.

Tyndall, J., 1856, Comparative view of the cleavage of crystal and slate rocks: Phil. Mag., 4th ser., v. 12, p. 35.

Van Bemmelen, R. W., 1936, The undation theory of the development of the earth's crust: 16 Int. Geol. Congr., Rept., Washington, v. 2, p. 965-982.

Veatch, A. C., 1907, Geography and geology of a portion of southwestern Wyoming: U.S. Geol. Survey Prof. Paper 56, 178 p.

Voight, B., 1973, The mechanics of retrogressive block-gliding, with emphasis on evolution of the Turnagain Heights Landslide, Anchorage, Alaska: p. 97-121 *in* De Jong, K. A., and R. Scholten, eds., Gravity and tectonics, Wiley-Interscience, New York, 501 p.

——, 1974, Architecture and mechanics of the Heart Mountain and South Fork rockslides, p. 26-36 *in* Voight, B. and M. A., eds., Rock mechanics: the American Northwest, 3rd Congr. Int. Soc. Rock Mech., Exped. Guidebook, Spec. Publ. Exp. Sta., College Earth Min. Sci., Penn. State Univ., 292 p.

——, and Cady, W. C., in press, Transported rocks of the Taconide Zone, eastern North America: *in* Voight, B., ed., Rockslides and Avalanches, Vol. 1, Elsevier, Amsterdam.

——, and Dahl, H. D., 1969, Numerical continuum approaches to analysis of nonlinear rock deformation: Can. Jour. Earth Sci., v. 7, p. 814-830.

Wang, Y.-J., and Voight, B., 1969, A discrete element stress analysis model for discontinuous materials, p. 111-115 *in* Brekke, T. L., and Jørstad, F. A., eds., Large permanent underground openings, Conf. Proc. Int. Soc. Rock Mech., Universitetsforlaget, 1970, 372 p.

Williams, H., 1975, Structural succession, nomenclature, and interpretation of transported rocks in western Newfoundland: Can. Jour. Earth Sci., v. 12, no. 11, p. 1874-1894.

Willis, B., 1902, Stratigraphy and structure, Lewis and Livingston ranges, Montana: Geol. Soc. America Bull., v. 13, p. 305-352.

Young, D. M., 1957, Deep drilling through Cumberland overthrust block in southwestern Virginia: Amer. Assoc. Petrol. Geol. Bull., v. 41, p. 2567-2573.

Zaruba, Q., and Mencl, V., 1969, Landslides and their control: Elsevier, Amsterdam.

Zen, E-an, 1961, Stratigraphy and structure at north end of Taconic Range in west-central Vermont: Geol. Soc. America Bull., v. 72, p. 293-338.

——, 1967, Time and space relationships of the Taconic allochthon and autochthon: Geol. Soc. America Spec. Paper 97, 107 p.

——, 1972, The Taconide Zone and the Taconic Orogeny in the western part of the Northern Appalachian Orogen: Geol. Soc. America Spec. Paper 135, 72 p.

Zienkiewicz, O. C., 1967, The finite element method in structural and continuum mechanics: McGraw-Hill, London, 272 p. (2nd ed., 1970).

AUTHOR CITATION INDEX

445

SUBJECT INDEX

451

469

About the Editor

One of a family of naturalists from Yonkers, New York, BARRY VOIGHT began geologic research under the guidance of Raymond Gutschick and his colleagues at the University of Notre Dame. Formal studies led to three degrees in geology and civil engineering from Notre Dame, and subsequently a Ph.D. in geology from Columbia University. Actively involved in structural geologic surveys and rock mechanics investigations in the Appalachians and Rocky Mountains for the past fifteen years, Professor Voight is co-editor of *Rock Mechanics: The American Northwest*, and editor of *Rockslides and Avalanches*. He has served as a visiting professor at the Delft Technical University in The Netherlands and at the University of Toronto.